Natural Antimicrobials in Food Safety and Quality

Natural Antimicrobials in Food Safety and Quality

Edited by

Mahendra Rai

SGB Amravati University, India

and

Michael Chikindas

Cook College Rutgers University, USA

www.cabi.org

CABI is a trading name of CAB International

CABI
Nosworthy Way
Wallingford
Oxfordshire, OX10 8DE
UK

Tel: +44 (0)1491 832111
Fax: +44 (0)1491 833508
E-mail: cabi@cabi.org
Website: www.cabi.org

CABI
875 Massachusetts Avenue
7th Floor
Cambridge, MA 02139
USA

Tel: +1 617 395 4056
Fax: +1 617 354 6875
E-mail: cabi-nao@cabi.org

A catalogue record for this book is available from the British Library, London, UK.

Library of Congress Cataloging-in-Publication Data

Natural antimicrobials in food safety and quality / edited by Mahendra Rai, Michael Chikindas.
p. ; cm.
Includes bibliographical references and index.
ISBN 978-1-84593-769-0 (alk. paper)
1. Anti-infective agents. 2. Food–Microbiology. 3. Food–Safety measures.
4. Food preservatives.
I. Rai, Mahendra. II. Chikindas, Michael. III. C.A.B. International.
[DNLM: 1. Anti-Infective Agents. 2. Biological Products. 3. Food Microbiology. 4. Food Preservatives. QV 250]
RM267.N38 2011
615.7'92--dc23

2011021521

Commissioning editor: Sarah Mellor
Editorial assistant: Gwenan Spearing
Production editor: Shankari Wilford

Typeset by Columns Design XML Ltd, Reading, UK.
Printed and bound in the UK by the MPG Books Group.

Contents

Contributors

Wolf-Rainer Abraham, Helmholtz Center for Infection Research, Chemical Microbiology, Inhoffenstrasse 7, 38124 Braunschweig, Germany; e-mail: Wolf-Rainer.Abraham@helmholtz-hzi.de

Hikmate Abriouel, Área de Microbiología, Departamento de Ciencias de la Salud, Facultad de Ciencias Experimentales, Universidad de Jaén, Campus Las Lagunillas s/n. 23071-Jaén, Spain.

Deepak Acharya, Abhumka Herbal Pvt Ltd, 502, 5th Floor, Shreeji Chambers, Behind Cargo Motors, CG Road, Ahmedabad- 380 006, Gujarat, India.

Emilio Alvarez-Parrilla, Universidad Autónoma de Ciudad Juárez, Instituto de Ciencias Biomédicas, Departamento de Ciencias Básicas, Ciudad Juárez, Chihuahua, México. Anillo Envolvente del PRONAF y Estocolmo s/n, Ciudad Juárez, Chihuahua, México, 32310.

Bassam A. Annous, Eastern Regional Research Center, NAA, ARS, US Department of Agriculture, Wyndmoor, PA 19038, USA; e-mail: bassam.annous@ars.usda.gov

Jesús Fernando Ayala-Zavala, Centro de Investigación en Alimentación y Desarrollo, A. C., Coordinación de Tecnología de Alimentos de Origen Vegetal. Carretera de la Victoria km. 0.6, Apartado Postal 1735, Hermosillo, Sonora, México 83000.

Nicoletta Belletti, CENTA Centre de Noves Tecnologies i processos Alimentaris, Finca Camps i Armet s/n Monells, Girona, Spain; e-mail: nicoletta@centa.cat

Jeannette Ben Hamida, Unité de Protéomie Fonctionnelle & Biopréservation Alimentaire, Institut Supérieur des Sciences Biologiques Appliquées de Tunis, Université El Manar, Tunisie.

Paolo Bertolini, Criof – DiProVal, University of Bologna, Italy.

Paula Bourke, School of Food Science and Environmental Health, Dublin Institute of Technology, Cathal Brugha Street, Dublin 1, Ireland.

Maria do Carmo de Freire Bastos, Departamento de Microbiologia Geral, Instituto de Microbiologia Prof. Paulo de Góes, Universidade Federal do Rio de Janeiro, Rio de Janeiro, RJ, 21941-590, Brazil; e-mail: mcbastos@micro.ufrj.br or mcbastos2@yahoo.com.br

Hilana Ceotto, Departamento de Microbiologia Geral, Instituto de Microbiologia Prof. Paulo de Góes, Universidade Federal do Rio de Janeiro, Rio de Janeiro, RJ, 21941-590, Brazil.

Patrick Cullen, School of Food Science and Environmental Health, Dublin Institute of Technology, Cathal Brugha Street, Dublin 1, Ireland; e-mail: pjcullen@dit.ie

Mohamed Faid, Department of Food Science, Hassan II Institute of Agronomy and Veterinary Medicine, PO Box 6202 Rabat-Inst; Morocco; e-mail: faidmohamed@yahoo.fr

Isabel M.P.L.V.O. Ferreira, REQUIMTE- Serviço de Bromatologia, Faculdade de Farmácia da Universidade do Porto, Rua Anibal Cunha 164, 4099-030 Porto; Portugal; e-mail: isabel.ferreira@ff.up.pt

William F. Fett, Eastern Regional Research Center, NAA, ARS, U.S. Department of Agriculture, Wyndmoor, PA 19038, USA.

Djordje Fira, Laboratory for Molecular Genetics of Industrial Microorganisms, Institute of Molecular Genetics and Genetic Engineering, University of Belgrade, Vojvode Stepe 444/a, P.O. Box 23, 11010 Belgrade, Serbia.

Ismail Fliss, STELA Dairy Research Centre, Nutraceuticals and Functional Foods Institute, Université Laval, G1K 7P4 Québec, QC, Canada; e-mail: ismail.fliss@fsaa.ulaval.ca

Mendel Friedman, Produce Safety and Microbiology Research Unit, Western Regional Research Center, Agricultural Research Service, U. S. Department of Agriculture, 800 Buchanan St., Albany, California 94710, USA; e-mail: Mendel.Friedman@ars.usda.gov

Antonio Gálvez, Área de Microbiología, Departamento de Ciencias de la Salud, Facultad de Ciencias Experimentales, Universidad de Jaén, Campus Las Lagunillas s/n. 23071-Jaén, Spain; e-mail: agalvez@ujaen.es

Pilar García, Instituto de Productos Lácteos de Asturias (IPLA-CSIC). Apdo. 85. 33300-Villaviciosa, Asturias, Spain; e-mail: pgarcia@ipla.csic.es

Fausto Gardini, Dipartimento di Scienze degli Alimenti, Università degli Studi di Bologna, Sede di Cesena, Piazza G. Goidanich, 60, 47023 Cesena, Italy.

Natasa Golic, Laboratory for Molecular Genetics of Industrial Microorganisms, Institute of Molecular Genetics and Genetic Engineering, University of Belgrade, Vojvode Stepe 444/a, P.O. Box 23, 11010 Belgrade, Serbia.

Gustavo Adolfo González-Aguilar, Centro de Investigación en Alimentación y Desarrollo, A. C., Coordinación de Tecnología de Alimentos de Origen Vegetal. Carretera de la Victoria km. 0.6, Apartado Postal 1735, Hermosillo, Sonora, México 83000; e-mail: gustavo@ciad.mx

María José Grande Burgos, Área de Microbiología, Departamento de Ciencias de la Salud, Facultad de Ciencias Experimentales, Universidad de Jaén, Campus Las Lagunillas s/n. 23071-Jaén, Spain.

Riadh Hammami, STELA Dairy Research Centre, Nutraceuticals and Functional Foods Institute, Université Laval, G1K 7P4 Québec, QC, Canada; e-mail: riadh.hammami@fsaa.ulaval.ca

Basilio Heredia, Ciencia y Tecnología de Alimentos, Coordinación CIAD Culiacán, Mexico.

Ana Maria Herrero, Department of Meat and Fish Science and Technology, ICTAN-Instituto del Frío, (CSIC),C/ Jose Antonio Novais, 10; Ciudad Universitaria, 28040 Madrid, Spain.

Francisco Jimenez-Colmenero, Department of Meat and Fish Science and Technology, ICTAN-Instituto del Frío, (CSIC),C/ Jose Antonio Novais, 10; Ciudad Universitaria, 28040 Madrid, Spain.

Vijay K. Juneja, Microbial Food Safety Research Unit, Eastern Regional Research Center, Agricultural Research Service, US Department of Agriculture, 600 E. Mermaid Lane, Wyndmoor, Pennsylvania 19038, USA.

Sylvain Sado Kamdem, Laboratoire de Microbiologie, Department of Biochemistry, University of Yaoundé I, P.O. Box 812, Yaounde, Cameroon.

Govind J. Kapadia, Department of Pharmaceutical Sciences, School of Pharmacy, Howard University, Washington, D.C. 20059, USA.

Khaoula Khwaldia, Institut National de Recherche et d'Analyse Physico-chimique, INRAP, Pôle Technologique de Sidi Thabet 2020 Sidi Thabet, Tunisia; e-mail: khaoula_khwaldia@yahoo.fr

Milan Kojic, Laboratory for Molecular Genetics of Industrial Microorganisms, Institute of Molecular Genetics and Genetic Engineering, University of Belgrade, Vojvode Stepe 444/a, P.O. Box 23, 11010 Belgrade, Serbia.

Rosalba Lanciotti, Dipartimento di Scienze degli Alimenti, Università degli Studi di Bologna, Sede di Cesena, Piazza G. Goidanich, 60, 47023 Cesena, Italy.

Christophe Le Lay, STELA Dairy Research Centre, Nutraceuticals and Functional Foods Institute, Université Laval, G1K 7P4 Québec, QC, Canada.

Yin Li, Institute of Microbiology, Chinese Academy of Sciences, China.

Ching-Hsing Liao, Eastern Regional Research Center, NAA, ARS, U.S. Department of Agriculture, Wyndmoor, PA 19038, USA.

Rosario Lucas, Área de Microbiología, Departamento de Ciencias de la Salud, Facultad de Ciencias Experimentales, Universidad de Jaén, Campus Las Lagunillas s/n. 23071-Jaén, Spain.

Mauro Malnati, Unità di Virologia Umana, DIBIT, Istituto Scientifico San Raffaele, Via Olgettina, 58, 20132 Milan, Italy.

Marta Mari, Criof – DiProVal, University of Bologna; e-mail: marta.mari@unibo.it

Beatriz Martínez, Instituto de Productos Lácteos de Asturias (IPLA-CSIC). Apdo. 85. 33300-Villaviciosa, Asturias, Spain.

Caterina Morcia, CRA-GPG, Genomic Research Center, Via San Protaso 302, 29017-Fiorenzuola d'Arda (PC), Italy.

María Muy-Rangel, Ciencia y Tecnología de Alimentos, Coordinación CIAD Culiacán, Mexico.

Fiorella Neri, Criof – DiProVal, University of Bologna, Italy.

Guadalupe Isela Olivas, Centro de Investigación en Alimentación y Desarrollo, A.C. Fisiología y Tecnología de Alimentos de Zona Templada, Cuauhtémoc, Chihuahua, México.

Olusegun Victor Oyetayo, Department of Microbiology, Federal University of Technology, P.M.B 704, Akure, Nigeria; e-mail: ovofuta@yahoo.com

Maira Peres de Carvalho, Helmholtz Center for Infection Research, Chemical Microbiology, Inhoffenstrasse 7, 38124 Braunschweig, Germany.

Olívia Pinho, Faculdade de Ciências da Nutrição e Alimentação da Universidade do Porto, Rua Dr. Roberto Frias, 4200-465 Porto, Portugal.

Mahendra Rai, Department of Biotechnology, SGB Amravati University, Amravati-444 602, Maharashtra, India; e-mail: mkrai123@rediffmail.com

Mehdi Razzaghi-Abyaneh, Department of Mycology, Pasteur Institute of Iran, Tehran 13164, Iran; e-mail: mrab442@yahoo.com & mrab442@pasteur.ac.ir

G. Subba Rao, Department of Pharmaceutical Sciences, School of Pharmacy, Howard University, Washington, D.C. 20059, USA.

Jose Luis Rios, Departamento de Farmacologia, Facultad de Farmacia, Universidad de Valencia 46100 Burjassot, Valencia, Spain.

Ana Rodríguez, Instituto de Productos Lácteos de Asturias (IPLA-CSIC). Apdo. 85. 33300-Villaviciosa, Asturias, Spain.

Lorena Rodríguez, Instituto de Productos Lácteos de Asturias (IPLA-CSIC). Apdo. 85. 33300-Villaviciosa, Asturias, Spain.

Laura de la Rosa, Universidad Autónoma de Ciudad Juárez, Instituto de Ciencias Biomédicas, Departamento de Ciencias Básicas, Ciudad Juárez, Chihuahua, México. Anillo Envolvente del PRONAF y Estocolmo s/n, Ciudad Juárez, Chihuahua, México, 32310.

Claudia Ruiz-Capillas, Department of Meat and Fish Science and Technology, ICTAN-Instituto del Frío, (CSIC),C/ Jose Antonio Novais, 10; Ciudad Universitaria, 28040 Madrid, Spain; e-mail: claudia@if.csic.es

Masoomeh Shams-Ghahfarokhi, Department of Mycology, Faculty of Medical Sciences, Tarbiat Modares University, Tehran 14115-111, Iran.

Hui Song, Tianjing Institute of Industrial Biotechnology, Chinese Academy of Sciences, China.

Martina Spini, CRA-GPG, Genomic Research Center, Via San Protaso 302, 29017-Fiorenzuola d'Arda (PC), Italy.

A. Michele Stanca, CRA-GPG, Genomic Research Center, Via San Protaso 302, 29017-Fiorenzuola d'Arda (PC), Italy.

Ivana Strahinic, Laboratory for Molecular Genetics of Industrial Microorganisms, Institute of Molecular Genetics and Genetic Engineering, University of Belgrade, Vojvode Stepe 444/a, P.O. Box 23, 11010 Belgrade, Serbia.

Yuanxia Sun, Tianjing Institute of Industrial Biotechnology, Chinese Academy of Sciences, China; e-mail: sun_yx@tib.cas.cn

Valeria Terzi, CRA-GPG, Genomic Research Center, Via San Protaso 302, 29017-Fiorenzuola d'Arda (PC), Italy; e-mail: valeria.terzi@entecra.it

Brijesh Kumar Tiwari, Department of Food & Tourism Management, Manchester Metropolitan University, M14 6HR, UK; e-mail: b.tiwari@mmu.ac.uk

Ljubisa Topisirovic, Laboratory for Molecular Genetics of Industrial Microorganisms, Institute of Molecular Genetics and Genetic Engineering, University of Belgrade, Vojvode Stepe 444/a, P.O. Box 23,11010 Belgrade, Serbia; e-mail: lab6@EUnet.rs

Vasilis P. Valdramidis, School of Food Science and Environmental Health, Dublin Institute of Technology, Cathal Brugha Street, Dublin 1, Ireland.

Yang Zhu, Department of Biosciences, TNO Quality of Life, Netherlands.

Abdelmajid Zouhir, Unité de Protéomie Fonctionnelle & Biopréservation Alimentaire, Institut Supérieur des Sciences Biologiques Appliquées de Tunis, Université El Manar, Tunisie.

Foreword

The World Health Organization of the United Nations estimates that every year one in three people worldwide get sick from consuming food contaminated with human pathogens (bacteria, viruses or parasites). The US Centers for Disease Control and Prevention estimates that one in six Americans get sick from eating unsafe food every year. In 2011, researchers at the University of Florida's Emerging Pathogens Institute reported that our food is not safe enough. Some of the foodborne outbreaks are extremely severe resulting in large numbers of human deaths every year. However, the incidence of foodborne illness worldwide is probably much higher due to unreported cases of foodborne disease. Foodborne outbreaks cost the US in excess of US$14 billion annually in direct medical costs and lost wages and in excess of US$40 billion in revenue losses by the food industry.

Fresh fruits and vegetables are normally consumed raw or minimally processed and are considered to be an important part of a healthy diet as they provide needed nutrients, fibre and antioxidants. The greater awareness of the health benefits of increased intake of fresh and fresh-cut produce has led to a sharp increase in per capita consumption of fresh produce in the US and around the world. Unfortunately, along with the increase in produce consumption there has been a sharp increase in the number of foodborne outbreaks due to fresh fruits and vegetables contaminated with a variety of human pathogens. The outbreaks associated with fresh produce doubled between the periods 1973 to 1987 and 1988 to 1992 and accounted for 6% of all reported foodborne outbreaks in the 1990s compared to only 0.7% in the 1970s. A recent analysis of US foodborne outbreaks occurring between 1990 and 2003 indicated that contaminated fresh produce caused the most illnesses and the second highest number of outbreaks. A wide variety of human pathogens can be isolated from the surfaces of raw produce, where the pathogens often survive for extended periods of time. Conventional washing and sanitizing technologies are not very effective in reducing the populations of human pathogens on fresh produce. Since the complete elimination of sources of contamination on the farm is not feasible, more effective intervention strategies are needed. Such strategies may include the use of natural-based interventions for suppressing the populations of pathogens and/or inhibiting the outgrowth of survivors after other chemical or physical interventions are applied (the multiple hurdle approach).

The purpose of this book is to provide a comprehensive reference covering a variety of aspects of natural and naturally-derived antimicrobials for enhancing the safety, shelf life and quality of the food supply. This book contains valuable information on novel analytical technologies, strategies to reduce or eliminate human pathogens in food supply, regulatory policy and emerging technologies for the production and the use of natural and naturally-derived antimicrobials. Chapter 1 focuses on the naturally occurring antimicrobials in the food industry. Chapters 2–6 look at bacteriophages and bacterial antimicrobial compounds used in food preservation. Chapter 7 describes the antimicrobials produced by algae. Chapter 8 describes the secondary metabolite antimicrobials derived from fungi. Chapters 9 and 10 focus on the antimicrobials derived from animal by-products and their efficacy as preservatives or antimicrobials by the food industry. Chapters 11 and 12 look into the use of naturally-derived preservatives in enhancing the chemical safety by reducing biogenic compounds present in food products. Chapters 13–20 study the use of natural antimicrobials derived from plant and plant products for the preservation of quality and enhancing microbial safety of the food supply. Chapters 21 and 22 focus on database and predictive modeling of natural antimicrobials.

I hope that this book will serve as a valuable reference source for research scientists in the food industry, academia, and government as well as graduate students, regulatory agencies and individuals interested in learning more about food preservation and microbial food safety. We also anticipate that the information presented in the various chapters written by a distinguished international group of scientists will stimulate ideas for future interdisciplinary research efforts in natural substances for food preservation and control of non-desired microorganisms in food environments.

I am grateful to Drs Mahendra Rai and Michael Chikindas (editors) for providing me with the opportunity to write this Foreword. Finally, on behalf of the editors, I would like to extend heartfelt thanks to the chapter authors in this book for contributing their time, knowledge and expertise to this endeavour.

Bassam A. Annous
United States Department of Agriculture
Agricultural Research Service
Eastern Regional Research Center
USA

Preface

The demand of producing high-quality, safe (pathogen-free) food relies increasingly on natural sources of antimicrobials to inhibit food-spoilage organisms and food-borne pathogens and toxins. The discovery and development of new antimicrobials from natural sources for a wide range of applications requires that knowledge of traditional sources for food antimicrobials is combined with the latest technologies in identification, characterization and application. This book explores some novel, natural sources of antimicrobials as well as the latest developments in using well-known antimicrobials in food.

The book includes antimicrobials from microbial sources (bacteriophages, bacteria, algae and fungi), animal-derived products (milk proteins, chitosan and reduction of biogenic amines), plants and plant products (essential oils, phytochemicals and bioactive compounds). A separate section of the book covers natural antimicrobials for fresh products. New and emerging technologies (a database for bacteriocin and predictive modelling of antimicrobial efficacy) concerning antimicrobials have also been incorporated.

The book could be considered essential reading for food technologists, microbiologists, biotechnologists, pharmacologists, botanists, agriculture specialists and those who believe in eco-friendly natural products. It should also be useful for postgraduate students and researchers.

We are grateful to all the contributors for the submission of their valuable manuscripts. Mahendra Rai wishes to thank Aniket Gade, Alka Karwa, Ravindra Ade, Avinash Ingle, Dhyaneshwar Rathod, Alka Yadav, Vaibhav Tiwari, Jayendra Kesharwani, Swapnil Gaikwad, Shital Bonde, Sonal Birla and Shivaji Deshmukh for help in editing.

The editors wish to thank Sarah Mellor for her encouragement and help in bringing out the book in the present form. We also thank Dr Geetika Sareen for suggestions during publication of the book.

Mahendra Rai
Michael Chikindas

1 Naturally Occurring Biocides in the Food Industry

Deepak Acharya, Jose Luis Rios and Mahendra Rai*

Introduction

Microbial contamination in food is a matter of great concern and with an increasing human population it is difficult to find a solution to reduce the contamination problems. The contamination by many species of bacteria, fungi and algae has become a major threat to food and packaging industries. The issue is more sensitive when it comes to the health industry. It is very important to maintain hygiene and disinfection so as to avoid health complications due to contamination. Natural biocidal substances can be used to nullify the effects of contaminants. According to Block (1991) a biocide is a substance that kills both pathogenic and non-pathogenic microorganisms. Extensive research has been carried out on naturally occurring biocides (Varona *et al.*, 2009).

Naturally occurring biocides include plant- or microbial-derived secondary metabolites that kill microbes, particularly pathogenic microbes. Thus, natural biocides can be categorized into plant-based biocides and microbe-based biocides. The majority of natural biocides are derived generally from the extracts and oils of plants. Essential oils play a major role in the protection of food (Valero *et al.*, 2006).

Biocides are important cleaning and disinfectant weapons used frequently in the food industry to control pathogenic and spoilage microbes (Holah, 2000). They can mainly be divided into four groups. **Disinfectants** are used on inanimate objects or intact skin to reduce the number of microorganisms. Disinfectants may be classified as low, medium or high level on the basis of the power by which they kill the microorganism. **Antiseptics** are used to treat infections in surface wounds, whereas **antibiotics** are used to treat microbes within the body. **Preservatives** are added to products, such as cosmetics, food, animal feeds or household products, for avoiding any microbial contamination. Fraise (2002) categorized biocides into disinfectants, antiseptics and preservatives and did not include antibiotics. As the name indicates, a biocide may deter, render or prevent action of any harmful microbe by chemical or biological means. Biocides are now commonly used in medicine, agriculture, forestry and the industrial world. The demand for biocides in the food industry is growing as every consumer needs products that are preserved naturally. The use of chemicals for the protection of food, including vegetables, fruits and drinks, is harmful to health and also to the environment. Essential oils are

* Corresponding author.

now well known for their antimicrobial activity. Due to their antimicrobial activity these oils are used as safe, natural food preservatives.

This chapter focuses on plant and microbially derived biocides and their role in the food industry.

Plant-based Biocides

A review of the literature on the antimicrobial activity of plants and plant-derived products revealed that many significant contributions have been made on the bioactivities of medicinal plants (Rios and Recio, 2005; Yoshida *et al.*, 2010).

Plant-based biocides have been used since the beginning of civilization. Rios and Recio (2005) reviewed the antimicrobial efficacy of *Syzygium aromaticum, Cinnamomum aromaticum* and *Origanum vulgare* in paper packaging. *C. aromaticum* oil, which is used in paraffin coating, totally inhibited growth of *Aspergillus flavus.* Kozlowski and Walentowska (2008) reported the role of plant derived biocides in protection of natural fibres against bio-deterioration and found that thyme oil and grapefruit extract play important roles as biocides. Similarly, in the protection of linen-cotton fabric, the essential oil of *Thymus vulgaris* was studied and showed remarkable antifungal activity (Kozlowski *et al.* 2008). The role of biocides can be vast in various industries. Marino *et al.* (2001) worked on the role of essential oils against fungi commonly found in library and archival materials, i.e. *Aspergillus niger, Chaetomium globosum, Penicillium frequentans* and *Paecilomyces variotii.* Essential oils obtained from armoise, boldo, clove, eucalyptus, lavender, tea tree, thuja and wormseed plants were investigated. Among these, eucalyptus oil (*Eucalyptus globulus*) showed moderate antifungal activity, whereas wormseed oil (*Chenopodium ambrosioides*) was the most effective.

People have been preserving food from spoilage by microbes for thousands of years. Curing with salt, pickling, dehydrating, pasteurizing and freezing are all examples of traditional food preservation methods that have been used historically and are still being used today. The preservation of food is essential. Without it, the rich nutrition of foodstuff becomes a breeding ground for bacteria, mould and other microorganisms that can cause sickness and disease. In the current fast-moving lifestyle, there is a major concern about health and fitness. Consumers are now more concerned about hygiene and the quality of food products. The growing demand of producing high-quality, microbe-free foodstuff means the food industries are now turning back to natural biocides. Natural source of antimicrobials would inhibit food-spoiling microbes, food-borne pathogens and toxic substances. There is a great need to identify, characterize and apply techniques to find some novel natural sources of antimicrobials for processed and fresh food. The demand for biocides in the food and beverage industry is growing at the rate of 4.2% per annum and it was expected to reach US$274 million in the USA alone by the end of 2010, as reported by Freedonia (Joseph and Sujatha, 2011; Thomas, 2011).

Essential Oils as Natural Biocides in the Food Industry

Food and its quality has been a matter of major concern all over the world. People now are more conscious about their health and fitness. They have become more choosy and selective in picking up the food products from the shop shelves. Food-borne infections are major public health concerns as they account for many cases of illness among human and animals. There are more than 250 food-borne diseases known and the majority of the disorders are a result of microbial infection in foodstuff (De Smet, 2002; Rhee *et al.*, 2004; Ernst, 2005). Abdominal cramps, vomiting and dysentery are a few of the symptoms seen in food-borne disorders (Boon, 1999; Herr *et al.*, 2002). Researchers are deeply involved in finding natural alternatives to safeguard foodstuff against microbial infections (Nychas, 1995; Lopez-Malo *et al.*, 2000; Ernst, 2000; Bent and Ko, 2004; Olasupo *et al.*, 2004; Lopez-Malo *et al.*, 2005).

A variety of bacteria and fungi, including the common spoilage bacteria *Salmonella, Listeria, Enterobacter, Lactobacillus* and *Pseudomonas*, contaminate foodstuff. Many herbs and spices are known to have flavour and fragrance and possess antimicrobial activity. Hence, they have been widely used as biocides (Ceylan and Fung, 2004; De, 2004; Davidson *et al.*, 2005). Researchers are involved in searching for bio tools to suppress or eradicate microbial contamination. Many plants, such as *Ocimum sanctum, Elettaria cardamomum, Brassica campestris, O. vulgare, Rosmarinus officinalis, Salvia officinalis* and *T. vulgaris*, have been evaluated for antimicrobial activity against food-borne pathogens (Deans and Ritchie, 1987; Lafont *et al.*, 1998; Davidson, 2001; Lambert *et al.*, 2001; Peter, 2001; Parker and Parker, 2004). Naturally occurring biocides are in great demand. Essential oils obtained from plants such as *Mentha piperita, T. vulgaris, Citrus paradisi, S. officinalis, Artemisia absinthium* and *Lavandula angustifolia* are known for their antimicrobial properties and currently they are widely applied in the fields of medical, sanitary, cosmetic, food and packaging industries, agriculture (plant protection), and so on. Active substances, e.g. alkaloids, flavonoids and terpenes, present in essential oils are a great source of antimicrobials (Seoud *et al.*, 2005; Dolan *et al.*, 2007). Essential oils obtained from *T. vulgaris, O. vulgare, Syzygium aromaticum* and *Mentha sylvestris* are a few among the many that are currently in demand in the food industry and that are known for their antimicrobial properties (Rakotonirainy and Lavedrine, 2005). Gulluce *et al.* (2007) observed the antifungal activity of *M. arvensis* oil and it can be applied as a good source of biocides against food infections of *Aspergillus* spp. and *Penicillium* spp. There have been plenty of examples of the historical use of essential oils in the treatment of different infectious diseases and their therapeutic use has been validated by various investigations that support the use of these plants or essential oils as biocides.

The market is full of processed and packaged food that lasts from months to years. Many products available on supermarket shelves contain synthetic and chemically derived preservatives. Many chemical preservatives are found naturally in some food; however, they are being produced synthetically to fulfil the growing demand for preservatives in the food industry. Many of the chemical preservatives are known to cause allergic reactions among sensitive individuals. Therefore, there is a great deal of effort being put in by scientists all over the world to develop a variety of safer and natural methods of food preservation. Researchers from all across the globe are trying to discover better alternatives to synthetic preservatives. Many researchers have demonstrated the efficacy of thyme (*T. vulgaris*) and oregano (*O. vulgare*) against various food-borne microbes (Elgayyar *et al.*, 2001; Burt, 2004; Oussalah *et al.*, 2006). Various findings suggest that combining essential oils can have a synergistic effect and be more effective than single applied essential oils. The essential oil of clove (*S. aromaticum*) was investigated for its ability to inhibit the growth of *Listeria monocytogenes*, a pathogen in salmon, and it was found to be very active in checking the growth of *L. monocytogenes* (Hanene *et al.*, 2010).

Laciar *et al.* (2009), in a study, screened the essential oil of *Artemisia echegarayi* against seven Gram-positive and Gram-negative bacterial species significantly responsible for food degradation and the results showed a high level of bacterial growth inhibition. *Listeria monocytogenes* is a bacterium that grows on a variety of processed meat products at refrigeration temperatures (Glass and Doyle, 1989) and also causes listeriosis, which is a death-causing disorder. *L. monocytogenes* is a leading cause of death among food-borne bacterial pathogens, with fatality rates exceeding even *Salmonella* and *Clostridium botulinum*. There is plenty of research being carried out to find natural biocides that suppress its growth (Buchanan *et al.*, 1997; Fernandez *et al.*, 1997; Parente *et al.*, 1998; Nerbrink *et al.*, 1999). There are a number of herbs, spices and natural substances that have shown antilisterial activity, including cinnamic acid (Ramos-Nino *et al.*, 1996;

Kouassi and Shelef, 1998), hop extracts (Larson *et al.*, 1996), carvacrol (Kim *et al.*, 1995), furanocoumarins (Ulate-Rodriguez *et al.*, 1997), eugenol (Blaszyk and Holley, 1998; Hao *et al.*, 1998a), horseradish distillates (Ward *et al.*, 1998), pimento leaf (Hao *et al.*, 1998a), rosemary and cloves (Pandit and Shelef, 1994; Lis-Balchin and Deans, 1997). There were many more investigations, but the results were not consistent (Kim *et al.*, 1995, Lis-Balchin and Deans, 1997; Hao *et al.*, 1998a). Horseradish distillates work as a wonderful preservative on roasted beef (Ward *et al.*, 1998), and similarly, eugenol and pimento leaf on refrigerated cooked beef (Hao *et al.*, 1998b).

Fresh vegetables are also susceptible to microbial infection because of the high water and nutritional contents (Ippolito and Nigro, 2003; Ragaert *et al.*, 2007). Such vegetables are very prone to infection by microbes as soon as they begin their journey from the farm to the store. The main bacterial populations of the Pseudomonadaceae and Enterobateriaceae families and a few from the lactic acid bacteria (LAB) group are known for their hazardous effect on fresh and refrigerated vegetables. Even if the vegetables are maintained with proper refrigeration, it is never sufficient to just check the contamination or spoiling of the material (Sapers, 2001). Essential oil obtained from plants such as oregano (*O. vulgare*) and thyme (*T. vulgaris*) can act as good biocides (Gutierrez *et al.*, 2009). The demand for natural biocides is high as they are more efficient and fall under the 'generally recognised as safe' (GRAS) category. Essential oils and plant-derived products are widely used as biocides for vegetables and ready-to-eat green foodstuff. They are known to check the growth of many food-borne pathogens and spoilage bacteria associated with vegetables (Gutierrez *et al.*, 2008). Carvacrol obtained from oregano and thymol obtained from thyme oils are known for their strong antibacterial properties (Dorman and Deans, 2000; Elgayyar *et al.*, 2001; Burt, 2004; Oussalah *et al.*, 2006). Tayel and El-Tras (2009) performed a study to check the antimicrobial potential of 25 species of herb extracts against eight food-borne bacterial strains and the results showed a high level of elimination of food-borne bacteria. The authors recommended that herbal biocides can be applied for the elimination of microbes and we can have better alternatives to chemical and synthetic biocides. Similarly, a study by Mishra and Behal (2010) indicates the microbial growth suppressing activity of many spices.

Role of Microbes in Food Safety

Microbes play a crucial role in the protection of food and food products. Among these microbes, bacteria and bacteriophages are very important.

Bacteriophages (lytic viruses or phages) are ubiquitously distributed in the environment and are responsible for killing bacteria without any harmful effect on humans or animals. Bacteriophages were co-discovered in the early 1900s by Twort (1915) and d'Herelle (1917). There is a growing interest of researchers in the application of bacteriophages to kill food-borne bacteria, particularly after the development of a multidrug resistance problem. Bacteriophages reproduce after attacking bacteria. In this process, the viral DNA is injected into the cells of the bacterial hosts, where they are multiplied. Eventually, the bacterial host cells are killed and the multiplied phages infect other bacterial cells. The method of multiplication continues. These bacteriophages are very useful for the protection of fresh-cut fruit and vegetables.

The Food and Drug Administration (2006) approved mixture of bacteriophage (lytic cocktail) to inhibit the growth of *L. monocytogenes* bacteria in meat. There is noteworthy research on the use of bacteriophages to control food-borne pathogens such as *Salmonella, L. monocytogenes, E. coli* O157:H7 and *Campylobacter* on agricultural products.

LAB have been in use in fermented food since ancient times. They produce organic acids in general and lactic acid in particular. LAB are used in the processing of fermented foods due to their wide range of antimicrobial activities. Some strains of LAB are reported to produce formic acid, fatty acids, hydrogen peroxide diacetyl, reuterin and reutericyclin. According to De Vuyst and Vandamme (1994) many strains of LAB also produce bacteriocins and bacteriocin-like molecules. Bacteriocins are small, ribosomally synthesized peptides produced by LAB that possess antimicrobial activity (Klaenhammer, 1988; De Vuyst and Vandamme, 1994; Cotter *et al.*, 2005).

The use of natural antimicrobial coatings and films is an emerging technology for maintaining food quality and safety. The use of edible antimicrobial coatings and films is gaining ground because such coatings have a variety of advantages and provide an innovative packaging concept. Cooksey (2005) reported Nisin as an efficient film against *Staphylococcus aureus* and *L. monocytogenes*. The use of raw milk for the development of films in food packaging is also gaining importance. Serrano *et al.* (2006) used *Aloe vera* gel for retaining the functional properties of table grapes (*Vitis vinifera*). The grapes coated with *Aloe vera* gel showed considerably extended retention of ascorbic acid during cold storage.

Research Priorities and Future Perspectives

Recent advances in information technology have allowed biologists to document, disseminate and analyse data from medicinal plants in general and plants exhibiting bioactivity in particular. The Internet makes it possible to access, share and disseminate information more effectively, thus contributing to global knowledge. So, there is a need to develop a database of medicinal plants and microbes showing antimicrobial activity.

Babu and his colleagues (2009) have constructed a minimum inhibitory concentration (MIC) database using html. MIC is the lowest concentration of an antimicrobial agent (antibiotics or essential oils) that inhibits the growth of microorganisms after a period of incubation. MIC is an important test in diagnostic laboratories to evaluate the activity or resistance of microorganisms to an antimicrobial agent. The authors collected data from different journals, namely *Antimicrobial Agents and Chemotherapy, Applied and Environmental Microbiology, Arkivoc, Bioorganic and Medicinal Chemistry Letters, Microbiology and Immunology, Journal of Dairy Science, Indian Journal of Pharmaceutical Sciences, Journal of Antimicrobial Chemotherapy, Journal of Clinical Microbiology* and *Journal of Bacteriology*. Presently, the MIC database includes 500 records containing minimum inhibitory concentration and zone of inhibition values of various well-known and newly synthesized antibiotics against a specific microbe.

There is a need to search for new antimicrobial agents because infectious diseases are still a worldwide problem due to the development and spread of drug-resistant pathogens. The increasing world population faces alarming health problems including cancer, drug-resistant bacteria, parasitic protozoa and fungi. Tuberculosis infections are a particular problem. Gordien *et al.* (2010) studied extracts from Scottish plants, lichens and mycoendophytes, which were screened for activity against *Mycobacterium aurum* and *Mycobacterium tuberculosis*. The highest activity against *M. aurum* was shown by extracts of *Juniperus communis* roots, of the lichen *Cladonia arbuscula* and of a mycoendophyte isolated from *Vaccinium myrtillus* (Gordien *et al.*, 2010). Obviously, mycoendophytes are a source of potentially useful medicinal compounds. For example, 3-nitropropionic acid was isolated from *Phomopsis* species which inhibited *M. tuberculosis* (Copp and Pearce 2007). The first anticancer agent produced by endophytes was paclytaxel and its derivatives. Paclytaxel is a highly functionalized diterpenoid, found in yew (*Taxus*) species (Bacon and White, 1994). The mode of action of paclytaxel is to prevent tubulin molecules from depolymerization during the processes of cell division (Tan and Zou, 2001). More novel anti-cancer drugs are required worldwide to combat this scurge.

The medicinal plants and microbes possess a great potential for the development of new drugs. Keeping this in mind, extensive surveys should be undertaken to search higher and lower plants and microbes including bryophytes, pteridophytes, algae, fungi and actinomycetes. In the past, much attention has been paid to the bioactivity of higher plants. The lower plants have not been considered. However, now many scientists are focusing their research on screening of bryophytes, pteridophytes, algae and fungi.

Conclusion

Natural biocides based on essential oils and extracts from plants are used for maintaining the quality of food. Moreover, microbes such as bacteria, bacteriophages, algae and fungi can also be used for the protection of food against pathogenic microbes. There is a pressing need to screen and develop newer biocides of plant and microbial origin.

The need of the hour is to develop combinations of natural biocides for better efficacy. A thorough study on how these natural biocides can really improve the shelf life of any food product should be investigated.

References

Babu, P.A., Kumar P.S., Padmaja, P., Rao, T.K. and Chitti, S. (2009) MIC Database: a collection of antimicrobial compounds from literature. *Bioinformation* 4, 75–77.

Bacon, C.W. and White, J.F. (1994) *Biotechnology of Endophytic Fungi of Grasses.* CRC Press, Boca Raton, Florida, pp. 47–56.

Bent, S. and Ko, R. (2004) Commonly used herbal medicines in the United States: a review. *American Journal of Medicine* 116, 478–485.

Blaszyk, M. and Holley, R.A. (1998) Interaction of monolaurin, eugenol and sodium citrate on growth of common meat spoilage and pathogenic organisms. *International Journal of Food Microbiology* 39, 175–183.

Block, S.S. (1991) Historical review. In S. S. Block (ed.), *Disinfection, Sterilization, and Preservation*, 4th edn. Lea & Febiger, Philadelphia, Pennsylvania, pp. 3–17.

Boon, H. (1999) *The Botanical Pharmacy.* Quarry Press, Kingston, ON, Canada.

Buchanan, R.L., Golden, M.H. and Phillips, J.G. (1997) Expanded models for the non-thermal inactivation of *Listeria monocytogenes. Journal of Applied Microbiology* 82, 567–577.

Burt, S. (2004) Essential oils: their antibacterial properties and potential applications in foods. *International Journal of Food Microbiology* 94, 223–253.

Ceylan, E. and Fung, D.Y. (2004) Antimicrobial activity of spices. *Journal of Rapid Methods and Automation Microbiology* 12, 1–55.

Cooksey, K. (2005) Effectiveness of antimicrobial food packaging materials. *Food Additives and Contaminants* 22, 980–987.

Copp, B.R. and Pearce, A.N. (2007) Natural product growth inhibitors of *Mycobacterium tuberculosis. Royal Society of Chemistry* 24, 278–297.

Cotter, P.D., Hill, C. and Ross, R.P. (2005) Bacteriocins: developing innate immunity for food. *Nature Reviews Microbiology* 3, 777–788.

Davidson, P.M. (2001) Chemical preservatives and naturally antimicrobial compounds. In: Beuchat, M.P.and Montville, L.R., (eds) *Food Microbiology. Fundamentals and Frontiers*, 2nd edn. ASM Press, Washington DC, pp. 593–628.

Davidson, P.M., Sofos, J.N. and Branen, A.L. (2005) *Antimicrobials in Food.* 3rd edn. CRC Press, Taylor and Francis Group. Boca Raton, Florida.

De, A.K. (2004) *Spices: Traditional Uses and Medicinal Properties.* Asian Books Pvt. Ltd. Daryaganj, India, pp. 7–17.

Deans, S.G. and Ritchie, G. (1987) Antibacterial properties of plant essential oils. *International Journal of Food Microbiology* 5, 165–180.

De Smet, P.A.G.M. (2002) Herbal remedies. *New England Journal of Medicine* 347, 2046–2056.

De Vuyst, L. and Vandamme, E.J. (1994). *Bacteriocins of Lactic Acid Bacteria: Microbiology, Genetics and Applications.* Blackie Academic and Professional, London.

d'Herelle, F. (1917) Sur un microbe invisible antagonists des bacilles dysenteriques. *Comptes Rendus l'Academie des Sciences Paris* 165, 373–375.

Dolan M.C., Dietrich, G., Panella, N.A., Montenieri, J.A. and Karchesy, J.J. (2007). Biocidal activity of three wood essential oils against *Ixodes scapularis* (Acari: Ixodidae), *Xenopsylla cheopis* (Siphonaptera: Pulicidae), and *Aedes aegypti* (Diptera: Culicidae). *Journal of Econonic Entomology* 100, 622–625.

Dorman, H.J.D. and Deans, S.G. (2000) Antimicrobial agents from plants: antibacterial activity of plant volatile oils. *Journal of Applied Microbiology* 88, 308–316.

Elgayyar, M., Draughon, F.A., Golden, D.A. and Mount, J.R. (2001) Antimicrobial activity of essential oils from plants against selected pathogenic and saprophytic microorganisms. *Journal of Food Protection* 64, 1019–1024.

Ernst, E. (2000) *Herbal Medicine. A Concise Overview for Professionals.* Butterworth Heinemann, Oxford.

Ernst, E. (2005) The efficacy of herbal medicine – an overview. *Fundamental Clinical Pharmacology* 19, 405–409.

Food and Drug Administration (2006) Food additives permitted for direct addition to food for human consumption; bacteriophage preparation. 21 CFR Part 172, pp. 47729–47732.

Fernandez, P.S., George, S.M., Sills, C.C. and Peck, M.W. (1997) Predictive model of the effect of CO_2, pH, temperature and NaCl on the growth of *Listeria monocytogenes. International Journal of Food Microbiology* 37, 37–45.

Fraise, A.P. (2002) Biocide abuse and antimicrobial resistance — a cause for concern? *Journal of Antimicrobial Chemotherapy* 49, 11–12.

Glass, K.A. and Doyle, M.P. (1989) Fate of *Listeria monocytogenes* in processed meat products during refrigerated storage. *Applied Environmental Microbiology* 55, 1565–1569.

Gordien, A.Y., Ingleby, K., Franzblau, S.G. and Seidel, V. (2010) Activity of Scottish plant, lichen and fungal endophyte extracts against *Mycobacterium aurum* and *Mycobacterium tuberculosis. Phytotherapy Research* 24, 692–698.

Gulluce, M., Sahin, F., Sokmen, M., Ozer, H., Daferera, D., Sokmen, A., Polissiou, M., Adiguzel, A. and Ozkan, H. (2007) Antimicrobial and antioxidant properties of the essential oils and methanol extract from *Mentha longifolia* L. ssp. *longifolia. Food Chemistry* 103, 1449–1456.

Gutierrez, J., Rodriguez, G., Barry-Ryan, C. and Bourke, P. (2008) Efficacy of plant essential oils against food-borne pathogens and spoilage bacteria associated with ready to eat vegetables: antimicrobial and sensory screening. *Journal of Food Protection* 71, 1846–1854.

Gutierrez J., Bourke, P., Lonchamp, J. and Barry-Ryan, C. (2009) Impact of plant essential oils on microbiological, organoleptic and quality markers of minimally processed vegetables. *Innovative Food Science and Emerging Technologies* 10, 135–296.

Hanene, M., Chaieb, K., Ammar, E. and Bakhrouf, A. (2010) Inhibitory effect of clove oil (*Syzygium aromaticum*) against *Listeria monocytogenes* cells incubated in fresh-cut salmon. *Journal of Food Safety* 30, 432–442.

Hao, Y.Y., Brackett, R.E. and Doyle, M.P. (1998a) Efficacy of plant extracts in inhibiting *Aeromonas hydrophila* and *Listeria monocytogenes* in refrigerated, cooked poultry. *Food Microbiology* 15, 367–378.

Hao, Y.Y., Brackett, R.E. and Doyle, M.P. (1998b) Inhibition of *Listeria monocytogenes* and *Aeromonas hydrophila* by plant extracts in refrigerated cooked beef. *Journal of Food Protection* 61, 307–312.

Herr, S.M., Ernst, E. and Young, V.S.L. (2002) *Herb–Drug Interaction Handbook*, Church Street Books, Nassau, NY.

Holah, J.T. (2000) Cleaning and disinfection. In: Dennis, C. and Stringer M.F. (eds) *Chilled Foods: a Comprehensive Guide.* Woodhead Publishing, Cambridge, UK, pp. 397–428.

Ippolito, A. and Nigro, F. (2003) Natural antimicrobials in post-harvest storage of fresh fruits and vegetables. In: Roller S., (ed.) *Natural Antimicrobials for the Minimal Processing of Foods.* Woodhead Publishing, Cambridge, UK pp. 201–223.

Joseph, B. and Sujatha, S. (2011) Bioactive compounds and its autochthonous microbial activity of extract and Clove oil (*Syzygium aromaticum* L.) on some food borne pathogen. *Asian Journal of Biological Sciences* 4, 35–43.

Kim, J.M., Marshall, M.R. and Wei, C. (1995) Antibacterial activity of some essential oil components against five foodborne pathogens. *Journal of Agricultural Food Chemistry* 43, 2839–2845.

Klaenhammer, T.R. (1988) Bacteriocins of lactic acid bacteria. *Biochimie* 70, 337–349.

Kouassi, Y. and Shelef, L.A. (1998) Inhibition of *Listeria monocytogenes* by cinnamic acid – possible interaction of the acid with cysteinyl residues. *Journal of Food Safety* 18, 231–242.

Kozlowski, R. and Walentowska, J. (2008) *Role of biocides from plant origin in protection of natural fibers against biodeterioration.* International Conference on Flax and Other Bast Plants, Saskatchewan, Canada, pp. 326–331.

Kozlowski, R., Muzyczek, M., Mieleniak, B., Walentowska, J., Flaczyk, J. and Konczewicz, W. (2008) Textile processing: state of the art and future developments, *The 5th International Conference of Textile Research Division,* NRC, Cairo, Egypt, pp. 7–8.

Laciar, A., Vaca Ruiz, M.L., Carrizo Flores, R. and Saad, J.R. (2009) Antibacterial and antioxidant activities of the essential oil of *Artemisia echegarayi* Hieron. (Asteraceae). *Revista Argentina de Microbiología* 41, 226–231.

Lafont, J., Jacyuet, J., Lafont, P., Romand, A. and Sarfasi, J. (1998) Some biological effects of spice, aromatics and condiments and other plant products on bacteria and micro mycelia. *Microbiologie-Aliments-Numtiox* 2, 239–249.

Lambert, R.J.W., Skandamis, P.N., Coote, P.J. and Nychas, G.J.E. (2001) A study of the minimum inhibitory concentration and mode of action of Oregano essential oil, thymol and carvacrol. *Journal of Applied Microbiology* 91, 453–462.

Larson, A.E., Yu, R.R.Y., Lee, O.A., Price, S., Haas, G.J. and Johnson, E.A. (1996) Antimicrobial activity of hop extracts against *Listeria monocytogenes* in media and in food. *International Journal of Food Microbiology* 33, 195–207.

Lis-Balchin, M. and Deans, S.G. (1997) Bioactivity of selected plant essential oils against *Listeria monocytogenes. Journal of Applied Microbiology* 82, 759–762.

Lopez-Malo, A., Alzamora, S.M. and Guerrero, S. (2000) Natural antimicrobials from plants. In: Alzamora, S.M., Tapia, M.S., López-Malo and A. Minimally (eds), *Processed Fruits and Vegetables. Fundamentals Aspects and Applications.* Aspen Publishers. Gaithersburg, Maryland, pp. 237–264.

Lopez-Malo, A., Palou, E. and Alzamora, S.M. (2005) Naturally occurring compounds – plant sources. In: Davidson P.M., Sofos, J.N., Branen A.L (eds) *Antimicrobials in Food.* CRC Press, New York, pp. 429–451.

Marino, M., Bersani, C. and Comi, G. (2001) Impedance measurements to study the antimicrobial activity of essential oils from Lamiaceae and Compositae. *International Journal of Food Microbiology* 67, 187–195.

Mishra, N. and Behal, K.K.. (2010) Antimicrobial activity of some spices against selected microbes. *International Journal of Pharmacy and Pharmacological Sciences* 2, 187–196.

Nerbrink, E., Borch, E., Blom, H. and Nesbakken, T. (1999) A model based on absorbance data on the growth rate of *Listeria monocytogenes* and including the effects of pH, NaCl, Na-lactate and Na-acetate. *International Journal of Food Microbiology* 47, 99–109.

Nychas, G.J.E. (1995) Natural antimicrobials from plants. In: Gould, G.W. (ed), *New Methods of Food Preservation.* Blackie Academic and Professional. Glasgow, pp. 58–89.

Olasupo, N.A., Fitzgerald, D.J., Narbad, A. and Gasson, M.J. (2004) Inhibition of *Bacillus subtilis* and *Listeria innocua* by Nisin in combination with some naturally occurring organic compounds. *Journal of Food Protection* 67, 596–600.

Oussalah, M., Caillet, S., Saucier, L. and Lacroix, M. (2006) Inhibitory effects of selected plant essential oils on the growth of four pathogenic bacteria: *E. coli* O157:H7, *Salmonella typhimurium*, *Staphylococcus aureus* and *Listeria monocytogenes. Food Control* 18, 414–420.

Pandit, V.A. and Shelef, L.A. (1994) Sensitivity of *Listeria monocytogenes* to rosemary (*Rosmarinus officinalis* L.). *Food Microbiology* 11, 57–63.

Parente, E., Giglio, M.A., Ricciardi, A. and Clementi, F. (1998) The combined effect of nisin, leucocin f10, pH, NaCl and EDTA on the survival of *Listeria monocytogenes* in broth. *International Journal of Food Microbiology* 40, 65–75.

Parker, J.N. and Parker, P.M. (2004) *Herbal Medicine.* ICON Group International, Inc., San Diego, California.

Peter, K.V. (2001) *Handbook of Herbs and Spices.* CRC Press, LLC, Boca Raton, Florida.

Ragaert, P., Devlieghere, F. and Debevere, J. (2007) Role of microbiological and physiological spoilage mechanisms during storage of minimally processed vegetables. *Postharvest Biology and Technology* 44, 185–194.

Rakotonirainy, M.S. and Lavedrine, B. (2005) Screening for antifungal activity of essential oils and related compounds to control the biocontamination in libraries and archives storage areas. *International Biodeterioration and Biodegradation* 55, 141–147.

Ramos-Nino, M.E., Clifford, M.N. and Adams, M.R. (1996) Quantitative structure activity relationship for the effect of benzoic acids, cinnamic acids and benzaldehydes on *Listeria monocytogenes. Journal of Applied Bacteriology* 80, 303–310.

Rhee, S.M., Garg, V.K. and Hershey, C.O. (2004) Use of complementary and alternative medicines by ambulatory patients. *Archives of Internal Medicine* 164, 1004–1009.

Rios, J.L. and Recio, M.C. (2005) Medicinal plants and antimicrobial activity. *Journal of Ethnopharmacology* 100, 80–84.

Sapers, G. (2001) Efficacy of washing and sanitizing methods for disinfection of fresh fruit and vegetable products. *Food Technology and Biotechnology* 39, 305–311.

Seoud, M.A., Abo-El; Sarhan M.M., Omar A.E. and Helal, M.M. (2005) Biocides formulation of essential oils having antimicrobial activity. *Archives of Phytopathology and Plant Protection* 38, 175–184.
Serrano, M., Valverde, J.M., Guillén, F., Castillo, S., Martínez-Romero, D. and Valero, D. (2006) Use of *Aloe vera* gel coating preserves the functional properties of table grapes. *Journal of Agricultural Food Chemistry* 54, 3882–3886.
Tan, R.X. and Zou, W.X. (2001) Endophytes: a rich source of functional metabolites. *Natural Products Reports* 18, 448–459.
Tayel, A.A. and El-Tras, W.F. (2009) Possibility of fighting food borne bacteria with herbs and spices extracts. *Internet Journal of Food Safety* 10, 92–97.
Thomas, E. (2011) Perspectives for globalized natural medicines. *Chinese Journal of Natural Medicines* 9, 1–6.
Twort, F.W. (1915) An investigation on the nature of ultramicroscopic viruses. *Lancet* 2, 1241–1243.
Ulate-Rodriguez, J., Schafer, H.W., Zottola, E.A. and Davidson, P.M. (1997) Inhibition of *Listeria monocytogenes, Escherichia coli* O157-H7, and *Micrococcus luteus* by linear furano-coumarins in culture media. *Journal of Food Protection* 60, 1046–1049.
Valero, A., Hierro, I., González, P., Montilla, P. and Navarro, M.C. (2006) Activity of various essential oils and their main components against L3 larvae of *Anisakis simplex* s.l. In: Govil, J.N., Singh, V.K. and Arunachalam, P. (eds) *Recent Progress in Medicinal Plants, Drug Development from Molecules* vol. 11. Estudium Press, LLC, Houston, pp. 247–265.
Varona, S., Martín, A. and Cocero, M.J. (2009) Formulation of a natural biocide based on lavandin essential oil by emulsification using modified starches. *Chemical Engineering and Processing: Process Intensification* 48, 1121–1128.
Ward, S.M., Delaquis, P.J., Holley, R.A. and Mazza, G. (1998) Inhibition of spoilage and pathogenic bacteria on agar and pre-cooked roast beef by volatile horseradish distillates. *Food Research International* 31, 19–26.
Yoshida, T., Amakura, Y. and Yoshimura, M. (2010) Structural features and biological properties of ellagitannins in some plant families of the order myrtales. *International Journal of Molecular Sciences.* 11, 79–106.

2 Bacteriophages and Phage-encoded Proteins: Prospects in Food Quality and Safety

Pilar García,* Beatriz Martínez, Lorena Rodríguez and Ana Rodríguez

Introduction

Recent trends in global food production, processing, distribution and preparation are creating an increasing demand for food safety research in order to ensure a safer global food supply. Thus, food preservation technologies should be regarded as an important topic for both developing and industrialized countries. Much work is still necessary to address food safety issues along the entire food production chain, from production to consumption ('farm to fork'). Several goals should be aimed at the development of new food safety strategies: decreasing the economic losses due to food spoilage, lowering the food processing costs, keeping on government requirements, avoiding transmission of microbial pathogens through the food chain and satisfying the new consumer's preferences. In this regard, consumers are particularly aware of the health concerns regarding food additives, so the health benefits of 'natural' and 'traditional' foods processed without chemical preservatives are becoming more attractive. Meanwhile, the increasing demand of ready-to-eat, fresh-tasting, nutrient- and vitamin-rich, and minimally processed foods has prompted the major challenges in the current food industry.

The extent of microbiological problems and the trends in food safety are summarized in the World Health Organization strategy '*WHO Global Strategy for Food Safety: Safer Food for Better Health*' (http://www.who.int/foodsafety/publications/general/global_strategy/en/). The WHO Department of Food Safety and Zoonoses has estimated that food-borne and waterborne diarrhoeal diseases cause the death of approximately 2.2 million people annually.

In industrialized countries, there are three major problems to be addressed. First, new pathogens and pathogens not previously associated with food consumption have emerged, resulting from a sum of changes in food production and handling practices (Skovgaard, 2007). Second, pathogens are able to adapt to the new food production and preservation technologies, mediated by the response of bacteria to stresses and sublethal treatments (Wesche *et al.*, 2009). Third, the widespread use of antibiotics in food animal production systems has resulted in the emergence of antibiotic-resistant zoonotic bacteria that can be transmitted to humans through the food chain (Walsh and Fanning, 2008).

The sustainability in food production along with ethical aspects such as animal welfare plays an important role in the

* Corresponding author.

consumer's selection of food in the developed countries. Some food preservation technologies are therefore more likely to be accepted by consumers. Biopreservation, for instance, refers to the extension of the shelf-life and improvement of food safety using microorganisms and/or their metabolites. In this regard, there has been considerable recent interest in bacteriophages as biocontrol agents in foods (Hudson *et al.*, 2005).

Since their discovery almost a century ago, bacterial viruses (bacteriophages or phages) have been used to prevent and treat a multitude of bacterial infections (Sulakvelidze and Kutter, 2005). Recently, the Food and Drug Administration (FDA) and the United States Department of Agriculture (USDA) approved a commercial phage blend (ListShield^{TM}) to reduce *Listeria monocytogenes* contamination in several foods, which proves their potential as food biopreservatives. In addition, several studies have established phages and phage-encoded lytic proteins as promising antibacterials with a great potential in veterinary and human medicine (Matsuzaki *et al.*, 2005), and food safety (Hagens and Loessner, 2007). This chapter is focused on bacteriophages and phage-encoded lytic proteins as tools to be used in the improvement of food safety, showing advantages and the remaining obstacles to their widespread use in many parts of the world.

Bacteriophages: Discovery and Biology

Bacteriophages were independently discovered by Twort (1915) and d'Herelle (1917), although initial observations of these viruses date back to Hankin in 1896. D'Herelle was the first to appreciate the potential of phages as antimicrobials and he began to treat bacterial infections. In 1919, he performed the first successful phage therapy trial on a 12-year-old boy suffering from bacterial dysentery (Sulakvelidze and Kutter, 2005). The antibacterial activity of phages was exploited since then in the former Soviet Union, whereas in the West phage therapy declined as a result of the widespread use of antibiotics. Today, the research and practical applications of phage therapy continues mainly in the George Eliava Institute of Bacteriophage, Microbiology and Virology (Tbilisi, Georgia; http://eliava-institute.org/). The failures observed in the early phage therapy trials could be attributed to a variety of factors that have been already overcome. At present, phage biology is well known, a great number of phage genomes has been sequenced, and their analysis has led to the understanding of phage evolution (Abedon, 2009), phage–host interactions (Chibani-Chennoufi *et al.*, 2004), bacterial pathogenicity (Boyd and Brüssow, 2002), phage ecology (Weinbauer, 2004) and origin (Hendrix *et al.*, 2000). Bacteriophages are found in all those habitats where bacteria live. It has been estimated that there are about 10^{31} phages on Earth and approximately 5500 have been identified (Ackerman, 2007). The classical bacteriophage taxonomy is based on shape, size and nucleic acid type. They have been classified into 13 families (Table 2.1), three of them (*Myoviridae, Siphoviridae* and *Podoviridae*) comprise tailed phages that have double-stranded DNA and are classified within the order *Caudovirales* (Ackerman, 1998). These phages account for the 96% of the identified bacteriophages. The tailless phages have been classified into ten families although they only account for 4% of the total phages. They are cubic, filamentous, or pleomorphic and they have double- or single-stranded DNA or RNA genomes. However, recent bacteriophage research has provided new data which can accurately identify relationships between phages. Phage genera can be defined by establishing genomic relationships based on shared homologous/orthologous proteins (Lavigne *et al.*, 2008).

Phages particles (virions) contain a nucleic acid genome in a protein or lipoprotein coat or capsid (Fig. 2.1A). Phages are obligate parasites as they need the host bacteria to multiply but do not infect mammalian cells. Upon infection of the bacterial host, phages may have quite different fates. Some phages follow the lytic infection cycle (Fig. 2.1B) whereby they

Table 2.1. Bacteriophage families identified to date.

DNA	Single stranded	Inoviridae Microviridae	
	Double stranded	Non-enveloped	Myoviridae Fuselloviridae Siphoviridae Tectiviridae Podoviridae Corticoviridae Rudiviridae
		Enveloped	Plasmaviridae Lipothrixviridae
RNA	Single stranded	Leviviridae	
	Double stranded	Cystoviridae	

multiply into the bacterial cell and lyse it at the end of the cycle to release newly formed phage particles. Other phages may follow the lysogenic pathway (Fig. 2.1B) where the phage genome will integrate into the host genome, being replicated as part of it and remain in a dormant state (prophage) for extended periods of time. The following phases can be distinguished in the lytic bacteriophage developmental cycle (Fig. 2.1B):

1. Adsorption of the phage particle to the host bacterial cell surface. This step is facilitated by tail fibres that bind to specific molecules, termed phage receptors, on the bacterial cell surface. The specificity of bacterial receptors determines the host range; some phages are specific at the strain level, whereas others have a broader spectrum and are able to infect many bacterial strains.
2. Injection of the nucleic acid into the bacterium.
3. Expression of the phage early genes, synthesis of early proteins, some of them involved in the regulation of host metabolic machinery.
4. Replication of the phage genome in multiple copies.
5. Expression of the phage late genes involved in the formation of new phage particles and lysis of the host bacterium.
6. Assembly of the phage heads and tails and packaging of the nucleic acid inside.
7. Lysis of the host bacterium and release of the new phage progeny to complete the cycle.

During the lysogenic cycle, progeny phages are not produced and the bacterial host cell is not lysed. These phages are called **temperate** phages and the bacterial strain containing the phage DNA is termed lysogenic. When these lysogenic bacteria encounter adverse environmental conditions, the prophage may be activated. The lytic cycle is turned on and the newly formed phage particles will lyse the host cell.

Bacteriophages play an important role in the evolution of their hosts. Whole genome sequencing of bacteria has revealed that phage elements contribute significantly to the bacterial diversity and pathogenicity, virulence genes exchange and ecological adaptation (Brüssow *et al.*, 2004; Chen and Novick, 2009).

Phage-encoded Proteins and their Antimicrobial Potential

Bacteriophages have developed two basic ways to liberate their progeny from bacterial cells: **Filamentous phages** are continuously extruded from bacterial cells without killing the cell; **Non-filamentous phages** induce lysis of the host cell by means of specific proteins by two different systems: **inhibition** of peptidoglycan synthesis or **enzymatic**

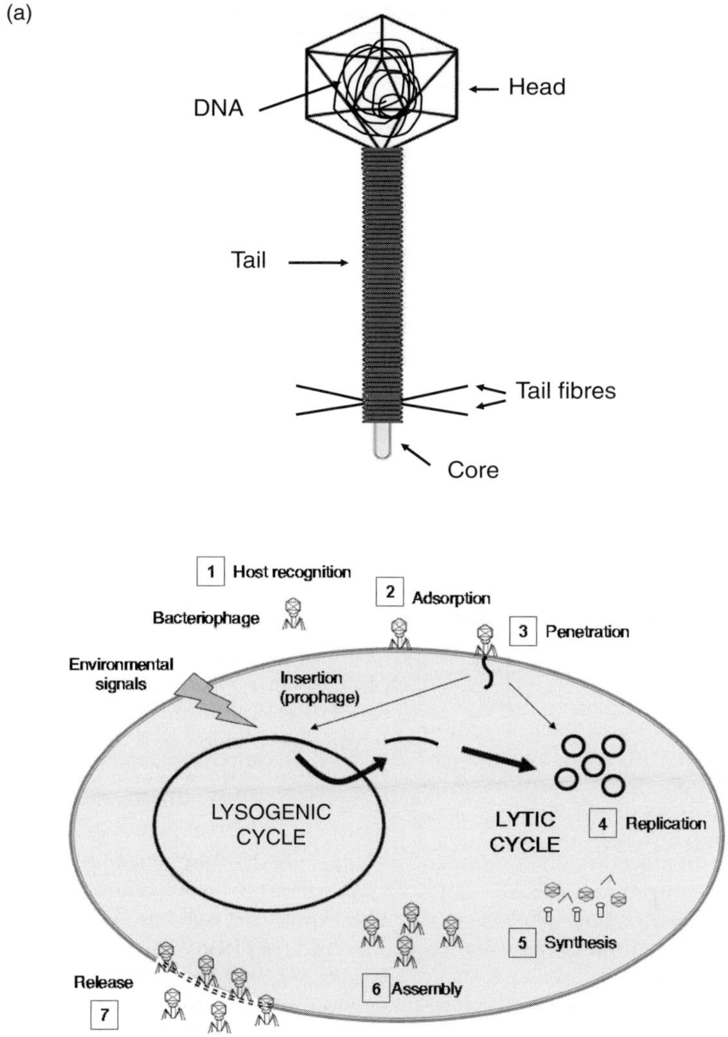

Fig. 2.1. Bacteriophage structure and life cycle: (a) schematic representation of a virion particle; and (b) main stages in the lytic and lysogenic cycles of bacteriophages.

cleavage of peptidoglycan by endolysins or holin-endolysin proteins.

Phage-encoded lytic enzymes, namely peptidoglycan cell wall hydrolases, can be divided into two groups: **endolysins** that accumulate in the bacterial cytoplasm during the lytic infection cycle; and **virion-associated peptidoglycan hydrolases**, responsible for 'lysis from without' at the initial infection step (Fig. 2.2). Both proteins cause bacteriolysis by degrading the peptidoglycan of bacterial wall (Fischetti, 2005). Endolysins act late during infection, undergoing holin-mediated translocation across the inner membrane into the peptidoglycan matrix, where they cleave cell-wall covalent bonds and cause bacterial lysis and progeny phage release. Mature virions are often endowed with peptidoglycan hydrolases involved in host cell-wall

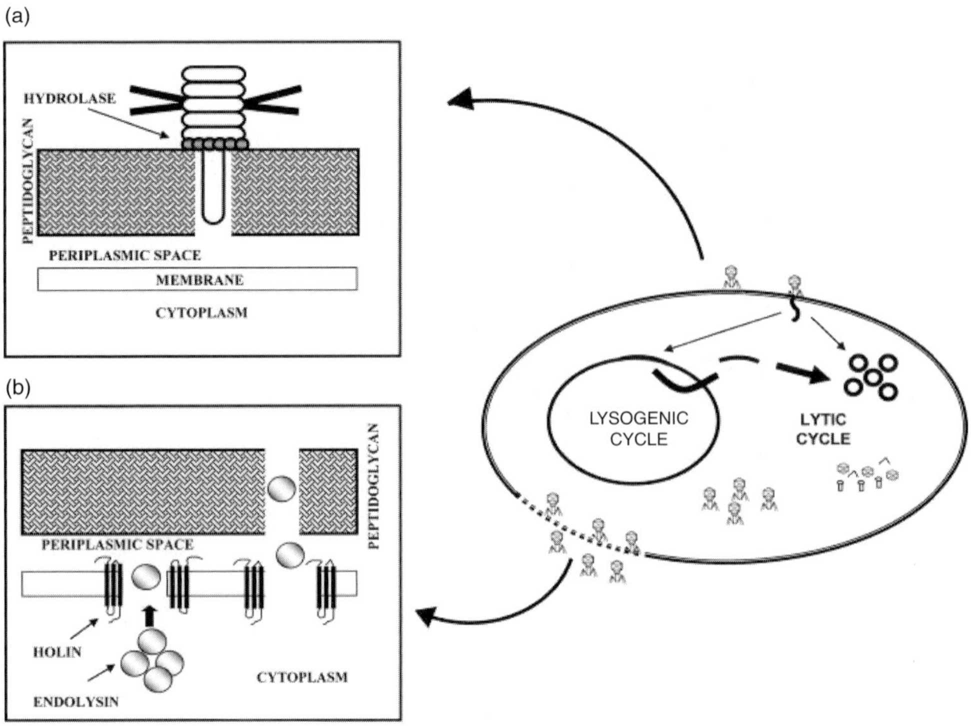

Fig. 2.2. Mode of action of the phage lytic enzymes: (a) virion-associated peptidoglycan hydrolase; and (b) holin-endolysin system.

degradation prior to injecting their genetic material during infection (Moak and Molineux, 2004). Their action generates a small hole through which the tail crosses the cell envelope.

Usually, the genes encoding the lytic proteins are clustered in a so-called lysis cassette, in which the holin gene is located immediately upstream of the endolysin gene (Young *et al.*, 2000). Genes encoding virion-associated peptidoglycan hydrolases are located in the morphogenetic module. However, four endolysins have been found to contain an N-terminal secretory signal and to be translocated across the cytoplasmic membrane by the host general secretion pathway. Signal peptides were identified in an *Oenococcus oeni* phage (São-José *et al.*, 2000), in *Lactobacillus plantarum* phage Φg1e (Kakikawa *et al.*, 2002), in *Escherichia coli* phage P1 (Xu *et al.*, 2004) and N4 endolysins (Stojkovic and Rothman-Denes, 2007). These endolysins bear an N-terminal signal arrest release (SAR) sequence by which they remain bound to the outer site of the inner membrane until the membrane potential is dissipated by the holin. Some lysins from phages of Gram-negative bacteria are capable of disturbing bacterial cells by means of a mechanism completely independent of their enzymatic activity. These endolysins contain sequences in the C terminus similar to those typical of cationic antimicrobial peptides (Orito *et al.*, 2004) that enable interactions with the negatively charged bacterial outer membrane components.

Most of endolysin and virion-associated peptidoglycan hydrolases proteins are composed of two structural domains, a C-terminal binding domain and one or two N-terminal catalytic domains (Fischetti, 2005). The catalytic domain expresses at least one of the six types of activity (Fig. 2.3): N-acetylmuramoyl-L-alanine amidase, inter-peptide bridge endopeptidase, N-acetyl-β-D-glucosaminidase, L-alanoyl-D-glutamate

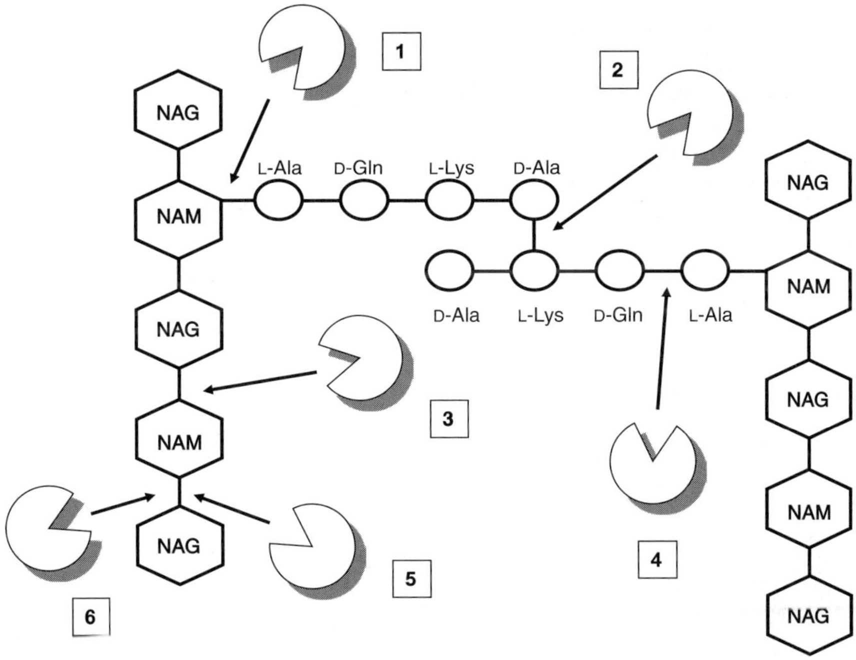

Fig. 2.3. Peptidoglycan hydrolase activities and their bond substrate in the cell wall. 1: N-acetylmuramoyl-L-alanine amidase; 2: interpeptide bridge endopeptidase; 3: N-acetyl-β-D-glucosaminidase; 4: L-alanoyl-D-glutamate endopeptidase; 5: N-acetyl- β-D-muramidase; 6: Transglycosylase.

endopeptidase, N-acetyl- β-D-muramidase and lytic transglycosylase.

Usually, one endolysin displays only one muralytic activity; however, several bifunctional lysins and virion-associated peptidoglycan hydrolases have also been reported so far. Phages B30 and NCTC 11261 from *Streptococcus agalactiae* display endopeptidase and lysozyme activities (Pritchard *et al.*, 2004); phages ΦI1, ΦH5 and ΦMR11 from *S. aureus* (Navarre *et al.*, 1999; Rashel *et al.*, 2007; Obeso *et al.*, 2008) and phage ΦWMY from *Staphylococcus warneri* M (Yokoi *et al.*, 2005) show endopeptidase and amidase activities. Finally, a virion-associated peptidoglycan hydrolase from *S. aureus* phiMR11 show amidase and lysozyme activities (Rashel *et al.*, 2008), The cell-wall-binding domain (CBD) at the C terminus binds to a specific substrate found in the cell wall of the host bacterium offering some degree of specificity to the enzyme (Loessner *et al.*, 2002). An important factor probably determining efficiency and specificity of lysins is the number and nature of their receptors on the cell wall. The endolysin receptors seem to be carbohydrate components that are present in $4–8 \times 10^4$ per cell and the affinity of the non-covalent binding is very high (affinity constant of $3–6 \times 10^8$) (Loessner *et al.*, 2002). Specificity of some endolysins is often restricted to the host bacterial species of the phage from which a certain endolysin was derived and, in some cases, it is genus specific (Loessner *et al.*, 2002).

Phage-encoded lytic enzymes are capable of degrading peptidoglycan when applied to Gram-positive bacterial cells, resulting in rapid lysis of the bacteria. It should be noticed that it is unlikely that bacteria become resistant to lysins as they target essential structures in the bacterial cell wall (Fischetti, 2005). The potential of many phage endolysins as therapeutics or biocontrol agents has already been demonstrated (Borysowski *et al.*, 2006; Obeso *et al.*, 2008). Additionally, the binding capacity of these enzymes could probably be exploited for sensitive diagnostic methods (Kretzer *et al.*, 2007). Endolysins could be used as therapeutic agents, particularly in external applications. Accordingly, several studies have been carried out to control infection by streptococci or by *Bacillus anthracis* in mice. Endolysins are also potentially useful for treating mucosal and other infections in animals and humans (Borysowski *et al.*, 2006).

Bacteriophages and Phage Lytic Proteins: Applications in Food Quality and Safety

The role of phages and phage lytic proteins to improve food quality and safety is focused on five main areas (Fig. 2.4).

Phage therapy and prophylaxis in primary production

Animals are the source of important zoonotic pathogens such as *E. coli*, *Salmonella* spp., *Campylobacter* spp. and *Listeria* spp. The use of phages as therapeutics to treat and/or prevent infections in animals will subsequently reduce the risk of food contamination. Since the 1980s when phage therapy trials began in animals, a number of papers confirm the success of phages to reduce pathogen bacteria. Phages did not completely eliminate bacteria, but numbers were reduced in treated animals compared with control animals (Johnson *et al.*, 2008; Atterbury, 2009). By contrast, the treatment of bovine mastitis caused by *S. aureus* with phages was far from successful as the cure rate of phage-treated quarters hardly differed from the control quarters (Gill *et al.*, 2006). However, the *S. aureus* bacteriophage phi11 endolysin has been proposed as an anti-mastitis agent that can be expressed in transgenic cow mammary glands (Donovan *et al.*, 2006a) in order to avoid *S. aureus* milk contamination. The ability of phages to prevent fish infections caused by *Lactococcus garviae* or *Pseudomonas plecoglossicida* was also evaluated. Phages were administered intraperitoneally or orally as phage-impregnated feed. The 80–100% of fish

Fig. 2.4. Bacteriophages and phage lytic enzymes: applications in food quality and safety.

inoculated with pathogenic bacteria survived following phage administration, compared with only 10% survival in the control group, where no phage was administered (Park and Nakai, 2003).

Decontamination of raw food

Raw food includes fresh fruit, vegetables, carcases and ready-to-eat foods, which may represent an important source of food-borne pathogens because most of them do not undergo any processing to kill pathogens before consumption. Different results were obtained after application of phages on diverse fruits. On fresh cuts of honey melons stored between 5°C and 10°C, *Salmonella enterica* serovar *enteritidis* counts were reduced by 3.5 log units but no significant decrease on apple slices was obtained due the low pH of this fruit (Leverentz *et al.*, 2001). On the other hand, two *Salmonella* bacteriophages were evaluated for pathogen control on experimentally contaminated alfalfa seeds and an approximately 1 log unit reduction was achieved 3 h after phage application (Kocharunchitt *et al.*, 2009). Likewise, a bacteriophage cocktail against *E. coli* O157:H7 was able to reduce the pathogen counts from 94% to 100% in experimental contaminated tomato, spinach, broccoli and ground beef (Abuladze *et al.*, 2008). Ready-to-eat products are prone to be contaminated by *L. monocytogenes* due to its ability to grow at refrigeration temperatures. Regarding this, *L. monocytogenes* contamination was controlled on surface-ripened red smear soft cheese with phage P100 (Carlton *et al.*, 2005), and reductions up to 5 log units were detected on solid foods (Guenther *et al.*, 2009).

Food environment decontamination

Phages and phage lytic proteins may also be used for sanitizing food-processing equipment. One of the main problems in the food industry is the contamination of surfaces with biofilm-producing pathogens as these structures are difficult to remove. Some phages have been shown to possess enzymes that can degrade bacterial polysaccharide and they have been recently used in biofilm removal of clinically relevant organisms (Azeredo and Sutherland, 2008). In the industrial environment, the treatment against *L. monocytogenes* on stainless steel and polypropylene surfaces gave a significant reduction of bacteria numbers (Roy *et al.*, 1993). Differences in phage effectiveness to remove *Pseudomonas fluorescens* biofilms were observed depending on the biofilm age (Sillankorva *et al.*, 2008). Additionally, a staphylococcal phage endolysin has been also shown to be effective in disrupting *S. aureus* biofilms (Sass and Bierbaum, 2007). Recently, phages have been successfully engineered to express an enzyme, DspB, which hydrolyses the biofilm formed by *Staphylococcus* and *E. coli* (Lu and Collins, 2007).

Food biopreservation

The addition of phages and endolysins as biopreservatives in food has been shown to be effective to inhibit pathogen and spoilage bacteria development and, thus, increase the shelf life of food products. Inhibition by phages of food-borne pathogens such as *S. enterica* serovar *typhimurium* DT104 on frankfurters (Whichard *et al.*, 2003), *S. aureus* in dairy products (Garcia *et al.*, 2007, 2009), and *Enterobacter sakazakii* in reconstituted infant formula milk (Kim *et al.*, 2007) has been assessed. Likewise, phage SJ2 was able to kill *Salmonella* in Cheddar cheese (Modi *et al.*, 2001). A shelf-life extension of pork adipose tissue from 4 to 8 days occurred by treating *Brochothrix thermosphacta* contamination with phages, indicating that phages could be used to extending the storage quality of chilled meats (Greer and Dilts, 2002). As lysins isolated so far are remarkably stable and relatively easy to produce in a purified form and in large quantities, they are amenable to preventing pathogen growth in the product (Obeso *et al.*, 2008). An alternative approach is the *in situ* production of endolysins in fermented products by using genetically modified starter cultures able to synthesize and secrete them (Gaeng *et al.*, 2000; Turner *et al.*, 2007).

Food-borne bacteria detection

Phages and phage lytic proteins can be used as tools for detecting pathogens in foods or in the food manufacturing environment (Rees and Dodd, 2006; Hagens and Loessner, 2007). Their low-cost and readily available production systems, added to their specificity for a target bacterial species, make phages and endolysins ideal for bacterial detection. Another important advantage of phage-based systems is that usually they only detect living bacteria, thereby reducing the number of false positives. Several methods have been developed to detect bacteria in foods by using phages. Phages have been covalently labelled with a fluorescent dye to the phage coat and used for specific binding to their host as a means of bacteria detection (Goodridge *et al.*, 1999; Oda *et al.*, 2004; Tanji *et al.*, 2004). In other systems, phages are used to promote bacterial lysis and bacterial components such adenylate kinase are detected (Corbitt *et al.*, 2000). Detection of pathogenic bacteria has also been approached by using phages to deliver reporter genes that are expressed after infection such *lux* or the green fluorescent protein gene (Kodikara *et al.*, 1991; Funatsu *et al.*, 2002). Finally, phages can be attached to specific peptides, antibody fragments or gold surfaces that will bind specific bacterial pathogens or toxins (Petrenko and Vodyanoy, 2003; Singh *et al.*, 2009). Regarding endolysins, the amino acid residues critical for specific binding of the lysin PlyG from *B. anthracis* gamma phage have been used for the effective identification of this pathogen (Fujinami *et al.*, 2007). The complete *B. anthracis* endolysin was also exploited as part of a rapid method able to detect as few as 100 spores in 5 min after enzyme addition (Schuch *et al.*, 2002). Recently, synthetic peptides based on PlyG coupled with Qdot-nanocrystals enhanced detection of this pathogen (Sainathrao *et al.*, 2009). Recombinant CBDs of *L. monocytogenes* bacteriophage endolysins coated to magnetic beads enabled immobilization and recovery of more than 90% of *L. monocytogenes* cells from foods (Kretzer *et al.*, 2007).

Advantages and Drawbacks in the Use of Bacteriophage-derived Biocontrol Agents

It is generally accepted that phages and phage lytic proteins have advantages over other antimicrobial technologies that might be exploited to improve food quality and safety. The main advantages are:

1. *Specificity and self-replicating capacity*. Phages will only actively replicate as long as susceptible hosts are available and phage number will increase in their presence. Phages and phage lytic enzymes can be targeted towards specific pathogens without interfering with the desirable natural microbiota or the starter cultures in fermented products.
2. *Effectiveness* in bacterial killing even against antibiotic-resistant bacteria, providing an alternative approach against pathogen bacteria causing infections in livestock.
3. *History of safe use*. The long-term use of phage to treat human infections in the former Soviet Union and Eastern Europe countries with hardly any negative side effects supports the safe use of phages. In addition, the animal studies performed in Western countries show that the oral administration of bacteriophages is harmless (Carlton *et al.*, 2005; Denou *et al.*, 2009), and human volunteers who ingested up to 10^5 PFU of phage T4 did not suffer any detectable negative side effects (Bruttin and Brüssow, 2005). Regarding lytic enzymes, the bonds that are targeted and hydrolysed are only present in bacterial cell walls but not in eukaryotic cells. No oral toxicity studies hitherto have been conducted with phage lytic proteins. However, lytic enzymes seem to be innocuous after both topical (Loeffler *et al.*, 2001) and systemic (Loeffler *et al.*, 2003) treatment in mice. Furthermore, even the repeated nasal or intravenous administration

of large amounts of the enzyme revealed no signs of toxicity, as assessed by observing the weight, aspect and behaviour of the treated mice for 4 weeks (Loeffler *et al.*, 2003).

4. *Phages do not alter organoleptic properties of food.* No changes in the colour, texture or taste were detected after the addition of phages (Greer, 2005).

5. *Endolysins do not generate bacterial resistance.* An important issue that should boost the applications of lytic enzymes and, in particular, their use as biopreservatives in food with respect to other antimicrobial treatments, is the lack of bacterial resistance to endolysins. It has been shown that pneumococcal phage lysins bind to essential molecules of these bacteria (Loeffler *et al.*, 2001). In fact, to date no bacterial resistance to endolysins has been reported even after repeated exposure of *S. pneumoniae* to endolysins (Loeffler *et al.*, 2001) or by stimulating mutant development in *B. cereus* (Schuch *et al.*, 2002).

However, there are other important characteristics of phages and phage lytic proteins that should be considered carefully as they may hamper its use as biocontrol agents in food:

1. *Phages may carry harmful genes.* Several temperate phages have been shown to carry bacterial toxin genes or virulence factors. Examples of phage-mediated virulence include virulent strains of *E. coli, Shigella* spp., *P. aeruginosa, Vibrio cholerae, S. aureus, Streptococcus pyogenes, Clostridium tetani, Cl. botulinum,* and *Corynebacterium diphtheriae* (Brüssow *et al.*, 2004). To overcome this drawback, full phage genome sequences and bioinformatic analysis are indispensable to minimize the risk of selecting phages which could confer pathogenic traits onto previously harmless bacteria.

2. *Narrow host range.* Besides being an advantage, the restricted host range of phages could become an obstacle to the successful use of phages as control agents of food-borne bacteria. This drawback could be, however, solved by designing phage cocktails with broader spectrum activity. Phages may also be genetically modified to alter their host binding profile (Mahichi *et al.*, 2009).

3. *Bacterial phage resistance.* Bacteria might develop resistance to phage infection by mutation or loss of the receptor site. When bacterial cells acquire the lysogeny status, they also become immune to external infection by another phage of the same type. The frequency of resistance may vary considerably depending on the phage–host combination. Recently, a new resistance mechanism has been reported (CRISPRs, (clustered regularly interspaced short palindromic repeats), which is widespread in the genomes of many bacteria (Sorek *et al.*, 2008). The use of phage cocktails could decrease the emergence of phage-resistant variants.

4. *Inactivation by food environmental conditions.* Another important issue is how a phage-derived agent will survive through processing and intrinsic chemical conditions of food. Specifically, the ratio phage:bacteria may determine the success of phage infection as a minimum density of host cells is required in order to support phage replication (Cairns *et al.*, 2009). This parameter may be crucial in food, where a low concentration of host bacteria is expected. The bacterial threshold is also determined by physico-chemical parameters such as the contact surface, the amount of fluid contained, adsorption capacity, proteases, etc. (Bigwood *et al.*, 2009). To overcome this difficulty, a deep knowledge of the dynamics of the phage infection process in different food matrices is required and should be determined empirically on a case-by-case basis. It might be also possible to design phage mutants able to resist specific food-processing conditions. Regarding phage lytic proteins, most of the endolysins require divalent cations for their activity (Donovan *et al.*, 2006a) and are susceptible to be inactivated by the chemical and physical conditions in food processes. Generally, endolysins seem to be rather thermostable proteins (Loeffler *et al.*, 2003). However, incubation of LysH5 for 30 min at 63°C or 1 min at 72°C fully inactivated the protein (Obeso *et al.*, 2008). Although the pH optimum for lysins is usually within the range of 4.0–6.0, low pH inactivates antibacterial activity (Obeso *et al.*, 2008).

5. *Phage lytic proteins do not lyse Gram-negative bacteria.* In these bacteria, the peptidoglycan layer is surrounded by the outer membrane that renders them resistant to endolysin activity. However, some endolysins are capable of killing Gram-negative bacteria despite the presence of the outer membrane, by means of their C-terminal peptide sequences (Orito *et al.*, 2004).

Prospects for the Use of Phages and Phage Lytic Enzymes in Food Safety

The ability of the phages to kill bacterial cells at the end of the infectious cycle is the basis of using phages as antimicrobial agents. As a result of studies on non-human applications of phages, some of which are already implemented in USA, it has gone a step further towards a wider use of these agents, including food safety, agriculture, animal veterinary, aquaculture, waste-water treatment, surface disinfection, bacteria detection and environmental remediation applications. In fact, a number of biotech companies have developed, or are in the process of developing, phage-based products, some of which have already received regulatory approval (Table 2.2). In spite of phage products not being commercially available in European countries, some companies such as Omnilytics Inc. commercializes a phage cocktail (AgriPhage^{TM}) that has received the Environmental Protection Agency (EPA) approval for application on tomato and pepper crops, and more recently, the USDA's Food Safety and Inspection Service has approved the use of a *E. coli* O157:H7 targeted bacteriophage product to treat live animals prior to slaughter. Other products such as ListShield^{TM} and EcoShield^{TM} targeting *L. monocytogenes* and *E. coli* O157:H7 contamination in food and food processing facilities, respectively, are commercialized by IntraLytix. ListShield^{TM} (formerly LMP-102) has received the FDA and USDA approval for direct application onto foods, and the EPA approval for application on surfaces in food facilities and other establishments. Likewise, Listex^{TM} P100, an antilisterial product manufactured by Ebi Food Safety,

Table 2.2. Main companies marketing phage-derived products.

Company	Country	Website	Target
Gangagen	USA	http://www.gangagen.com/	Human infections
Intralytix	USA	http://www.intralytix.com/	Food safety
Omnilytics	USA	http://www.phage.com/home5.html	Crops (Agriphage)
Phage Biotech	Israel	http://www.phage-biotech.com/	Human infections
Hexal Genentech	Germany	http://www.hexal-gentech.com/index.html	Human infections
Novolytics	UK	http://www.novolytics.co.uk/	Human infections
Biophage Inc.	Canada	http://www.biophagepharma.net/index.html	Environment, human and animal
Biopharm Pharmaceuticals	Georgia	http://www.biopharmservices.com/Pharma.aspx	Human infections
EBI Food Safety	Netherlands	http://www.ebifoodsafety.com	Food safety (Listex P100^{TM})
BigDNA	UK	(http://www.bigdna.com/)	Animal infections
JSC Biochimpharm	Georgia	(http://www.biochimpharm.ge/)	Human infections
Biocontrol	UK	(http://www.biocontrol-ltd.com/)	Human infections
Innophage	Portugal	(http://www.innophage.com/)	Human infections
Phico Therapeutics	UK	(http://www.phicotherapeutics.co.uk/)	Human infections
Phage International	USA, Georgia	(http://www.phageinternational.com/)	Human infections
Targanta Therapeutics	USA	(http://www.targanta.com/)	Human infections
Viridax	USA	(http://www.viridax.com/)	Human infections

has got FDA and USDA approval as 'generally recognized as safe' (GRAS) for use in cheese and, more recently, for all food products.

Most of these biotech companies are also investing in phage-based products for animal and human use. These studies will promote acceptance and future approval by regulatory government agencies of novel phage products.

Regarding phage lytic enzymes, the main characteristic of endolysins is their high lytic activity. A small amount of purified recombinant endolysin applied exogenously is sufficient to rapidly lyse a dense suspension (10^9–10^{10}cfu/ml) of cells within minutes. The high substrate specificity and the high activity of endolysins support a number of future applications in food science. Their activity could even be enhanced because they act synergistically with other antimicrobials (Loeffler and Fischetti, 2003). Current molecular biology techniques offer attractive options to develop new antimicrobials by protein engineering. It seems plausible that different catalytic domains could be swapped or put together resulting in proteins with different bacterial and catalytic specificities, as previously showed in pneumococcal lysins (Lopez *et al.*, 1997). In addition, several truncated proteins have allowed the determination of the activity of each domain. It has been shown that in phi11 and LysK endolysins the CHAP domain is sufficient to lyse untreated *S. aureus* cells without either the amidase or the SH3b domains (Donovan *et al.*, 2006b). A similar result was obtained with *S. agalactiae* bacteriophage B30 endolysin (Donovan *et al.*, 2006b,c). Deletion analysis of LambdaSa2 prophage endolysin indicates that the endopeptidase domain can lyse *Streptococcus* strains with a higher specific activity than the full length protein, while the truncated constructs harbouring the glycosidase domain are virtually inactive (Donovan and Foster-Frey, 2008). Thus, the domains can function independently and maintain their activity when fused to create novel recombinant fusion hydrolases. Furthermore, new catalytic domains can be obtained from virion-associated peptidoglycan hydrolase activities responsible for 'lysis from without' (Moak and Molineux, 2004; Manoharadas *et al.*, 2009). A growing number of conserved domains are even being recognized on other phage structural proteins that play roles in degradation of the polysaccharides on the cell surface. The MT3 motif in the tape measure protein (TMP) of the mycobacteriophage TM4 genome has been reported to be responsible for peptidoglycan-hydrolysing activity, enabling the entry of phage DNA through the thickened peptidoglycan layer of *Mycobacterium smegmatis* at the stationary phase (Dusthackeer *et al.*, 2008). Protein Pb2 from phage T5 straight fibre was predicted to consist of three domains. The Pb2 C-terminal region caused peptidoglycan hydrolysis (Boulanger *et al.*, 2008). Future bioinformatic studies to identify conserved hydrolytic domains found in phage and bacterial proteins will be essential. In addition, more clinical trials have to be carried out to assess the recent data from animal models that indicate that these enzymes are safe and effective.

Conclusions

During the past decade, knowledge on the molecular biology of phages has led to the rational development of both phages and their products for their use in a number of applications to prevent unwanted bacterial growth. Since bacteriophages are the most abundant biological entities on Earth, they constitute a very rich natural source of new phages and proteins. Furthermore, bacteriophages and derived proteins can be redesigned by protein-engineering techniques to improve and adapt their killing activity to specific requirements and to modify its target specificity. Taking in account these characteristics, phage-derived products could be used along the food chain as components of antibacterial prophylaxis treatments in animal reservoirs of pathogenic bacteria, as disinfectants in cleaning systems, as sanitizers of raw products and as biopreservatives. Complementary to this, lytic phages can specifically lyse bacteria to release cell-specific marker molecules such as enzymes to be monitored. In addition, phages and lytic enzymes have good chemical and thermal stability, and can be conjugated with nanomaterials and immobilized on a transducer surface in an analytical device. These could be used in specific detection of pathogenic or spoilage bacteria along the food chain.

Regarding their use as antimicrobials, the most recent studies and clinical trials indicate that the development of phage-derived products will be successful. The authorization by the FDA of a cocktail of phages that could be used as a food additive to kill *Listeria* opens the way for other phage-based products. In fact, the European Food Safety Authority (EFSA) had positively evaluated the use of phages as biocontrol agents in foods (EFSA, 2009). However, despite of the number of studies supporting the effectiveness of phages and lytic proteins as antibacterials especially in clinical applications, the most important hurdle to the wide use of phages in the food industry is the acceptance by the consumer. The consumers' perception of adding 'viruses' to foods will be probably very critical, although phages have been present in the human environment. In this context, phage lytic enzymes would have a better acceptance as food additives. The scientific community is aware that many studies are necessary to confirm all the safety issues for each new product. At the same time, the enormous advance in phage genomics and proteomics support their use as biological and biotechnological tools. The success of phages and phage products in food will be determined by the investment on thorough studies on safety, the transmission of correct information about the products, the regulatory status and governments' approval, and the final decision of the consumer.

References

Abedon, S.T. (2009) Phage evolution and ecology. *Advances in Applied Microbiology* 67, 1–45.

Abuladze, T., Li, M., Menetrez, M.Y., Dean, T., Senecal, A. and Sulakvelidze, A. (2008) Bacteriophages reduce experimental contamination of hard surfaces, tomato, spinach, broccoli, and ground beef by *Escherichia coli* O157:H7. *Applied and Environmental Microbiology* 74, 6230–6238.

Ackerman, H.W. (1998) Tailed bacteriophages: the Caudovirales. *Advances in Virus Research* 51, 135–201.

Ackerman, H.W. (2007) 5500 Phages examined in the electron microscope. *Archives of Virology* 152, 227–243.

Atterbury, R.J. (2009) Bacteriophage biocontrol in animals and meat products. *Microbial Biotechnology* doi :10.1111/j.1751-7915.2009.00089.

Azeredo, J. and Sutherland, I. (2008) The use of phages for the removal of infectious biofilms. *Current Pharmaceutical Biotechnology* 9, 261–266.

Bigwood, T., Hudson, J.A. and Billington, C. (2009) Influence of host and bacteriophage concentrations on the inactivation of food-borne pathogenic bacteria by two phages. *FEMS Microbiology Letters* 291, 59–64.

Borysowski, J., Weber-Dabrowska, B. and Górski, A. (2006) Bacteriophage endolysins as a novel class of antibacterial agents. *Experimental Biology and Medicine* 231, 366–377.

Boulanger, P., Jacquot, P., Plançon, L., Chami, M., Engel, A., Parquet, C., Herbeuval, C. and Letellier, L. (2008) Phage T5 straight tail fiber is a multifunctional protein acting as a tape measure and carrying fusogenic and muralytic activities. *Journal of Biological Chemistry* 283, 13556–13564.

Boyd, E.F. and Brüssow, H. (2002) Common themes among bacteriophage-encoded virulence factors and diversity among the bacteriophages involved. *Trends in Microbiology* 10, 521–529.

Brüssow, H., Canchaya, C. and Hardt, W.D. (2004) Phages and the evolution of bacterial pathogens: from genomic rearrangements to lysogenic conversion. *Microbiology and Molecular Biology Reviews* 68, 560–602.

Bruttin, A. and Brüssow, H. (2005) Human volunteers receiving *Escherichia coli* phage T4 orally: a safety test of phage therapy. *Antimicrobial Agents and Chemotherapy* 49, 2874–2878.

Cairns, B.J., Timms, A.R., Jansen, V.A., Connerton, I.F. and Payne, R.J. (2009) Quantitative models of *in vitro* bacteriophage-host dynamics and their application to phage therapy. *PLoS Pathogens* 5(1), e1000253.

Carlton, R.M., Noordman, W.H., Biswas, B., de Meester, E.D. and Loessner, M.J. (2005) Bacteriophage P100 for control of *Listeria monocytogenes* in foods: Genome sequence, bioinformatic analyses, oral toxicity study, and application. *Regulatory Toxicology and Pharmacology* 43, 301–312.

Chen, J. and Novick, R.P. (2009) Phage-mediated intergeneric transfer of toxin genes. *Science* 323, 139–141.

Chibani-Chennoufi, S., Bruttin, A., Dillmann, M.L. and Brüssow, H. (2004) Phage–host interaction: an ecological perspective. *Journal of Bacteriology* 186, 3677–3686.

Corbitt, A.J., Bennion, N. and Forsythe, S.J. (2000) Adenylate kinase amplification of ATP bioluminescence for hygiene monitoring in the food and beverage industry. *Letters in Applied Microbiology* 30, 443–447.

Denou, E., Bruttin, A., Barretto, C., Ngom-Bru, C., Brüssow, H. and Zuber, S. (2009) T4 phages against *Escherichia coli* diarrhea: potential and problems. *Virology* 388, 21–30.

d'Herelle, F. (1917) Sur un microbe invisible antagoniste des bacilles dysentériques. *Comptes Rendus l'Academie des Sciences Paris* 165, 373–375.

Donovan, D.M. and Foster-Frey, J. (2008) LambdaSa2prophage endolysin requires Cpl-7-binding domains and amidase-5 domain for antimicrobial lysis of streptococci. *FEMS Microbiology Letters* 287, 22–33.

Donovan, D.M., Lardeo, M. and Foster-Frey, J. (2006a) Lysis of staphylococcal mastitis pathogens by bacteriophage phi11 endolysin. *FEMS Microbiology Letters* 265, 133–139.

Donovan, D.M., Dong, S., Garrett, W., Rousseau, G.M., Moineau, S. and Pritchard, D.G. (2006b) Peptidoglycan hydrolase fusions maintain their parental specificities. *Applied and Environmental Microbiology* 72, 2988–2996.

Donovan, D.M., Foster-Frey, J., Dong, S., Rousseau, G.M., Moineau, S. and Pritchard, D.G. (2006c) The cell lysis activity of the *Streptococcus agalactiae* bacteriophage B30 endolysin relies on the cysteine, histidine-dependent amidohydrolase/peptidase domain. *Applied and Environmental Microbiology* 72, 5108–5112.

Dusthackeer, A., Hassan, V.N. and Kumar, V. (2008) Tape measure protein having MT3 motif facilitates phage entry into stationary phase cells of *Mycobacterium tuberculosis*. *Computational Biology and Chemistry* 32, 367–369.

EFSA (2009) Scientific Opinion of the Panel on Biological Hazards on a request from European Commission on the use and mode of action of bacteriophages in food production. *EFSA Journal* 1076, 1–26.

Fischetti, V.A. (2005) Bacteriophage lytic enzymes: novel anti-infectives. *Trends in Microbiology* 13, 491–496.

Fujinami, Y., Hirai, Y., Sakai, I., Yoshino, M. and Yasuda, J. (2007) Sensitive detection of *Bacillus anthracis* using a binding protein originating from gamma-phage. *Microbiology and Immunology* 51, 163–169.

Funatsu, T., Taniyama, T., Tajima, T., Tadakuma, H. and Namiki, H. (2002) Rapid and sensitive detection method of a bacterium by using a GFP reporter phage. *Microbiology and Immunology* 46, 365–369.

Gaeng, S., Scherer, S., Neve, H. and Loessner, M.J. (2000) Gene cloning and expression and secretion of *Listeria monocytogenes* bacteriophage-lytic enzymes in *Lactococcus lactis*. *Applied and Environmental Microbiology* 66, 2951–2958.

García, P., Madera, C., Martínez, B. and Rodríguez, A. (2007) Biocontrol of *Staphylococcus aureus* in curd manufacturing processes using bacteriophages. *International Dairy Journal* 17, 1232–1239.

García, P., Madera, C., Martínez, B., Rodríguez, A. and Suárez, J.E. (2009) Prevalence of bacteriophages infecting *Staphylococcus aureus* in dairy samples and their potential as biocontrol agents. *Journal of Dairy Science* 92, 3019–3026.

Gill, J.J., Pacan, J.C., Carson, M.E., Leslie, K.E., Griffiths, M.W. and Sabour, P.M. (2006) Efficacy and pharmacokinetics of bacteriophage therapy in treatment of subclinical *Staphylococcus aureus* mastitis in lactating dairy cattle. *Antimicrobial Agents and Chemotherapy* 50, 2912–2918.

Goodridge, L., Chen, J. and Griffiths, M. (1999) Development and characterization of a fluorescent-bacteriophage assay for detection of *Escherichia coli* O157:H7. *Applied and Environmental Microbiology* 65, 1397–1404.

Greer, G.G. (2005) Bacteriophage control of foodborne bacteria. *Journal of Food Protection* 68, 1102–1111.

Greer, G.G. and Dilts, B.D. (2002) Control of *Brochothrix thermosphacta* spoilage of pork adipose tissue using bacteriophages. *Journal of Food Protection* 65, 861–863.

Guenther, S., Huwyler, D., Richard, S. and Loessner, M.J. (2009) Virulent bacteriophage for efficient biocontrol of *Listeria monocytogenes* in ready-to-eat foods. *Applied and Environmental Microbiology* 75, 93–100.

Hagens, S. and Loessner, M.J. (2007) Application of bacteriophages for detection and control of foodborne pathogens. *Applied Microbiology and Biotechnology* 76, 513–519.

Hendrix, R.W., Lawrence, J.G., Hatfull, G.F. and Casjens, S. (2000) The origins and ongoing evolution of viruses. *Trends in Microbiology* 8, 504–508.

Hudson, J.A., Billington, C., Carey-Smith, C. and Greening, G. (2005) Bacteriophages as biocontrol agents in food. *Journal of Food Protection*. 68, 426–437.

Johnson, R.P., Gyles, C.L., Huff, W.E., Ojha, S., Huff, G.R., Rath, N.C. and Donoghue, A.M. (2008) Bacteriophages for prophylaxis and therapy in cattle, poultry and pigs. *Animal Health Research Reviews* 9, 201–215.

Kakikawa, M., Yokoi, K.J., Kimoto, H., Nakano, M., Kawasaki, K., Taketo, A. and Kodaira, K. (2002) Molecular analysis of the lysis protein Lys encoded by *Lactobacillus plantarum* phage Φg1e. *Gene* 299, 227–234.

Kim, K.P., Klumpp, J. and Loessner, M.J. (2007) *Enterobacter sakazakii* bacteriophages can prevent bacterial growth in reconstituted infant formula. *International Journal of Food Microbiology* 115, 195–203.

Kocharunchitt, C., Ross, T. and McNeil, D.L. (2009) Use of bacteriophages as biocontrol agents to control *Salmonella* associated with seed sprouts. *International Journal of Food Microbiology* 128, 453–459.

Kodikara, C.P., Crew, H.H. and Stewart, G.S. (1991) Near on-line detection of enteric bacteria using *lux* recombinant bacteriophage. *FEMS Microbiology Letters* 67, 261–265.

Kretzer, J.W., Lehmann, R., Schmelcher, M., Banz, M., Kim, K., Korn, C. and Loessner, M.J. (2007) Use of high-affinity cell wall-binding domains of bacteriophage endolysins for immobilization and separation of bacterial cells. *Applied and Environmental Microbiology* 73, 1992–2000.

Lavigne, R., Seto, D., Mahadevan, P., Ackermann, H.W. and Kropinski, A.M. (2008) Unifying classical and molecular taxonomic classification: analysis of the *Podoviridae* using BLASTP-based tools. *Research in Microbiology* 159, 406–414.

Leverentz, B., Conway, W.S., Alavidze, Z., Janisiewicz, W.J., Fuchs, Y., Camp, M.J., Chighladze, E. and Sulakvelidze, A. (2001) Examination of bacteriophage as a biocontrol method for *Salmonella* on fresh-cut fruit: a model study. *Journal of Food Protection* 64, 1116–1121.

Loeffler J.M. and Fischetti, V.A. (2003) Synergistic lethal effect of a combination of phage lytic enzymes with different activities on penicillin-sensitive and -resistant *Streptococcus pneumoniae* strains. *Antimicrobial Agents and Chemotherapy* 47, 375–377.

Loeffler, J.M., Nelson, D. and Fischetti, V.A. (2001) Rapid killing of *Streptococcus pneumoniae* with a bacteriophage cell wall hydrolase. *Science* 294, 2170–2172.

Loessner, M.J., Kramer, K., Ebel, F. and Scherer, S. (2002) C-terminal domains of *Listeria monocytogenes* bacteriophage murein hydrolases determine specific recognition and high-affinity binding to bacterial cell wall carbohydrates. *Molecular Microbiology* 44, 335–349.

Loeffler, J.M., Djurkovic, S. and Fischetti, V.A. (2003) Phage lytic enzyme Cpl-1 as a novel antimicrobial for pneumococcal bacteremia. *Infection and Immunity* 71, 6199–6204.

Lopez, R., Garcia, E., Garcia, P. and Garcia, J.L. (1997) The pneumococcal cell wall degrading enzymes: a modular design to create new lysins? *Microbial Drug Resistance* 3, 199–211.

Lu, T. and Collins, J. (2007) Dispersing biofilms with engineered enzymatic bacteriophage. *Proceedings of the National Academy of Sciences of the United States of America* 104, 11197–11202.

Mahichi, F., Synnott, A.J., Yamamichi, K., Osada, T. and Tanji, Y. (2009) Site-specific recombination of T2 phage using IP008 long tail fiber genes provides a targeted method for expanding host range while retaining lytic activity. *FEMS Microbiology Letters* 295, 211–217.

Manoharadas, S., Witte, A. and Bläsi, U. (2009) Antimicrobial activity of a chimeric enzybiotic towards *Staphylococcus aureus*. *Journal of Biotechnology* 139, 118–123.

Matsuzaki, S., Rashel, M., Uchiyama, J., Sakurai, S., Ujihara, T., Kuroda, M., Ikeuchi, M., Tani, T., Fujieda, M., Wakiguchi, H. and Imai, S. (2005) Bacteriophage therapy: a revitalized therapy against bacterial infectious diseases. *Journal of Infection and Chemotherapy* 11, 211–219.

Moak, M. and Molineux, I.J. (2004) Peptidoglycan hydrolytic activities associated with bacteriophage virions. *Molecular Microbiology* 51, 1169–1183.

Modi, R., Hirvi, Y., Hill, A. and Griffiths, M.W. (2001) Effect of phage on survival of *Salmonella* Enteritidis during manufacture and storage of Cheddar cheese made from raw and pasteurized milk. *Journal of Food Protection* 64, 927–933.

Navarre, W.W., Ton-That, H., Faull, K.F. and Schneewind, O. (1999) Multiple enzymatic activities of the murein hydrolase from staphylococcal phage Φ11. *Journal of Biological Chemistry* 274, 15847–15856.

Obeso, J.M., Martínez, B., Rodríguez, A. and García, P. (2008) Lytic activity of the recombinant staphylococcal bacteriophage phiH5 endolysin active against *Staphylococcus aureus* in milk. *International Journal of Food Microbiology* 128, 212–218.

Oda, M., Morita, M., Unno, H. and Tanji, Y. (2004) Rapid detection of *Escherichia coli* O157:H7 by using green fluorescent protein labeled PP01 bacteriophage. *Applied and Environmental Microbiology* 70, 527–534.

Orito, Y., Morita, M., Hori, K., Unno, H. and Tanji, Y. (2004) *Bacillus amyloliquefaciens* phage endolysin can enhance permeability of *Pseudomonas aeruginosa* outer membrane and induce cell lysis. *Applied Microbiology and Biotechnology* 65, 105–109.

Park, S.C. and Nakai, T. (2003) Bacteriophage control of *Pseudomonas plecoglossicida* infection in ayu *Plecoglossus altivelis*. *Diseases of Aquatic Organisms* 53, 33–39.

Petrenko, V.A. and Vodyanoy, V.J. (2003) Phage display for detection of biological threat agents. *Journal of Microbiological Methods* 53, 253–262.

Pritchard, D.G., Dong, S., Baker, J.R. and Engler, J.A. (2004) The bifunctional peptidoglycan lysin of *Streptococcus agalactiae* bacteriophage B30. *Microbiology* 150, 2079–2087.

Rashel, M., Uchiyama, J., Ujihara, T., Uehara, Y., Kuramoto, S., Sugihara, S., Yagyu, K., Muraoka, A., Sugai, M., Hiramatsu, K., Honke, K. and Matsuzaki, S. (2007) Efficient elimination of multidrug-resistant *Staphylococcus aureus* by cloned lysin derived from bacteriophage ΦMR11. *Journal of Infectious Diseases* 196, 1237–1247.

Rashel, M., Uchiyama, J., Takemura, I., Hoshiba, H., Ujihara, T., Takatsuji, H., Honke, K. and Matsuzaki, S. (2008) Tail-associated structural protein gp61 of *Staphylococcus aureus* phage ΦMR11 has bifunctional lytic activity. *FEMS Microbiology Letters* 284, 9–16.

Rees, C.E.D. and Dodd, C.E.R. (2006) Phage for rapid detection and control of bacterial pathogens in food. *Advances in Applied Microbiology* 59, 159–186.

Roy, B., Ackermann, H.W., Pandian, S., Picard, G. and Goulet, J. (1993) Biological inactivation of adhering *Listeria monocytogenes* by listeria phages and a quaternary ammonium compound. *Applied and Environmental Microbiology* 59, 2914–2917.

Sainathrao, S., Mohan, K.V. and Atreya, C. (2009) Gamma-phage lysin PlyG sequence-based synthetic peptides coupled with Qdot-nanocrystals are useful for developing detection methods for *Bacillus anthracis* by using its surrogates, *B. anthracis*-Sterne and *B. cereus*-4342. *BMC Biotechnology* 22, 67.

São-José, C., Parreira, R., Vieira, G. and Santos, M.A. (2000) The N-terminal region of the *Oenococcus oeni* bacteriophage fOg44 lysin behaves as a bona fide signal peptide in *Escherichia coli* and as a cis-inhibitory element, preventing lytic activity on oenococcal cells. *Journal of Bacteriology* 182, 5823–5831.

Sass, P. and Bierbaum, G. (2007) Lytic activity of recombinant bacteriophage Φ11 and Φ12 endolysins on whole cells and biofilms of *Staphylococcus aureus*. *Applied and Environmental Microbiology* 73, 347–352.

Schuch, R., Nelson, D. and Fischetti, V.A. (2002) A bacteriolytic enzyme that detects and kills *Bacillus anthracis*. *Nature* 418, 884–889.

Sillankorva, S., Neubauer, P. and Azeredo, J. (2008) *Pseudomonas fluorescens* biofilms subjected to phage phiIBB-PF7A. *BMC Biotechnology* 27, 79.

Singh, A., Glass, N., Tolba, M., Brovko, L., Griffiths, M. and Evoy, S. (2009) Immobilization of bacteriophages on gold surfaces for the specific capture of pathogens. *Biosensors and Bioelectronics* 24, 3645–3651.

Skovgaard, N. (2007) New trends in emerging pathogens. *International Journal of Food Microbiology* 120, 217–224.

Sorek, R., Kunin, V. and Hugenholtz, P. (2008) CRISPR – a widespread system that provides acquired resistance against phages in bacteria and archeae. *Nature Reviews Microbiology* 6, 181–186.

Stojković, E.A. and Rothman-Denes, L.B. (2007) Coliphage N4 N-acetylmuramidase defines a new family of murein hydrolases. *Journal of Molecular Biology* 366, 406–419.

Sulakvelidze, A. and Kutter, E. (2005) Bacteriophage therapy in humans. In: Kutter, E. and Sulakvelidze, A. (eds) *Bacteriophages: Biology and Application.* Boca Raton CRC Press, Palm Beach, Florida, pp. 381–436.

Tanji, Y., Furukawa, C., Na, S.H., Hijikata, T., Miyanaga, K. and Unno, H. (2004) *Escherichia coli* detection by GFP-labeled lysozyme-inactivated T4 bacteriophage. *Journal of Biotechnology* 114, 11–20.

Turner, M.S., Waldherr, F., Loessner, M.J. and Giffard, P.M. (2007) Antimicrobial activity of lysostaphin and a *Listeria monocytogenes* bacteriophage endolysin produced and secreted by lactic acid bacteria. *Systematic and Applied Microbiology* 30, 58–67.

Twort, F. W. (1915) An investigation on the nature of ultramicroscopic viruses. *Lancet* ii, 1241.

Walsh, C. and Fanning, S. (2008) Antimicrobial resistance in foodborne pathogens – a cause for concern? *Current Drug Targets* 9, 808–815.

Weinbauer, M.G. (2004) Ecology of prokaryotic viruses. *FEMS Microbiology Reviews* 28, 127–181.

Wesche, A.M., Gurtler, J.B., Marks, B.P. and Ryser, E.T. (2009) Stress, sublethal injury, resuscitation, and virulence of bacterial foodborne pathogens. *Journal of Food Protection* 72, 1121–1138.

Whichard, J.M., Sriranganathan, N. and Pierson, F. W. (2003) Suppression of *Salmonella* growth by wild-type and large-plaque variants of bacteriophage Felix O1 in liquid culture and on chicken frankfurters. *Journal of Food Protection* 66, 220–225.

Xu, M., Struck, D.K., Deaton, J., Wang, I.N. and Young, R.Y. (2004) A signal-arrest release sequence mediates export and control of the phage P1 endolysin. *Proceedings of the National Academy of Sciences of the United States of America* 101, 6415–6420.

Yokoi, K., Kawahigashi, N., Uchida, M., Sugahara, K., Shinohara, M., Kawasaki, K.I., Nakamura, S., Taketo, A. and Kodaira, K.I. (2005) The two component cell lysis genes holWMY and lysWMY of the *Staphylococcus warneri* M phage varΦWMY: cloning, sequencing, expression, and mutational analysis in *Escherichia coli*. *Gene* 351, 97–108.

Young, R.Y., Wang, I.N. and Roof, W.D. (2000) Phages will out: strategies of host cell lysis. *Trends in Microbiology* 8, 120–128.

3 A Survey of Antimicrobial Activity in Lactic Acid Bacteria of Different Origin

Ljubisa Topisirovic,* Milan Kojic, Ivana Strahinic, Djordje Fira and Natasa Golic

Introduction

Lactic acid bacteria (LAB) represent a heterogeneous group of bacteria that includes several genera: *Aerococcus, Carnobacterium, Enterococcus, Lactobacillus, Lactococcus, Leuconostoc, Oenococcus, Pediococcus, Streptococcus, Tetragenococcus, Vagococcus* and *Weissella*. These genera are commonly defined as Gram-positive, non-sporulating, catalase-negative, anaerobic but aero tolerant, acid tolerant, nutritionally fastidious, strictly fermentative organisms that lack cytochromes and produce lactic acid as a major end product of carbohydrate metabolism (Axelsson, 2004). They can synthesize lactic acid from lactose and are used for the production of a wide range of fermented dairy, meat and vegetable products. In addition, LAB are part of the indigenous microflora of human and animal gastrointestinal (GIT) and urogenital (UGT) tracts.

Some metabolic properties of LAB have been used traditionally to improve the flavour development and ripening of fermented products (Beresford *et al.*, 2001), and to prevent rapid spoilage of dairy foods and meat, as well as vegetables and silages. The ability of LAB to synthesize and secrete antimicrobial substances has long been used to preserve food, and is mainly connected with the formation of lactic acid and the concurrent reduction of pH during their metabolic activity. These bacteria produce two types of inhibitory substances: metabolic end products such as organic acids, hydrogen peroxide and diacetyl, as well as inhibitory peptides called bacteriocins (Klaenhammer, 1988; Holzapfel *et al.*, 1995). Bacteriocins are a group of proteinaceous compounds that generally show antimicrobial activity towards closely related bacteria (for a review see Nes *et al.*, 1996). However, some LAB secrete bacteriocins with a wider antibacterial spectrum (Jack *et al.*, 1995). It was found that such LAB can inhibit not only the growth of Gram-positive pathogenic and spoilage bacteria (Cintas *et al.*, 1995; Atanassova *et al.*, 2003), but also that of Gram-negative bacteria (Stevens *et al.*, 1991; Cardi, 2002). The producers of bacteriocins are not affected themselves due to specific protection mechanisms. The structure, biosynthesis, genetics and food application of LAB bacteriocins have been reviewed (Cotter *et al.*, 2005; Deegan *et al.*, 2006; Drider *et al.*, 2006; Gálvez *et al.*, 2007).

The Balkan Peninsula harbours a large variety of traditional, spontaneously fermented foods, produced from cow's, ewe's and goat's milk, such as white-pickled, soft,

* Corresponding author.

semi-hard cheeses and 'kajmak' (an autochthonous product made by the fermentation of milk fat) (Jokovic *et al.*, 2008). Artisanal fermented milk products are part of the heritage of Balkan countries. Manufacture of these homemade milk-fermented products has been conducted in a traditional way for centuries throughout the Balkan region. Thus, processes of preparation were passed down from generation to generation and have not been significantly influenced by modern food technology. Large quantities of different artisanal cheeses and other specific fermented milk products are made in households of this region without adding any known starter culture. In Serbia, these products are still manufactured in small rural households immediately after milking without any special treatment. Moreover, farmers very often prepare rennin by themselves. Since most of these products are made from non-pasteurized milk, the composition of the 'natural starter' depends on the presence of LAB in the raw milk and environment. Accordingly, these fermented products contain specific LAB that exist in the ecological localities where they were originally produced.

Bearing in mind the importance of LAB in many aspects of human activity, research on the molecular genetics and manipulation of these microorganisms has greatly expanded in the past ten years. However, so far, the main attention has been paid to genetic investigation of the LAB strains routinely used in industrial processes. Much less is known about the genetic organization of LAB isolated from specific natural niches. During the past 20 years it has been demonstrated that sources for new strains of LAB, so-called non-starter natural isolates or wild-type strains, could be artisanal milk products (McSweeney *et al.*, 1993; Cogan *et al.*, 1997; Randazzo *et al.*, 2002; Terzic-Vidojevic *et al.*, 2007). Therefore, study of the genetic organization of LAB isolated from traditionally produced homemade fermented products, as well as those of human origin, would be very interesting. Such LAB could be a potential source of genes encoding new variants of, for instance, bacteriocins, proteinases or exopolysaccharides.

The use of either bacteriocin-producing LAB or their bacteriocins in food production might be useful for food preservation and safety. A population of adventitious microflora, otherwise known as non-starter LAB, can proliferate during the ripening period and often constitutes the dominant cheese microflora. These bacteria represent the local, geographically specific microflora, and it is believed that differences between cheese qualities arise from the presence of non-starter microorganisms. The precise role of non-starter strains in flavour development is still unclear. However, whether exerting a positive or negative effect, they certainly contribute unpredictably to cheese quality.

One of the main goals of the Laboratory for Molecular Genetics of Industrial Microorganisms in the past 20 years has been the isolation and characterization of natural isolates of LAB with industrially important phenotypes like production of proteinases, bacteriocins and exopolysaccharides, aggregation ability, and probiotics, from homemade artisanal milk products (cheese, yoghurt, kajmak and butter) produced at different ecological localities of Serbia and neighbouring countries.

Antimicrobial Activity of Lactococci

The genus *Lactococcus* belongs to the group of lactic acid bacteria (LAB). *Lactococcus* species have frequently been isolated from milk, but also from other sources, indicating that they are widespread in the environment and not strictly dairy related. The genus *Lactococcus* currently comprises five species: *L. lactis*, *L. gravie*, *L. plantarum*, *L. piscium* and *L. raffinolactis* (Mundt, 1986; Cogan, 1996). *L. lactis* can be divided into two subspecies, *L. lactis* subsp. *lactis* and *L. lactis* subsp. *cremoris*.

Lactococci produce a wide variety of bacteriocins that belong to class I and class II based on their structure (De Vuyst, 1994). The distribution of bacteriocin production in lactococci was evaluated by Geis and co-authors (1983), who showed that about

5% of 280 strains surveyed produced proteinaceous inhibitors. Nisin, the most well-known and best characterized bacteriocin produced by *Lactococcus lactis,* was discovered in 1928 (Hurst, 1967) and has been used as a food preservative for more than 50 years. Nisin is a class I bacteriocin, a lantibiotic, which are small, temperature-stable bacteriocins containing post-translationally modified amino acids. Genes for nisin synthesis, immunity and regulation of expression are located on conjugative transposons (Tn*5301*, Tn*5276*) of 70 kb (Horn *et al.*, 1991; Rauch and de Vos, 1992). On the basis of structural and functional features lantibiotics are subdivided into type-A and type-B.

The class II bacteriocins consist of small (<10 kDa) heat-stable, unmodified membrane-active peptides. Although a small number of these class II bacteriocins are translated with a *sec*-dependent leader sequence, most possess a double glycine-type leader peptide. According to some of their features, this class of bacteriocins can be further divided into at least six subclasses: pediocin-like bacteriocins, two peptide bacteriocins, *sec*-dependent bacteriocins, bacteriocins without a leader sequence, cyclic peptide bacteriocins and other unmodified bacteriocins (Diep and Nes, 2002). Class II antimicrobial peptides of LAB can either be plasmid or chromosomally encoded. The synthesis and export of these bacteriocins involve four or more different genes organized in a cluster that usually consists of a number of operons that are not necessarily transcribed in the same direction (Holo *et al.*, 1991; van Belkum *et al.*, 1991). This gene cluster comprises one or two structural genes encoding the pre-bacteriocin(s), and a specific immunity gene that is usually located next to the structural gene(s). Furthermore, it includes genes encoding an ABC transporter and an accessory protein (except for *sec*-dependent bacteriocins), both of which are needed for externalization of the bacteriocin concomitant with processing of the leader peptide (Håvarstein *et al.*, 1995). In addition to the four basic genes, regulatory genes are associated with the genetic determinants of several class II bacteriocins. They specify three-component regulatory mechanisms consisting of a small induction peptide, a histidine protein kinase and a response regulator (Kleerebezem *et al.*, 1997). This so-called cell-density-dependent regulation (quorum sensing) of bacteriocin production enables the producing organisms to switch on bacteriocin production when competition for nutrients is likely to become more severe. Bacteriocin production provides a selective advantage of the producing organisms over their closely related natural competitors.

As earlier reported by Geis and co-authors (1983), we found that 7–30% of LAB isolates produce antimicrobial compounds (Kojic *et al.*, 2006; Terzic-Vidojevic *et al.*, 2007; Veljovic *et al.*, 2007). Among the lactococci, this includes different bacteriocins. Strains isolated from semi-hard cheese manufactured in Zagradje village (BGMN) and from kefir (BGKF) predominantly produce lactococcin B (Kojic *et al.*, 2006; Kojic *et al.*, 2007). In both groups the genes responsible for lactococcin B production are located on plasmids of different size, as confirmed by plasmid curing experiments. Besides lactococcin B, some BGMN isolates (BGMN1-3 and BGMN1-5), produce two more bacteriocins, designated LsbA and LsbB. The genes responsible for lactococcin B are located on plasmid pMN80 (80kb) and for LsbA and LsbB on pMN5, a small rolling circle replicating 5.6 kb plasmid (Gajic *et al.*, 2003; Kojic *et al.*, 2006). The cognate genes for LsbA and LsbB are located within a gene cluster specifying LmrB, an ATP-binding cassette-type multidrug resistance transporter protein. LsbA is a hydrophobic peptide initially synthesized with a leader peptide, while LsbB is a relatively hydrophilic protein synthesized without an N-terminal leader sequence or signal peptide. The secretion of both polypeptides was shown to be mediated by LmrB (Gajic *et al.*, 2003). We also found that the strain *Lactococcus lactis* subsp. *lactis* BGSM1-19, isolated from traditionally home-made white cheese, produces two bacteriocins: a lactococcin B-like bacteriocin named bacteriocin BacSMa and bacteriocin BacSMb that show similarity with lacticin

RM. Plasmid curing indicated that genes coding for bacteriocin synthesis and immunity seem to be located on plasmids. Strain BGSM1-19 could be used in food preservation because it exhibited anti-microbial activity against some pathogenic bacteria, such as *Salmonella paratyphi*, *Micrococcus flavus*, *Pseudomonas aeruginosa* and *Staphylococcus aureus* (Strahinic *et al.*, 2007b). Production of two or more bacteriocins by one lactococcal strain has been demonstrated elsewhere (e.g. LcnA, LcnB and LcnM/N; van Belkum *et al.*, 1992). It was found that in BGMN1-3 and BGMN1-5 the *prtP* gene encoding proteinase is collocated with *lcnB* on plasmid pMN80. Collocations of proteinase and bacteriocin genes on different plasmids were also found in other lactococcal strains (Kojic *et al.*, 2005; Kojic *et al.*, 2007). In *L. lactis* subsp. *lactis* bv. diacetylactis S50, bacteriocin production is correlated with the largest self-conjugal plasmid isolated from lactococci (pS140) of 140 kb (Kojic *et al.*, 2005). Strain S50 most probably produces two bacteriocins, lactococcin A-like and bacteriocin S50. In isolates that exhibited both bacteriocin and proteinase production, it was found that secretion of some bacteriocins is dependent on the concentration of casitone or triptone in the growth medium, as well as on the activity of PrtP proteinase, indicating that some lactococcal bacteriocins are regulated at the transcriptional and/or post-translational level (Gajic *et al.*, 1999; Kojic *et al.*, 2005; Kojic *et al.*, 2006; Kojic *et al.*, 2007). These findings can be exploited for controlled expression of bacteriocin and also other genes of interest.

Among bacteriocin producers isolated from artisanal Zlatar cheese, we also found nisin producers (Veljovic *et al.*, 2007). All of them belonged to the nisin Z variant, confirmed by PCR and sequencing. It is interesting that some bacteriocin-producing strains showed a high level of resistance to nisin (Kojic *et al.*, 2007). Natural isolate BGKF26, which produces lactococcin B, also showed resistance to the antimicrobial activity of nisin. Curing experiments indicated that nisin resistance is collocated with lactococcin B production on the plasmid of BGKF26. The nisin-resistant isolates are interesting because they can be suitable candidates for combination with nisin-producing starter cultures or nisin itself for the production of fermented milk products.

Although bacteriocins of LAB are already applied in the food industry as natural preservatives (nisin and pediocin PA-1), there is still a need for isolation of strains that produce bacteriocin(s) with broad-spectrum activity against food-spoilage bacteria. Moreover, narrow-spectrum bacteriocins can be used more specifically to inhibit certain high-risk bacteria in foods (such as *Listeria monocytogenes*) selectively without affecting harmless microbiota.

Antimicrobial Activity of Lactobacilli

The genus *Lactobacillus* contains over 110 species widely distributed in the environment (Rodas *et al.*, 2005). Lactobacilli produce a number of antimicrobial substances that might be important for food and feed fermentation and preservation, for which they would be of economic significance. Production of antimicrobial substances may be a mechanism that allows *Lactobacillus* to dominate the ecosystem by suppressing not only other bacteria but also other lactobacilli. Most strains producing antimicrobial substances in our collection were predominantly isolated from traditionally fermented food (Topisirovic *et al.*, 2006).

Analysis of autochthonous lactobacilli from this collection revealed that about 7% of them produced bacteriocins. For instance, mountain Zlatar is a specific ecological region in Western Balkans with high plant biodiversity. It was confirmed that 37 *Lactobacillus* strains among 253 isolates examined produced antimicrobial bacteriocins (Terzic-Vidojevic *et al.*, 2007). The isolates belong to subspecies *L. paracasei* subsp. *paracasei*. Eight strains separated from 1- and 10-day-old cheeses gave clear zones of growth inhibition only on the indicator strains *L. plantarum* A112 and *L. paracasei* subsp. *paracasei* BGBUK2-16/K4. On the other hand, two isolates from milk, showed 2 mm turbid zones of growth inhibition on *L. lactis* BGMN1-596, as well as on BGBUK2-16/K4 and A112 indicator strains. The remaining isolated lactobacilli exposed turbid zones of growth inhibition only on the BGBUK2-16/K4 indicator strain and all of them were isolated from 20-day-old cheese. Interestingly, no isolates producing antimicrobials were isolated from 60-day-old cheese (Topisirovic *et al.*, 2006).

Furthermore, Bukuljac cheese is traditionally homemade from heat-treated goat's milk without the addition of any bacterial starter culture. *L. paracasei* subsp. *paracasei* was the dominant strain in the microflora. Other *Lactobacillus* strains isolated from Bukuljac cheese displayed a broad spectrum of antimicrobial activity (Nikolic *et al.*, 2008). The representative isolates BGAR88-2 and BGAR89-2 showed mutual cross growth inhibition and affected the growth of *L. lactis* subsp. *lactis* strains used in the test. They also influenced the growth of *L. paracasei* subsp. *paracasei* BGSJ2-8, a producer of the bacteriocin SJ (Lozo *et al.*, 2007), but had no effect on BGBUK2-16 and BGUB9, both bacteriocin producers. Interestingly, BGAR88-2 and BGAR89-2 inhibited the growth of *Salmonella enteritidis* (clinical isolate). This is important since *Salmonella enteritidis* is associated with food-borne diseases (Notermans and Hoogenboom-Verdegaal, 1992). All BGAR lactobacilli isolates exhibited a continual clear zone of inhibition around the well even in the presence of proteinases, indicating the non-proteinaceous nature of the antimicrobials. Moreover, the antimicrobial effect was not obtained if samples were adjusted to pH 7. When isolates of *L. paracasei* subsp. *paracasei* BGAR88-2 and BGAR89-2 were further tested for hydrogen peroxide production it was found that they were intensive H_2O_2 producers, as blue pigment appeared on the colonies resulting from tetramethylbenzidine dihydrochloride oxidation by peroxidase. Hydrogen peroxide is an antimicrobial substance, with a strong oxidizing effect on bacterial cell surface proteins and lipids (Kamau *et al.*, 1990).

The strain *L. paracasei* subsp. *paracasei* BGBUK2-16, one of 23 LAB isolated from traditionally homemade white-pickled cheese produced in Bukovica village, Montenegro, exhibited the highest antibacterial activity against closely related microorganisms and some pathogenic bacteria, such as *Staphylococcus aureus, Bacillus cereus, Salmonella* sp. and *Pseudomonas aeruginosa*. It was found that strain BGBUK2-16 was able to control the overgrowth of pathogenic *S. aureus* in mixed culture with a bactericidal mode of action. Biochemical characterization of BGBUK2-16 bacteriocin, named Bac217, showed that it is active within the range of pH 3 to 12, after thermal treatment at 100°C for 15 min and after storage at 4°C for 6 months or –20°C for up to 12 months. The plateau of Bac217 production was detected in the early stationary phase of growth at 37 and 30°C. Therefore, Bac217 was classified as a class II bacteriocin, which are small (<10 kDa), heat-stable, unmodified membrane-active peptides. Additionally, plasmid curing experiments indicated that genes coding for Bac217 synthesis and immunity seem to be located on an 80 kb plasmid present in the strain. Further analysis showed that strain BGBUK2-16 was resistant to high concentrations of food-grade bacteriocin nisin. Therefore, it might be chosen together with nisin-producing *Lactococcus* strains in starter cultures for fermented food production (Lozo *et al.*, 2004).

Another bacteriocin producer was selected among 55 natural isolates of LAB separated from semi-hard homemade cheese produced on the Sjenica high plateau, Serbia. The strain was determined as *L. paracasei* subsp. *paracasei* BGSJ2-8 and its bacteriocin named BacSJ. In contrast to Bac217 this bacteriocin has a narrow antimicrobial spectrum inhibiting only the growth of the closely related species, *L. paracasei* and *L. lactis*. After three-step chromatographic purification and mass spectrometric analysis, a molecular mass of 5372 Da was demonstrated. The biochemical characteristics of BacSJ suggest it also belongs to class II bacteriocins (Lozo *et al.*, 2007).

Several species, including *L. acidophilus* and *L. rhamnosus*, are members of the normal intestinal and vaginal flora of healthy humans (Begovic *et al.*, 2009). As part of the natural vaginal flora, *Lactobacillus* plays an important role in the reproductive health of women by maintaining an acidic vaginal pH and through antagonistic interaction with pathogenic bacteria, it maintains the vaginal ecosystem in a healthy state (Marelli *et al.*, 2004; Falagas *et al.*, 2006). The acidic pH itself acts as a natural defence against sexually transmitted diseases and AIDS (Garg *et al.*, 2001). Other species, such as *L. paracasei* and *L. plantarum*, are commonly isolated from dairy products as well as from fruit and vegetables and they play an important role in human and animal nutrition. During the last two decades, screening of lactobacilli from different human sources for antibacterial activity is usually one of the first approaches in the search for putative probiotic strains. The antibacterial compound should be effective against indigenous and transient undesirable bacteria that colonize humans.

There is evidence that lactobacilli can inhibit the growth and attachment of pathogens to epithelial cells. Hydrogen peroxide and bacteriocins produced by lactobacilli can kill pathogenic microorganisms in the human body (Falagas *et al.*, 2006; Atassi *et al.*, 2006). There is increasing evidence that lactobacilli among the gastrointestinal microflora develop antimicrobial activities that participate in the host's gastrointestinal system of defence. From our collection of human intestinal isolates, *L. helveticus* BGRA43 showed a strong inhibitory effect on the growth of some closely related *Lactococcus* and *Lactobacillus* strains. Moreover, antagonistic activity against *Staphylococcus aureus*, *Escherichia coli*, *Bacillus mycoides*, *Pseudomonas* sp. and especially *Clostridium sporogenes* was confirmed. It was demonstrated that strain BGRA43 does not produce bacteriocin(s) or hydrogen peroxide. According to results indicating rapid lowering of pH during growth in skimmed milk, it could be hypothesized that organic acids are responsible for the growth inhibition (Banina *et al.*, 1998). Besides its antimicrobial activity, the rapid acidification, short lag growth phase, pleasant aroma and adequate viscosity formation have made strain BGRA43 very attractive as a starter for the production of fermented milk products.

Human isolates of *Lactobacillus* species were found to have more antagonistic activity against other pathogenic microorganisms. A strain isolated from human faeces produced a substance with potent bacteriostatic activity against a wide range of bacterial species. It inhibited anaerobic bacteria (*Clostridium* sp., *Bacteroides* sp., *Bifidobacterium* sp.) and members of the *Enterobacteriaceae*, *Pseudomonas* sp., *Staphylococcus* sp. and *Streptococcus* sp. but it did not inhibit other lactobacilli. The inhibitory activity occurred between pH 3 and pH 5 and was heat stable (Silva *et al.*, 1987). *L. gasseri* is considered to be a dominant species inhabiting the human intestine (Fujisawa *et al.*, 1992) and was found to produce a bacteriocin with a wide spectrum of bactericidal activity against enteric pathogens (Kawai *et al.*, 1994; Itoh *et al.*, 1995). McGroarty and Reid (1988) purified antibacterial proteins from urethral *L. acidophilus* and *L. casei* that were active against *E. coli*. A heat-resistant peptide was extracted from a vaginal isolate of *L. salivarius* that inhibited growth of *Enterococcus faecalis*, *Enterococcus faecium* and *Neisseria gonorrhoeae* (Ocana *et al.*, 1999). Barefoot and Klaenhammer (1983) found that 63% of the *L. acidophilus* strains examined produced bacteriocin. Interest in *L. acidophilus* resulted from its ability to colonize the human intestinal tract (Kleeman and Klaenhammer, 1982). Purified bacteriocin from intestinal *L. acidophilus* was found to be inhibitory to other lactobacilli as well as *E. faecalis* (Muriana and Klaenhammer, 1991).

There are reports that several lactobacilli strains isolated from different parts of the oral cavity produce antibacterial compounds (Koll-Klais *et al.*, 2005; Pangsomboon *et al.*, 2009). Among lactobacilli isolated from the surface of healthy teeth, oral mucous, surface and deep tooth decay, five different antimicrobial producers were found. The isolates *L. salivarius* BGHO1 and BGHO64, *L. fermentum* BGHO36, *L. gasseri* BGHO89 and *L. delbrueckii* subsp. *lactis* BGHO99 acted antagonistically on the growth of several LAB strains and pathogenic strains, such as *S. aureus, E. faecalis, Micrococcus flavus, Salmonella enteritidis, Streptococcus pneumoniae* and *Streptococcus mutans*. Additional testing of these strains showed that BGHO1, BGHO36 and BGHO89 produced bacteriocins, while BGHO64 and BGHO99 produced organic acids. Furthermore, 49% of nucleotide sequence homology with the structural acidocin LF221 A gene (Majhenic *et al.*, 2004) suggested that the isolate *L. gasseri* BGHO89 produces a variant of acidocin A (Strahinic *et al.*, 2007a). Considering that isolate *L. salivarius* BGHO1 exhibited the strongest antimicrobial activity against the growth of pathogens, its bacteriocin, named LS1 (molecular mass 10 kDa) was purified (Busarcevic *et al.*, 2008). It was shown that bacteriocins identified and characterized from strains *L. salivarius* subsp. *salivarius* UCC118 (Flynn *et al.*, 2002) and *L. salivarius* NRRL B-30514 (Stern *et al.*, 2006) had molecular masses of 4096.69 Da and 5132 Da, respectively. The results obtained for bacteriocin LS1 indicated it could be a new bacteriocin with different antimicrobial spectra. Anti *Helicobacter pylori* activity of a metabolite produced by an *L. plantarum* group isolated from white cabbage was confirmed by Rokka *et al.* (2006).

Antimicrobial Activity of Enterococci

Bacteria of the genus *Enterococcus* play a significant role in environmental, clinical and food microbiology. They inhabit different ecological niches, and can be found in the gastrointestinal tract of humans and animals, or as constituents of the microflora of raw milk and different types of fermented foods like meat and dairy products (Giraffa, 2003; Hugas *et al.*, 2003). Enterococci also contribute to the development of organoleptic properties during the ripening of cheese and sausages (Centeno *et al.*, 1999; Ogier and Serror, 2008), due to their proteolytic and lipolytic activities. *E. faecalis* and *E. faecium* were the most frequently occurring species isolated from artisanal dairy products from different countries (Psoni *et al.*, 2006; Valenzuela *et al.*, 2009). In addition, enterococci have also been used as probiotics (Franz *et al.*, 1999).

Many strains of enterococci produce bacteriocins generally designated as enterocins, and they have been isolated, purified and genetically characterized over the years. Most of them were obtained from *E. faecium* and *E. faecalis* strains separated from various samples of food and also from animals and humans (Nes *et al.*, 2007). The classification of enterocins basically follows the classification of bacteriocins in general. Thus, four classes of enterocins are recognized (Franz *et al.*, 2007). Among them, enterolysin A has a wide antimicrobial spectrum, since it is active against members of other genera of LAB (*Lactobacillus, Lactococcus* and *Pediococcus*), as well as against some strains of *Listeria, Bacillus* and *Staphylococcus* spp.

The search for new strains of enterococci with different antimicrobial activity and other features of interest for application has been intensified during the last decade. In a recent study, a collection of natural isolates of enterococci was tested for the presence of genes responsible for bacteriocin production and virulence factors and also for antibiotic resistance (Valenzuela *et al.*, 2009). The strains were isolated mostly from cheese samples of different origin (21 out of 25), and from milk, meat and ham manufactured throughout Serbia. Screening of the collection by PCR amplification revealed the presence of genes encoding for enterocins A, B, P, L50, and 1071 in six isolates. Incomplete sets of genes involved in the expression of cytolysin were detected in several natural isolates of *E. faecalis* and *E. faecium*, too.

The investigation of bacteriocin production in natural isolates of enterococci illustrates their potential for application in the food industry. In another study, the antimicrobial and proteolytic activities of enterococci isolated from artisanal fermented dairy and meat products of different origin, manufactured traditionally in Serbian households, were screened (Veljovic *et al.*, 2009). The most frequent species among 26 natural isolates of enterococci were *E. faecium* and *E. faecalis*, and 11 isolates produced antimicrobial compounds. Ten out of eleven enterococci synthesized enterocins with antimicrobial activity against food-borne pathogenic species, such as *L. monocytogenes* and *S. aureus*. The broadest spectrum of antimicrobial activity was detected in *E. faecalis* BGPT1-10P and BGPT1-78, since they inhibited the growth of *L. monocytogenes, Listeria innocua, Candida pseudotropicalis* and plant pathogenic bacteria such as *Erwinia carotovora* and *Burkholderia plantarii*. The natural isolate *E. faecalis* BG221 showed antimicrobial activity that was not enterocin, H_2O_2 or organic acid. A majority of the tested strains of enterococci showed strong or moderate proteolytic activity towards β-casein. Two isolates, BGPT1-10P and BGPT1-78, hydrolysed casein and gelatine highly efficiently. The extracellular proteinases produced by strains BGPT1-10P and BGPT1-78 have a molecular mass of about 29 kDa. Together with bacteriocin production, the proteolytic activity of natural isolates of enterococci may be of additional technological interest for application of these strains in the manufacture of fermented dairy and meat products.

Conclusion

The application of LAB as natural inhibitors of food-spoilage bacteria in the construction of bioactive-producing starters could result in sufficient product stability during processing and storage, and may imply a key role for enhancing food functionality. On the other hand, bioactive producers introduced into the organism with a fermented product may lead to *in situ* production of antimicrobials in the gastrointestinal tract inhibiting the growth of pathogenic microorganisms. In addition, natural isolates of LAB of human origin with antimicrobial activity could be used as probiotics. Finally, extensive research on LAB and identification of new antimicrobial producing strains should contribute to future progress in related scientific and biotechnological areas, especially because LAB are generally regarded as safe (GRAS) microorganisms from the aspect of human health.

Acknowledgements

We would like to thank our collaborators Jelena Begovic, Jelena Lozo, Amarela Terzic-Vidojevic, Branko Jovcic, Katarina Veljovic, Maja Tolinacki and Milica Nikolic for their excellent work in the field of molecular genetics and the application of LAB. This chapter would be impossible to write without their results. The authors are grateful to Dr Anna Nikolic, a native English Scientific Counsellor for editing the language.

The Ministry of Education and Science, Republic of Serbia, Grant No. 173019 funded this work.

References

Atanassova, M., Choiset, Y., Dalgalarrondo, M., Chobert, J-M., Dousset, X., Ivanova, I. and Haertlé, T. (2003) Isolation and partial characterization of a proteinaceous anti-bacteria and anti-yeast compound produced by *Lactobacillus paracasei* subsp. *paracasei* strain M3. *International Journal of Food Microbiology* 87, 63–73.

Atassi, F., Brassart, D., Grob, P., Graf, F. and Servin, A.L. (2006) *Lactobacillus* strains isolated from the vaginal microbiota of healthy women inhibit *Prevotella bivia* and *Gardnerella vaginalis* in coculture and cell culture. *FEMS Immunology and Medical Microbiology* 48, 424–432.

Axelsson, L. (2004) Lactic acid bacteria: classification and physiology. In: Salminen, S., von Wright, A. and Ouwehand, A. (eds) *Lactic Acid Bacteria: Microbiology and Functional Aspects*. Marcel Dekker Inc., New York, pp. 1–66.

Banina, A., Vukasinovic, M., Brankovic, S., Fira, D., Kojic, M. and Topisirovic, L. (1998) Characterization of natural isolate *Lactobacillus acidophilus* BGRA43 useful for acidophilus milk production. *Journal of Applied Microbiology* 84, 593–599.

Barefoot, S.F. and Klaenhammer, T.R. (1983) Detection and activity of lactacin B, a bacteriocin produced by *Lactobacillus acidophilus*. *Applied and Environmental Microbiology* 45, 1808–1815.

Begovic, J., Huys, G., Mayo, B., D'Haene, K., Florez, A.B., Lozo, J., Kojic, M., Strahinic, I. and Topisirovic, L. (2009) Human vaginal *Lactobacillus rhamnosus* harbor mutation in 23S rRNA associated with erythromycin resistance. *Research in Microbiology* 160, 421–426.

Beresford, T.P., Fitzsimons, N.A., Brennan, N.L. and Cogan, T.M. (2001) Recent advances in cheese microbiology. *International Dairy Journal* 11, 259–274.

Busarcevic, M., Kojic, M., Dalgalarrondo, M., Chobert, J.M., Haertlé, T. and Topisirovic, L. (2008) Purification of bacteriocin LS1 produced by human oral isolate *Lactobacillus salivarius* BGHO1. *Oral Microbiology and Immunology* 23, 254–258.

Cardi A. (2002) Selection of *Escherichia coli*-inhibiting strains of *Lactobacillus paracasei* subsp. *paracasei*. *Journal of Industrial Microbiology and Biotechnology* 29, 303–308.

Centeno, J.A., Menendez, S., Hermida, M. and Rodriguez-Otero, J.L. (1999) Effect of the addition of *Enterococcus faecalis* in Cebreiro cheese manufacture. *International Journal Food Microbiology* 48, 97–111.

Cintas, L., Cintas, M., Rodriguez, J.M., Fernandez, M.F., Sletten, K., Nes, I.F., Hernandez, P.E. and Holo, H. (1995) Isolation and characterization of pediocin L50, a new bacteriocin from *Pediococcus acidilacti* with a broad inhibitory spectrum. *Applied and Environmental Microbiology* 61, 2643–2648.

Cogan T.M. (1996) History and taxonomy of starter cultures. In: Cogan, T.M. and Accolas, J.P. (eds) *Dairy Starter Cultures.* VCH Publishers, Inc., New York, USA, pp. 1–20.

Cogan, T.M., Barbosa, M., Beuvier, E., Bianchi-Salvadori, B., Cocconcelli, P.S., Fernandes, I., Gomez, J., Gomez, R., Kalantzopoulos, G., Ledda, A. and Rodriguez, E. (1997) Characterization of the lactic acid bacteria in artisanal dairy products. *Journal of Dairy Research* 64, 409–421.

Cotter, P.D., Hill, C. and Ross, R.P. (2005) Bacteriocins: developing innate immunity for food. *Nature Reviews Microbiology* 3, 777–788.

Deegan, L.H., Cotter, P.D., Hill, C. and Ross, P. (2006) Bacteriocins: biological tools for bio-preservation and shelf-life extension. *International Dairy Journal* 16, 1058–1071.

De Vuyst, L. (1994) Bacteriocins produce by *Lactococcus lactis* subsp. *lactis*. In: De Vuyst, L. and Vandamme, E.J. (eds) *Bacteriocins of Lactic Acid Bacteria: Microbiology, Genetics and Application* Blackie Academic and Professional, London and New York, pp. 143–149.

Diep, D.B. and Nes, I.F. (2002) Ribosomally synthesised antibacterial peptides in Gram positive bacteria. *Current Drug Targets* 3, 107–122.

Drider, D., Fimland, G., Héchard, Y., McMullen, L.M. and Prévost, H. (2006) The continuing story of class IIa bacteriocins. *Microbiology and Molecular Biology Reviews* 70, 564–582.

Falagas, M.E., Betsi, G.I., Tokas, T. and Athaanasiou, S. (2006) Probiotics for prevention of recurrent urinary tract infections in women: a review of the evidence from microbiological and clinical studies. *Drugs* 66, 1253–1261.

Flynn, S., van Sinderen, D., Thornton, G.M., Holo, H., Nes, I.F. and Collins, J.K. (2002) Characterization of the genetic locus responsible for the production of ABP-118, a novel bacteriocin produced by the probiotic bacterium *Lactobacillus salivarius* subsp. *salivarius* UCC118. *Microbiology* 148, 973–984.

Franz, C.M., Holzapfel, W.H. and Stiles, M.E. (1999) Enterococci at the crossroads of food safety? *International Journal of Food Microbiology* 47, 1–24.

Franz, C.M., van Belkum, M.J., Holzapfel, W.H., Abriouel, H. and Galvez, A. (2007) Diversity of enterococcal bacteriocins and their grouping in a new classification scheme. *FEMS Microbiology Reviews* 31, 293–310.

Fujisawa, T., Benno, Y., Yaeshima, T. and Mitsuoka, T. (1992) Taxonomic study of the *Lactobacillus acidophilus* group, with recognition of *Lactobacillus gallinarum* sp. nov. and *Lactobacillus johnsonii* sp. nov. and synonymy of *Lactobacillus acidophilus* group A3 (Johnson *et al.* 1980) with the type strain of *Lactobacillus amylovorus* (Nakamura 1981) *International Journal of Systematic Bacteriology* 42, 487–491.

Gajic, O., Kojic, M., Banina, A. and Topisirovic, L. (1999) Characterization of natural isolate *Lactococcus lactis* subsp. *lactis* BGMN1-5, a strain producing two bacteriocins, cell wall-associated proteinase and showing clumping phenotype. *Archives of Biological Sciences* 51, 69–78.

Gajic, O., Buist, G., Kojic, M., Topisirovic, L., Kuipers, O.P and Kok, J. (2003) Novel mechanism of bacteriocin secretion and immunity carried out by lactococcal MDR proteins. *Journal of Biological Chemistry* 278, 34291-34298.

Gálvez, A., Abriouel, H., López, R.L. and Omar, N.B. (2007) Bacteriocin-based strategies for food biopreservation. *International Journal of Food Microbiology* 120, 51–70.

Garg, S., Anderson, R.A., Chany, C.J., Waller, D.P., Diao, X.H., Vermani, K. and Zaneveld, L.J. (2001) Properties of a new acid-buffering bioadhesive vaginal formulation (ACIDFORM) *Contraception* 64, 67–75.

Geis, A., Singh, J. and Teuber, M. (1983) Potential of lactic streptococci to produce bacteriocin. *Applied and Environmental Microbiology* 45, 205–211.

Giraffa, G. (2003) Functionality of enterococci in dairy products. *International Journal of Food Microbiology* 88, 215–222.

Håvarstein, L.S., Diep, D.B. and Nes, I.F. (1995) A family of bacteriocin ABC transporters carry out proteolytic processing of their substrates concomitant with export. *Molecular Microbiology* 16, 229–240.

Holo, H., Nilssen, O. and Nes, I.F. (1991) Lactococcin A, a new bacteriocin from *Lactococcus lactis* subsp. *cremoris*: isolation and characterization of the protein and its gene. *Journal of Bacteriology* 173, 3879–3887.

Holzapfel, W.H., Geisen, R. and Schillinger, U. (1995) Biological preservation of foods with reference to protective cultures, bacteriocins and food-grade enzymes. *International Journal of Food Microbiology* 24, 343–362.

Horn, N., Swindell, S., Dodd, H. and Gasson, M. (1991) Nisin biosynthesis genes are encoded by novel conjugative transposon. *Molecular and General Genetics* 228, 129–135.

Hugas, M., Garriga, M. and Aymerich, T. (2003) Functionality of enterococci in meat products. *International Journal Food Microbiology* 88, 223–233.

Hurst, A. (1967) Function of nisin and nisin-like basic proteins in the growth cycle of *Streptococcus lactis*. *Nature* 214, 1232–1234.

Itoh, T., Fujimoto, Y., Kawai, Y., Toba, T. and Saito, T. (1995) Inhibition of food-borne pathogenic bacteria by bacteriocins from *Lactobacillus gasseri*. *Letters in Applied Microbiology* 21, 137–141.

Jack, R.W., Tagg, J.R. and Ray, B. (1995) Bacteriocins of Gram positive bacteria. *Microbiological Reviews* 59, 171–200.

Jokovic, N., Nikolic, M., Begovic, J., Jovcic, B., Savic, D. and Topisirovic, L. (2008) A survey of the lactic acid bacteria isolated from Serbian artisanal dairy product kajmak. *International Journal of Food Microbiology* 127, 305–311.

Kamau, D.N., Doores, S. and Pruitt, K.M. (1990) Enhanced thermal destruction of *Listeria monocytogenes* and *Staphylococcus aureus* by the lactoperoxidase system. *Applied and Environmental Microbiology* 56, 2711–2716.

Kawai, Y., Saito, T., Toba, T., Samant, S.K. and Itoh, T. (1994) Isolation and characterization of a highly hydrophobic new bacteriocin (gassericin A) from *Lactobacillus gasseri* LA39. *Bioscience, Biotechnology, and Biochemistry* 58, 1218–1221.

Klaenhammer, T.R. (1988) Bacteriocins of lactic acid bacteria. *Biochimie* 70, 337–349.

Kleeman, E.G. and Klaenhammer, T.R. (1982) Adherence of *Lactobacillus* species to human fetal intestinal cells. *Journal of Dairy Science* 65, 2063–2069.

Kleerebezem, M., Beerthuyzen, M.M., Vaughan, E.E., de Vos, W.M. and Kuipers, O.P. (1997) Controlled gene expression systems for lactic acid bacteria: transferable nisin-inducible expression cassettes for *Lactococcus*, *Leuconostoc*, and *Lactobacillus* spp. *Applied and Environmental Microbiology* 63, 4581–4584.

Kojic, M., Strahinic, I. and Topisirovic, L. (2005) Proteinase PI and lactococcin A genes are located on the largest plasmid in *Lactococcus lactis* subsp. *lactis* bv. diacetylactis S50. *Canadian Journal of Microbiology* 51, 305–314.

Kojic, M., Strahinic, I., Fira, D., Jovcic, B. and Topisirovic, L. (2006) Plasmid content and bacteriocin production by five strains of *Lactococcus lactis* isolated from semi-hard homemade cheese. *Canadian Journal of Microbiology* 52, 1110–1120.

Kojic, M., Lozo, J., Begovic, J., Jovcic, B. and Topisirovic, L. (2007) Characterisation of lactococci isolated from homemade kefir. *Archives of Biological Sciences* 59, 13–22.

Koll-Klais, P., Mandar, R., Leibur, E., Marcotte, H., Hammarstrom, L. and Mikelsaar, M. (2005) Oral lactobacilli in chronic periodontitis and periodontal health: species composition and antimicrobial activity. *Oral Microbiology and Immunology* 20, 354–361.

Lozo, J., Vukasinovic, M., Strahinic, I. and Topisirovic, L. (2004) Characterization and antimicrobial activity of bacteriocin 217 produced by natural isolate *Lactobacillus paracasei* subsp. *paracasei* BGBUK2-16. *Journal of Food Protection* 67, 2727–2734.

Lozo, J., Jovcic, B., Kojic, M., Dalgalarrondo, M., Chobert, J., Haertlé, T. and Topisirovic, L. (2007) Molecular characterization of a novel bacteriocin and an unusually large aggregation factor of *Lactobacillus paracasei* subsp. *paracasei* BGSJ2-8, a natural isolate from home-made cheese. *Current Microbiology* 55, 266–271.
Majhenic, A.C., Venema, K., Allison, G.E., Matijasic, B.B., Rogelj, I. and Klaenhammer, T.R. (2004) DNA analysis of the genes encoding acidocin LF221 A and acidocin LF221 B, two bacteriocins produced by *Lactobacillus gasseri* LF221. *Applied Microbiology and Biotechnology* 63, 705–714.
Marelli, G., Papaleo, E. and Ferrari, A. (2004) Lactobacilli for prevention of urogenital infections: a review. *European Review for Medical and Pharmacological Sciences* 8, 87–95.
McGroarty, J.A. and Reid, G. (1988) Detection of a *Lactobacillus* substance that inhibits *Escherichia coli*. *Canadian Journal of Microbiology* 34, 974–978.
McSweeney, P.L.H., Fox, P.F., Lucey, J.A., Jordan, K.N. and Cogan, T.M. (1993) Contribution of the indigenous microflora to the maturation of cheddar cheese. *International Dairy Journal* 3, 613–634.
Mundt, J.O. (1986) Lactic acid streptococci. In: Sneath, P.H.A., Mair, N.S., Sharpe, M.E. and Holt, J.G. (eds) *Bergey's Manual of Systematic Bacteriology.* Vol. 2, Williams and Wilkins Co., Baltimore, Maryland, USA, pp. 1065–1071.
Muriana, P.M. and Klaenhammer, T.R. (1991) Purification and partial characterization of lactacin F, a bacteriocin produced by *Lactobacillus acidophilus* 11088. *Applied and Environmental Microbiology* 57, 114–121.
Nes, I.F., Diep, D.B., Havarstein, L.S., Brurberg, M.B., Eijsink, V. and Holo, H. (1996) Biosynthesis of bacteriocins in lactic acid bacteria. In: Venema, G., Huis Veld, J.H.J. and Hugenholtz J. (eds) *Lactic Acid Bacteria: Genetics, Metabolism and Applications.* Kluwer Academic Publishers, The Netherlands, pp. 17–32.
Nes, I.F., Diep, D.B. and Holo, H. (2007) Bacteriocin diversity in *Streptococcus* and *Enterococcus*. *Journal of Bacteriology* 189, 1189–1198.
Nikolic, M., Terzic-Vidojevic, A., Jovcic, B., Begovic, J., Golic, N. and Topisirovic, L. (2008) Characterization of lactic acid bacteria isolated from Bukuljac, a homemade goat's milk cheese. *International Journal of Food Microbiology* 122, 162–170.
Notermans, S. and Hoogenboom-Verdegaal, A. (1992) Existing and emerging foodborne diseases. *International Journal of Food Microbiology* 15, 197–205.
Ocana, V.S., Pesce De Ruiz Holgado, A.A. and Nader-Macias, M.E. (1999) Characterization of a bacteriocin-like substance produced by a vaginal *Lactobacillus salivarius* strain. *Applied and Environmental Microbiology* 65, 5631–5635.
Ogier, J. and Serror, C.P. (2008) Safety assessment of dairy microorganisms: The *Enterococcus* genus. *International Journal of Food Microbiology* 126, 291–301.
Pangsomboon, K., Bansal, S., Martin, G.P., Suntinanalert, P., Kaewnopparat, S. and Srichana, T. (2009) Further characterization of a bacteriocin produced by *Lactobacillus paracasei* HL32. *Journal of Applied Microbiology* 106, 1928–1940.
Psoni, L., Kotzamanides, C., Andrighetto, C., Lombardi, A., Tzanetakis, N. and Litopoulou-Tzanetaki, E. (2006) Genotypic and phenotypic heterogeneity in *Enterococcus* isolates from Batzos, a raw goat milk cheese. *International Journal of Food Microbiology* 109, 109–120.
Randazzo, C.L., Torriani, S., Akkermans, A.D.L., de Vos, W.M. and Vaughan, E.E. (2002) Diversity, dynamics, and activity of bacterial communities during production of an artisanal Sicilian cheese as evaluated by 16S rRNA analysis. *Applied and Environmental Microbiology* 68, 1882–1892.
Rauch, P.J. and de Vos, W.M. (1992) Characterization of the novel nisin-sucrose conjugative transposon Tn*5276* and its insertion in *Lactococcus lactis*. *Journal of Bacteriology* 174, 1280–1287.
Rodas, A.M., Ferrer, S. and Pardo, I. (2005) Polyphasic study of wine *Lactobacillus* strains: taxonomic implications. *International Journal of Systematic and Evolutionary Microbiology* 55, 197–207.
Rokka, S., Pihlanto, A., Korhonen, H. and Joutsjoki, V. (2006) *In-vitro* growth inhibition of *Helicobacter pylori* by lactobacilli belonging to the *Lactobacillus plantarum* group. *Letters in Applied Microbiology* 43, 508–513.
Silva, M., Jacobus, N.V., Deneke, C. and Gorbach, S.L. (1987) Antimicrobial substance from a human *Lactobacillus* strain. *Antimicrobial Agents and Chemotherapy* 31, 1231–1233.
Stern, N.J., Svetoch, E.A., Eruslanov, B.V., Perelygin, V.V., Mitsevich, E.V., Mitsevich, I.P., Pokhilenko, V.D., Levchuk, V.P., Svetoch, O.E. and Seal, B.S. (2006) Isolation of a *Lactobacillus salivarius* strain and purification of its bacteriocin, which is inhibitory to *Campylobacter jejuni* in the chicken gastrointestinal system. *Antimicrobial Agents and Chemotherapy* 50, 3111–3116.

Stevens, K.A., Sheldon, B.W., Klapes, N.A. and Klaenhammer, T.R. (1991) Nisin treatment for inactivation of *Salmonella* species and other Gram-negative bacteria. *Applied and Environmental Microbiology* 57, 3613–3615.

Strahinic, I., Busarcevic, M., Pavlica, D., Milasin, J., Golic, N. and Topisirovic, L. (2007a) Molecular and biochemical characterizations of human oral lactobacilli as putative probiotic candidates. *Oral Microbiology and Immunology* 22, 111–117.

Strahinic, I., Cvetanovic, D., Kojic, M., Fira, D., Tolinacki, M. and Topisirovic, L. (2007b) Characterization and antimicrobial activity of natural isolate *Lactococcus lactis* subsp. *lactis* BGSM1-19. *Acta Veterinaria* 57, 509–521.

Terzic-Vidojevic, A., Vukasinovic, M., Veljovic, K., Ostojic, M. and Topisirovic, L. (2007) Characterization of microflora in homemade semi-hard white Zlatar cheese. *International Journal of Food Microbiology* 114, 36–42.

Topisirovic, L., Kojic, M., Fira, D., Golic, N., Strahinic I. and Lozo, J. (2006) Potential of lactic acid bacteria isolated from specific natural niches in food production and preservation. *International Journal of Food Microbiology* 112, 230–235.

Topisirovic, L., Veljovic, K., Terzic-Vidojevic, A., Strahinic, I. and Kojic, M. (2007) Comparative analysis of antimicrobial and proteolytic activity of lactic acid bacteria isolated from Zlatar cheese. *Genetika* 39, 125–138.

Valenzuela, A.S., Ben Omar, N., Abriouel, H., Lópes, R.L., Veljovic, K., Cañamero, M.M., Kojic, M., Topisirovic, L. and Galvez, A. (2009) Virulence factors, antibiotic resistance, and bacteriocins in enterococci from artisan foods of animal origin. *Food Control* 20: 381–385.

van Belkum, M.J., Hayema, B.J., Jeeninga, R.E., Kok, J. and Venema, G. (1991) Organization and nucleotide sequences of two lactococcal bacteriocin operons. *Applied and Environmental Microbiology* 57, 492–498.

van Belkum, M.J., Kok, J. and Venema, G. (1992) Cloning, sequencing, and expression in *Escherichia coli* of *lcnB*, a third bacteriocin determinant from the lactococcal bacteriocin plasmid p9B4-6. *Applied and Environmental Microbiology* 58, 572–577.

Veljovic, K., Terzic-Vidojevic, A., Vukasinovic, M., Strahinic, I., Begovic, J., Lozo, J., Ostojic, M. and Topisirovic, L. (2007) Preliminary characterization of lactic acid bacteria isolated from Zlatar cheese. *Journal of Applied Microbiology* 103, 2142–2152.

Veljovic, K., Fira, D., Terzic-Vidojevic, A., Abriouel, H., Topisirovic, L. and Galvez, A. (2009) Evaluation of antimicrobial and proteolytic activity of enterococci isolated from fermented products. *European Food Research and Technology* 230, 63–70.

4 Bacteriocins for Bioprotection of Foods

Antonio Gálvez,* Hikmate Abriouel, Rosario Lucas, María José Grande Burgos

Introduction

Bacteriocins are ribosomally synthesized antimicrobial peptides or proteins (Jack *et al.*, 1995). Because of their association with foods, the lactic acid bacteria (LAB) are so far the bacterial group in which bacteriocin production has been studied more extensively and in which more bacteriocins have been described.

LAB bacteriocins and their producer strains can be exploited to improve the quality and safety of foods (Thomas *et al.*, 2000; Gálvez *et al.*, 2007, Gálvez *et al.*, 2008: Settani and Corsetti, 2008). One of the main purposes of bacteriocin addition is to reduce the microbial load of food-borne pathogenic bacteria. Many LAB bacteriocins have bactericidal activity against pathogenic or toxigenic Gram-positive bacteria such as *Listeria monocytogenes*, *Bacillus cereus*, *Clostridium botulinum* or *Staphylococcus aureus*. Some bacteriocins can also inhibit Gram-negative bacteria, especially when tested in combination with other antimicrobials as will be discussed later in this chapter. The inhibitory activity of bacteriocins can also be exploited as a hurdle against proliferation of pathogenic bacteria during food storage. Food processing conditions and refrigeration storage of processed foods create new opportunities for proliferation of psychrotrophic bacteria such as *L. monocytogenes*. In the absence of a competing microbiota (such as in many processed foods), *Listeria* from cross contamination can proliferate easily if no other hurdles are imposed. Furthermore, accidental cold-chain-breakage events also increase the risks for the proliferation of bacteria in refrigerated foods. Incorporation of bacteriocins in refrigerated foods can therefore provide extra protection.

The shelf life of foods depends on several parameters such as nutritional value, organoleptic properties, physical aspects, and microbiological quality. Since microbial proliferation is one of the main factors responsible for modification of these parameters and making food unacceptable, one of the main interests in the application of bacteriocins has been to prolong the shelf life of foods. Several examples will be described in the text where bacteriocins can selectively inhibit food-spoilage bacteria such as those involved in gas formation, slime formation or overacidification. Extending the shelf life of foods is most frequently achieved by addition of chemical preservatives or by application of physico-chemical treatments that often have an impact on the food quality and sometimes are not fully effective against all bacteria, like for example endospore formers. Since many bacteriocins are active

* Corresponding author.

against endospore-forming bacteria, they can be applied as complementary hurdles to physico-chemical treatments. In addition, many bacteriocins act synergistically with other antimicrobial agents. The microbial reductions achieved by application of bacteriocins in combination with other antimicrobials are often much greater compared with the single treatments. For this reason, the concentrations or intensity of antimicrobial treatments in foods can be decreased when applied in combination with bacteriocins. This results in a better preservation of the food nutritional value and organoleptic properties and a reduction in the content of chemical preservatives. The resulting lightly preserved food products are better accepted by consumers and are considered to be healthier because of their reduced content of substances known to be detrimental to human health such as nitrites, sulfites or salt. This change of focus towards biopreservation of foods could also be exploited to develop 'novel' foods, such as foods that are less acidic, with a lower salt content and with higher water content. The application of bacteriocins in foods could satisfy, to a lesser or greater extent, some of the trends of the food market, such as the need to eliminate the use of artificial ingredients and additives, the demands for minimally processed and fresher foods, as well as for ready-to-eat food or the request for functional foods and nutraceuticals (Robertson *et al.*, 2004).

The purpose of this chapter is to summarize two main aspects related to application of bacteriocins as food bioprotectants: (i) how to improve the efficacy of bacteriocins in combination with other barriers or hurdles; and (ii) the main results and conclusions of work carried out on the specific application of bacteriocins for the preservation of different types of foods and beverages.

The Application of Bacteriocins as Part of Hurdle Technology

Food preservation relies on the simultaneous or consecutive application of barriers or hurdles that microbes need to overcome in order to proliferate successfully on the food substrate. The concept of hurdle technology includes bacteriocins as part of the multiple antimicrobial factors or hurdles that can be applied (Leistner, 2000). Because of the amount of energy required to repair the multiple damages caused by different barriers or antimicrobial agents, the cells will ultimately die as a consequence of energy exhaustion and failure to repair cell damage. Since most bacteriocins act on the bacterial cytoplasmic membrane, they interfere with the generation of energy required to repair bacterial cell damage. Bacteriocins can exhibit either additive or synergistic effects with other antimicrobial substances or treatments used in hurdle technology. The additive effect of two antimicrobial agents results from the sum of the single killing effects for each antimicrobial separately. In this way, bacteriocin addition can compensate for a reduction in the concentration of other antimicrobials being added to the food. The second, and most interesting interaction is synergy, whereby the killing or inhibitory effect resulting from the combined action of two antimicrobials is much greater than the sum of the single effects for each antimicrobial. This phenomenon is reported very often for combinations of bacteriocins and chemical food preservatives as well as physical treatments, and can be exploited to reduce the amount of bacteriocin and chemical preservatives or the intensity of treatments. It can also be exploited to extend the spectrum of bacteriocins against intrinsically resistant bacteria (such as Gram-negative bacteria) or increase the inactivation of resting forms (such as bacterial endospores). Table 4.1 summarizes some of the treatment combinations reported to increase the effectiveness of bacteriocins in food systems.

Chemical preservatives and natural antimicrobials

Nitrates and nitrites are the main chemical barrier in meat products against *C. botulinum*. Positive interactions between

Table 4.1. Effects of bacteriocins in combination with chemical preservatives and other antimicrobial compounds.

Antimicrobial	Effects
Nitrates, nitrites	Increased antibacterial activity in meats
	Reduced addition of nitrites
Organic acids and salts	Increased inactivation of bacteria and growth inhibition
	Increased solubility and activity of bacteriocin molecules
Chelating agents	Permeation of the outer cell membrane in Gram-negative bacteria: increased sensitivity to bacteriocins
	Decreased availability of essential mineral nutrients: increased activity of bacteriocins on Gram-positive bacteria
Essential oils and their bioactive components	Increased inactivation of Gram-positive and Gram-negative bacteria Reduction of impact on food organoleptic properties
Other antimicrobials of chemical or proteinaceous nature	Increased inactivation of Gram-positive / Gram-negative bacteria
CO_2 atmosphere	Inhibition of strictly aerobic bacteria (complementary action with bacteriocins)
	Synergism with bacteriocins against food-borne pathogens

nisin and nitrites have been demonstrated in meats against *C. botulinum*, *Leuconostoc mesenteroides* and *L. monocytogenes* (Rayman *et al.*, 1981, 1983; Taylor *et al.*, 1985; Gill and Holley, 2003). Nitrites increased the antimicrobial activities of enterocins EJ97 (García *et al.*, 2003, 2004a,b) and AS-48 (Abriouel *et al.*, 2002). Nitrites also increased the anti-listeria effects of bacteriocin-producing lactobacilli in meat in one example (Hugas *et al.*, 1996), but had the opposite effect in another (Kouakou *et al.*, 2009).

Organic acids and their salts can enhance the solubility and activity of bacteriocin molecules, while at the same time increasing the sensitivity of target cells (Jack *et al.*, 1995; Stiles, 1996). The scientific literature abounds with examples of positive interactions between bacteriocins (such as nisin, pediocin PA-1/AcH, enterocin AS-48 or lacticin 3147) and lactic acid or lactate, sorbic acid or sorbate, citric acid or citrate, acetic and peracetic acids, and others (Scannell *et al.*, 1997, 2000a,b; Nykänen *et al.*, 2000; Long and Phillips, 2003; Ukuku and Fett, 2004; Uhart *et al.*, 2004; Cobo Molinos *et al.*, 2005; Grande *et al.*, 2006). Organic acids are widely used as food preservatives or produced naturally in fermented foods, hence the great interest and relevance of their positive interactions with bacteriocins. Furthermore, due to their chelating effect, organic acids sensitize Gram-negative bacteria to bacteriocins (such as nisin or enterocin AS-48), widening their inhibitory spectra (Scannell *et al.*, 1997; Long and Phillips, 2003; Cobo Molinos *et al.*, 2008c).

Ethylenediaminetetraacetic acid (EDTA) and other food-grade chelators (such as polyphosphates) sensitize Gram-negative bacteria to bacteriocins by destabilizing the bacterial outer membrane (Stevens *et al.*, 1991; Vaara, 1992; Helander *et al.*, 1997). This interaction is highly relevant to overcoming the natural resistance of Gram-negative bacteria to most bacteriocins and to increasing the efficacy of combined treatments in foods. EDTA also increases the sensitivity of bacteria to other antimicrobials (such as lysozyme) which in turn may act synergistically with bacteriocins. Combinations of nisin–lysozyme–EDTA show increased inhibitory activity against Gram-positive and Gram-negative bacteria such as *Brochothrix thermosphacta, Lactobacillus curvatus, L. mesenteroides, L. monocytogenes,* and *Escherichia coli* O157:H7 (Gill and Holley, 2000a,b).

Bacteriocin activity can be potentiated by other chemicals such as ethanol, sulfites, monolaurin, sucrose fatty acid esters, 2-nitropropanol, reuterin, and several other miscellaneous compounds (Brewer *et al.*, 2002; Mansour *et al.*, 1999; Thomas *et al.*, 1998; Arqués *et al.*, 2004; Rojo-Bezares *et al.*, 2007; Cobo Molinos *et al.*, 2008a, 2009b; Abriouel *et al.*, 2010).

Essential oils and the phenolic compounds they contain are strong, natural antimicrobials. However, their application in foods is limited by the strong impact they have on the food organoleptic properties. Recent research works carried out with nisin and enterocin AS-48 suggest that, based on their positive interactions, essential oils or preparations containing phenolic compounds could be applied in certain foods in combination with bacteriocins at concentrations where they have a much lower impact on the food (Pol *et al.*, 2001a,b; Yuste and Fung, 2004; Grande *et al.*, 2007; Cobo Molinos *et al.*, 2009a).

The antimicrobial activity of bacteriocins can be improved by combination with other antimicrobial peptides of proteins, including other bacteriocins, lysozyme, lactoferrin, the milk lactoperoxidase system (LPS) and others. For example, the combinations of nisin or pediocin PA-1/AcH and lactacins showed greater antibacterial activity than each bacteriocin separately (Mulet-Powell *et al.*, 1998). Nisin and lysozyme act synergistically against Gram-positive as well as Gram-negative bacteria (Gill and Holley, 2000a,b; Nattress and Baker, 2003), and the combination of nisin and lactoferrin showed increased antilisterial activity (Branen and Davidson, 2004). Nisin, in combination with the LPS system, had a stronger antilisterial effect in skimmed milk compared to the single nisin addition (Boussouel *et al.*, 2000). Positive effects have been reported for several other combinations such as divercin V41 with nisin and/or polymixins (Rihakova *et al.*, 2009), nisin and polymixin A (Nagmouchi *et al.*, 2010) or the bacteriocins pediocin PA-1/AcH, sakacin P and curvacin A with the fish antimicrobial peptide pleurocidin (Lüders *et al.*, 2003).

Modified atmosphere packaging

Many perishable food products are often stored under modified atmosphere packaging (MAP). MAP is 'the enclosure of food products in gas-barrier materials, in which the gaseous environment has been changed' (Young *et al.*, 1988). Selected gaseous environments (usually based on reduced O_2 and increased CO_2 levels) are chosen so as to retard intrinsic food changes and inhibit spoilage microbiota. Growth inhibition of microorganisms is mediated by the low oxygen levels (in the case of aerobic bacteria) and by the antimicrobial action of dissolved CO_2 (Devlieghere *et al.*, 1998). Gram-negative bacteria are generally more sensitive to CO_2 (Farber, 1991; Church, 1994), although they are usually resistant to bacteriocins, while the opposite holds true for most Gram-positive bacteria including LAB. Nisin and a CO_2 atmosphere act synergistically on *L. monocytogenes* by enhancing membrane permeabilization (Nilsson *et al.*, 2000). Nisin in combination with CO_2 MAP increased the inhibition of *L. monocytogenes* in pork (Fang and Lin, 1994a,b) and in cold smoked salmon (Nilsson *et al.*, 1997; Szabo and Cahill, 1999). Similar effects were reported for pediocin PA-1/AcH (Szabo and Cahill, 1999).

Heat treatments

Heat treatments cause multiple cell damages, including denaturation of enzymes and structural proteins and changes in membrane fluidity and structure. Many bacteriocins have been shown to increase the bactericidal effects of heat treatments both in culture media and in food systems. This positive interaction can be exploited to reduce the intensity of heat treatments applied to food (Table 4.2). For example, the addition of nisin reduced the heat resistance of *L. monocytogenes* in milk (Maisnier-Patin *et al.*, 1995) and in cold-pack lobster meat (Budu-Amoako *et al.*, 1999). Heat treatments also sensitize Gram-negative bacteria to

Table 4.2. Effects of bacteriocins in combination with physical treatments.

Treatment	Effect(s)	Reference(s)
Heat	Increased inactivation of food-poisoning and pathogenic Gram-positive bacteria cells	Budu-Amoako *et al.*, 1999; Ananou *et al.*, 2004
	Sensitization of Gram-negative bacteria to bacteriocins	Kalchayanand *et al.*, 1992; Boziaris *et al.*, 1998; Ananou *et al.*, 2005; Bakes *et al.*, 2004
	Reduced the heat resistance of bacterial endospores	Beard *et al.*, 1999; Wandling *et al.*, 1999; Grande *et al.*, 2006
PEF	Synergistic effect against Gram-positive bacteria in liquid foods	Calderón-Miranda *et al.*, 1999a,b; Ulmer *et al.*, 2002; Sobrino-López and Martin-Belloso, 2006; Pol *et al.*, 2001
	Decreased proliferation of survivors during food storage	Martínez Viedma *et al.*, 2009
	Increased inactivation of Gram-negative bacteria in fruit juices and/or in milk	Liang *et al.*, 2002; Terebiznik *et al.*, 2000; Martínez Viedma *et al.*, 2008
HHP	Inactivation of Gram-positive and Gram-negative bacteria in milk and in liquid egg products	Ponce *et al.*, 1998; Garcia-Graells *et al.*, 1999; Black *et al.*, 2005
	Inactivation of *L. monocytogenes*, aerobic mesophilic bacteria, and bacterial endospores in cheese	Capellas *et al.*, 2000; López-Pedemonte *et al.*, 2003; Arqués *et al.*, 2005
	Greater inactivation of *L. monocytogenes* in meat and inhibition of survivor proliferation	Morgan *et al.*, 2000; Garriga *et al.*, 2002; Marcos *et al.*, 2008
Irradiation	Increased inactivation of *L. monocytogenes* and protection against post-process contamination of RTE meats	Chen *et al.*, 2004
Pulsed light	Reduction of post-process contamination in RTE meats	Uesugi and Moraru, 2009

bacteriocins, because of the alterations induced on the bacterial outer membrane. In one example, inactivation of *E. coli* O157:H7 in apple juice by enterocin AS-48 increased dramatically in combination with heat (Ananou *et al.*, 2005). Bacterial endospores are in general resistant to bacteriocins, and they are also much more resistant to heat treatments than vegetative cells. Application of heat treatments to inactivate bacterial endospores often has an impact on the food nutritional and organoleptic properties. However, the combination of bacteriocin treatments decreases the heat intensity required for inactivation of endospores (Beard *et al.*, 1999; Grande *et al.*, 2006). The residual bacteriocin after treatment is also an additional hurdle against the surviving endospores, acting as an inhibitor during germination and outgrowth (Wandling *et al.*, 1999).

Pulsed electric fields

Several bacteriocins have been tested in combination with high-intensity pulsed electric fields (PEF), with promising results (Table 4.2). PEF is a non-thermal food-processing technology based on the repeated application of short-duration electric pulses to a liquid food as it flows between two electrodes (Martín-Belloso and Elez-Martínez, 2005; Mittal and Griffiths, 2005). However, PEF treatments have no protective effect against survivors and do not inactivate bacterial endospores. The application of bacteriocins in combination with PEF increases the efficacy of treatments both against Gram-positive and Gram-negative bacteria. The observed synergistic effects between bacteriocins and PEF are predictable, since the two hurdles act on the bacterial cytoplasmic membrane. Furthermore, disorganization of the outer cell membrane of Gram-negative bacteria by PEF should facilitate penetration of the cell envelope by bacteriocin molecules, allowing them to reach the cytoplasmic membrane more easily. Additionally, bacteriocins may increase the lethal effects of PEF treatments by inhibition of membrane repair in sublethally injured cells. They may also provide an additional hurdle against proliferation of survivors after PEF treatment during storage of samples, increasing the product shelf life. This second barrier effect of bacteriocins may be important to inhibit outgrowth of bacterial endospores in foods treated by PEF.

The scientific literature includes many examples of application of bacteriocins in combination with PEF in food systems such as milk and fruit juices. In dairy substrates such as skimmed milk, whey or simulated milk ultrafiltrate, the combined treatments nisin–PEF increased the microbial inactivation of *L. monocytogenes, S. aureus, B. cereus* and *E. coli* (Sobrino-López and Martín-Belloso, 2008). Application of enterocin AS-48 in skimmed milk with or without nisin and PEF treatment increased the inactivation of *S. aureus* (Sobrino *et al.*, 2009). The combined treatment extended the shelf life of milk by at least one week compared to a conventional pasteurization treatment.

In freshly squeezed, unpasteurized and preservative-free apple juice, the addition of nisin in combination with PEF treatment increased the inactivation of *E. coli* O157:H7 compared to PEF treatment alone (Iu *et al.*, 2001). Similar effects were reported using nisin or nisin/lysozyme and PEF against *Salmonella typhimurium* in pasteurized and freshly squeezed orange juices (Liang *et al.*, 2002). The synergistic effect seemed to be dependent on treatment temperature (which was held at 42°C for *E. coli* and required at least 45°C or above for *Salmonella*). Inactivation of *Salmonella enterica* cells in apple juice by enterocin AS-48 in combination with PEF was observed at a treatment temperature of 40°C (Martínez-Viedma *et al.*, 2008). Altogether, these results indicate that the temperature of treatment is a critical parameter for inactivation of Gram-negative bacteria in fruit juices. Similarly, inactivation of naturally occurring microorganisms (including yeasts and moulds) in several substrates such as freshly squeezed apple cider, red and white grape juices and tomato juice was observed for combined treatments of nisin and PEF at 50°C (Wu *et al.*, 2005; Liang *et al.*, 2006; Nguyen and Mittal, 2007). At this treatment temperature, the vitamin C content of tomato juice remained unaltered after treatment (Nguyen and Mittal, 2007). Synergistic inactivation of *Listeria innocua* and *E. coli* K12 by nisin and PEF (maximum outlet temperature of 56°C) was also reported recently (McNamee *et al.*, 2010). By contrast, inactivation of Gram-positive bacteria such as the cider spoilage LAB strains *Lactobacillus diolivorans* and *Pediococcus parvulus* was achieved easily by bacteriocin–PEF treatments carried out at room temperature (Martínez-Viedma *et al.*, 2009, 2010). Application of enterocin AS-48 and PEF inhibited the proliferation of the surviving fraction after treatment for at least 15 days in the case of *L. diolivorans* and for up to 30 days for *P. parvulus* under storage temperatures of 4°C as well as 22°C (Martínez-Viedma *et al.*, 2009, 2010).

High hydrostatic pressure

The application of bacteriocins in combination with high hydrostatic pressure (HHP) has shown promising results. HHP treatments in the range of 300 to 700 MPa induce lethal cell injury on bacteria in food systems. More sophisticated equipment reaching up to 1200 MPa is now available, some of which offer the possibility to control the temperature of treatments between below 0°C and 100°C. Pressurization adversely affects H-bonds, ionic bonds and hydrophobic interactions of the macromolecules (Hoover, 1993) and induces membrane-phase transitions, affecting mainly ATP-generating and transport proteins (Kato and Hayashi, 1999). The loss of structure and function of proteins involved in critical cellular processes such as energy metabolism is per se enough to explain cell death induced by HHP treatments. It is believed that cell death occurs as a consequence of multiple events or cumulative cell damage (Kalchayanand *et al.*, 1998). Currently, there are already different products on the market treated by HHP, such as ready-to-eat (RTE) chicken meat, sliced ham, fresh whole oysters, jams, fruit juices or guacamole. However, higher pressure treatments induce greater texture changes that may be unacceptable for some types of foods, while lower pressure treatments increase the risk for selection of pressure-resistant strains. Bacterial endospores are still a pending issue, since they are intrinsically resistant to HHP and can be induced to germinate by the treatment, increasing the risk for outgrowth and proliferation in the treated food. For all these reasons, bacteriocins seem attractive as additional hurdles to increase the efficacy of HHP treatments (Table 4.2).

Several bacteriocins (such as nisin, pediocin PA-1/AcH, lacticin 3147, and enterocin AS-48) have been shown to enhance the lethal effects of HHP treatments against bacteria of concern in food systems (Table 4.2). Interestingly, HHP transiently sensitized Gram-negative bacteria to nisin (Masschalck *et al.*, 2001). In milk, nisin or nisin/lysozyme in combination with HHP treatment increased the microbial inactivation of *E. coli* cells. Lacticin 3147 also increased the efficacy of HHP against *S. aureus* and *L. monocytogenes* in milk and in whey (Morgan *et al.*, 2000), and residual lacticin still showed an inhibitory effect in the food, inactivating and preventing growth of sub-lethally injured cells. In cheeses, the combination of nisin and HHP increased the inactivation of *L. monocytogenes* (Arqués *et al.*, 2005), aerobic mesophilic bacteria, as well as *Bacillus* and *Clostridium* endospores (Roberts and Hoover, 1996; Stewart *et al.*, 2000; Capellas *et al.*, 2000). The combined treatment of nisin–HHP could improve the microbial stability and safety of cheeses.

In meat systems, the combination of nisin and HHP achieved a greater inactivation of *E. coli* compared to HHP treatment alone. The combined treatment also reduced viable counts of staphylococci significantly and completely inactivated slime-producing LAB. However, it was not effective against *L. monocytogenes* proliferation during storage (probably due to the lower efficacy of nisin in meat systems). Nevertheless, other bacteriocins (such as sakacin K, enterocins A and B as well as pediocin AcH) were much more effective against the listeria (Garriga *et al.*, 2002). Similar results were reported recently for enterocin AS-48 (Ananou *et al.*, 2010). It is expected that combinations of different bacteriocins may increase the bactericidal effects of HHP treatments against a wider spectrum of bacteria compared to single bacteriocin–HHP treatments.

High-pressure homogenization (HPH) is becoming widely used during processing of certain foods. In spite of the lower pressure reached (which usually does not exceed 350 MPa), this process has also shown to sensitize bacterial cells to antibacterial peptides and enzymes (Diels *et al.*, 2004, 2005). Addition of nisin increased the bactericidal effect of HPH treatment against *L. innocua* in apple and carrot juices, reducing to some extent the intensity of the HPH treatment (Pathanibul *et al.*, 2009). However, addition of nisin did not increase the effect of HPH treatment on *E. coli* cells.

Other physical treatments

The combination of bacteriocins and low-dose irradiation has shown promising results for inactivation of *L. monocytogenes* on meat products (Table 4.2). Low-dose gamma irradiation has less adverse effects on food and is accepted better than high-dose irradiation by consumers. Irradiation (2.3 kGy) of packaged frankfurters treated with pediocin showed a greater inhibition of *L. monocytogenes* compared to the single treatments without adverse effects on the product sensory quality (Chen *et al.*, 2004). Similarly, irradiation at 2 kGy of RTE turkey meat vacuum-packaged with a pectin–nisin film increased the inactivation of *L. monocytogenes* (Jin *et al.*, 2009). It was proposed that combined treatments of bacteriocins and low-dose irradiation on the surface of RTE meat products could help to prevent listeriosis due to post-processing contamination of irradiated RTE meats.

Technologies based on UV light or pulsed light are gaining interest for surface application on foods. There is scarce work concerning the application of bacteriocins in combination with these technologies (Table 4.2). According to a recent study, application of a Nisaplin dip followed by exposure to pulsed light (9.4 J/cm^2) reduced the population of *L. innocua* on sausages by 4 log cycles and inhibited the microbial growth during refrigeration storage for 24 to 48 days (Uesugi and Moraru, 2009). The combined treatment could be applied as a post-processing step to reduce surface contamination and increase the safety of RTE meat products.

Food Applications of Bacteriocins

Bacteriocins for the preservation of meat and poultry products

Bacteriocins (either as partially purified concentrates or produced *in situ* by selected strains) have been tested at every different step of the meat-processing chain (Table 4.3).

In raw meats, treatment with nisin alone has several limitations derived from the interaction of nisin molecules with food components and poor solubility of nisin preparations at pH close to neutrality (Thomas *et al.*, 2000; Stergiou *et al.*, 2006). However, surface application of combinations containing nisin and other barriers (organic acids, chelators, lysozyme, vacuum packaging or MAP) improved the bactericidal effects against *L. monocytogenes*, *B. thermosphacta* and *E. coli* O157:H7. The results improved with immobilization of nisin in combination with other antimicrobials (mainly organic acids and EDTA) (Chen and Hoover, 2003; Galvez *et al.*, 2007, 2008; Aymerich *et al.*, 2008). The application of pediocin PA-1/AcH on meats reduced the populations of spoilage and pathogenic bacteria (*B. thermosphacta*, *L. monocytogenes*, *Clostridium perfringens*) (Rodríguez *et al.*, 2002; Nieto-Lozano *et al.*, 2006). Other bacteriocins tested (such as sakacins, carnobacteriocins, bifidocins, lactocins, lactococcins or pentocins) showed variable inhibitory effects against spoilage or pathogenic bacteria in raw meats or poultry meats (Aymerich *et al.*, 2008; Gálvez *et al.*, 2008).

In cooked and partially processed meat products, bacteriocins have been tested to prevent growth of *L. monocytogenes* and spoilage bacteria that may recontaminate the products during handling and slicing. Nisin, pediocin PA-1/AcH, enterocins and lacticin 3147 were tested alone or in combination with organic acids, lysozyme and EDTA on cooked meat products, such as hot dogs, frankfurters or ham (Gálvez *et al.*, 2008). Surface application of bacteriocins in combination with other antimicrobials or post-processing treatments such as HHP, irradiation or pulsed light is an attractive approach for the preservation of cooked RTE meat products (Aymerich *et al.*, 2008; Gálvez *et al.*, 2008).

In fermented meats, addition of bacteriocins such as nisin, enterocins (CCM 4231, A, B and AS-48) or leucocins reduce the populations of *L. monocytogenes* or *S. aureus* (Rodríguez *et al.*, 2002; Chen and Hoover, 2003; Aymerich *et al.*, 2008; Gálvez *et al.*, 2008). Organic acid production during fermentation improves the bactericidal action of bacteriocins. Nevertheless, it is important to avoid interference of the added bacteriocin with growth and performance of meat starter cultures.

In poultry products such as liquid egg, bacteriocins have been tested as hurdles against *L. monocytogenes* and other bacteria. Nisin added to pasteurized liquid whole egg reduced viable cell counts of *L. monocytogenes* and prevented its growth and that of *B. cereus* during storage (Delves-Broughton *et al.*, 1992; Knight *et al.*, 1999; Schuman and Sheldon, 2003). Pediocin PA-1/

Table 4.3. Potential applications of bacteriocins or bacteriocin-producing strains in foods of animal origin and seafood.

Raw meats
Decontamination of raw meats before processing
Inhibition of foodborne pathogens (*L. monocytogenes, C. perfringens, E. coli, S. enterica, C. jejuni*)
Inhibition of spoilage bacteria (*B. thermosphacta*, spoilage LAB, *C. estertheticum*)
Processed meat products
Inhibition of recontaminating pathogenic bacteria
Inhibition of slime formation
Fermented meats
Inhibition of *L. monocytogenes* and/or *S. aureus*
Improving meat fermentation
Egg products
Inhibition of *L. monocytogenes* and *B. cereus*
Increased inactivation of *L. monocytogenes* and *Salmonella enteritidis*
Raw milk
Inactivation of *L. monocytogenes* and *S. aureus*
Reduction of pasteurization intensity
Inactivation of pathogenic and spoilage bacteria in combination with other treatments (such as PEF or HHP)
Cheeses and dairy products
Control of late blowing caused by *C. tyrobutyricum*
Control of food-borne pathogens (*L. monocytogenes, S. aureus, B. cereus, C. botulinum, E. coli*)
Control of NSLAB
Acceleration of cheese ripening
Control of *L. monocytogenes* in yogurt
Seafood
Retard spoilage of raw fish
Control of *L. monocytogenes* in slightly processed seafood products (caviar, cold-smoked fish products)
Control of *L. monocytogenes* in processed seafood products (such as crabmeat or lobster meat)
Improving fish fermentation

AcH acted synergistically with heat against *L. monocytogenes* (Muriana, 1996), whereas nisin increased the heat sensitivity of both *L. monocytogenes* and *Salmonella enteritidis* in egg products (Boziaris *et al.*, 1998; Knight *et al.*, 1999). The effectiveness of nisin in liquid egg increased in combination with PEF and HHP treatments (Ponce *et al.*, 1998; Calderón-Miranda *et al.*, 1999a).

Many bacteriocin-producing strains have been tested for bioprotection of meat and meat products. Inoculation of raw meats with bacteriocin-producing strains (mainly *Lactobacillus sakei* and *Lactobacillus curvatus* strains) showed promising results against *L. monocytogenes* or *B. thermosphacta*. Other positive effects reported recently include inhibition of blown-pack spoilage caused by *Clostridium estertheticum,* reduction of *Campylobacter jejuni* survival or inhibition of *S. enteritidis* growth (Jones *et al.*, 2009; Maragkoudakis *et al.*, 2009). In processed meat products, strains producing sakacins, pediocins, leucocins, plantaricins, enterocins, bavaricins or curvaticins can inhibit growth of *Listeria* and sometimes avoid spoilage due to slime formation (Aymerich *et al.*, 1998; Hugas *et al.*, 1998). Bacteriocin-producing LAB cultures are already in the market, for improvement of the microbiological safety of semi-processed and cooked meats (Aymerich *et al.*, 2008).

In fermented meat products, inoculation with bacteriocin-producing strains with anti-listeria activity has been proposed by different studies (Työppönen *et al.*, 2003; Leroy *et al.*, 2006; Aymerich *et al.*, 2008). Bacteriocin-producing strains of *L. sakei* and *L. curvatus*, and to a less extent *Lactobacillus rhamnosus* and *Lactobacillus plantarum* have shown variable anti-listerial effects in sausage or salami fermentations (Erkkilä *et al.*, 2001; Dicks *et al.*, 2004; Leroy *et al.*, 2005; Benkerroum *et al.*, 2005; Todorov *et al.*, 2007). Pediocin-producing strains can reduce *L. monocytogenes* populations in fermented meats (Amezquita and Brashears, 2002; Rodríguez *et al.*, 2002; Aymerich *et al.*, 2008). Performance of bacteriocin producers greatly depends on the environmental factors influencing bacteriocin production, such as sausage ingredients, salt, fat and nitrite content, acidification level and temperature (Leroy *et al.*, 2006).

Bacteriocin applications in milk and dairy products

Commercial preparations of nisin and pediocin PA-1/AcH are widely used in the preservation of dairy products (Table 4.3). Pediocin is used mainly for control of *L. monocytogenes,* while nisin is used mainly to control *C. botulinum* in processed cheese products, and secondarily to prevent a gas-blowing defect in cheeses caused by *Clostridium tyrobutyricum* and to inhibit *L. monocytogenes* (Thomas *et al.*, 2000; Thomas and Delves-Broughton, 2001; Rodríguez *et al.*, 2002; Deegan *et al.*, 2006; Sobrino-López and Martín-Belloso, 2008). In raw milk, there is a growing interest for the application of nisin in combination with novel food-processing technologies (such as PEF) in order to decrease the intensity of pasteurization treatments (Sobrino-López and Martín-Belloso, 2008).

Commercial preparations obtained by the cultivation of propionibacteria are currently being used for the preservation of dairy products. Microgard™ is a lyophilized preparation from cultured broths of *Propionibacterium freudenreichii* ssp. *shermanii,* containing organic acids and an anti-microbial peptide (Weber and Broich, 1986). Microgard™ is approved in certain countries as an ingredient in dairy products such as cottage cheese and yogurt. Other bacteriocin preparations in the form of lyophilized powders have been proposed as candidates for the preservation of dairy products, such as lacticin 3147, enterocin AS-48, or variacin (Morgan *et al.*, 2001; Mollet *et al.*, 2004; Ananou *et al.*, 2010). In addition, crude or partially purified preparations of several other bacteriocins (including lacticins, enterocins, pediocins and propionicins) have been shown to inhibit bacteria of concern in milk and dairy products (Gálvez *et al.*, 2008).

Bacteriocin-producing lactococci can inhibit food-borne pathogens (such as *L. monocytogenes* or *S. aureus*) in fermented milk products and also *C. tyrobutyricum* in cheese (Martínez-Cuesta *et al.*, 2010). However, the main difficulty for application of such strains is often their lack of suitable technological properties (such as fast acidification capacity or proteolytic activity). This drawback has been solved in the case of lacticin 3147 by transfer of the lacticin plasmid to suitable recipient strains (O'Sullivan *et al.*, 2006). However, in the case of nisin, co-inoculation of nisin-producing strains in combination with nisin-resistant starters has been proposed (Bouksaim *et al.*, 2000; Rilla *et al.*, 2003).

Enterococci are frequently found in cheeses, not only in those made from raw milk, but also in many other cheeses due to recontamination during cheese-making operations and also because of the higher thermal resistance of these bacteria. Enterocin-producing strains have shown to perform satisfactorily in cheeses and to produce bacteriocins which inhibit the growth of bacteria such as *L. monocytogenes, B. cereus* or *S. aureus* (Muñoz *et al.*, 2004, 2007). Nevertheless, there is a strong concern about recommending enterococci for application in foods due to their implication in nosocomial infections and their role as reservoirs of antimicrobial resistance genes. However, strains naturally occurring in foods and in the gastro-intestinal tract of animals and humans also carry genetic determinants for virulence factors and antibiotic-resistance traits, adding confusion to the debate about their safe use in foods.

Inoculation with thermophilic streptococci has been proposed for the control of gas-blowing defects caused by *C. tyrobutyricum* in some cheeses such as hard cheese. Bacteriocin-producing strains such as *Streptococcus thermophilus* ST580 (Mathot *et al.*, 2003) and *Streptococcus macedonicus* ACA-DC 198 seem to be particularly useful for this purpose (Anastasiou *et al.*, 2007).

An emerging field of interest is the application of bacteriocin-producing strains for acceleration of cheese ripening. This is a complex process driven by enzymes present in the starting materials as well as those released by bacteria during fermentation and by intracellular enzymes released by dead cells. Bacteriocin production in cheese may accelerate the death of starter and adventitious microbiota, with the concomitant release of intracellular enzymes (Lortal and Chapot-Chartier, 2005; Peláez and Requena, 2005; Deegan *et al.*, 2006). Furthermore, bacteriocins can increase the cell-membrane permeability to small extracellular compounds (such as amino acids and organic acids), enhancing their access to intracellular enzymes of bacteriocin-injured cells, with their concomitant transformation into intermediates or final products involved in cheese flavour. These effects have been observed for several bacteriocin-producing strains, including lacticin 3147, lacticin 481, lactococcins A, B and M, enterocin AS-48 or nisin (Gálvez *et al.*, 2008).

Another suggested application of bacteriocin-producing strains is the control of adventitious non-starter LAB (NSLAB) in cheeses. NSLAB microbiota are often responsible for cheese defects and batch-to-batch changes during cheese manufacture. In one example, the application of lacticin 3147-producing starters achieved a more homogeneous microbiota during cheese ripening, through inhibition of the lacticin-sensitive adventitious NSLAB (Ryan *et al.*, 2001; Deegan *et al.*, 2006).

Bacteriocins for the preservation of seafood products

Bacteriocins have been tested on seafood products with the aims of avoiding spoilage of fresh seafood and to inhibit or inactivate human pathogenic bacteria and prolong the product shelf life in slightly processed products (Table 4.3).

Addition of nisin in combination with other antimicrobials (such as the lactoperoxidase system or with headspace CO_2 levels and EDTA) achieved an increased inactivation or growth delay of spoilage microbiota in sardines and in fish muscle extract (Gálvez *et al.*, 2008). Similarly, the combination of nisin and Microgard^{TM} reduced the total aerobic bacteria populations and delayed growth of *L. monocytogenes* in fresh-chilled salmon (Calo-Mata *et al.*, 2008; Gálvez *et al.*, 2008).

Several bacteriocin preparations (nisin, pediocin and enterocin 1083) reduced the viable counts of *L. monocytogenes* on crabmeat during refrigeration storage (Degnan *et al.*, 1994). In cold-pack lobster, nisin addition increased the microbial inactivation of *L. monocytogenes*, reducing the energy input of heat treatments (Budu-Amoako *et al.*, 1999). Nisin addition was also reported to reduce viable cell counts of listeria and total mesophilic bacteria in caviars (Al-Holy *et al.*, 2004).

Cold-smoked seafood products (such as cold-smoked salmon; CSS) may become contaminated frequently with *L. monocytogenes*. Proliferation of this bacterium during the product shelf life period under refrigeration is a matter of concern. This safety issue has driven many investigations on the application of bacteriocin treatments to combat *L. monocytogenes* in CSS and other similar products such as cold-smoked salmon-trout. Promising results have been reported for sakacin P, carnobacteriocins, nisin and pediocins (Gálvez *et al.*, 2008). The application of nisin as a coating on plastic films inhibited the proliferation of *L. monocytogenes* strains and background microbiota (aerobic, anaerobic and LAB) in

CSS during storage at 4°C (Neetoo *et al.*, 2008).

LAB strains isolated from seafoods such as CSS can be an important source of bacteriocins, with better adaptation to seafood substrates and more suitability for application as protective cultures. Antagonistic strains of *C. piscicola, C. divergens, L. sakei, Lactobacillus casei, L. curvatus, Lactobacillus delbrueckii, L. plantarum, Pediococcus acidilactici* or *Enterococcus faecium* have been tested for the biocontrol of *L. monocytogenes* on cold-smoked seafood products (Leisner *et al.*, 2007; Calo-Mata *et al.*, 2008; Gálvez *et al.*, 2008; Tomé *et al.*, 2008; Rihakova *et al.*, 2009; Tahiri *et al.*, 2009). Other strains producing antimicrobial substances with broader antibacterial spectra (such as *Leuconostoc gelidum, Lactococcus piscium, Lactobacillus fuchuensis* and *Carnobacterium alterfunditum*) are currently under investigation as bio-protective cultures in fish preservation (Matamoros *et al.*, 2009). Strains with broader spectra of inhibition may be of value in the control of fish-spoilage bacteria as well as human pathogens associated with seafood or reaching seafood products by cross-contamination. Selected LAB strains may also find application to improve some types of fish fermentation (Diop *et al.*, 2009).

Bacteriocins for the preservation of vegetable foods and fruit juices

The control of human pathogenic bacteria on vegetable foods is difficult because of the limited types of treatments that can be applied without altering the quality of the raw materials. Bacteriocins can be applied as mild treatments compatible with the processing of raw vegetables and fruits (Table 4.4). This is particularly significant for fruits and vegetables consumed without cooking. Washing treatments or surface application of bacteriocin solutions (nisin, pediocin or enterocin AS-48) can reduce the microbial load of Gram-positive bacteria

Table 4.4. Potential applications of bacteriocins or bacteriocin-producing strains in vegetable foods and beverages.

Raw vegetables and fruits
Decontamination of fruit and vegetable surfaces
Inhibition of pathogenic bacteria during storage of processed fruits
Fruit juices
Control of spoilage bacteria (*Alicyclobacillus* sp., *G. stearothermophilus*, LAB)
Increased inactivation of pathogenic bacteria (*E. coli, S. enterica*)
Ready-to-eat vegetable foods
Inactivation of pathogenic and/or toxigenic bacteria (*B. cereus* and related, *C. botulinum, S. aureus, L. monocytogenes*, enterobacteriaceae)
Canned vegetables
Control of endospore formers (*C. botulinum, G. stearothermophilus*)
Fermented vegetables
Control of fermentation
Control of product over-ripening
Inhibition of rope-forming bacilli in bread
Inhibition of pathogenic bacteria in cereal gruels
Fermented beverages
Inhibition of spoilage bacteria in ciders
Inhibition of spoilage LAB in beer
Control of wine malolactic fermentation
Inhibition of wine-spoilage LAB and biogenic amine producers

such as *L. monocytogenes, B. cereus* and *Bacillus weihenstephanensis* (Gálvez *et al.*, 2008; Randazzo *et al.*, 2009; Abriouel *et al.* 2010). The efficacy of treatments increases considerably in combination with other antimicrobials, such as organic acids, extending the inhibitory spectrum of treatments to Gram-negative bacteria as well (Bari *et al.*, 2005; Cobo Molinos *et al.*, 2008b,c). This is an important finding because Gram-negative bacteria are the main bacteria of concern in raw-vegetable foods, either as pathogens (such as members of Enterobacteriaceae) or as food spoilers (such as Pseudomonads and related bacteria). Similar washing or decontamination treatments can be applied on whole-fruit surfaces to avoid transmission of pathogenic bacteria to the processed fruits (Ukuku *et al.*, 2005) or to inactivate the pathogenic bacteria on whole fruits (such as strawberries) or sliced fruits (Cobo Molinos *et al.*, 2008b).

Inoculation with live cultures is another alternative for the inhibition of pathogenic bacteria on fresh-produce surfaces. Bacterial strains (including species of genera such as *Bacillus, Pseudomonas, Enterococcus, Lactococcus, Leuconostoc, Weissella* or *Lactobacillus*) isolated from raw vegetables may produce antagonistic substances against food-borne pathogens (Gálvez *et al.*, 2008; Trias *et al.*, 2008a,b). These strains may be better adapted to vegetable substrates and growth under cold or moderate temperatures.

In fruit juices and drinks, bacteriocins (nisin and enterocin AS-48) can inhibit spoilage caused by endospore-forming bacteria such as *Alicyclobacillus acidoterrestris, Bacillus licheniformis* or *Geobacillus stearothermophilus* (Gálvez *et al.*, 2008; Abriouel *et al.*, 2010). Other bacteria involved in spoilage such as lactobacilli, pediococci and *Propionibacterium cyclohexanicum* can also be inhibited by bacteriocins. For the Gram-negative pathogenic bacteria (such as *E. coli* and *S. enterica*), an increased inactivation has been reported in fruit juices when bacteriocins were tested in combination with other hurdles, as described previously in the text.

Reports on the application of bacteriocins for biocontrol of pathogenic bacteria in RTE foods such as salads and sauces and in canned foods include some promising results for bacteriocins such as nisin, pediocin PA-1/AcH, and enterocins AS-48 and EJ-97 (Gálvez *et al.*, 2008). Some of these bacteriocins could be applied in soups, purées and canned foods in order to reduce the intensity of heat treatments (increasing the heat sensitivity of endospores) and to protect against the outgrowth of bacterial endospores surviving the heat treatments.

Bacteriocins in fermented vegetable foods and beverages

There are few examples on the addition of bacteriocins to fermenting vegetables (Table 4.4), such as nisin addition to control cabbage fermentation (Breidt *et al.*, 1995) and to avoid over-ripening of kimchi (Choi and Park, 2000). Bacteriocin-producing strains are often isolated from fermenting vegetables and some of them have been tested with satisfactory results in vegetable fermentations. For example, plantaricin-producing strains are available commercially as starter cultures for table olive fermentation (Gálvez *et al.*, 2008). These plantaricin producers may be used to speed up fermentation, ensure a homogeneous fermentation process in newly operating plants that still lack the appropriate resident LAB microbiota or to resolve stuck fermentations.

Bacteriocin-producing LAB strains isolated from cereal-fermented foods may be interesting for several purposes. In the bread-making industry, some LAB strains (mainly *L. plantarum* and *L. mesenteroides*) have shown to inhibit proliferation of *B. subtilis*, the main causative agent of bread ropiness (Pepe *et al.*, 2003; Valerio *et al.*, 2008). Wheat flours often contain other bacilli, which may produce enterotoxins. Addition of bacteriocin preparations (such as enterocin AS-48) to wheat dough reduced the microbial load of enterotoxigenic bacilli (Martínez-Viedma *et al.*, unpublished results).

In fermented beverages such as beer, several bacteriocins (such as nisin, lacticin M30, and enterocins L50A and L50B) can inhibit the main LAB strains involved in beer spoilage (Thomas *et al.*, 2000; Basanta *et al.*, 2008). The application of nisin in the beer-manufacturing process has been proposed with several purposes including cleaning of equipment, washing of pitching yeasts to eliminate contaminating LAB, and also as a preservative in unpasteurised beers. Although LAB do not play an important role in beer fermentation (except for some specific beers), the heterologous production of bacteriocins by yeast cells has been proposed in several occasions (Schoeman *et al.*, 1999; Du Toit and Pretorius, 2000; Van Reenen *et al.*, 2002; Gutiérrez *et al.*, 2005; Sánchez *et al.*, 2008; Basanta *et al.*, 2009). Some of these yeast strains producing bacteriocins could be applied to beer fermentation.

In apple cider, the control of LAB causing spoilage such as acrolein producers and exopolysaccharide producers can be achieved by adding enterocin AS-48 even at very low concentrations (Abriouel *et al.*, 2010). This bacteriocin could be added at the end of fermentation to inhibit proliferation of spoilage LAB at the expense of residual sugars.

There is a growing interest in the application of bacteriocins in the wine industry. Nisin shows interesting synergistic activity against wine bacteria in combination with sulfites and with ethanol (Rojo-Bezares *et al.*, 2007). Nisin could be exploited to reduce the levels of sulfites in wine, since sulfites are known to cause adverse reactions in some consumers. Certain LAB such as the oenococci are recognized as essential components in the fermentation of many wines, being responsible for malolactic fermentation. Bacteriocin-producing bacteria from wine could be applied to improve the malolactic fermentation and to control other bacteria causing undesirable effects during or after fermentation, such as spoilage bacteria or histamine producers.

Concluding Remarks

Much research has been done on the possible applications of bacteriocins and/or their producer strains for the bioprotection of foods, including drinks, as well as fermented beverages. This has led to the commercialization of bacteriocin preparations (such as Nisaplin^{TM}, Alta^{TM} or Microgard^{TM}) as well as bioprotective cultures. The promising results reported for other bacteriocins (such as lacticin 3147 or enterocin AS-48) suggest that they could also become available commercially in the future. Bacteriocins perform best in combination with other hurdles, of which modern food-processing technologies such as HHP, PEF, irradiation or pulsed light technology seem very promising. The development of bioactive packaging based on immobilized bacteriocin preparations seems a more rational way for protecting certain types of foods, such as many RTE food products, decreasing bacteriocin expense and avoiding the problems of cross-contamination during or after processing. The application of bacteriocins as part of hurdle technology can improve food safety and shelf life while preserving nutritional and organoleptic properties of the food and reducing the amounts of added chemical preservatives. The application of bacteriocin-producing strains also seems an attractive and maybe cheaper approach for the bioprotection of foods, especially those based on LAB strains naturally occurring in the food. Such strains could serve not only as bioprotectants but also to improve the fermentation processes by controlling adventitious microbiota, resolving stuck fermentation, avoiding food overacidification, or reducing the risks for spoilage and biogenic amine production in fermented beverages. Bacteriocin-producing strains with specific technological properties may find additional applications as enhancers of ripening in fermented foods.

References

Abriouel, H., Maqueda, M., Gálvez, A., Martínez-Bueno, M. and Valdivia, E. (2002) Inhibition of bacterial growth, enterotoxin production, and spore outgrowth in strains of *Bacillus cereus* by bacteriocin AS-48. *Applied and Environmental Microbiology* 68, 1473–1477.

Abriouel, H., Lucas, R., Ben Omar, N., Valdivia, E. and Gálvez, A. (2010) Potential applications of the cyclic peptide enterocin AS-48 in the preservation of vegetable foods and beverages. *Probiotics and Antimicrobial Proteins* 2, 77–89.

Al-Holy, M., Ruiter, J., Lin, M., Kang, D.H. and Rasco, B. (2004) Inactivation of *Listeria innocua* in nisin-treated salmon (*Oncorhynchus keta*) and sturgeon (*Acipenser transmontanus*) caviar heated by radio frequency. *Journal of Food Protection* 67, 1848–1854.

Amezquita, A. and Brashears, M.M. (2002) Competitive inhibition of *Listeria monocytogenes* in ready-to-eat meat products by lactic acid bacteria. *Journal of Food Protection* 65, 316–325.

Ananou, S., Valdivia, E., Martínez Bueno, M., Gálvez, A. and Maqueda, M. (2004) Effect of combined physico-chemical preservatives on enterocin AS-48 activity against the enterotoxigenic *Staphylococcus aureus* CECT 976 strain. *Journal of Applied Microbiology* 97, 48–56.

Ananou, S., Gálvez, A., Martínez-Bueno, M., Maqueda, M. and Valdivia, E. (2005) Synergistic effect of enterocin AS-48 in combination with outer membrane permeabilizing treatments against *Escherichia coli* O157:H7. *Journal of Applied Microbiology* 99, 1364–1372.

Ananou, S., Garriga, M., Joffré, A., Aymerich, T., Gálvez, A., Maqueda, M., Martínez-Bueno, M. and Valdivia, E. (2010) Combined effect of enterocin AS-48 and high hydrostatic pressure treatment to control food-borne pathogens in slightly fermented sausages. *Meat Science* 84, 594–600.

Anastasiou, R., Georgalaki, M., Manolopoulou, E., Kandarakis, I., De Vuyst, L. and Tsakalidou, E. (2007) The performance of *Streptococcus macedonicus* ACA-DC 198 as starter culture in Kasseri cheese production. *International Dairy Journal* 17, 208–217.

Arqués, J.L., Fernández, J., Gaya, P., Nuñez, M., Rodríguez, E. and Medina, M. (2004) Antimicrobial activity of reuterin in combination with nisin against food-borne pathogens. *International Journal of Food Microbiology* 95, 225–229.

Arqués, J.L., Rodríguez, E., Gaya, P., Medina, M. and Nuñez, M. (2005) Effect of combinations of high-pressure treatment and bacteriocin-producing lactic acid bacteria on the survival of *Listeria monocytogenes* in raw milk cheese. *International Dairy Journal* 15, 893–900.

Aymerich, M.T., Hugas, M. and Monfort, J.M. (1998) Review: bacteriocinogenic lactic acid bacteria associated with meat products. *Food Science and Technology International* 4, 141–158.

Aymerich, T., Picouet, P.A. and Monfort, J.M. (2008) Decontamination technologies for meat products. *Meat Science* 78, 114–129.

Bakes, S.H., Kitis, F.Y.E., Quattlebaum, R.G. and Barefoot, S.F. (2004) Sensitization of Gram-negative and Gram-positive bacteria to jenseniin G by sublethal injury. *Journal of Food Protection* 67, 1009–1013.

Bari, M.L., Ukuku, D.O., Kawasaki, T., Inatsu, Y., Isshiki, K. and Kawamoto, S. (2005) Combined efficacy of nisin and pediocin with sodium lactate, citric acid, phytic acid, and potassium sorbate and EDTA in reducing the *Listeria monocytogenes* population of inoculated fresh-cut produce. *Journal of Food Protection* 68, 1381–1387.

Basanta, A., Sánchez, J., Gómez-Sala, B., Herranz, C., Hernández, P.E. and Cintas, L.M. (2008) Antimicrobial activity of *Enterococcus faecium* L50, a strain producing enterocins L50 (L50A and L50B), P and Q, against beer-spoilage lactic acid bacteria in broth, wort (hopped and unhopped), and alcoholic and non-alcoholic lager beers. *International Journal of Food Microbiology* 125, 293–307.

Basanta, A., Herranz, C., Gutiérrez, J., Criado, R., Hernández, P.E. and Cintas, L.M. (2009) Development of bacteriocinogenic strains of *Saccharomyces cerevisiae* heterologously expressing and secreting the leaderless enterocin L50 peptides L50A and L50B from *Enterococcus faecium* L50. *Applied and Environmental Microbiology* 75, 2382–2392.

Beard, B.M., Sheldon, B.W. and Foegeding, P.M. (1999) Thermal resistance of bacterial spores in milk-based beverages supplemented with nisin. *Journal of Food Protection* 62, 484–491.

Benkerroum, N., Daoudi, A., Hamraoui, T., Ghalfi, H., Thiry, C., Duroy, M., Evrart, P., Roblain, D. and Thonart, P. (2005) Lyophilized preparations of bacteriocinogenic *Lactobacillus curvatus* and *Lactococcus lactis* subsp. *lactis* as potential protective adjuncts to control *Listeria monocytogenes* in dry-fermented sausages. *Journal of Applied Microbiology* 98, 56–63.

Black, E.P., Kelly, A.L. and Fitzgerald, G.F. (2005) The combined effect of high pressure and nisin on inactivation of microorganisms in milk. *Innovative Food Science and Emerging Technologies* 6, 286–292.

Bouksaim, M., Lacroix, C., Audet, P. and Simard, R.E. (2000) Effects of mixed starter composition on nisin Z production by *Lactococcus lactis* subsp. *lactis* biovar. *diacetylactis* UL 719 during production and ripening of Gouda cheese. *International Journal of Food Microbiology* 59, 141–156.

Boussouel, N., Mathieu, F., Revol-Junelles, A.M. and Millière, J.B. (2000) Effects of combinations of lactoperoxidase system and nisin on the behaviour of *Listeria monocytogenes* ATCC 15313 in skim milk. *International Journal of Food Microbiology* 61, 169–175.

Boziaris, I.S., Humpheson, L. and Adams, M.R. (1998) Effect of nisin on heat injury and inactivation of *Salmonella enteritidis* PT4. *International Journal of Food Microbiology* 43, 7–13.

Branen, J.K. and Davidson, P.M. (2004) Enhancement of nisin, lysozyme, and monolaurin antimicrobial activities by ethylenediaminetetraacetic acid and lactoferrin. *International Journal of Food Microbiology* 90, 63–74.

Breidt, F., Crowley, K.A. and Fleming, H.P. (1995) Controlling cabbage fermentations with nisin and nisin-resistant *Leuconostoc mesenteroides*. *Food Microbiology* 12, 109–116.

Brewer, R., Adams, M.R. and Park, S.F. (2002) Enhanced inactivation of *Listeria monocytogenes* by nisin in the presence of ethanol. *Letters in Applied Microbiology* 34, 18–21.

Budu-Amoako, E., Ablett, R.F., Harris, J. and Delves-Broughton, J. (1999) Combined effect of nisin and moderate heat on destruction of *Listeria monocytogenes* in cold-pack lobster meat. *Journal of Food Protection* 62, 46–50.

Calderón-Miranda, M.L., Barbosa-Cánovas, G.V. and Swanson, B.G. (1999a) Inactivation of *Listeria innocua* in liquid whole egg by pulsed electric fields and nisin. *International Journal of Food Microbiology* 51, 7–17.

Calderón-Miranda, M.L., Barbosa-Canovas, G.V. and Swanson, B.G. (1999b) Inactivation of *Listeria innocua* in skim milk by pulsed electric fields and nisin. *International Journal of Food Microbiology* 51, 19–30.

Calo-Mata, P., Arlindo, S., Boehme, K., de Miguel, T., Pascoal, A. and Barros-Velazquez, J. (2008) Current applications and future trends of lactic acid bacteria and their bacteriocins for the biopreservation of aquatic food products. *Food Bioprocess Technology* 1, 43–63.

Capellas, M., Mor-Mur, M., Gervilla, R., Yuste, J. and Guamis, B. (2000) Effect of high pressure combined with mild heat or nisin on inoculated bacteria and mesophiles of goats' milk fresh cheese. *Food Microbiology* 17, 633–641.

Chen, H. and Hoover, D.G. (2003) Bacteriocins and their food applications. *Comprehensive Reviews in Food Science and Food Safety* 2, 82–100.

Chen, C.M., Sebranek, J.G., Dickson, J.S. and Mendonca, A.F. (2004) Combining pediocin with post-packaging irradiation for control of *Listeria monocytogenes* on frankfurters. *Journal of Food Protection* 67, 1866–1875.

Choi, M.H. and Park, Y.H. (2000) Selective control of lactobacilli in kimchi with nisin. *Letters in Applied Microbiology* 30, 173–177.

Church, N. (1994) Developments in modified-atmosphere packaging and related technologies. *Trends in Food Science and Technology* 5, 345–352.

Cobo Molinos, A., Abriouel, H., Ben Omar, N., Valdivia, E., Lucas, R., Maqueda, M., Martínez Cañamero, M. and Gálvez, A. (2005) Effect of immersion solutions containing enterocin AS-48 on *Listeria monocytogenes* in vegetable foods. *Applied and Environmental Microbiology* 71, 7781–7787.

Cobo Molinos, A., Abriouel, H., Ben Omar, N., Lucas, R., Valdivia, E. and Gálvez, A. (2008a) Inactivation of *Listeria monocytogenes* in raw fruits by enterocin AS-48. *Journal of Food Protection* 71, 2460–2467.

Cobo Molinos, A., Abriouel, H., Lucas, R., Ben Omar, N., Valdivia, E. and Gálvez, A. (2008b) Inhibition of *Bacillus cereus* and *B. weihenstephanensis* in raw vegetables by application of washing solutions containing enterocin AS-48 alone and in combination with other antimicrobials. *Food Microbiology* 25, 762–770.

Cobo Molinos, A., Abriouel, H., Lucas, R., Valdivia, E., Ben Omar, N. and Gálvez, A. (2008c) Combined physico-chemical treatments based on enterocin AS-48 for inactivation of Gram-negative bacteria in soybean sprouts. *Food and Chemical Toxicology* 46, 2912–2921.

Cobo Molinos, A., Abriouel, H., Ben Omar, N., Lucas, R., Valdivia, E. and Gálvez, A. (2009a) Enhanced bactericidal activity of enterocin AS-48 in combination with essential oils, natural bioactive compounds, and chemical preservatives against *Listeria monocytogenes* in ready-to-eat salads. *Food and Chemical Toxicology* 47, 2216–2223.

Cobo Molinos, A., Lucas, R., Abriouel, H., Ben Omar, N., Valdivia, E. and Gálvez, A. (2009b) Inhibition of *Salmonella enterica* cells in deli-type salad by enterocin AS-48 in combination with other antimicrobials. *Probiotics and Antimicrobial Proteins* 1, 85–90.

Deegan, L.H., Cotter, P.D., Hill, C. and Ross, P. (2006) Bacteriocins: biological tools for bio-preservation and shelf-life extension. *International Dairy Journal* 16, 1058–1071.

Degnan, A.J., Kaspar, C.W., Otwell, W.S., Tamplin, M.L. and Luchansky, J.B. (1994) Evaluation of lactic acid bacterium fermentation products and food-grade chemicals to control *Listeria monocytogenes* in blue crab (*Callinectes sapidus*) meat. *Applied and Environmental Microbiology* 60, 3198–3203.

Delves-Broughton, J., Williams, G.C. and Wilkinson, S. (1992) The use of bacteriocin, nisin, as a preservative in pasteurized liquid whole egg. *Letters in Applied Microbiology* 15, 133–136.

Devlieghere, F., Debevere, J. and van Impe, J. (1998) Concentration of carbon dioxide in the water-phase as a parameter to model the effect of a modified atmosphere on microorganisms. *International Journal of Food Microbiology* 43, 105–113.

Dicks, L.M.T., Mellett, F.D. and Hoffman, L.C. (2004) Use of bacteriocin-producing starter cultures of *Lactobacillus plantarum* and *Lactobacillus curvatus* in production of ostrich meat salami. *Meat Science* 66, 703–708.

Diels, A.M., De Taeye, J. and Michiels, C.W. (2004) High-pressure homogenisation sensitises *Escherichia coli* to lysozyme and nisin. *Community for Agricultural and Applied Biological Sciences* 69, 115–117.

Diels, A.M.J., De Taeye, J. and Michiels, C.W. (2005) Sensitisation of *Escherichia coli* to antibacterial peptides and enzymes by high-pressure homogenisation. *International Journal of Food Microbiology* 105, 165–175.

Diop, M.B., Dubois-Dauphin, R., Destain, J., Tine, E. and Thonart, P. (2009) Use of a nisin-producing starter culture of *Lactococcus lactis* subsp. *lactis* to improve traditional fish fermentation in Senegal. *Journal of Food Protection* 72, 1930–1934.

Du Toit, M. and Pretorius, I.S. (2000) Microbial spoilage and preservation of wine: using weapons from nature's own arsenal. *South African Journal of Enology and Viticulture* 21, 74–96.

Erkkilä, S., Suihko, M.L., Eerola, S., Petäjä, E. and Mattila-Sandholm, T. (2001) Dry sausage fermented by *Lactobacillus rhamnosus* strains. *International Journal of Food Microbiology* 28, 205–210.

Fang, T.J. and Lin, L.W. (1994a) Growth of *Listeria monocytogenes* and *Pseudomonas fragi* on cooked pork in a modified atmosphere packaging/ nisin combination system. *Journal of Food Protection* 57, 479–485.

Fang, T.J. and Lin, L.W. (1994b) Inactivation of *Listeria monocytogenes* on raw pork treated with modified atmosphere packaging and nisin. *Journal of Food and Drug Analysis* 2, 189–200.

Farber, J.M. (1991) Microbiological aspects of modified-atmosphere packaging technology-a review. *Journal of Food Protection* 54, 58–70.

Gálvez, A., Abriouel, H., Lucas López, R. and Ben Omar, N. (2007) Bacteriocin-based strategies for food biopreservation. *International Journal of Food Microbiology* 120, 51–70.

Galvez, A., Lopez, R.L., Abriouel, H., Valdivia, E. and Ben Omar, N. (2008) Application of bacteriocins in the control of foodborne pathogenic and spoilage bacteria. *Critical Reviews in Biotechnology* 28, 125–152.

García, M.T., Ben Omar, N., Lucas, R., Pérez-Pulido, R., Castro, A., Grande, M.J., Martínez-Cañamero, M. and Gálvez, A. (2003) Antimicrobial activity of enterocin EJ97 on *Bacillus coagulans* CECT 12. *Food Microbiology* 20, 533–536.

García, M.T., Lucas, R., Abriouel, H., Ben Omar, N., Pérez, R., Grande, M.J., Martínez-Cañamero, M. and Gálvez, A. (2004a) Antimicrobial activity of enterocin EJ97 against '*Bacillus macroides/Bacillus maroccanus*' isolated from zucchini purée. *Journal of Applied Microbiology* 97, 731–737.

García, M.T., Martínez Cañamero, M., Lucas, R., Ben Omar, N., Pérez Pulido, R. and Gálvez, A. (2004b) Inhibition of *Listeria monocytogenes* by enterocin EJ97 produced by *Enterococcus faecalis* EJ97. *International Journal of Food Microbiology* 90, 161–170.

Garcia-Graells, C., Masschalck, B. and Michiels, C.W. (1999) Inactivation of *Escherichia coli* in milk by high-hydrostatic-pressure treatment in combination with antimicrobial peptides. *Journal of Food Protection* 62, 1248–1254.

Garriga, M., Aymerich, M.T., Costa, S., Monfort, J.M. and Hugas, M. (2002) Bactericidal synergism through bacteriocins and high pressure in a meat model system during storage. *Food Microbiology* 19, 509–518.

Gill, A.O. and Holley, R.A. (2000a) Inhibition of bacterial growth on ham and bologna by lysozyme, nisin and EDTA. *Food Research International* 33, 83–90.

Gill, A.O. and Holley, R.A. (2000b) Surface application of lysozyme, nisin, and EDTA to inhibit spoilage and pathogenic bacteria on ham and bologna. *Journal of Food Protection* 63, 1338–1346.

Gill, A.O. and Holley, R.A. (2003) Interactive inhibition of meat spoilage and pathogenic bacteria by lysozyme, nisin and EDTA in the presence of nitrite and sodium chloride at 24°C. *International Journal of Food Microbiology* 80, 251–259.

Grande, M.J., Lucas, R., Abriouel, H., Valdivia, E., Ben Omar, N., Maqueda, M., Martínez-Bueno, M., Martínez-Cañamero, M. and Gálvez, A. (2006) Inhibition of toxicogenic *Bacillus cereus* in rice-based foods by enterocin AS-48. *International Journal of Food Microbiology* 106, 185–194.

Grande, M.J., Lucas, R., Abriouel, H., Valdivia, E., Ben Omar, N., Maqueda, M., Martínez-Cañamero, M. and Gálvez, A. (2007) Treatment of vegetable sauces with enterocin AS-48 alone or in combination with phenolic compounds to inhibit proliferation of Staphylococcus aureus. *Journal of Food Protection* 70, 405–411.

Gutiérrez, J., Criado, R., Martín, M., Herranz, C., Cintas, L.M. and Hernández, P.E. (2005) Production of enterocin P, an antilisterial pediocin-like bacteriocin from *Enterococcus faecium* P13, in *Pichia pastoris*. *Antimicrobial Agents and Chemotherapy* 49, 3004–3008.

Helander, I.M., von Wright, A. and Mattila-Sandholm, T.-M. (1997) Potential of lactic acid bacteria and novel antimicrobials against Gram-negative bacteria. *Trends in Food Science and Technology* 8, 146–150.

Hoover, D.G. (1993) Pressure effects on biological systems. *Food Technology* 47, 150–155.

Hugas, M., Neumeyer, B., PagCs, F., Garriga, F. and Hammes, W.P. (1996) Die antimikrobielle Wirkung von Bakteriozin-bildenden Kulturen in Fleischwaren: IT) Vergleich des Effektes unterschiedlicher bakteriozinbildender Laktobazillen auf Listerien in Rohwurst. *Fleischwirtsch* 76, 649–652.

Hugas, M., Pagés, F., Garriga, M. and Monfort, J.M. (1998) Application of the bacteriocinogenic *Lactobacillus sakei* CTC494 to prevent growth of *Listeria* in fresh and cooked meat products packaged with different atmospheres. *Food Microbiology* 15, 639–650.

Iseppi, R., Pilati, F., Marini, M., Toselli, M., Niederhäusern, S., Guerrieri, E., Messi, P., Sabia, C., Manicardi, G., Anacarso, I. and Bondi, M. (2008) Anti-listerial activity of a polymeric film coated with hybrid coatings doped with Enterocin 416K1 for use as bioactive food packaging. *International Journal of Food Microbiology* 123, 281–287.

Iu, J., Mittal, G.S. and Griffiths, M.W. (2001) Reduction in levels of *Escherichia coli* O157:H7 in apple cider by pulsed electric fields. *Journal of Food Protection* 64, 964–969.

Jack, R.W., Tagg, J.R. and Ray, B. (1995) Bacteriocins of Gram positive bacteria. *Microbiology Reviews* 59, 171–200.

Jin, T., Liu, L., Sommers, C.H., Boyd, G. and Zhang, H. (2009) Radiation sensitization and postirradiation proliferation of *Listeria monocytogenes* on ready-to-eat deli meat in the presence of pectin-nisin films. *Journal of Food Protection* 72, 644–649.

Jones, R.J., Zagorec, M., Brightwell, G. and Tagg, J.R. (2009) Inhibition by *Lactobacillus sakei* of other species in the flora of vacuum packaged raw meats during prolonged storage. *Food Microbiology* 26, 876–881.

Kalchayanand, N., Hanlin, M.B. and Ray, B. (1992) Sublethal injury makes Gram-negative and resistant Gram-positive bacteria sensitive to the bacteriocins, pediocin AcH and nisin. *Letters in Applied Microbiology* 16, 239–243.

Kalchayanand, N., Sikes, A., Dunne, C.P. and Ray, B. (1998) Factors influencing death and injury of foodborne pathogens by hydrostatic pressure pasteurization. *Food Microbiology* 15, 207–214.

Kato, M. and Hayashi, R. (1999) Effects of high pressure on lipids of biomembranes for understanding high-pressure-induced biological phenomena. *Bioscience, Biotechnology and Biochemistry* 68, 1321–1328.

Klaenhammer, T.R. (1993) Genetics of bacteriocins produced by lactic acid bacteria. *FEMS Microbiology Reviews* 12, 39–86.

Knight, K.P., Bartlett, F.M., McKellar, R.C. and Harris, L.J. (1999) Nisin reduces the thermal resistance of *Listeria monocytogenes* Scott A in liquid whole egg. *Journal of Food Protection* 62, 999–1003.

Leisner, J.J., Laursen, B.G., Prévost, H., Drider, D. and Dalgaard, P. (2007) *Carnobacterium*: positive and negative effects in the environment and in foods. *FEMS Microbiology Reviews* 31, 592–613.

Leistner, L. (2000) Basic aspects of food preservation by hurdle technology. *International Journal of Food Microbiology* 55, 181–186.

Leroy, F., Lievens, K. and De Vuyst, L. (2005) Modeling bacteriocin resistance and inactivation of *Listeria innocua* LMG 13568 by *Lactobacillus sakei* CTC 494 under sausage fermentation conditions. *Applied and Environmental Microbiology* 71, 7567–7570.

Leroy, F., Verluyten, J. and De Vuyst, L. (2006) Functional meat starter cultures for improved sausage fermentation. *International Journal of Food Microbiology* 106, 270–285.

Liang, Z., Mittal, G.S. and Griffiths, M.W. (2002) Inactivation of *Salmonella typhimurium* in orange juice containing antimicrobial agents by pulsed electric field. *Journal of Food Protection* 65, 1081–1087.

Liang, Z., Cheng, Z. and Mittal, G.S. (2006) Inactivation of spoilage microorganisms in apple cider using a continuous flow pulsed electric field system. *LWT – Food Science and Technology* 39, 351–357.

Long, C. and Phillips, C.A. (2003) The effect of sodium citrate, sodium lactate and nisin on the survival of *Arcobacter butzleri* NCTC 12481 on chicken. *Food Microbiology* 20, 495–502.

López-Pedemonte, T., Roig-Sagués, A.X., Trujillo, A.J., Capellas, M. and Guamis, B. (2003) Inactivation of spores of *Bacillus cereus* in cheese by high hydrostatic pressure with the addition of nisin or lysozyme. *Journal of Dairy Science* 86, 3075–3081.

Lortal, S. and Chapot-Chartier, M.P. (2005) Role, mechanisms and control of lactic acid bacteria lysis in cheese. *International Dairy Journal* 15, 857–871.

Lüders, T., Birkemo, G.A., Fimland, G., Nissen-Meyer, J. and Nes, I.F. (2003) Strong synergy between a eukaryotic antimicrobial peptide and bacteriocins from lactic acid bacteria. *Applied and Environmental Microbiology* 69, 1797–1799.

Maisnier-Patin, S., Tatini, S.R. and Richard, J. (1995) Combined effect of nisin and moderate heat on destruction of *Listeria monocytogenes* in milk. *Lait* 75, 81–91.

Mansour, M., Amri, D., Bouttefroy, A., Linder, M. and Milliere, J.B. (1999) Inhibition of *Bacillus licheniformis* spore growth in milk by nisin, monolaurin, and pH combinations. *Journal of Applied Microbiology* 86, 311–324.

Maragkoudakis, P.A., Mountzouris, K.C., Psyrras, D., Cremonese, S., Fischer, J., Cantor, M.D. and Tsakalidou, E. (2009) Functional properties of novel protective lactic acid bacteria and application in raw chicken meat against *Listeria monocytogenes* and *Salmonella enteritidis. International Journal of Food Microbiology* 130, 219–226.

Marcos, B., Aymerich, T., Monfort, J.M. and Garriga, M. (2008) High-pressure processing and antimicrobial biodegradable packaging to control *Listeria monocytogenes* during storage of cooked ham. *Food Microbiology* 25, 177–182.

Martín-Belloso, O. and Elez-Martínez, P. (2005) Food safety aspects of pulsed electric fields. In: Sun, D.W. (ed.) *Emerging Technologies for Food Processing.* Elsevier, Amsterdam. pp. 183–217.

Martínez-Cuesta, M., Bengoechea, J., Bustos, I., Rodríguez, B., Requena, T., Peláez, C. (2010) Control of late blowing in cheese by adding lacticin 3147-producing *Lactococcus lactis* IFPL 3593 to the starter. *International Dairy Journal* 20, 18–24.

Martínez-Viedma, P., Sobrino, A., Ben Omar, N., Abriouel, H., Lucas López, R., Valdivia, E., Martín Belloso, O. and Gálvez, A. (2008) Enhanced bactericidal effect of High-Intensity Pulsed-Electric Field treatment in combination with enterocin AS-48 against *Salmonella enterica* in apple juice. *International Journal of Food Microbiology* 128, 244–249.

Martínez-Viedma, P., Abriouel, H., Sobrino, A., Ben Omar, N., Lucas López, R., Valdivia, E., Martín Belloso, O. and Gálvez, A. (2009) Effect of enterocin AS-48 in combination with High-Intensity Pulsed-Electric Field treatment against the spoilage bacterium *Lactobacillus diolivorans* in apple juice. *Food Microbiology* 26, 491–496.

Martínez Viedma, P., Sobrino López, A., Ben Omar, N., Abriouel, H., Lucas López, R., Martín Belloso, O. and Gálvez, A. (2010) Increased inactivation of exopolysaccharide-producing *Pediococcus parvulus* in apple juice by combined treatment with enterocin AS-48 and high-intensity pulsed electric field. *Journal of Food Protection* 73, 39–43.

Masschalck, B., Van Houdt, R. and Michiels, C.W. (2001) High pressure increases bactericidal activity and spectrum of lactoferrin, lactoferricin and nisin. *International Journal of Food Microbiology* 64, 325–332.

Matamoros, S., Pilet, M.F., Gigout, F., Prévost, H. and Leroi, F. (2009) Selection and evaluation of seafood-borne psychrotrophic lactic acid bacteria as inhibitors of pathogenic and spoilage bacteria. *Food Microbiology* 26, 638–644.

Mathot, A.G., Beliard, E. and Thuault, D. (2003) *Streptococcus thermophilus* 580 produces a bacteriocin potentially suitable for inhibition of *Clostridium tyrobutyricum* in hard cheese. *Journal of Dairy Science* 86, 3068–3074.

McNamee, C., Noci, F., Cronin, D.A., Lyng, J.G., Morgan, D.J. and Scannell, A.G. (2010) PEF based hurdle strategy to control *Pichia fermentans, Listeria innocua* and *Escherichia coli* k12 in orange juice. *International Journal of Food Microbiology* 138, 13–18.

Mittal, G.S. and Griffiths, M.W. (2005) Pulsed electric field processing of liquid foods and beverages. In: Sun, D.W. (ed.), *Emerging Technologies for Food Processing.* Elsevier, Amsterdam, pp. 99–139.

Mollet, B., Peel, J., Pridmore, D., Rekhif, N. and Suri, B. (2004) Bactericide compositions prepared and obtained from *Micrococcus varians*. US Patent 6,689,750 (Appl. No.: 08/693,353).

Morgan, S.M., Ross, R.P., Beresford, T. and Hill, C. (2000) Combination of hydrostatic pressure and lacticin 3147 causes increased killing of *Staphylococcus* and *Listeria*. *Journal of Applied Microbiology* 88, 414–420.

Morgan, S.M., Garvin, M., Ross, R.P. and Hill, C. (2001) Evaluation of a spray-dried lacticin 3147 powder for the control of *Listeria monocytogenes* and *Bacillus cereus* in a range of food systems. *Letters in Applied Microbiology* 33, 387–391.

Mulet-Powell, N., Lacoste-Armynot, A.M., Vinas, M. and Simeon De Buochberg, M. (1998) Interactions between pairs of bacteriocins from lactic bacteria. *Journal of Food Protection* 61, 1210–1212.

Muñoz, A., Maqueda, M., Gálvez, A., Martínez-Bueno, M., Rodriguez, A. and Valdivia, E. (2004) Biocontrol of psychrotrophic enterotoxigenic *Bacillus cereus* in a non fat hard type cheese by an enterococcal strain-producing enterocin AS-48. *Journal of Food Protection* 67, 1517–1521.

Muñoz, A., Ananou, S., Gálvez, A., Martínez-Bueno, M., Rodríguez, A., Maqueda, M. and Valdivia, E. (2007) Inhibition of *Staphylococcus aureus* in dairy products by enterocin AS-48 produced in situ and ex situ: bactericidal synergism through heat and AS-48. *International Dairy Journal* 17, 760–769.

Muriana, P.M. (1996) Bacteriocins for control of *Listeria* spp. in food. *Journal of Food Protection* 59, S54–S63.

Naghmouchi, K., Drider, D., Baah, J. and Teather, R. (2010) Nisin A and polymyxin B as synergistic inhibitors of Gram-positive and Gram-negative bacteria. *Probiotics and Antimicrobial Proteins* 2, 98–103.

Nattress, F.M. and Baker, L.P. (2003) Effects of treatment with lysozyme and nisin on the microflora and sensory properties of commercial pork. *International Journal of Food Microbiology* 85, 259–267.

Neetoo, H., Ye, M., Chen, H., Joerger, R.D., Hicks, D.T. and Hoover, D.G. (2008) Use of nisin-coated plastic films to control *Listeria monocytogenes* on vacuum-packaged cold-smoked salmon. *International Journal of Food Microbiology* 122, 8–15.

Nes, I.F., Yoon, S. and Diep, D.B. (2007) Ribosomally synthesized antimicrobial peptides (bacteriocins) in lactic acid bacteria: A review. *Food Science and Biotechnology* 5, 675–690.

Nguyen, P. and Mittal, G.S. (2007) Inactivation of naturally occurring microorganisms in tomato juice using pulsed electric field (PEF) with and without antimicrobials. *Chemical Engineering and Processing* 46, 360–365.

Nieto-Lozano, J.C., Reguera-Useros, J.I., Peláez-Martínez, M.C. and Hardisson de la Torre, A. (2006) Effect of a bacteriocin produced by *Pediococcus acidilactici* against *Listeria monocytogenes* and *Clostridium perfringens* on Spanish raw meat. *Meat Science* 72, 57–61.

Nilsson, L., Huss, H.H. and Gram, L. (1997) Inhibition of *Listeria monocytogenes* on cold-smoked salmon by nisin and carbon dioxide atmosphere. *International Journal of Food Microbiology* 38, 217–227.

Nilsson, L., Chen, Y., Chikindas, M.L., Huss, H.H., Gram, L. and Montville, T.J. (2000) Carbon dioxide and nisin act synergistically on *Listeria monocytogenes*. *Applied and Environmental Microbiology* 66, 769–774.

Nykänen, A., Weckman, K. and Lapvetelainen, A. (2000) Synergistic inhibition of *Listeria monocytogenes* on cold-smoked rainbow trout by nisin and sodium lactate. *International Journal of Food Microbiology* 61, 63–72.

O´Sullivan, L., O´Connor, E.B., Ross, R.P. and Hill, C. (2006) Evaluation of live-culture-producing lacticin 3147 as a treatment for the control of *Listeria monocytogenes* on the surface of smear-ripened cheese. *Journal of Applied Microbiology* 100, 135–143.

Pathanibul, P., Taylor, T.M., Davidson, P.M. and Harte, F. (2009) Inactivation of *Escherichia coli* and *Listeria innocua* in apple and carrot juices using high pressure homogenization and nisin. *International Journal of Food Microbiology* 129, 316–320.

Peláez, C. and Requena, T. (2005) Exploiting the potential of bacteria in the cheese ecosystem. *International Dairy Journal* 15, 831–844.

Pepe, O., Blaiotta, G., Moschetti, G., Greco, T. and Villani, F. (2003) Rope-producing strains of *Bacillus* spp. from wheat bread and strategy for their control by lactic acid bacteria. *Applied and Environmental Microbiology* 69, 2321–2329.

Pol, I.E., Mastwijk, H.C., Slump, R.A., Popa, M.E. and Smith, E.J. (2001a) Influence of food matrix on inactivation of *Bacillus cereus* by combinations of nisin, pulsed electric field treatment, and carvacrol. *Journal of Food Protection* 64, 1012–1018.

Pol, I.E., van Arendonk, W.G., Mastwijk, H.C., Krommer, J., Smid, E. and Moezelaar, R. (2001b) Sensitivities of germinating spores and carvacrol-adapted vegetative cells and spores of *Bacillus cereus* to nisin and pulsed-electric-field treatment. *Applied and Environmental Microbiology* 67, 1693–1699.
Ponce, E., Pla, R., Sendra, E., Guamis, B. and Mor-Mur, M. (1998) Combined effect of nisin and high hydrostatic pressure on destruction of *Listeria innocua* and *Escherichia coli* in liquid whole egg. *International Journal of Food Microbiology* 43, 15–19.
Randazzo, C.L., Pitino, I., Scifo, G.O. and Caggia, C. (2009) Biopreservation of minimally processed iceberg lettuces using a bacteriocin produced by *Lactococcus lactis* wild strain. *Food Control* 20, 756–763.
Rayman, K., Aris, B. and Hurst, A. (1981) Nisin: possible alternative or adjunct to nitrite in the preservation of meats. *Applied and Environmental Microbiology* 41, 375–380.
Rayman, K., Malik, N. and Hurst, A. (1983) Failure of nisin to inhibit outgrowth of *Clostridium botulinum* in a model cured meat system. *Applied and Environmental Microbiology* 46, 1450–1452.
Rihakova, J., Belguesmia, Y., Petit, V.W., Pilet, M,F., Prévost, H., Dousset, X. and Drider, D. (2009) Divercin V41 from gene characterization to food applications: 1998–2008, a decade of solved and unsolved questions. *Letters in Applied Microbiology* 48, 1–7.
Rilla, N., Martínez, B., Delgado, T. and Rodríguez, A. (2003) Inhibition of *Clostridium tyrobutyricum* in Vidiago cheese by *Lactococcus lactis* ssp. *lactis* IPLA 729, a nisin Z producer. *International Journal of Food Microbiology* 85, 23–33.
Roberts, C.M. and Hoover, D.G. (1996) Sensitivity of *Bacillus coagulans* spores to combinations of high hydrostatic pressure, heat, acidity and nisin. *Journal of Applied Bacteriology* 81, 363–368.
Robertson, A., Tirado, C., Lobstein, T., Jermini, M., Knai, C., Jensen, J.H., Ferro-Luzzi, A. and James, W.P.T. (2004) Food and health in Europe: a new basis for action. WHO Regional Publications, European Series, No. 96, Geneva.
Rodríguez, J.M., Martinez, M.I. and Kok, J. (2002) Pediocin PA-1, a wide-spectrum bacteriocin from lactic acid bacteria. *Critical Reviews in Food Science and Nutrition* 42, 91–121.
Rojo-Bezares, B., Sáenz, Y., Zarazaga, M., Torres, C. and Ruiz-Larrea, F. (2007) Antimicrobial activity of nisin against *Oenococcus oeni* and other wine bacteria, *International Journal of Food Microbiology* 116, 32–36.
Ryan, M.P., Ross, R.P. and Hill, C. (2001) Strategy for manipulation of cheese flora using combinations of lacticin 3147-producing and -resistant cultures. *Applied and Environmental Microbiology* 67, 2699–2704.
Sánchez, J., Borrero, J., Gómez-Sala, B., Basanta, A., Herranz, C., Cintas, L.M. and Hernández, P.E. (2008) Cloning and heterologous production of hiracin JM79, a Sec-dependent bacteriocin produced by *Enterococcus hirae* DCH5, in lactic acid bacteria and *Pichia pastoris*. *Applied and Environmental Microbiology* 74, 2471–2479.
Scannell, A.G.M., Hill, C., Buckley, D.J. and Arendt, E.K. (1997) Determination of the influence of organic acids and nisin on shelf-life and microbiological safety aspects of fresh pork sausage. *Journal of Applied Microbiology* 83, 407– 412.
Scannell, A.G., Ross, R.P., Hill, C. and Arendt, E.K. (2000b) An effective lacticin biopreservative in fresh pork sausage. *Journal of Food Protection* 63, 370–375.
Schoeman, H., Vivier, M.A., du Toit, M., Dicks, L.M.T. and Pretorius, I.S. (1999) The development of bactericidal yeast strains by expressing the *Pediococcus acidilactici* pediocin gene (*pedA*) in *Saccharomyces cerevisiae*. *Yeast* 15, 647–656.
Schuman, J.D. and Sheldon, B.W. (2003) Inhibition of *Listeria monocytogenes* in pH-adjusted pasteurized liquid whole egg. *Journal of Food Protection* 66, 999–1006.
Settanni, L. and Corsetti, A. (2008) Application of bacteriocins in vegetable food biopreservation. *International Journal of Food Microbiology* 121, 123–138.
Sobrino-Lopez, A. and Martin-Belloso, O. (2006) Enhancing inactivation of *Staphylococcus aureus* in skim milk by combining high-intensity pulsed electric fields and nisin. *Journal of Food Protection* 69, 345–353.
Sobrino-López, A. and Martín-Belloso, O. (2008) Use of nisin and other bacteriocins for preservation of dairy products. *International Dairy Journal* 18, 329–343.
Sobrino, A., Martínez Viedma, P., Abriouel, H., Valdivia, E., Gálvez, A. and Martín-Belloso, O. (2009) The impact of adding antimicrobial peptides to milk inoculated with *Staphylococcus aureus* and processed by high-intensity pulsed electric field. *Journal of Dairy Science* 92, 2514–2523.

Stergiou, V.A., Thomas, L.V. and Adams, M.R. (2006) Interactions of nisin with glutathione in a model protein system and meat. *Journal of Food Protection* 69, 951–956.
Stevens, K.A., Sheldon, B.W., Klapes, N.A. and Klaenhammer, T.R. (1991) Nisin treatment for inactivation of *Salmonella* species and other gram-negative bacteria. *Applied and Environmental Microbiology* 57, 3613–3615.
Stewart, C.M., Dunne, C.P., Sikes, A. and Hoover, D.G. (2000) Sensitivity of spores of *Bacillus subtilis* and *Clostridium sporogenes* PA3679 to combinations of high hydrostatic pressure and other processing parameters. *Innovative Food Science and Emerging Technologies* 1, 49–56.
Stiles, M.E. (1996) Biopreservation by lactic acid bacteria. *Antonie van Leeuwenhoek* 70, 331–345.
Szabo, E.A. and Cahill, M.E. (1999) Nisin and ALTATM 2341 inhibit the growth of *Listeria monocytogenes* on smoked salmon packaged under vacuum or 100% CO_2. *Letters in Applied Microbiology* 28, 373–377.
Tahiri, I., Desbiens, M., Kheadr, E., Lacroix, C. and Fliss, I. (2009) Comparison of different application strategies of divergicin M35 for inactivation of *Listeria monocytogenes* in cold-smoked wild salmon. *Food Microbiology* 26, 783–793.
Taylor, J.I., Somer, E.B. and Kruger, L.A. (1985) Antibotulinal effectiveness of nisin–nitrite combinations in culture medium and chicken frankfurter emulsions. *Journal of Food Protection* 48, 234–249.
Terebiznik, M.R., Jagus, R.J., Cerrutti, P., De Huergo, M. and Pilosof, A.M. (2000) Combined effect of nisin and pulsed electric fields on the inactivation of *Escherichia coli. Journal of Food Protection* 63, 741–746.
Thomas, L.V. and Delves-Broughton, J. (2001) New advances in the application of the food preservative nisin. *Advances in Food Science* 2, 11–22.
Thomas, L.V., Davies, E.A., Delves-Broughton, J. and Wimpenny, J.W.T. (1998) Synergistic effect of sucrose fatty acid ester on nisin inhibition of Gram-positive bacteria. *Journal of Applied Microbiology* 85, 1013–1022.
Thomas, L.V., Clarkson, M.R. and Delves-Broughton, J. (2000) Nisin. In: Naidu, A.S. (ed.), *Natural Food Antimicrobial Systems.* CRC Press, Boca-Raton, Florida, pp. 463–524.
Todorov, S.D., Koep, K.S.C., Van Reenen, C.A., Hoffman, L.C., Slinde, E. and Dicks L.M.T. (2007) Production of salami from beef, horse, mutton, Blesbok (*Damaliscus dorcas phillipsi*) and Springbok (*Antidorcas marsupialis*) with bacteriocinogenic strains of *Lactobacillus plantarum* and *Lactobacillus curvatus. Meat Science* 77, 405–412.
Tomé, E., Gibbs, P.A. and Teixeira, P.C. (2008) Growth control of *Listeria innocua* 2030c on vacuum-packaged cold-smoked salmon by lactic acid bacteria. *International Journal of Food Microbiology* 121, 285–294.
Trias, R., Bañeras, L., Badosa, E. and Montesinos, E. (2008a) Bioprotection of Golden Delicious apples and Iceberg lettuce against foodborne bacterial pathogens by lactic acid bacteria. *International Journal of Food Microbiology* 123, 50–60.
Trias, R., Badosa, E., Montesinos, E. and Bañeras, L. (2008b) Bioprotective *Leuconostoc* strains against *Listeria monocytogenes* in fresh fruits and vegetables. *International Journal of Food Microbiology* 127, 91–98.
Työppönen, S., Petäjä, E. and Mattila-Sandholm, T. (2003) Bioprotectives and probiotics for dry sausages. *International Journal of Food Microbiology* 83, 233–244.
Uesugi, A.R. and Moraru, C.I. (2009) Reduction of *Listeria* on ready-to-eat sausages after exposure to a combination of pulsed light and Nisin. *Journal of Food Protection* 72, 347–353.
Uhart, M., Ravishankar, S. and Maks, N.D. (2004) Control of *Listeria monocytogenes* with combined antimicrobials on beef franks stored at 4°C. *Journal of Food Protection* 67, 2296–2301.
Ukuku, D.O. and Fett, W.F. (2004) Effect of nisin in combination with EDTA, sodium lactate, and potassium sorbate for reducing *Salmonella* on whole and fresh-cut cantaloupe. *Journal of Food Protection* 67, 2143–2150.
Ukuku, D.O., Bari, M.L., Kawamoto, S. and Isshiki, K. (2005) Use of hydrogen peroxide in combination with nisin, sodium lactate and citric acid for reducing transfer of bacterial pathogens from whole melon surfaces to fresh-cut pieces. *International Journal of Food Microbiology* 104, 225–233.
Ulmer, H.M., Heinz, V., Gänzle, M.G., Knorr, D. and Vogel, R.F. (2002) Effects of pulsed electric fields on inactivation and metabolic activity of *Lactobacillus plantarum* in model beer. *Journal of Applied Microbiology* 93, 326–335.
Vaara, M. (1992) Agents that increase the permeability of the outer membrane. *Microbiological Reviews* 56, 395–411.

Valerio, F., De Bellis, P., Lonigro, S.L., Visconti, A. and Lavermicocca, P. (2008) Use of *Lactobacillus plantarum* fermentation products in bread-making to prevent *Bacillus subtilis* ropy spoilage. *International Journal of Food Microbiology* 122, 328–332.

Van Reenen, C.A., Chikindas, M.L., van Zyl, W.H. and Dicks, L.M.T. (2002) Characterisation and heterologous expression of a class IIa bacteriocin, plantaricin 423 from *Lactobacillus plantarum* 423, in *Saccharomyces cerevisiae. International Journal of Food Microbiology* 81, 29–40.

Wandling, L.R., Sheldon, B.W. and Foegeding, P.M. (1999) Nisin in milk sensitizes *Bacillus* spores to heat and prevents recovery of survivors. *Journal of Food Protection* 62, 492–498.

Weber, G.H. and Broich, W.A. (1986) Shelf-life extension of cultured dairy foods. *Culture Dairy Products Journal* 21, 1–9.

Wu, Y., Mittal, G.S. and Griffiths, M.W. (2005) Effect of Pulsed Electric Field on the inactivation of microorganisms in grape juices with and without antimicrobials. *Biosystems Engineering* 90, 1–7.

Young, L.L., Reverie, R.D. and Cole, A.B. (1988) Fresh red meats: A place to apply modified atmospheres. *Food Technology* 42, 64–66, 68–69.

Yuste, J. and Fung, D.Y. (2004) Inactivation of *Salmonella typhimurium* and *Escherichia coli* O157:H7 in apple juice by a combination of nisin and cinnamon. *Journal of Food Protection* 67, 371–377.

5 Bacterial Antimicrobial Peptides and Food Preservation

Maria do Carmo de Freire Bastos* and Hilana Ceotto

Introduction

There are already many control measures within the food industry to prevent or minimize bacterial contamination which can result in food spoilage or food-borne diseases. However, despite these precautions, food-borne outbreaks do occur alarmingly frequently (Deegan *et al.*, 2006). *Listeria monocytogenes*, the bacterial agent of listeriosis, is of particular concern to the food industry since it is a difficult food-borne pathogen to control because of its ubiquitous distribution, tolerance to high levels of salt, and its ability to grow at a relatively low pH and at refrigeration temperatures. Other pathogens can also cause food-borne diseases, including *Staphylococcus aureus* and many Gram-negative bacteria such as *Escherichia coli* O157:H7, *Campylobacter* spp., and *Salmonella* spp., among others (Deegan *et al.*, 2006). Therefore, food safety and quality have become an increasingly important international concern. Amelioration of economic losses due to food spoilage, lowering the food-processing costs and avoiding transmission of microbial pathogens through the food chain, while satisfying the growing consumer demands for foods that are ready-to-eat, fresh-tasting, nutrient and vitamin rich, and minimally processed and preserved, became the major challenges for the current food industry. As a result, there has been a great interest in naturally produced antimicrobial agents as biopreservatives that target food pathogens and food-spoilage microorganisms without toxic or other adverse effects.

Biopreservation refers to the extension of the shelf-life and improvement of the safety of food using microorganisms and/or their metabolites (Ross *et al.*, 2002). Bacteriocins (Bac) are ribosomally synthesized bacterial peptides with inhibitory activity against other bacteria (Heng *et al.*, 2007). Gram-positive bacteria are the major source of bacteriocins that have been examined for applications in microbial food safety and quality. This chapter will therefore address different aspects related to food preservation by bacteriocins.

Bacteriocin Biology

Bacteriocin production could be considered as advantageous to the producing bacterial cells as these peptides can kill or inhibit bacteria competing for the same ecological niche or the same nutrient pool. The bacteriocin-producing strains have developed a protection system against their own product, named immunity. Each bacteriocin has its immunity system, which is generally

* Corresponding author.

expressed concomitantly with the bacteriocin structural genes (Heng *et al.*, 2007).

Bacteriocins produced by Gram-positive bacteria are the most studied ones. They form a heterogeneous group of peptides and proteins. They can be active against other bacteria, belonging to only closely related species (narrow spectrum) or to different bacterial genera (broad spectrum). Their genetic determinants are generally arranged in the form of operons, which may be encoded on the bacterial chromosome, although they are usually found on plasmids.

Recently, Bierbaum and Sahl (2009) and Nissen-Meyer *et al.* (2009) proposed modifications in the classification of the bacteriocins produced by Gram-positive bacteria to accommodate data related to the annual description of a repertoire of bacteriocins, some of them with distinctive characteristics. According to the current classification, bacteriocins produced by Gram-positive bacteria can be grouped into three main classes, all of them with subdivisions. Bacteriocins from classes I and II are the most studied and have a better elucidated mode of action, since they occur more frequently and have a potential industrial application as food preservatives (Gálvez *et al.*, 2007; Settanni and Corsetti, 2008). Members of the two classes are clearly different, both in terms of the structure of the bacteriocin itself and in terms of the machinery involved in production and processing. Knowledge of all aspects of bacteriocin biology, including the elucidation of their structure–function relationships, production, immunity, regulation and mode of action, is required when considering bacteriocin applications. However, for a more comprehensive review on both class I and class II bacteriocins, the reader is referred to recent publications (Bierbaum and Sahl, 2009; Nissen-Meyer *et al.*, 2009). Examples of bacteriocins of both classes with potential application in food preservation are described in Table 5.1.

The three classes of bacteriocins produced by Gram-positive bacteria are:

- Class I bacteriocins (lantibiotics): The lantibiotics (lanthionine-containing antibiotics) are generally small peptides (< 5 kDa) that contain uncommon amino acids, such as lanthionine (Lan) and methyl-lanthionine (MeLan), among others, which are formed by post-translational modifications of the peptide. The post-translational modifications generally involve three amino acids: serine, threonine and cysteine. The dehydration of serine leads to the formation of didehydroalanine (Dha), while the dehydration of threonine leads to the formation of didehydrobutyrine (Dhb). Lan and MeLan are formed by the addition of a thiol group of a cysteine near to these dehydrated amino acids, which generates thiol bridges between them. As a result of the presence of these intramolecular bonds, the lantibiotics are polycyclic structures containing rings of Lan and MeLan. The conformation of polycyclic lantibiotics seems to confer rigidity and heat resistance to them and contributes to their antimicrobial activity (Bierbaum and Sahl, 2009). Lantibiotics are only produced by Gram-positive bacteria and most of them present bactericidal activity mainly against this group (Bierbaum and Sahl, 2009).
- Class II bacteriocins: Class II comprises peptides smaller than 10 kDa without modified amino acids.
- Class III bacteriocins: Class III is composed of peptides larger than 10 kDa, generally thermolabile.

All bacteriocins that have been tested for application in food preservation belong to either class I or class II, acting by permeabilization of bacterial membranes. Most of them are produced by lactic acid bacteria (LAB). The LAB group is composed of Gram-positive, non-sporulating micro-aerophilic bacteria, including a number of genera such as *Lactococcus, Lactobacillus, Enterococcus, Streptococcus, Leuconostoc, Pediococcus* and *Carnobacterium*, among others, which are strictly fermentative, generally producing lactic acid as the major end product. They occur as natural microbiota in raw milk (de Arauz *et al.*, 2009).

Table 5.1. Examples of bacteriocins with potential application in biopreservation.

Bacteriocin	Bacteriocin class/subclass	Producing bacteria	Microorganisms inhibited	Application / potential application in food preservation (reference)
Nisin	I/AI	*Lactococcus lactis* subsp. *lactis*	*Listeria monocytogenes* *Staphylococcus aureus*	Dairy products (de Arauz *et al.*, 2009)
Variacin	I/AI	*Kokuria varians*	*Bacillus cereus*	Vanilla chilled dessert Chocolate mousse (O'Mahony *et al.*, 2001)
Warnericin RB4	I/AII	*Staphylococcus warneri* RB4	*Alyciclobacillus acidoterrestris*	Acidic drinks (Minamikawa *et al.*, 2005)
Lacticin 3147	I/B	*L. lactis* DPC3147	NSLAB *L. monocytogenes*	Dairy products (Ryan *et al.*, 1996; McAuliffe *et al.*, 1999)
Pediocin PA1/AcH	II/a	*Pediococcus acidilactici*	*L. monocytogenes* *Clostridium* spp.	Dairy products (Rodriguez *et al.*, 2002)
Lactocin AL705	IIa	*Lactobacillus curvatus*	*Listeria* spp. *Brochothrix thermosphacta*	Vacuum-packaged fresh meat (Castellano *et al.*, 2008)
Lactocin 705	IIb	*Lactobacillus curvatus*	*Listeria* spp. *Brochothrix thermosphacta*	Vacuum-packaged fresh meat (Castellano *et al.*, 2008)
Enterocin AS-48	II/c	*Enterococcus faecalis* S-48	*Geobacillus stearothermophilus* *B. cereus* *B. thermosphacta*	Fruit juices and vegetables (Maqueda *et al.*, 2004)

NSLAB, non-starter lactic acid bacteria.

Bacteriocins Produced by Lactic Acid Bacteria

LAB have been used for a variety of dairy, vegetable and meat fermentations because they significantly contribute to the flavour, texture and, in many cases, to the nutritional value of the food products. Since bacteriocins are produced by LAB during their growth, they are considered natural ingredients found in all fermented foods and dairy products, and have thus been consumed unknowingly by humans for thousands of years (Randazzo *et al.*, 2009).

Bacteriocins that are produced by LAB can be broad or narrow spectrum but, in general, activity is directed against low-C+G Gram-positive species. Bacteriocins produced by Gram-positive bacteria are generally inactive against Gram-negative bacteria due to resistance conferred by the outer membrane. However, activity against these microorganisms has been described, but usually only in situations where the integrity of the outer membrane has been compromised. There are also rare natural bacteriocins, such as some enterocins, that possess inherent activity against Gram-negative microorganisms (Maqueda *et al.*, 2004).

The bacteriocins produced by LAB offer several desirable properties that make them suitable for food preservation: (i) they have the status GRAS (generally recognized as safe); (ii) they are not active and are nontoxic on eukaryotic cells; (iii) they become inactivated by digestive proteases, having little influence on the gut microbiota; (iv) they are usually pH and heat tolerant; (v) they have a relatively broad antimicrobial spectrum against many food-borne pathogenic and spoilage bacteria; (vi) they generally exhibit a bactericidal mode of action and no cross-resistance with antibiotics; and (vii) their genetic determinants are usually plasmid encoded, facilitating genetic manipulation, including the transfer of the gene clusters involved in their production to other food-grade bacteria (Gálvez *et al.*, 2007).

In general, studies on food safety using bacteriocins as preservatives follow this scheme: (i) isolation of bacteria from raw materials or final products; (ii) screening for bacteriocin production; (iii) characterization of bacteriocins; (iv) production of bacteriocins in food model systems, and (v) *in situ* application. To date, the only commercially produced bacteriocins are nisin and pediocin PA-1.

Nisin

Undoubtedly, the most well-known and studied bacteriocin is the lantibiotic nisin, which has found application as a shelf-life extender in a broad range of dairy and non-dairy products worldwide. Nisin (N inhibitory substance), produced by certain strains of *L. lactis* subsp. *lactis*, was the first discovered LAB bacteriocin (Rogers and Whittier, 1928). It has the status GRAS and has been used as a direct human food ingredient in approximately 50 countries. Nisin is manufactured via fermentation of fluid milk or whey by strains of *L. lactis*. The resulting fermentation broth is subsequently concentrated and separated, spray dried, and milled to yield small particles. The final product is commercially available as Nisaplin™ (Danisco, Copenhagen, Denmark) and may be purchased from different suppliers as a preparation that contains nisin (1×10^6 IU/g) with NaCl (74.4%), 23.8% denatured milk solids and 1.7% moisture (Deegan *et al.*, 2006).

The nisin variants A and Z differ by a single amino acid substitution (histidine at position 27 in nisin A is replaced by asparagine in nisin Z). The structural modification has no effect on the antimicrobial activity, but it gives nisin Z higher solubility and diffusion compared to nisin A, which are important characteristics for food applications (de Vos *et al.*, 1993).

Nisin is an effective bactericidal agent against Gram-positive bacteria, including strains of *Streptococcus* spp., *Staphylococcus* spp. and *Listeria* spp. While fermented foods are generally considered to be relatively free of pathogens, *Listeria* spp. has been found to survive and possibly grow in fermented foods made with raw materials containing the organism (Guinane *et al.*, 2005).

Spores produced by *Bacillus* spp. and *Clostridium* spp., important food-spoilage and pathogenic bacteria, are particularly susceptible to nisin, with spores being more sensitive than vegetative cells especially after heat treatments (Delves-Broughton *et al.*, 1996). Probably, the ability of nisin to inhibit spore germination is related to the formation of pores in the membrane of cells in the early stages of germination (Bierbaum and Sahl, 2009).

Pediocin PA-1/AcH

Pediocin PA-1 is a class II bacteriocin with strong anti-listerial activity produced by some strains of pediococci, generally of meat origin. Pediocin AcH, produced by *P. acidilactici* H, is identical to pediocin PA-1. Pediocin PA-1/AcH is also approved for use in food. This bacteriocin has been commercially exploited in the form of ALTA™ 2431 (Kerry Bioscience, Carrigaline, Co. Cork, Ireland), which is based on a fermentate generated from its producing strain.

Enterocin AS-48

Enterocin AS-48 is a class II bacteriocin produced by *E. faecalis* S-48 (Maqueda *et al.*, 2004). Although enterocin AS-48 is not commercially available yet, it presents a great potential application in food systems, since it exhibits bactericidal activity against a wide variety of Gram-positive bacteria, including food-spoilage and pathogenic strains.

Bacteriocins Produced by Other Bacteria

Since the discovery of nisin and its potential applications, bacteriocins have attracted an intensive research interest over the last three decades, resulting in the discovery and characterization of a growing arsenal of novel bacteriocins produced not only by LAB but also by other bacteria.

For food applications, any bacteriocin-producing strain should meet important criteria: (i) it should preferably have a GRAS status; (ii) the bacteriocin should exhibit either a broad spectrum of activity that includes pathogens and/or spoilage microorganisms, or activity against a particular pathogen; (iii) the bacteriocin should be heat stable to withstand thermal processing; (iv) it should have no associated health risks; (v) its inclusion in products should lead to beneficial effects such as improved safety and quality; and (vi) it should have a high specific activity (Cotter *et al.*, 2005).

Although nisin and pediocin PA-1 are currently the only commercially produced bacteriocins, many other bacteriocins so far described do have a potential application in food products (Table 5.1).

Effectiveness of Bacteriocins in Food Systems

The application of bacteriocins in food preservation can offer several benefits: (i) an extended shelf-life of foods; (ii) provide extra protection during temperature abuse conditions; (iii) decrease the risk for transmission of food-borne pathogens through the food chain; (iv) ameliorate the economic losses due to food spoilage; (v) reduce the application of chemical preservatives; (vi) permit the application of less severe heat treatments without compromising food safety, better preserving food nutrients and vitamins, as well as organoleptic properties of foods; and (vii) they may serve to satisfy industrial and consumer demands (Gálvez *et al.*, 2007).

Although many bacteriocins have been isolated from food-associated LAB, they are not necessarily effective in all food systems. Therefore, results of inhibition of target organisms obtained from *in vitro* systems should be confirmed by applied studies performed in food systems.

Bacteriocins can be introduced into food in at least three different ways: (i) by adding the purified or semi-purified bacteriocin directly to food; (ii) by incorporating an ingredient based on a fermentate of a bacteriocin-producing strain into the food; or (iii) by producing the bacteriocin *in situ* by using the Bac$^+$ strains as starter or protective/adjunct cultures (Cotter *et al.*, 2005; Deegan *et al.*, 2006). The use of purified/semi-purified bacteriocins is not always attractive to the food industry, as in this form they may be labelled as additives and require regulatory approval. The two other alternatives (fermented ingredient/*in situ* production) do not require regulatory approval or preservative label declarations. These options are frequently regarded as more attractive routes through which bacteriocins can be incorporated into a food (Deegan *et al.*, 2006).

Comparative studies revealed that bacteriocins are much more efficient *in vitro* (culture medium) than in food systems and that sometimes at least tenfold higher bacteriocin concentrations are required in food to achieve the same inhibitory effect (Schillinger *et al.*, 1996). The efficacy of bacteriocins in foods will greatly depend on various food-related factors that in most cases involve: (i) interactions with food components (fat or protein particles); (ii) precipitation; (iii) inactivation by conditions that destabilize the biological activity, like proteolytic degradation or oxidation; and (iv) uneven distribution of bacteriocin molecules in the food matrix (Coma, 2008). The chemical composition and the physical conditions of food, e.g. pH, osmolarity and fat content, can have a significant influence on the activity of bacteriocins. For example, the solubility, stability and biological activity of nisin are dependent on the pH medium. Nisin is 228 times more soluble at pH 2.0 than at pH 8.0 (Liu and Hansen, 1990) and at pH 2.0, nisin can be autoclaved at 121°C for 15 min without inactivation. In fermentation processes, at pH <6.0, more than 80% of nisin produced is released into the medium. On the other hand, at pH >6.0, most of the nisin is associated with the cellular membrane. In neutral and alkaline conditions, nisin is almost insoluble, but solubility and stability increase drastically with the lowering of pH (Penna *et al.*, 2005). Lacticin 3147 (another lactococcal lantibiotic) retains activity at neutral pH and is particularly heat stable at low pH (McAuliffe *et al.*, 1998).

Foods are complex ecosystems with a range of microbial compositions. Microorganisms are seldom distributed homogeneously in the food matrix, tending to form microcolonies in a solid food, or grow in the form of slime-covered biofilms on food surfaces. The food microbiota can interfere in bacteriocin activity depending on microbial load and diversity, bacteriocin sensitivity and microbial interactions in the food system (Gálvez *et al.*, 2007).

Use of Bacteriocin-producing Bacteria *In Situ*

The use of Bac$^+$ cultures in the food industry requires careful selection of strains that are: (i) well adapted to the particular food environment in which they will be used; (ii) able to grow under the food processing and/or storage conditions; and (iii) able to produce enough bacteriocin to inhibit the target pathogenic or spoilage bacteria. Therefore, it is necessary to implement the right experimental approaches to select bacteriocin-producing strains that are suitable for use in food production (Gálvez *et al.*, 2007).

Bac$^+$ strains can be used either directly as starter cultures, as adjunct or co-cultures in combination with a starter culture, or as protective cultures, especially in the case of non-fermented foods (Gálvez *et al.*, 2007). The demonstration of antimicrobial activities may constitute a secondary effect for starter cultures, whereas it represents the unique characteristic requested for protective cultures or co-cultures.

When used as a starter culture, the Bac$^+$ strain must be able to carry out the desired fermentation process optimally besides being able to produce enough bacteriocin to afford protection. Adjunct cultures do not need to contribute to the fermentation, but they must not interfere with the primary function of the starter culture. For this reason, bacteriocin resistance of the starter culture may be required. Differences in inoculum density, a faster growth rate of the starter or a delayed bacteriocin production may also permit the starter to grow without interference from the Bac$^+$ adjunct culture.

A number of studies have been performed with bacteriocin-producing cultures, demonstrating the effectiveness of bacteriocins for the inhibition of *L. monocytogenes* in food. As an example, Nunez *et al.* (1997) found that counts of *L. monocytogenes* Ohio in Manchego cheese inoculated with a bacteriocin-producing *E. faecalis* strain decreased by 6 $\log_{10}$ in 7 days, whereas the survival of the organism in cheese made with the commercial starter culture was not affected.

Because pediococci do not have an application as cheese starter cultures, the plasmid encoding pediocin PA-1 was expressed heterologously in *L. lactis* to aid in the preservation of Cheddar cheese and to assure the microbial quality of the fermentation process. Rodríguez *et al.* (2005) reported that pediocin PA-1 produced heterologously by a *L. lactis* adjunct strain lowered *L. monocytogenes* and *S. aureus* counts by 2.97 and 0.98 $\log_{10}$ units, respectively, compared with control cheese. These results show that pediocin PA-1 production *in situ* by strains of LAB adapted to the dairy environment would extend the application of this bacteriocin in cheese manufacture. Pediocin PA-1 has also been expressed in the yeast *Saccharomyces cerevisiae* to improve the preservation of wine, bread and other food products where yeast is used (Schoeman *et al.*, 1999).

Lacticin 3147 is a two-component lantibiotic which has been shown to be an effective biopreservative in many food systems (Guinane *et al.*, 2005). As the genetic determinants for lacticin 3147 are encoded on the large conjugative plasmid pMRC01, they can be transferred in a food-grade manner to many commercially useful starter cultures. The ability of a lacticin 3147-producing starter culture, constructed by the conjugative transfer of pMRC01, to reduce numbers of *L. monocytogenes* (10^4 cfu/ml) artificially introduced into cottage cheese at 4, 18 and 30°C was tested over a period of one week. Reductions of 99.9% were seen in cottage cheese held at 4°C after only 5 days, whereas at the elevated temperatures the reductions occurred more promptly (McAuliffe *et al.*, 1999).

In another study, *L. curvatus* CRL705 proved able to grow, produce bacteriocins (lactocin 705 and lactocin AL705), and control the growth of *Listeria* spp. and *B. thermosphacta* (the predominant spoilage organism in meat products) in vacuum-packaged fresh meat stored at chill temperatures (2–8°C) with a negligible effect on the sensory characteristics of the meat (Castellano *et al.*, 2008).

Bac^+ protective cultures can be used to inhibit spoilage and pathogenic bacteria during the shelf-life period of non-fermented foods. A protective culture may grow and produce bacteriocin during refrigeration storage of the food and/or during temperature abuse conditions. In the first case, growth of the protective cultures must have a neutral impact on the physicochemical and organoleptic properties of the food, while under temperature abuse conditions the protective culture may even act as the predominant spoiler, ensuring that pathogenic bacteria do not grow and that the spoiled food is not consumed (Gálvez *et al.*, 2007).

In situ bacteriocin production offers several advantages compared to *ex situ* production (to be described below) regarding both legal aspects and costs. Lowering the costs of biopreservation processes may be highly attractive, especially for small economies and developing countries, where food safety may be seriously compromised (Gálvez *et al.*, 2007).

Bacteriocins as Food Additives (*Ex Situ* Bacteriocin Production)

Preparations of bacteriocins obtained by cultivation of the producer strain in a fermentor at industrial scale followed by adequate recovery and processing can be added to foods as partially purified or purified concentrates. *Ex situ* produced bacteriocins can also be added in the form of raw concentrates obtained by cultivation of the producer strain in a food-grade substrate such as milk or whey. The resulting preparations may be regarded as food additives or ingredients from the legal point of view, since some of their components may play a recognized function in the food, such as increase in protein content or thickening. They also may contain other cell-derived metabolites, such as lactic acid, affording an additional bioprotectant function (Gálvez *et al.*, 2007).

Bacteriocins have been directly added to foods such as cheese to prevent against *C. botulinum* and *L. monocytogenes.* As an example, in long-life cottage cheese spiked with 10^4 cfu/g *L. monocytogenes*, the addition of 2000 IU/g nisin resulted in a 1000-fold decrease in the bacterial counts after 7-day storage at 20°C, compared to a tenfold decrease in the control (Ferreira and Lund, 1996). Nisin has also been demonstrated to be effective in a range of food products that include processed cheese and cheese spreads, milk products, canned foods, fish and meat products, brewing, wine manufacture, liquid egg and confectionery (Ross *et al.*, 2002).

The effect of pediocin PA-1 on the growth of *L. monocytogenes* has been studied in cottage cheese, half-and-half cream, and cream sauce systems (Pucci *et al.*, 1988). Control counts of the bacteria in the half-and-half and cheese sauce increased by almost 4 $\log_{10}$ after 7 days at 4°C (5.4×10^6 cfu/ml and 1.7×10^7 cfu/g, respectively). When 100 AU/ml pediocin was added, cell counts had just reached the detection limit for half-and-half (10^2 cfu/ml) and were 5-$\log_{10}$ lower than the control in the cheese sauce.

Meat is an excellent substrate for bacterial growth; hence, if restriction methods are not used, it becomes easily spoiled. *Clostridium perfringens* is the leading cause of bacterial food-borne illness in countries where meat and poultry consumption is high (Castellano *et al.*, 2008). Nitrates are commonly used to prevent clostridial growth in meat. However, safety concerns regarding the presence of nitrites have prompted the food industry to look for alternative methods of preservation. The amphipathic nature of nisin limits widespread application of nisin to various products, including meat, because of its interaction with fat and other food components, which reduces the antimicrobial potential of nisin in a food matrix (Taylor *et al.*, 2007). Since there are difficulties using nisin, the use of other bacteriocins has been examined. LeucocinA, enterocins, sakacins and carnobacteriocins prolong the shelf life of fresh meat. However, the most promising results in meats were obtained using pediocin PA-1. Used alone (Coventry *et al.*, 1995) or in combination with diacetate (Schlyter *et al.*, 1993), pediocin PA-1 was active in turkey slurries against *L. monocytogenes* and *Lactobacillus curvatus*, a spoilage microorganism. In another study, pediocin AcH/PA-1 successfully controlled the growth of *L. monocytogenes* in raw chicken (Goff *et al.*, 1996). The use of 2400 AU/g pediocin resulted in 2.8 $\log_{10}$ cfu/g *L. monocytogenes* after 28 days of storage at 5°C, whereas the control chicken had 8.1 $\log_{10}$ cfu/g. Pediocin AcH/PA-1 binds to raw chicken but not to cooked. However, when raw chicken with applied pediocin was cooked, the antimicrobial activity was retained. The authors suggested that pediocin should be applied to chicken before cooking for maximum effectiveness.

Bacteriocins can also be employed for the preservation of foods of vegetable origin. Some examples are described in Table 5.2. However, for a more comprehensive review on this subject, the reader is referred to a recent publication written by Settanni and Corsetti (2008).

Use of Bacteriocins to Promote Quality

Bacteriocins can also be used to promote quality, rather than simply to prevent spoilage or safety problems. For example, bacteriocins can be used to control adventitious non-starter flora such as non-starter LAB (NSLAB) in cheese and wine. The uncontrolled growth of NSLAB, which are primarily lactobacilli, can cause major economic losses since these microorganisms are often associated with quality defects in the cheese, such as calcium-lactate crystal formation, flavour and slit defects, and the production of detrimental compounds in wine. Bacteriocin-producing starters have been found to reduce these problems significantly.

Table 5.2. Application of bacteriocins in preservation of vegetable foods

Bacteriocin	Target bacteria	Food	Reference
Fermentable vegetables			
Plantaricins S and T	LAB	Green olives	Leal-Sánchez *et al.*, 2003
Nisin	*Bacillus* spp.	Soybean paste	Kato *et al.*, 1999
Nisin	Lactobacilli	Kimchi	Choi and Park, 2000
Enterocin AS-48	*Bacillus licheniformis*	Apple juice and apple cider	Grande *et al.*, 2006
Non-fermentable vegetables			
Nisin	*Bacillus* spp., *Clostridium* spp.	Mashed potatoes	Thomas *et al.*, 2002
Enterocin CCM 4231	*L. monocytogenes* and *S. aureus*	Soy milk	Lauková and Czikková, 1999
Nisin, Enterocin AS-48	*A. acidoterrestris*	Fruit juices	Komitopoulou *et al.*, 1999 Grande *et al.*, 2005
Enterocin AS-48	*Bacillus coagulans*	Canned vegetable foods	Lucas *et al.*, 2006

Lacticin 3147 has been investigated for use in the dairy industry to control NSLAB during fermentation of dairy products. The most efficient method of introducing lacticin to a fermented dairy product is to use a lacticin-producing culture as the starter or as an adjunct in the fermentation process. DPC3147, the natural lacticin 3147 producer, is unsuitable as a starter culture as this strain is associated with an 'off' flavour. However, the conjugative plasmid which carries the lacticin 3147 gene cluster (pMRC01) can be directly transferred to commercially used lactococcal starters. Using one of such transconjugant starters in Cheddar cheese manufacture, Ryan *et al.* (1996) showed that the level of lacticin 3147 produced was constant throughout the ripening process and that, at the end of cheese production, the NSLAB population had been reduced by at least 100-fold.

As some NSLAB can improve flavour in some foods, their complete elimination is not always desirable. This problem was overcome through the use of a system in which an adjunct *Lactobacillus paracasei* strain with reduced lacticin 3147 sensitivity (obtained on repeated exposure to increasing concentrations of the bacteriocin) was used with a lacticin 3147-producing starter in Cheddar cheese manufacture. While other NSLAB were inhibited during ripening, the resistant mutant could tolerate the levels of lacticin 3147 and became the dominant microbiota in the cheese (Ryan *et al.*, 2001).

Another advantage of using lacticin 3147 transconjugants as starters is that the plasmid pMRC01 encodes bacteriophage resistance. The susceptibility of lactococcal starter cultures to phage attack is a serious problem in the dairy industry, affecting cheese starter performance and resulting in production and economic losses. Thus, the introduction of the lacticin 3147 genetic determinants to a starter culture has the added benefit of conferring protection from lactococcal phage infection (O'Sullivan *et al.*, 1998). This is in contrast to nisin-producing starters which are associated with phage susceptibility.

Bacteriocins can also be applied in other ways to enhance food fermentation. Cheese usually has a maturation time of at least a few months, during which gradual autolysis of the starter cultures occurs. Lysis results in the release of intracellular enzymes which break down the casein in the cheese to small peptides and amino acids. The amino acids released are the precursor compounds responsible for flavour development in cheese. As cell lysis is a slow process and often a limiting step in cheese maturation, controlled early lysis is known to be advantageous for improved flavour development (Guinane *et al.*, 2005). The lytic abilities of many bacteriocins may have an application in the acceleration of cheese ripening. This has been shown during semi-hard and hard cheese manufacture in which bacteriocin production brought about the controlled lysis of starter LAB, which resulted in the release of intracellular proteinases and peptidases and ultimately accelerated ripening and even improved flavour (Oumer *et al.*, 2001; O'Sullivan *et al.*, 2003).

A problem that may be encountered when using bacteriocins to lyse starters is that the rate of acidification may be compromised if the target lactococcal strain is also the primary acidifier. To overcome this problem, a three-strain system can be used, which includes a bacteriocin-producing adjunct, a bacteriocin-sensitive starter as a target and a bacteriocin-resistant species as an acidifier (Morgan *et al.*, 2002).

Use of Bacteriocin in Bioactive Packaging

In addition to the three approaches described above as commonly used in the application of bacteriocins for biopreservation, a fourth recent food application of bacteriocins is their use in bioactive packaging, in which bacteriocins can be incorporated into packaging destined to be in contact with food. This system combines the preservation function of bacteriocins with conventional packaging materials, which protects the food from external contaminants. Spoilage of refrigerated foods usually begins with microbial growth on the surface, which reinforces the attractive use of bacteriocins being used in conjunction with packaging to improve food safety and quality and to prolong shelf life of food products (Coma, 2008).

Bioactive packaging can be prepared by directly immobilizing bacteriocin to the food packaging, or by addition of a sachet containing the bacteriocin into the packaged food. The bacteriocin carrier acts as a reservoir and diffuser of the concentrated bacteriocin molecules to the food during storage of the food product, ensuring a gradient-dependent continuous supply of bacteriocin. The carrier may also protect the bacteriocin from inactivation by interaction with food components and enzymatic inactivation. Moreover, the precise localized application of bacteriocin on the food surface requires much lower amounts of bacteriocin when compared to application in whole food volume, decreasing the processing costs (Gálvez *et al.*, 2007). Moreover, immobilization has the advantages of increased bacteriocin productivity and improved long-term stability over free-cell fermentations.

A variety of methods have been proposed for bacteriocin immobilization, including silica particles, corn starch powder, liposome encapsulation, and incorporation on gel coatings and films of different materials, such as calcium alginate, gelatine, cellulose, soy protein, corn zein, collagen casings, polysaccharide-based films, cellophane, silicon coatings, polyethylene, nylon or other polymer plastic films (Gálvez *et al.*, 2007). In most cases, immobilized bacteriocin preparations are applied on the surface of the processed food to avoid post-process contamination and surface proliferation of unwanted bacteria. To date, this kind of bacteriocin application has been reported only for preserving foods of animal origin and some examples are shown in Table 5.3. However, it has a great potential as a biopreservation technology for vegetable origin foods as well.

Another possibility for bioactive packaging is to incorporate the antimicrobial compound into an edible coating applied by dipping or spraying onto the food. Selection of the incorporated bioactive agents is limited to edible compounds, because they have to be consumed with the coating layers and foods together. For example, the chitosan- or cellulose-derivatives-based films associated with bacteriocins may also be included in these packaging concepts. Edible coatings can improve the quality of fresh, frozen and processed meat and poultry products by, for instance, delaying moisture loss, reducing lipid oxidation and discolouration, enhancing product appearance and functioning as carriers of food additives (Coma, 2008).

Dykes, G.A. and Hastings, J.W. (1998) Fitness costs associated with class IIa bacteriocin resistance in *Listeria monocytogenes* B73. *Letters in Applied Microbiology* 26, 5–8.
Ercolini, D., Storia, A., Villani, F. and Mauriello, G. (2006) Effect of a bacteriocin-activated polythene film on *Listeria monocytogenes* as evaluated by viable staining and epifluorescence microscopy. *Journal of Applied Microbiology* 100, 765–772.
Ferreira, M.A. and Lund, B.M. (1996) The effect of nisin on *Listeria monocytogenes* in culture medium and long-life cottage cheese. *Letters in Applied Microbiology* 22, 433–438.
Gálvez, A., Abriouel, H., Lopez, R.L. and Ben Omar, N. (2007) Bacteriocin-based strategies for food preservation. *International Journal of Food Microbiology* 120, 51–70.
Goff, J.H., Bhunia, A.K. and Johnson, M.G. (1996) Complete inhibition of low levels of *Listeria monocytogenes* on refrigerated chicken meat with pediocin AcH bound to heat-killed *Pediococcus acidilactici* cells. *Journal of Food Protection* 59, 1187–1192.
Grande, M.J., Lucas, R., Abriouel, H., Ben Omar, N., Maqueda, M., Martínez-Bueno, M., Martínez-Cañamero, M., Valdivia, E. and Gálvez, A. (2005) Control of *Alicyclobacillus acidoterrestris* in fruit juices by enterocin AS-48. *International Journal of Food Microbiology* 104, 289–297.
Grande, M.J., Lucas, R., Abriouel, H., Valdivia, E., Ben Omar, N., Maqueda, M., Martínez-Cañamero, M. and Gálvez, A. (2006) Inhibition of *Bacillus licheniformis* LMG 19409 from ropy cider by enterocin AS-48. *Journal of Applied Microbiology* 101, 422–428.
Gravesen, A., Axelsen, J., Mendes da Silva, A.M., Hansen, T.B. and Knochel, S. (2002) Frequency of bacteriocin resistance development and associated fitness costs in *Listeria monocytogenes*. *Applied and Environmental Microbiology* 68, 756–764.
Guinane, C.M., Cotter, P.D., Hill, C. and Ross, R.P. (2005) Microbial solutions to microbial problems; latococcal bacteriocins for the control of undesirable biota in food. *Journal of Applied Microbiology* 98, 1316–1325.
Héchard, Y., Pelletier, C., Cenatiempo, Y. and Frère, J. (2001) Analysis of σ^{54}-dependent genes in *Enterococcus faecalis*: a mannose PTS permease (EII^{Man}) is involved in sensitivity to a bacteriocin, mesentericin Y105. *Microbiology* 147, 1575–1580.
Heng, N.C.K., Wescombe, P.A., Burton, J.P., Jack, R.W. and Tagg, J.R. (2007) The diversity of bacteriocins in Gram-positive bacteria. In: Riley, M.A. and Chavan, M.A. (eds) *Bacteriocins: Ecology and Evolution*. Springer, Berlin, pp. 45–92.
Iseppi, R., Pilati, F., Marini, M., Toselli, M., de Niederhäusern, S., Guerrieri, E., Messi, P., Sabia, C., Manicardi, G., Anacarso, I. and Bondi, M. (2008) Anti-listerial activity of a polymeric film coated with hybrid coatings doped with enterocin 416K1 for use as bioactive food packaging. *International Journal of Food Microbiology* 123, 281–287.
Jarvis, B. (1967) Resistance to nisin and production of nisin-inactivating enzymes by several *Bacillus* species. *Journal of General Microbiology* 47, 33–48.
Kato, T., Maeda, K., Kasuya, H. and Matsuda, T. (1999) Complete growth inhibition of *Bacillus subtilis* by nisin-producing lactococci in fermented soybeans. *Bioscience, Biotechnology and Biochemistry* 63, 642–647.
Komitopoulou, E., Boziaris, I.S., Davies, E.A., Delves-Broughton, J. and Adam, M.R. (1999) *Alicyclobacillus acidoterrestris* in fruit juices and its control by nisin. *International Journal of Food Science & Technology* 34, 81–85.
Kramer, N.E., van Hijum, S.A., Knol, J., Kok, J. and Kuipers, O.P. (2006) Transcriptome analysis reveals mechanisms by which *Lactococcus lactis* acquires nisin resistance. *Antimicrobial Agents and Chemotherapy* 50, 1753–1761.
Lauková, A. and Czikková, S. (1999) The use of enterocin CCM 4231 in soy milk to control the growth of *Listeria monocytogenes* and *Staphylococcus aureus*. *Journal of Applied Microbiology* 87, 182–186.
Leal-Sánchez, M.V., Ruiz-Barba, J.L., Sánchez, A.H., Rejano, L., Jiménez-Díaz, R. and Garrido, A. (2003) Fermentation profile and optimization of green olive fermentation using *Lactobacillus plantarum* LPCO10 as a starter culture. *Food Microbiology* 20, 421–430.
Liu, W. and Hansen, J.N. (1990) Some chemical and physical properties of nisin, a small-protein antibiotic produced by *Lactococcus lactis*. *Applied and Environmental Microbiology* 56, 2551–2558.
Lucas, R., Grande, M.J., Abriouel, H., Maqueda, M., Ben Omar, N., Valdivia, E., Martínez-Cañamero, M. and Gálvez, A. (2006) Application of the broad-spectrum bacteriocin enterocin AS-48 to inhibit *Bacillus coagulans* in canned fruit and vegetable foods. *Food and Chemical Toxicology* 44, 1774–1781.

Maisnier-Patin, S. and Richard, J. (1996) Cell wall changes in nisin-resistant variants of *Listeria innocua* in the presence of high nisin concentrations. *FEMS Microbiology Letters* 140, 29–35.

Mantovani, H.C. and Russell, J.B. (2001) Nisin resistance of *Streptococcus bovis. Applied and Environmental Microbiology* 67, 808–813.

Maqueda, M., Gálvez, A., Sánchez-Barrena, M.J., Gonzáles, C., Albert, A., Rico, M. and Valdivia, E. (2004) Peptide AS-48: prototype of a new class of cyclic bacteriocins. *Current Protein and Peptide Science* 5, 399–416.

Marcos, B., Aymerich, T., Monfort, J.M. and Garriga, M. (2007) Use of antimicrobial biodegradable packaging to control *Listeria monocytogenes* during storage of cooked ham. *International Journal of Food Microbiology* 120, 152–158.

Mauriello, G., Ercolini, D., La Storia, A., Casaburi, A. and Villani, F. (2004) Development of polythene films for food packaging activated with an antilisterial bacteriocin from *Lactobacillus curvatus* 32Y. *Journal of Applied Microbiology* 97, 314–322.

McAuliffe, O., Ryan, M.P., Ross, R.P., Hill, C., Breeuwer, P. and Abee, T. (1998) Lacticin 3147, a broad spectrum bacteriocin which selectively dissipates the membrane potential. *Applied and Environmental Microbiology* 64, 439–445.

McAuliffe, O., Hill, C. and Ross, R.P. (1999) Inhibition of *Listeria monocytogenes* in cottage cheese manufactured with a lacticin 3147-producing starter culture. *Journal of Applied Microbiology* 86, 251–256.

Millette, M., Le Tien, C., Smoragiewicz, W. and Lacroix, M. (2007) Inhibition of *Staphylococcus aureus* on beef by nisin-containing modified alginate films and beads. *Food Control* 18, 878–884.

Minamikawa, M., Kawai, Y., Inoue, N. and Yamazaki, K. (2005) Purification and characterization of warnericin RB4, anti-*Alicyclobacillus* bacteriocin, produced by *Staphylococcus warneri* RB4. *Current Microbiology* 51, 22–26.

Morgan, S.M., O'Sullivan, L., Ross, R.P. and Hill, C. (2002) The design of a three strain starter system for Cheddar cheese manufacture exploiting bacteriocin-induced starter lysis. *International Dairy Journal,* 12, 985–993.

Nissen-Meyer, J., Rogne, P., Oppegård, C., Haugen, H.S. and Kristiansen, P.E. (2009) Structure-function relationships of the non-lanthionine-containing peptide (class II) bacteriocins produced by Gram-positive bacteria. *Current Pharmaceutical Biotechnology* 10, 10–37.

Nunez, M., Rodriguez, J.L., Garcia, E., Gaya, P. and Medina, M. (1997) Inhibition of *Listeria monocytogenes* by enterocin 4 during the manufacture and ripening of Manchego cheese. *Journal of Applied Microbiology* 83, 671–677.

O'Mahony, T., Rekhif, N., Cavadini, C. and Fitzgerald, G.F. (2001) The application of a fermented food ingredient containing 'variacin', a novel antimicrobial produced by *Kokuria varians,* to control the growth of *Bacillus cereus* in chilled dairy products. *Journal of Applied Microbiology* 90, 106–114.

O'Sullivan, D., Coffey, A., Fitzgerald, G.F., Hill, C. and Ross, R.P. (1998) Design of a phage-insensitive lactococcal dairy starter via sequential transfer of naturally occurring conjugative plasmids. *Applied and Environmental Microbiology* 64, 4618–4622.

O'Sullivan, L., Ross, R.P. and Hill, C. (2003) A lacticin 481 producing adjunct culture increases starter lysis while inhibiting nonstarter lactic acid bacteria proliferation during cheddar cheese ripening. *Journal of Applied Microbiology* 95, 1235–1241.

Oumer, A., Garde, S., Gaya, P., Medina, M. and Nuñez, M. (2001) The effects of cultivating lactic acid starter cultures with bacteriocin-producing lactic acid bacteria. *Journal of Food Protection* 64, 81–86.

Penna, T.C.V., Jozala, A.F., Novaes, L.C.L., Pessoa Jr., A. and Cholewa, O (2005) Production of nisin by *Lactococcus lactis* in media with skimmed milk. *Applied Biochemistry and Biotechnology,* 122, 619–637.

Pucci, M.J., Vedamuthu, E.R., Kunka, B.S. and Vandenbergh, P.A. (1988) Inhibition of *Listeria monocytogenes* by using bacteriocin PA-1 produced by *Pediococcus acidilactici* PAC 1.0. *Applied and Environmental Microbiology* 54, 2349–2353.

Randazzo, C.L., Pitino, I., Scifò, G.O. and Caggia, C. (2009) Biopreservation of minimally processed iceberg lettuces using a bacteriocin produced by *Lactococcus lactis* wild strain. *Food Control* 20, 756–763.

Rodríguez, E., Calzada, J., Arqués, J.L., Rodríguez, J.M., Nuñez, M. and Medina, M. (2005) Antimicrobial activity of pediocin-producing *Lactococcus lactis* on *Listeria monocytogenes, Staphylococcus aureus* and *Escherichia coli* O157:H7 in cheese. *International Dairy Journal* 15, 51–57.

Rodriguez, J.M., Martinez, M.I. and Kok, J. (2002) Pediocin PA-1, a wide-spectrum bacteriocin from lactic acid bacteria. *Critical Reviews in Food Science and Nutrition* 42, 91–121.
Rogers, L.A. and Whittier, E.D. (1928) Limiting factors in lactic acid fermentation. *Journal of Bacteriology* 16, 211–229.
Ross, R.P., Morgan, S. and Hill, C. (2002) Preservation and fermentation: past, present and future. *International Journal of Food Microbiology* 79, 3–16.
Ryan, M.P., Rea, M.C., Hill, C. and Ross, R.P. (1996) An application in cheddar cheese manufacture for a strain of *Lactococcus lactis* producing a novel broad spectrum bacteriocin lacticin 3147. *Applied and Environmental Microbiology* 62, 612–619.
Ryan, M.P., Ross, R.P. and Hill, C. (2001) Strategy for manipulation of cheese flora using combinations of lacticin 3147-producing and -resistant cultures. *Applied and Environmental Microbiology* 67, 2699–2704.
Scannell, A.G.M., Hill, C., Ross, R.P., Marx, S., Hartmeier, W. and Arendt, E.K. (2000) Development of bioactivre packaging materials using immobilised bacteriocins lacticin 3147 and Nisaplin®. *International Journal of Food Microbiology* 60, 241–249.
Schillinger, U., Geisen, R. and Holzapfel, W.H. (1996) Potential of antagonistic microorganisms and bacteriocins for the biological preservation of foods. *Trends in Food Science & Technology* 71, 58–64.
Schlyter, J.H., Glass, K.A., Loeffelholz, J., Degnan, A.J. and Luchansky, J.B. (1993) The effects of diacetate with nitrate, lactate, or pediocin on the viability of *Listeria monocytogenes* in turkey slurries. *International Journal of Food Microbiology* 19, 271–281.
Schoeman, H., Vivier, M.A., Du, T.M., Dicks, L.M. and Pretorius, I.S. (1999) The development of bactericidal yeast strains by expressing the *Pediococcus acidilactici* pediocin gene (*pedA*) in *Saccharomyces cerevisiae*. *Yeast* 15, 647–656.
Settanni, L. and Corsetti, A. (2008) Application of bacteriocins in vegetable food biopreservation. *International Journal of Food Microbiology* 121, 123–138.
Siragusa, G.R., Cutter, C.N. and Willet, J.L. (1999) Incorporation of bacteriocin in plastic retains activity and inhibits surface growth of bacteria on meat. *Food Microbiology* 16, 229–235.
Taylor, T.M., Davidson, P.M. and Zhong, Q. (2007) Extraction of nisin from a 2.5% commercial nisin product using methanol and ethanol solutions. *Journal of Food Protection* 70, 1272–1276.
Thomas, L.V., Ingram, R.E., Bevis, H.E., Davies, E.A., Milne, C.F. and Delves-Broughton, J. (2002) Effective use of nisin to control *Bacillus* and *Clostridium* spoilage of a pasteurized mashed potato product. *Journal of Food Protection* 65, 1580–1585.
Vadyvaloo, V., Hasting, J.W., van der Merwe, M.J. and Rautenbach, M. (2002) Membranes of class IIa bacteriocin-resistant *Listeria monocytogenes* cells contain increased levels of desaturated and short-acyl-chain phosphatidylglycerols. *Applied and Environmental Microbiology* 68, 5223–5230.
Vadyvaloo, V., Arous, S., Gravensen, A., Héchard, Y., Chauhan-Haubrock, R., Hastings, J.W. and Rautenbach, M. (2004) Cell-surface alterations in class IIa bacteriocin-resistant *Listeria monocytogenes* strains. *Microbiology* 150, 3025–3033.

6 Microbial Fermentation for Food Preservation

Yuanxia Sun,* Yin Li, Hui Song and Yang Zhu

Introduction

Food preservation is the process of treating and handling food to stop or greatly slow down spoilage caused or accelerated by microorganisms. It involves the action taken to maintain foods with the desired properties or nature for as long as possible. Preservation using chemicals and microbes comprises comprehensive methods based on additives of a chemical or microbiological nature, including fermentations, antimicrobials, antioxidants, pH-lowering agents and nitrides. However, consumers demand a reduced use of chemical preservatives or additives in food. As a result, naturally produced antimicrobial agents are requested. The development of a new product and approach is leading to the reduction or even displacement of heat treatments and traditional preservatives using treatments capable of assuring the sensory and nutritional properties of the product without reducing food safety (Brul and Coote, 1999; López and Belloso, 2008). For example, a variety of microorganisms, such as lactic acid bacteria (LAB), acetic acid bacteria, yeasts and moulds, are frequently used in fermented foods to add the specific qualities and extend the shelf life of the product.

Food preservation by fermentation is a common practice and ancient technology. Fermentation as a food-preservation technique can be traced back to thousands of years ago. Indeed, it is thought that the art of cheese making was invented 8000 years ago in Iraq (Fox, 1993). Fermentation has been exploited for preservation of food and beverages for thousands of years, and the knowledge was enriched later by the development of leavened bread, wine, beer, mouldy soybean curds, whisky and yogurt. However, it was from advances in the past 150 years that microorganisms were recognized as responsible for the fermentation process. Pasteurization was discovered in 1861 and, for the first time, the essential role of microorganisms in the fermentation process was realized (Ross *et al.*, 2002). This breakthrough greeted the development of large-scale fermentation processes for the commercial production of fermented foods and alcoholic beverages, with the most widely used microorganisms including yeast for the production of beer, wine and spirits, and LAB for a variety of dairy, vegetable and meat fermentations. In each case, the food or beverage raw material provides the substrates for the formation of a range of microbial metabolites that contribute to the

* Corresponding author.

shelf-life extension and the enhanced quality of the product.

Fermentation is a process dependent on the biological activity of microorganisms for producing a range of metabolites that can suppress the growth of undesirable microflora in foodstuffs (Blom and Mörtvedt, 1991). Fermentation as a food-preservation method can be effective at extending the shelf life of foods and can often be carried out with relatively inexpensive, basic equipment. Such microbial fermentation can produce a wide variety of metabolites including organic acids, diacetyl and bacteriocins. Some metabolites have been used as a bio-preservative in food products because their unique metabolic characteristics are involved in many fermentation processes of milk, meats, cereals, vegetables and teas. This chapter will therefore focus mainly on introducing the characteristics of antimicrobial metabolites in food fermentation processes, their mechanisms of bio-preservatives, and their potential application in different food systems.

Microorganisms Used in Food Fermentation

Lactic acid bacteria

With respect to the natural antimicrobial activity associated with microorganisms, the most promising ongoing development in food preservation is the use of LAB. These bacteria have a long and safe tradition in food fermentation, and many potent applications as food preservatives have been established. LAB can produce a number of antimicrobial substances with economic significance for food fermentation and preservation.

LAB are subdivided based on their products from glucose fermentation: (i) homofermenters producing lactic acid; (ii) heterofermenters producing equimolar amounts of lactate, carbon dioxide and ethanol. Their most effective mechanism is to grow readily in most foods, producing acid that lowers the pH rapidly to a point where other competing organisms can no longer grow (Steinkraus, 1983). Lactobacilli have the ability to produce hydrogen peroxide that inhibits spoilage organisms, while lactobacilli themselves are relatively resistant to hydrogen peroxide (Hurst and Collins-Thompson, 1979; Banks *et al.*, 1986).

Bacteriocins produced by LAB as food additives also effectively inhibit food-borne pathogens (Berry *et al.*, 1991; O'Sullivan *et al.*, 2002). Bacteriocins are ribosomally synthesized bactericidal peptides or proteins that are usually inhibitory to species closely related to the producer. Nisin, lactococcin and pediocin are bacteriocins that are produced by a food-grade organism, LAB. Therefore, they could be classified as 'natural' and are considered to be safe for consumers as they have been consumed in fermented foods for generations.

Acetic acid bacteria

The acetic acid bacteria with importance in food fermentations are acid tolerant, growing well at pH levels below pH 5.0. The acetic acid bacteria consist of two genera, *Acetobacter* and *Gluconobacter*. They are found in nature where ethanol is produced from the fermentation of carbohydrates by yeasts, such as in plant nectars and damaged fruits. Other good sources are alcoholic beverages such as fresh cider and unpasteurized beer.

Yeasts

Yeasts are widely distributed in natural habitats that are nutritionally rich and high in carbohydrates (Suomalainen and Oura, 1971; Adams and Moss, 2000). Yeasts have been classified into about 500 species (Kreger-van Rij, 1984). However, only a small number are regularly used to make alcoholic beverages, of which *Saccharomyces cerevisiae* is the most frequently used and many variants of which are available. Yeasts are used to produce ethanol, CO_2, flavour and aroma.

Moulds

Moulds are important to the food industry, both as spoilers and preservers. They are used in food fermentations to produce specific enzymes, antibiotics and flavours (Gourama and Bullerman, 1995; Nout, 1995). Some moulds, such as *Aspergillus oryzae* and *Rhizopus oligosporus,* are often used in traditional food fermentations of soybeans to make miso, soy sauce and tempeh.

Antimicrobial Metabolites in LAB Fermentation

Microorganisms naturally produce an arsenal of antimicrobial agents to improve their competitiveness. The use of the generally recognized as safe (GRAS) organisms or the antimicrobial compounds produced by microbial fermentation has been successfully achieved in many types of foods. Most prominently, bacteriocins produced by LAB have been under investigation worldwide for food-preservation purposes (Cleveland *et al.*, 2001; O'Sullivan *et al.*, 2002).

LAB produce a variety of antibacterial substances in the process of growth and metabolism. This ability to produce antimicrobial substances has historically long been used to preserve food. These substances are mainly a range of small-molecular-mass organic molecules, such as organic acids (lactic acid, butyric acid, acetic acid, etc.) (Caplice and Fitzgerald, 1999), which are usually divided into two groups: proteins and non-proteins. Proteins of antibacterial nature are mainly bacteriocins. These bacteriocins are synthesized and secreted by the ribosome into the environment in bacterial metabolism. Antibacterial non-proteins include organic acids, hydrogen peroxide, diacetyl and other compounds. These substances have a wide range of inhibition of food-spoilage and pathogenic bacteria.

Organic acids

The most common organic acids are lactic, acetic and propionic acids. Acetic acid is the strongest inhibitor and has a wide range of inhibitory activity, inhibiting yeasts, moulds and bacteria (Blom and Mörtvedt, 1991), while propionic acid has been observed to exert a strong antimicrobial effect, in particular towards yeasts and moulds (Suomalainen and Mäyrä-Mäkinen, 1999). This stronger antimicrobial activity of acetic and propionic acids can be explained in part by their higher pKa (4.87 and 4.75) compared to lactic acid (3.08). At pH 4, only 11% of lactic acid is undissociated, whereas 85% of acetic acid and 92% of propionic acid is undissociated (Eklund, 1983). In the mixture of acids, propionic and acetic acid may be the actual antimicrobial agents, whereas lactic acid contributes mainly to the reduction in pH. In addition to reducing the pH, lactic acid has been observed to also permeabilize membranes, thereby further enhancing the activity of other antimicrobial substances (Alakomi *et al.*, 2000). Acids are generally thought to exert their antimicrobial effect by interfering with the maintenance of cell membrane potential, inhibiting active transport, reducing intracellular pH and inhibiting a variety of metabolic functions (Doores, 1993). Indeed, the low intracellular pH leads to protein denaturation, resulting in an imbalance of the cell structure and function of other components, and thereby interfering with cell growth and survival.

Low molecular weight antimicrobial substances

Reuterin, produced by *Lactobacillus reuteri*, has a very broad spectrum of antimicrobial activities. It was found to have antibacterial, antifungal, antiprotozoal and antiviral activity (Axelsson *et al.*, 1989; Dobrogosz *et al.*, 1989). Reuterin is an equilibrium mixture of monomeric, hydrated monomeric and cyclic dimeric forms of β-hydroxypropionaldehyde. No reports on possible negative effects of reuterin on human cells could be found, although reuterin has been observed to be able to cross link biological tissues in a similar way to glutaraldehyde (Sung *et al.*, 2003). Reuterin is produced by stationary-phase cultures during anaerobic growth on a mixture of glucose and glycerol or glyceraldehydes. Consequently, in order to use reuterin-producing *L. reuteri* for biopreservation in a food product, it would be beneficial to include glycerol. Reuterin is very stable under acidic conditions, but in alkaline conditions it loses activity irreversibly. Although the mechanism of antimicrobial activity is not yet clear, it is thought that reuterin acts against sulfhydryl enzymes. Reuterin inhibits the substrate-binding subunit of ribonucleotide reductase, thereby interfering with the DNA synthesis (Dobrogosz *et al.*, 1989).

Reutericyclin, a tetramic acid antibiotic with a broad antimicrobial spectrum also produced by *L. reuteri,* is believed to be responsible for the stability of certain German sourdoughs. It has a molecular weight of 349 Da, is negatively charged and is a highly hydrophobic antagonist (Höltzel *et al.*, 2000). The spectrum of inhibition of the antibiotic is confined to Gram-positive bacteria including *Lactobacillus* spp., *Bacillus subtilis, Bacillus cereus, Enterococcus faecalis, Staphylococcus aureus* and *Listeria innocua.* Interestingly, the inhibitory activity of reutericyclin against *Escherichia coli* and *Salmonella* is increased dramatically at higher salt concentration (2%) and low pH (4.5). The minimum inhibitory concentration was found to be approximately 0.05–1 mg/l for Gram-positive bacteria. Gram-negative bacteria and yeasts were not found to be sensitive (>100 mg/l) (Gänzle *et al.*, 2000). Reutericyclin was not found to form pores in the membrane of the target cells, but rather works as a proton ionophore. Since nisin was observed to kill Gram-negative bacteria under conditions which disturb the outer membrane (Stevens *et al.*, 1992), there are similarities in the mode of action of nisin and reutericyclin.

Pyroglutamic acid, also known as 2-pyrrolidone-5-carboxylic acid, was found to contribute to the antimicrobial activity. This antimicrobial substance was produced by *Lactobacillus casei* ssp. *casei, L. casei* ssp. *Pseudoplantarum* and *Streptococcus bovis* (Chen and Russell, 1989; Huttunen *et al.*, 1995), though it is also discovered in fruits, vegetables and grasses. Pyroglutamic acid inhibits pathogenic bacteria and spoilage bacteria such as *B. subtilis, Enterobacter cloacae, Pseudomonas putida* and *Pseudomonas fluorescens.* Although pyroglutamic acid has a stronger antimicrobial activity than lactic acid at the same concentration (Yang *et al.*, 1997), it is likely that its mechanism of action is similar to that of organic acids. It is also heat stable at 121°C for 20 min, but loses its activity when the pH is raised above 2.5, depending on the target strain used. Obviously, each antibacterial substance produced during fermentation provides an additional hurdle for pathogens and spoilage bacteria before they can survive and proliferate in a food or beverage, from manufacture to consumption.

Bacteriocins and their effectiveness in food systems

Microorganisms can also produce a range of antimicrobial peptides and proteins that are collectively referred to as bacteriocins. In particular, almost all the different species which make up the LAB group have been reported to produce these inhibitory proteins. The bacteriocins offer potential applications in food preservation, and the use of bacteriocins in the food industry can help to reduce the addition of chemical preservatives as well as the intensity of heat treatments, resulting in foods which are more naturally preserved and richer in organoleptic and nutritional properties. This can be an alternative to satisfy the increasing consumer demands for safe, fresh-tasting, ready-to-eat, minimally processed foods and also to develop 'novel' food products (e.g. less acidic or with a lower salt content).

Bacteriocins of LAB, according to the classification procedure proposed by Klaenhammer (1993) and modified by Nes *et al.* (1996), are divided into four major subclasses (Table 6.1). The majority of those produced by bacteria associated with food belong to classes I and II. Class I bacteriocins are composed of one or two small peptides of approximately 3 kDa. An unusual feature of this class is that they are post-translationally modified to contain lanthionine, β-methyllanthionine and dehydrated amino acids (Sahl and Bierbaum, 1998; Ryan *et al.*, 1999). On the other hand, small and heat-stable, non-lantibiotic peptides are class II bacteriocins (molecular weight ≤ 10 kDa) (Hastings *et al.*, 1991; Bennik *et al.*, 1997; Nes *et al.*, 2000). This class is further subdivided into three groups (Table 6.1).

Nisin belongs to the lantibiotic family (class I), and is a heat-stable bacteriocin produced by *Lactococcus lactis* ssp. (Ray, 1992; Delves-Broughton *et al.*, 1996). It is composed of 34 amino acids and has a pentacyclic structure with one lanthionine residue (ring A) and four β-methyllanthionine residues (rings B, C, D and E) (Cleveland *et al.*, 2001). It can inhibit a broad range of Gram-positive bacteria including *Listeria monocytogenes* and *S. aureus*, and prevent the outgrowth of spores of many *Clostridium* and *Bacillus* spp. (Joerger and Klaenhammer, 1990; Jack and Sahl, 1995; Ryan *et al.*, 1999). It has been recommended as an efficient and safe natural preservative by the United Nations Food and Agriculture Organization (FAO) in 1969. Thus, nisin is approved for use as a component of the preservation procedure for processed and fresh cheese, canned foods, processed vegetables and baby foods, in up to 50 countries (Delves-Broughton *et al.*, 1996; O'Sullivan *et al.*, 2002).

One of the principal applications of nisin-producing strains is in the manufacture of cheese. From a food safety point of view, the pathogen of primary concern in a number of cheeses is *L. monocytogenes*, which is capable of growing at refrigeration temperatures and has the ability to survive the acidic conditions of cheese manufacture (Ferreira and Lund, 1996). A number of studies have been performed with both nisin-producing cultures and Nisaplin®, demonstrating the effectiveness of nisin for the inhibition of *L. monocytogenes* in cheese (Maisnier-Patin *et al.*, 1996). The other example is the use of nisin in meat products. It was indicated that nisin could prevent the growth of *Clostridium* in meat under certain conditions, and lower fat content correlated with higher nisin activity in sausage meat (Rayman *et al.*, 1981; Davies *et al.*, 1999).

Pediocin is produced by LAB of the genus *Pediococcus*, such as *P. pentosaceus* and *P. acidilactici*. The most promising results in meats were obtained using pediocin PA-1. It was found that pediocin PA-1 immediately reduces the number of target organisms (Nielsen *et al.*, 1990; Luchansky *et al.*, 1992). The many studies had indicated that pediocins could be more effective than nisin in some meat products, but it is not yet an approved food additive in the USA. The effect of pediocin PA-1 on the growth of *L. monocytogenes* has also been studied in cottage cheese, half-and-half cream and cheese sauce systems (Bhunia *et al.*, 1988; Pucci *et al.*, 1988). In addition, pediococci are the main starter cultures used in the manufacture of American-style fermentation of many vegetables (Bennik *et al.*, 1999). Three isolates obtained from different fresh vegetables were found to have the required characteristics: one strain of *Enterococcus mundtii* and two strains of *Pediococcus parvulus*. Both types produced a bacteriocin that effectively controlled growth of *L. monocytogenes in vitro*. The bacteriocin produced by both strains was characterized and appeared to be identical to pediocin PA-1. The findings implied that pediocin production might be a favourable asset of starter cultures in the fermentation of sausages, cheese and vegetables. Pediocins therefore have potential in food applications when used under the proper conditions.

Enterocin AS-48 is a broad-spectrum cyclic antimicrobial peptide that is active against several food spoilage and pathogenic bacteria (Gálvez *et al.*, 1991; Abriouel *et al.*, 2002; Mendoza *et al.*, 2002; Maqueda *et al.*, 2004). Enterocin AS-48 is produced by *E. faecalis*, and its antimicrobial activity against *B. cereus* has been established in conventional culture media (Abriouel *et al.*, 2002; Muñoz *et al.*, 2004). For example, the efficacy of enterocin AS-48 against *B. cereus* inoculated in cheese has been demonstrated by inoculation with an enterococcal strain that produced the bacteriocin *in situ* (Muñoz *et al.*, 2004). Its antimicrobial activity against a toxicogenic psychrotrophic strain of *B. cereus* has also been reported in a model food system consisting of boiled rice and in a rice-

Table 6.1. Classification of bacteriocins from lactic acid bacteria.

Type	Structure	Description	Example	References
Class I	Small (≤ 5kDa)	Heat stable	Nisin	Joerger and Klaenhammer, 1990
Lantibiotic family	Unusual amino acids	Antibacterial spectrum: medium to broad	Lacticin 3147	Piard *et al.*, 1992
	e.g. lanthionine, β-methyllanthionine, dehydrated amino acids		Mersacidin	De Vuyst and Vandamme, 1994
			Actagardine	Ryan *et al.*, 1996, 1999
			Lacticin 481	Sahl and Bierbaum, 1998
Class II	Small (≤10 kDa)	Heat stable	Sakacin A	Hastings *et al.*, 1991
Small non-modified peptide	(30–100 amino acids)	Antibacterial spectrum: medium to broad	Pediocin PA-1	Foegeding *et al.*, 1992
	No lanthionine	a: pediocin-like bacteriocins with anti-listerial effects	Pediocin AcH	Holck *et al.*, 1992
			Leucocin UAL 187	Bhunia *et al.*, 1991
		b: two peptide bacteriocins		Nes *et al.*, 2000
		c: sec-dependent secretion of bacteriocins		Gao *et al.*, 2010
Class III	Large (≥ 30 kDa)	Heat labile	Helveticin J	Joerger and Klaenhammer, 1986
Large heat-labile protein	No lanthionine	Antibacterial spectrum: narrow	Cascicin 80	Rammelsberg *et al.*, 1990
			Enterolysin	Nilsson *et al.*, 1999
Class IV	Glyco and/or lipid moieties	Heat stable	Leuconocin S	Lewus *et al.*, 1992
Complex bacteriocins		Antibacterial spectrum: medium	Pediocin SJ-1	Schved *et al.*, 1993

based infant formula dissolved in whole milk (Grande *et al.*, 2006). The highest activity of enterocin AS-48 against *B. cereus* was detected in the pH range of 6–8 (Abriouel *et al.*, 2002). Enterocin AS-48 would be suitable to inhibit growth of *B. cereus* in non-acid foods such as cooked rice or rice gruel.

Broad-spectrum bacteriocins present potential wider uses, while narrow-spectrum bacteriocins can be used more specifically to selectively inhibit certain high-risk bacteria such as *L. monocytogenes* without affecting harmless microbiota. Bacteriocins can be added to foods in the form of concentrated preparations as food preservatives, shelf-life extenders, additives or ingredients, or produced *in situ* by bacteriocinogenic starters, adjunct or protective cultures. In addition, the activity of bacteriocins in foods is greatly influenced by different factors including food composition, interaction with food components, bacteriocin stability, pH and storage temperature. Some bacteriocins also show synergistic effects when used in combination with other antimicrobial agents. Thus, the combined use of different bacteriocins may be an attractive approach to avoid development of resistant strains and influence of different factors.

Application of Microbial Fermentation in Food Preservation

Chemical food additives are commonly applied in food preservation. With the trend of increasing use of naturally produced biopreservatives in food products, natural antibacterial agents from food fermentation may offer an innovative and interesting measure for such applications. In fact, the antimicrobial compounds produced by LAB strains or other natural antimicrobial substances from microbially fermented food can help to combat microbial contamination and reduce health risks without changing the sensory quality of the product. Therefore, some potential and unexplored food fermentation process certainly presents new possibilities for use in food preservation to meet consumer demands relating to health, nutrition, safety and convenience.

Such potential biopreservatives originate from traditional food fermentations. Fermented foods are now regarded as part of our staple diet. The main substrates used in the commercial production of the most familiar fermented products are milk, meat, cucumber and cabbage. These fermentations are classified in a number of different ways, according to the microorganisms, the biochemistry, the type of fermentation and the product type. In this topic, the fermentations are grouped in terms of the product type used (Table 6.2), and the application of biopreservative is discussed according to the different fermentation processes including cheese, sausages, sourdough, cabbage and tea.

Dairy products

Cheese is a concentrated milk product obtained after coagulation and whey separation of milk, cream or a mixture of these products. There are over 400 varieties of cheeses, and they have been classified according to the different methods of preparation (Jay, 1998). The milk receives a treatment equivalent to pasteurization at the beginning of the processing, and is then inoculated with an appropriate lactic starter. The starter organisms used for cheese production are mostly mesophilic starters, strains of *Lactococcus lactis* and its subspecies (Table 6.2). Thermophilic starters are used in the production of cheeses where a higher incubation temperature is employed. Propionic bacteria, moulds and red- or yellow-smearing cultures are also added, depending on the type of cheese to be manufactured (Radke-Mitchell and Sandine, 1984; Jin and Park, 1995).

The most ripened cheeses are the products of metabolic activities of LAB that produce lactic acid. The lactic acid plays a major role in the suppression of pathogenic and spoilage microorganisms and in the production of volatile flavour compounds. In addition, several well-known cheeses owe their particular character to other related

Table 6.2. The application of microbial fermentation in different food systems.

Type	Microorganisms	Substrates and products	References
Dairy products	Lactic acid bacteria	Milk	Jay, 1998
	Lactobacillus (lactis, casei, helveticus, delbruckii, bulgaricus)	Cheese	Radke-Mitchell and Sandine, 1984
		Yoghurt	Jin and Park, 1995
	S. thermophilus		Ferreira and Lund, 1996
	Sometimes moulds		Maisnier-Patin *et al.*, 1996
	Penicillium (camemberti, candium, roqueforti)		Steffen *et al.*, 1993
			Munoz *et al.*, 2004
	Carnobacterium piscicola		
	Brevibacterium linens		
Cereal-based products	Lactic acid bacteria	Wheat, rye, other grains	Sugihara *et al.*, 1971
	Lactobacillus (sanfranciscensis, reuteri, brevis, pontis, plantarum)	Sourdough	Corsetti *et al.*,1996, 1998
		Bread	Rosenquist and Hansen, 1998
	Yeast		Lavermicocca *et al.*, 2000
	Saccharomyces cerevisiae		Gänzle *et al.*, 2000
	Saccharomyces exiguous		Messens and De Vuyst, 2002
	Candida milleri		Katina *et al.*, 2002
Animal products	Lactic acid bacteria	Mammalian and fish meat	Nielson *et al.*, 1990
	Pediococcus (cerevisiae, acidilactici)	(pork, beef, fish)	Schillinger *et al.*, 1991
	Staphylococcus carnosus	Sausages	Luchansky *et al.*, 1992

	L. plantarum	Fish sauces	Foegeding *et al.*, 1992
	Carnobacterium (piscicola, divergens)	Fish pastes	Paludan-Muller *et al.*, 1999, 2002
	Yeast		Waites *et al.*, 2001
	Debaryomyces hansenii		Adams and Moss, 2000; Adams and Nout, 2001
	Mould		Ariyapitipun *et al.*, 2000
	Penicillium spp.		Elotmani and Assobhei, 2004
Vegetable products	Lactic acid bacteria	Vegetable	Russell, 1992
	Lb. (plantarum, curvatus, brevis, sake)	Cucumbers	Buckenhüskes,1993
		Sauerkraut	Cheigh and Park, 1994
	Ln. mesenteroides	Olives	Harris, 1998
	Pediococcus cerevisiae		Gardner *et al.*, 2001
	Enterococcus (mundtii, faecium)		Savard *et al.*, 2002
Tea products	Lactic acid bacteria	Tea	Greewalt *et al.*, 1998
	Acetobacter (xylinum, xylinoids)	Puer tea	Sreeramulu *et al.*, 2000, 2001
	Bacterium gluconicum	Fuzhuan brick-tea	Chen and Liu, 2000
	Aspergillus spp.	Kombucha	Teoh *et al.*, 2004
	Penicillium spp.		Mo *et al.*, 2005
	Eurotium spp.		Chu and Chen, 2006
	Yeast		Wu *et al.*, 2007
	Schizosaccharomyces pombe		Xu *et al.*, 2007
	Saccharomycodes ludwigii		Aidoo *et al.*, 2006
	Saccharomyces cerevisiae		

organisms. For example, *Propionibacterium shermanii* is added to the lactic bacterial *Lactobacillus bulgaricus* and *Streptococcus thermophilus*. Propionibacteria contribute to the typical flavour and texture of Swiss-type cheese (Steffen *et al.*, 1993).

Yogurt is also a coagulated milk product obtained by lactic acid fermentation through the action of *Streptococcus thermophilus* and *Lactobacillus delbrueckii* subsp. *Bulgaricus*. The symbiotic growth of the two organisms of the yogurt starter culture has been reviewed by many authors (Radke-Mitchell and Sandine, 1984). Streptococci produce lactic acid, formic acid and carbon dioxide. Formic acid stimulates the growth of lactobacilli. The lactobacilli liberate some amino acids needed for the growth of the streptococci, and produce acetaldehyde and more lactic acid to bring the pH to 4.4–4.6. Furthermore, the type of yogurt starter used can change the physical characteristics of the final yogurt product.

Cereal-based products

Cereals are a good medium (dry matter basis of 70–80% polysaccharides) for microbial fermentation. Cereal grains normally carry an indigenous microbial flora composed of a variety of microbes, such as moulds, enterobacteria and aerobic sporeformers, all competing for nutrients. Because no pasteurization can be applied without affecting the technological properties of starch and protein, a vigorous starter flora of LAB and yeast is required for successful fermentation.

A number of cereal-based foods have been characteristically fermented by LAB, such as the European sour rye bread, various Asian flat breads and numerous types of fermented sour porridge and dumplings. Among cereal-based foods, most scientific research and technological development with respect to LAB and yeast has been associated with the sourdough breadmaking process. The fermentation combines the metabolic activity of LAB for souring and yeast for leavening. The dominant yeast strain in sourdough starter cultures was classified as *Saccharomyces exiguous* (Sugihara *et al.*, 1971) and later reclassified as *Candida milleri*. In San Francisco sourdough culture, the ratio of yeast to bacteria is about 1:100. The most common LAB are members of the genus *Lactobacillus*, shown in Table 6.2. To obtain a stable symbiotic relationship, the fermentation conditions must encourage metabolic activity of both yeasts and LAB.

The benefits of sourdough technology contribute to both quality of the end products and inhibition of contaminating or spoiling flora during fermentation. Sour bread is characterized by better resistance to microbiological spoilage by moulds and rope-forming bacilli (Corsetti *et al.*, 1998; Rosenquist and Hansen, 1998). The major antimicrobial compound in sourdough is acetic acid (Katina *et al.*, 2002). However, other compounds may play a role too. Corsetti *et al.* suggested that caproic acid formed by sourdough lactobacilli contributed to antifungal function. They also identified a bacteriocin-like substance from a *Lactobacillus* strain and found it active against a *B. subtilis* strain (Corsetti *et al.*, 1996). Lavermicocca *et al.* found new antifungal compounds produced by a *Lactobacillus plantarum* strain (Lavermicocca *et al.*, 2000). A heat-labile antibiotically acting compound, reutericyclin, was formed by a strain of *L. reuteri* isolated from sourdough (Gänzle *et al.*, 2000). A review of the inhibitory substances produced by lactobacilli isolated from sourdoughs was presented by Messens and De Vuyst (2002). Sourdough technology provides possibilities for the elongation of shelf life without added anti-mould or anti-rope or staling-inhibiting agents (Gobbetti and Corsetti, 1997).

Animal products

A variety of procedures for producing fermented animal products have been developed, including fermented sausages, fish sauces and fish pastes (Schillinger and Lucke, 1989; 1991; Ordonez *et al.*, 1999; Elotmani and Assobhei, 2004). The primary reason was to extend the shelf life of these highly prized, perishable foods. In general, the preservation of the meat keeps a low water content, achieved by the addition of salts and the generation of lactic acid by bacteria. In addition, the meat processing may include curing, smoking, drying and aging to improve both the flavour and the shelf life.

Fermented sausages can remain shelf stable when the moisture content of product is lower than 50%. Bacteria responsible for the fermentation need to tolerate both low water activity and salt. These environmental conditions encourage the generation of LAB, resulting in a decrease in pH and in the amount of available oxygen. The most common bacteria involved in meat fermentations are *Pediococcus cerevisiae, Pediococcus acidilactici, Staphylococcus carnosus* and *L. plantarum* (Nielsen *et al.*, 1990; Waites *et al.*, 2001). LAB and nitrate-reducing bacteria are also important members of starter cultures because of their greater reliability. To achieve the expected colour, flavour and antimicrobial property particular to all cured meats, either nitrate or nitrite must be added as a curing agent. However, it is the nitrite form, and not the nitrate, that actually reacts with the meat pigments and provides the curing effect. If the sausage formulation contains nitrate, it must first be converted to nitrite. Then its conversion to nitrite will depend on the presence of nitrate-reductase-producing strains. Thus, the nitrate-reducing bacteria are important to perform this function (Morot-Bizot *et al.*, 2004; Tang and Gillevet, 2003). In addition to fermentation by LAB, meat fermentation can be achieved using high-salt-tolerant yeasts such as *Debaryomyces hansenii* and moulds such as *Penicillium* spp. (Table 6.2) (Adams and Moss, 2000; Adams and Nout, 2001).

The fermented fish products known as fish sauces and fish pastes are protein rich with all the essential amino acids. They are produced by microbial fermentation and by the degrading activity of autolytic fish enzymes. The carbohydrate content of fish is low; therefore, for microbial fermentation an additional source of carbohydrate is required (Paludan-Muller, 1999). The supplementation of carbohydrates enables the microorganisms to ferment, and an acidic and stable product to be made (Adam and Moss, 2000). Some streptococci, micrococci, staphylococci and *Bacillus* spp. have been isolated from fish sauces. In the fermented products, LAB play an important role in the organoleptic properties and shelf life of these products.

Fermented vegetable products

The fermentation of vegetables is an ancient preservation method; the origins have been traced to Asia (Buckenhüskes, 1993). The most important commercially fermented vegetables in Europe and USA are cabbage (sauerkraut), cucumbers and olives. Others include carrots, cauliflower, celery, okra, onions and peppers. In Korea, kimchi is a traditional fermented vegetable food (Cheigh and Park, 1994). Typically, these fermentations do not involve the use of starter cultures and rely on the natural flora. Brine solutions are prepared in the fermentation of vegetables. Due to the desired metabolites such as lactic acids produced in fermentation, the pH of the fermenting material drops quickly. The organic acids and low pH will inhibit the growth of undesirable Gram-negative organisms (Harris, 1998). Consequently, a high level of hygiene can be achieved by repressing the growth of pathogenic bacteria.

Gardner *et al.* (2001) evaluated development of various LAB during fermentation and storage phases in vegetable mixtures of carrot, beet and cabbage. The selected starter consisting of *L. plantarum* NK 312, *P. acidilactici* AFERM 772 and *Leuconostoc mesenteroides* BLAC produced good sensory quality and repressed the growth of yeasts. By increasing the ratio of *L. mesenteroides*, the proportion of acetic acid increased and lactic acid decreased (Savard *et al.*, 2002). These results suggest that the development of starter cultures can aid in the economic improvement of fermentation processes as well as the safety and health aspects of the vegetable products.

As raw vegetables have a high microbial load and cannot be pasteurized without compromising product quality, most vegetable fermentations occur as a consequence of providing growth conditions that favour the LAB. Although the amounts of LAB in fresh vegetables are very low, accounting for only 0.15–1.5% of total population, the primary fermentation of vegetable is dominated by LAB, belonging to the genera *Lactobacillus, Leuconostoc* and *Pediococcus* (Valdez *et al.*, 1990; Nout and Rombouts, 1992). The antimicrobial effect of fermentation acids is based on the concentration of the undissociated form of the acids in synergy with a low pH (Russell, 1992). Upon entering the cell, the undissociated acid dissociates into its anion and proton because of the somewhat neutral intracellular pH. This reduces the intracellular pH to a level that will rapidly kill the cell unless the ions are excreted again by active transport, a process that requires energy. Anions have antimicrobial activity as well. In addition, the accumulation of CO_2 in fermented vegetable products is the result of an endogenous respiration of the plant cells combined with microbial activities (Clark and Takács, 1980). The overall effect of CO_2 on microorganisms is an extension of the lag phase of growth and a decrease in the growth rate during the logarithmic phase. LAB in vegetable fermentations are beneficial for the sensory and hygienic quality of the final products.

Tea products

Recently, several microbial fermented teas have become noted worldwide, not only because of their beneficial health properties, but also because of their antimicrobial activities. A few studies indicate that the antimicrobial components of fermented tea have an inhibitory effect against several food-borne, spoilage and pathogenic bacteria (Sreeramulu *et al.*, 2000; 2001; Mo *et al.*, 2005). Traditional fermented teas, such as Puer tea, Fuzhuan brick-tea and Kombucha, have shown obvious antibacterial effects (Wu *et al.*, 2007; Mo *et al.*, 2005).

Both Puer tea and Fuzhuan brick-tea are the unique Chinese microbial fermented black teas obtained through indigenous tea fermentation (Xu *et al.*, 2007; Mo *et al.* 2008). A microbiological analysis of Puer tea revealed that *Aspergillus niger* was the dominating microorganism during the fermentation. Its antimicrobial activity shows an inhibitory effect on the spore-forming bacteria *B. cereus, B. subtilis, Clostridium perfringens* and *Clostridium sporogenes.* The microbiological composition and the antimicrobial activity of extracts from the fermented Fuzhuan brick-tea have also been analysed. The results show that *Aspergillus* spp., *Penicillium* spp. and *Eurotium* spp. were the main microorganisms isolated from the Fuzhuan brick-tea during fermentation. Its antibacterial tests showed inhibitory effect on several food-borne bacteria, including the spore-forming bacteria *B. cereus, B. subtilis, C. perfringens* and *C. sporogenes.*

Kombucha is a fermented drink of tea extract supplemented with sucrose and fermented with yeasts and acetic acid bacteria (Teoh *et al.*, 2004; Chu and Chen, 2006; Adioo *et al.*, 2006). It originated in northeast China and later spread to Russia and the rest of the world. The antimicrobial activity of Kombucha was tested against a number of pathogenic microorganisms (Sreeramulu *et al.*, 2000). *Staphylococcus aureus, Shigella sonnei, E. coli, Aeromonas hydrophila, Yersinia enterolitica, Pseudomonas aeruginosa, E. cloacae, Staphylococcus epidermis, Campylobacter jejuni, Salmonella enteritidis, Salmonella trphimurium, B. cereus, Helicobacter pylori,* and *L. monocytogenes* were found to be sensitive to Kombucha. Kombucha proved to exert antimicrobial activities against *E. coli, S. sonnei, S. typhimurium, S. enteritidis* and *C. jejuni,* even at neutral pH and after thermal denaturation, which suggests the presence of antimicrobial compounds other than acetic acid and large proteins in Kombucha (Sreeramulu *et al.*, 2001).

The antimicrobial activities in the microbial fermentation process of tea leaves are attributed to some organic acids (depending on the source of the culture: acetic acid, butyric acid, gluconic acid, glucuronic acid, lactic acid, malic acid, oxalic acid and usnic acid), proving the major role of these acids in the microorganism's growth inhibition. The activity against pathogenic microorganisms was proved largely attributable to acetic acid, which is known to inhibit and destroy a number of Gram-positive and Gram-negative microorganisms (Steinkraus *et al.*, 1996; Greenwalt *et al.*, 1998). In addition, the active antimicrobial components were substances other than organic acid and their synergistic effect, ethanol, proteins or tannins originally present in tea or their derivatives. The active components are very likely to be microbial metabolites produced by bacteria and yeast during fermentation with tea and sugar as substrates.

Conclusion

The preservation of food resulting from fermentation has been an effective form of extending the shelf life of food for millennia. Traditionally, foods were preserved through naturally occurring fermentation; however, modern large-scale production now exploits the use of defined-strain starter systems to ensure consistency and quality of the final product. Current research trends in food preservation focus on the use of natural antimicrobial compounds produced by the fermentation process. The identification and characterization of the antimicrobial compounds are of further importance for understanding how compounds in food systems are transformed by the metabolic pathways into antimicrobial agents. The production of one or more active antimicrobials is part of the complex mechanisms. Furthermore, understanding these mechanisms and antimicrobial properties of a number of metabolites from microbial fermentation are also important for biological approaches in food preservation.

Food preservatives of natural origin are generally considered as potential, safe sources of antimicrobials, but their effective use in practice is still rare. For example, knowing that other bacteriocins exist and can work at least as effectively as nisin with respect to particular foods/target bacteria, the question is often posed why more have not been exploited to the same extent as nisin. In addition, the application of bioengineered/modified bacteriocins or other antimicrobial metabolites may be considered counterproductive to the marketing of the metabolites as natural products. Continued research on antimicrobial metabolites will undoubtedly lead to an increased understanding, and with the emergence of new antimicrobial metabolites, new potential biopreservatives.

We are now entering the post-genomic age of microbiology at a time when many microorganisms used for food production have already been sequenced. This offers a new knowledge-based approach to the exploitation of bacteria for food production, from metabolic engineering of microorganisms to produce antimicrobial or nutritional compounds, to the molecular mining of activities as yet unknown but which could benefit food production. In addition, the availability of the genomes of many pathogenic and food-spoilage bacteria may open up new possibilities for the design of novel antimicrobials that target essential functions of these problematic bacteria. Such techniques could improve the stability, efficacy, and production of antimicrobial compounds so they may be more applicable in food preservation.

References

Abriouel, H., Maqueda, M., Gálvez, A., Martínez-Bueno, M. and Valdivia, E. (2002) Inhibition of bacterial growth, enterotoxin production, and spore outgrowth in strains of *Bacillus cereus* by bacteriocin AS-48. *Applied Environmental Microbiology* 68, 1473–1477.

Adams, M.R. and Nout, M.J.R. (2001) *Fermentation and Food Safety.* Aspen Publishers Inc., Maryland.

Adams,M.R. and Moss, M.O. (2000) *Food Microbiology,* 2nd edn. The Royal Society of Chemistry, Cambridge, UK.

Adioo, K.E., Nout, M.J.R. and Sarkar, P.K. (2006) Occurrence and fuction of yeasts in Asian indigenous fermented foods. *FEMS Yeast Research* 6, 30–39.

Alakomi, H-L., Skyttä, E., Saarela, M., Mattila-Sandholm, T., Latva-Kala, K. and Helander, I.M. (2000) Lactic acid permeabilizes gram-negative bacteria by disrupting the outer membrane. *Applied Environmental Microbiology* 66, 2001–2005.

Ariyapitipun, T., Mustapha, A. and Clarke, A.D. (2000) Survival of *Listeria monocytogenes* Scott A on vacuum-packaged raw beef treated with polylactic acid, lactic acid, and nisin. *Journal of Food Protection* 63, 131–136.

Axelsson, L.T., Chung, T.C., Dobrogosz, W.J. and Lindgren, S.E. (1989) Production of a broad spectrum antimicrobial substance by *Lactobacillus reuteri. Microbial Ecology in Health and Disease* 2, 131–136.

Banks, J.G., Board, R.G. and Sparks, N.H.C. (1986) Natural antimicrobial systems and their potential in food preservation of the future. *Biotechnology Applied Biochemistry* 8, 103–111.

Bennik, M.H.J., Smid, E.J. and Gorris, L.G.M. (1997) Vegetable associated *Pediococcus parvulus* produces pediocin PA-1. *Applied Environmental Microbiology* 63, 2074–2071.

Bennik, M.H.J., Van Overbeek, W., Smid, E.J. and Gorris, L.G.M. (1999) Biopreservation in modified atmosphere stored mungbean spouts: the use of vegetable-associated bacteriocinogenic lactic acid bacteria to control the growth of *Listeria monocytogenes. Letters in Applied Microbiology* 28, 226–232.

Berry, E.D., Hutkins, R.W. and Mandigo, R.W. (1991) The use of a bacteriocin-producing *Pediococcus acidilactici* to control postprocessing *Listeria monocytogenes* contamination of frankfurters. *Journal Food Protein* 54, 681–690.

Bhunia, A.K., Johnson, M.C. and Ray, B. (1988) Purification, characterization and antimicrobial spectrum of a bacteriocin produced by *Pediococcus acidilactici. Journal of Applied Bacteriology* 65, 261–268.

Bhunia, A.K., Johnson, M.C., Ray, B. and Kalchayanand, N. (1991) Mode of action of pediocin AcH from *Pediococcus acidilactici* H on sensitive bacterial strains. *Journal of Applied Bacteriology* 70, 25–30.

Blom, H. and Mörtvedt, C. (1991) Anti-microbial substances produced by food associated microorganisms. *Biochemical Society Transactions* 19, 694–698.

Brul, S. and Coote, P. (1999) Preservative agents in foods. Mode of action and microbial resistance mechanisms. *International Journal of Food Microbiology* 50, 1–17.

Buckenhüskes, H.J. (1993) Selection criteria for lactic acid bacteria to be used as starter cultures for various food commodities. *FEMS Microbiology Reviews* 12, 253–272.

Caplice, E. and Fitzgerald, G.F. (1999) Food fermentations: role of microorganisms in food production and preservation. *International Journal of Food Microbiology* 50, 131–149.

Cheigh, H-S. and Park, K-Y. (1994) Biochemical, microbiological and nutritional aspects of kimchi (Korean fermented vegetable products). *Critical Reviews in Food Science and Nutrition* 34, 175–203.

Chen, C. and Liu, B.Y. (2000) Changes in major components of tea fungus metabolites during prolonged fermentation. Journal of Applied Microbiology, 89, 834–839.

Chen, G. and Russell, J.B. (1989) Transport of glutamine by *Streptococcus bovis* and conversion of glutamine to pyroglutamic acid and ammonia. *Journal of Bacteriology* 171, 2981–2985.

Chu, S.C. and Chen, C.S. (2006) Effects of origins and fermentation time on the antioxidant activities of Kombucha. *Food Chemistry* 98, 502–507.

Clark, D.S. and Takács, J. (1980) Gases as preservatives. In: Silliker, J.H. (ed.) *Microbial Ecology of Foods*, Academic Press, London, pp.70–180.

Cleveland, J., Montville, T.J., Nes, I.F. and Chikindas, M.L. (2001) Bacteriocins: safe, natural antimicrobials for food preservation. *International Journal of Food Microbiology* 71, 1–20.

Corsetti, A., Gobbetti, M. and Smacchi, E. (1996) Antibacterial activity of sourdough lactic acid bacteria: isolation of a bacteriocin-like inhibitory substance from *Lactobacillus sanfrancisco* C57. *Food Microbiology* 13, 447–456.

Corsetti, A., Gobbetti, M., Rossi, J. and Damiani, P. (1998) Antimould activity of sourdough lactic acid bacteria: identification of a mixture of organic acids produced by *Lactobacillus sanfrancisco* CB1. *Applied Microbiology and Biotechnology* 50, 253–256.

Davies, E.A., Milne, C.F., Bevis, H.E., Potter, R.W., Harris, J.M., Williams, G.C., Thomas, L.V. and Delves-Broughton, J. (1999) Effective use of nisin to control lactic acid bacterial spoilage in vacuum-packed bologna-type sausage. *Journal of Food Protection* 62, 1004–1010.

De Vuyst, L. and Vandamme, E. (1994) Nisin, a lantibiotic produced by *Lactococcus lactis* subsp. *lactis*: properties, biosynthesis and applications. In: de Vuyst, L. and Vandamme, E. (eds) *Bacteriocins of Lactic Acid Bacteria. Microbiology, Genetics and Applications*. Blackie Academic and Professional, London, pp. 151–221.

Delves-Broughton, J., Blackburn, P., Evans, R.J. and Hugenholtz, J. (1996) Applications of the bacteriocin nisin. *Antonie van Leeuwenhoek* 69, 193–202.

Dobrogosz, W.J., Casas, I.A., Pagano, G.A., Talarico, T.L., Sjöberg, B-M. and Karlsson, M. (1989) *Lactobacillus reuteri* and the enteric microbiota. In: Norin, E. (ed.) The *Regulatory and Protective Role of the Normal Microflora,* Stockton Press, New York, pp. 283–292.

Doores, S. (1993) Organic acids. In: Davidson, P.M. and Branen, A.L. (eds) *Antimicrobials in Foods.* Marcel Dekker, New York, pp. 95–136.

Eklund, T. (1983) The antimicrobial effect of dissociated and undissociated sorbic acid at different pH levels. *Journal of Applied Bacteriology* 54, 383–389.

Elotmani, F. and Assobhei, O. (2004) *In vitro* inhibition of microbial flora of fish by nisin and lactoperoxidase system. *Letters in Applied Microbiology* 38, 60–65.

Ferreira, M.A. and Lund, B.M. (1996) The effect of nisin on *Listeria monocytogenes* in culture medium and long-life cottage cheese. *Letters in Applied Microbiology* 22, 433–438.

Foegeding, P.M., Thomas, A.B., Pilkington, D.H. and Klaenhammer, T.R. (1992) Enhanced control of *Listeria monocytogenes* by in situ-produced pediocin during dry fermented sausage production. *Applied Environmental Microbiology* 58, 884–890.

Fox, P.F. (1993) Cheese: an overview. In: Fox, P.F. (ed.), *Cheese; Chemistry, Physics and Microbiology.* Chapman and Hall, London, pp. 1–36.

Gálvez, A., Maqueda, M., Martínez-Bueno, M. and Valdivia, E. (1991) Permeation of bacterial cells, permeation of cytoplasmic and artificial membrane vesicles, and channel formation on lipid bilayers by peptide antibiotic AS-48. *Journal of Bacteriology* 173, 886–892.

Gänzle, M.G., Höltzel, A., Walter, J., Jung, G. and Hammes, W.P. (2000) Characterization of reutericyclin produced by *Lactobacillus reuteri* LTH2584. *Applied Environmental Microbiology* 66, 4325–4333.

Gao, Y., Jia, S., Gao, Q. and Tan, Z. (2010) A novel bacteriocin with a broad inhibitory spectrum produced by *Lactobacillus sake* C2, isolated from traditional Chinese fermented cabbage. *Food Control* 21, 76–81.

Gardner, N.J., Savard, T., Obermeier, P., Caldwell, G. and Champagne, C.P. (2001) Selection and characterization of mixed starter cultures for lactic acid fermentation of carrot, cabbage, beet and onion vegetable mixtures. *International Journal of Food Microbiology* 64, 261–275.

Gobbetti, M. and Corsetti. A. (1997) *Lactobacillus sanfrancisco*, a key sourdough lactic acid bacterium: a review. *Food Microbiology* 14, 175–187.

Gourama, H. and Bullerman, L.B. (1995) Antimycotic and antiaflatoxigenic effect of lactic acid bacteria, a review. *Journal of Food Protein* 58, 1275–1280.

Grande, M.J., Lucas, R., Abriouel, H., Valdivia, E., Omar, N.B., Maqueda, M., Martínez-Bueno, M., Martínez-Canamero, M. and Gálvez, A. (2006) Inhibition of toxicogenic *Bacillus cereus* in rice-based foors by enterocin AS-48. *International Journal of Food Microbiology* 106, 185–194.

Greenwalt, C.J., Ledford, R.A. and Steinkraus, K.H. (1998) Determination and characterization of the antimicrobial activity of the fermented tea Kombucha. *LWT-Food Science and Technology* 31, 291–296.

Harris, L.J. (1998) The microbiology of vegetable fermentations. In: Wood, B.J.E. (ed.) *Microbiology of Fermented Foods,* 2nd edn. Blackie Academic and Professional, London, pp. 45–72.

Hastings, J.W., Sailer, M., Johnson, K., Roy, K.L., Vederase, J.C. and Stiles, M.E. (1991) Characterization of leucocin A-UAL 187 and cloning of the bacteriocin gene from *Leuconostoc gelidum. Journal of Bacteriology* 173, 7491–7500.

Holck, A., Axelsson, L., Birkeland, S.E., Aukrust, T. and Blom, H. (1992) Purification and amino acid sequence of sakacin A, a bacteiociocin from *Lactobacillus sake* Lb706. *Journal of General Microbiology* 38, 2715–2720.

Höltzel, A., Gänzle, M.G., Nicholson, G.J., Hammes, W.P. and Jung, G. (2000) The first low molecular weight antibiotic from lactic acid bacteria: reutericyclin, a new tetramic acid. *Angewandte Chemie International Edition* 39, 2766–2768.

Hurst, A. and Collins-Thompson, D.L. (1979) Food as a bacterial habitat, In: Alexander, M. (ed.) *Advances in Microbial Ecology,* Vol. 3, Plenum Publishing Corporation, New York, pp. 79–134.

Huttunen, E., Noro, K. and Yang, Z. (1995) Purification and identification of antimicrobial substances produced by two *Lactobacillus casei* strains. *International Dairy Journal* 5, 503–513.

Jack, R.W. and Sahl, H.G. (1995) Unique peptide modifications involved in the biosynthesis of lantibiotics. *Trends in Biotechnology* 13, 269–278.

Jay, J.M. (1996) *Modern Food Microbiology*, 5th edn. Chapman & Hall, New York.
Jin, Y.K. and Park, Y.W. (1995) Effects of aging time and temperature on proteolysis of commercial goat milk cheeses produced in the United States. *Journal of Dairy Science* 78, 2598.
Joerger, M.C. and Klaenhammer, T.R. (1986) Characterization and purification of helveticin J and evidence for a chromosomally determined bacteriocin produced by *Lactobacillus helveticus* 481. *Journal of Bacteriology* 167, 439–446.
Joerger, M.C. and Klaenhammer, T.R. (1990) Cloning, expression and nucleotide sequence of the *Lactobacillus helveticus* 481 gene encoding the bacteriocin helveticin. *Journal of Bacteriology* 171, 6339–6347.
Katina, K., Sauri, M., Alakomi, H.-L. and Mattila-Sandholm, T. (2002) Potential of lactic acid bacteria to inhibit rope spoilage in wheat sourdough bread. *Lebensmittel-Wissenschaft und Technologie* 35, 38–45.
Klaenhammer, T.R. (1993) Genetics of bacteriocins produced by lactic acid bacteria. *FEMS Microbiology Reviews* 12, 39–85.
Kreger-van Rij, N.J.W. (1984) *The Yeasts. A Taxonomic Study.* Elsevier, Amsterdam.
Lavermicocca, P., Valerio, F., Evidente, A., Lazzaroni, S., Corsetti, A. and Gobbetti, M. (2000) Purification and characterization of novel antifungal compounds from the sourdough *Lactobacillus plantarum* strain 21B. *Applied Environmental Microbiology* 66, 4084–4090.
Lewus, C.B., Sun, S. and Montville, T.J. (1992) Production of an amylase-sensitive bacteriocin by an atypical *Leuconostoc paramesenteroides* strain. *Applied Environmental Microbiology* 58, 143–149.
López A.S. and Belloso, O.M. (2008) Use of nisin and other bacteriocins for preservation of dairy products. *International Dairy Journal*, 329–343.
Luchansky, J.B., Glass, K.A., Harsono, K.D., Degnan, A.J., Faith, N.G., Cauvin, B., Bater, B., Maurer, A.J. and Cassens, R.G. (1992) Genomic analysis of *Pediococcus* starter cultures used to control *Listeria monocytogenes* in turkey summer sausage. *Applied Environmental Microbiology* 58, 3053–3059.
Maisnier-Patin, S., Forni, E. and Richard, J. (1996) Purification, partial characterization and mode of action of enterococcin EFS2, an antilisterial bacteriocin produced by a strain of *Enterococcus faecalis* isolated from a cheese. *International Journal of Food Microbiology* 30, 255–270.
Maqueda, M., Gálvez, A., Martínez-Bueno, M., Sánches-Barrena, J., González, C., Albert, A., Rico, M. and Valdivia, E. (2004) Peptide AS-48: prototype of a new class of cyclic bacteriocins. *Current Protein Peptide Science* 5, 399–416.
Mendoza, F., Maqueda, M., Gálvez, A., Martínez-Bueno, M. and Valdivia, E. (2002) Antilisterial activity of peptide AS-48 and study of changes induced in the cell envelope properties of an AS-48-adapted strain of *Listeria monocytogenes. Applied Environmental Microbiology* 68, 1473–1477.
Messens, W. and De Vuyst, L. (2002) Inhibitory substances produced by lactobacilli isolated from sourdoughs-a review. *International Journal of Food Microbiology* 72, 31–43.
Mo, H.Z., Xu, X.O., Yan, M.C. and Zhu, Y. (2005) Microbiological analysis and antibacterial effects of the indigenous fermented Puer tea. *Agro Food Industry Hi-Tech,* 16, 16–18.
Mo, D., Zhu, Y. and Chen, Z. (2008) Microbial fermented tea – a potential source of natural food preservatives. *Trends in Food Science and Technology* 19, 124–130.
Morot-Bizot, S.C., Talon, R. and Leroy, S. (2004) Development of a multiplex PCR for the identification of *Staphylococcus* genus and four *staphylococcal* species isolated from food. *Journal of Applied Microbiology* 97, 1087–1094.
Munõz, A., Maqueda, M., Gálvez, A., Martínez-Bueno, M., Rodriguez, A. and Valdivia, E. (2004) Biocontrol of psychrotrophic enterotoxigenic *Bacillus cereus* in a non fat hard type cheese by an enterococcal strain-producing enterocin AS-48. *Journal of Food Protection* 67, 1517–1521.
Nes, I.F., Diep, D.B., Havarstein, L.S., Brurberg, M.B., Eijsink, V. and Holo, H. (1996) Biosynthesis of bacteriocins in lactic acid bacteria. *Antonie van Leeuwenhoek* 70, 113–128.
Nes, I.F., Holo, H., Fimland, G., Hauge, H.H. and Nissen-Meyer, J. (2000) Unmodified peptide-bacteriocins (class II) produced by lactic acid bacteria. In: Dutton, C.J., Haxell, M.A., McArthur, H.A.I. and Wax, R.G. (eds), *Peptide Antibiotics: Discovery, Modes of Action and Applications.* Marcel Dekker, New York/Basel.
Nielsen, J.W., Dickson, J.S. and Crouse, J.D. (1990) Use of a bacteriocin produced by *Pediococcus acidilactici* to inhibit *Listeria monocytogenes* associated with fresh meat. *Applied Environmental Microbiology* 56, 2142–2145.
Nilsson, L., Gram, L. and Huss, H.H. (1999) Growth control of *Listeria monocytogenes* on cold-smoked salmon using a competitive lactic acid bacteria flora. *Journal of Food Protection* 62, 336–342.

Nout, M.J.R. (1995) Fungal interactions in food fermentations. *Canadian Journal of Botany* 73, 1291–1300.
Nout, M.J.R. and Rombouts, F.M. (1992) Fermentative preservation of plant foods. *Journal of Applied Bacteriology Symposium Supplement* 73, 136S–147S.
Ordonez, J.A., Hierro, E.M., Bruna, J.M. and de la Hoz, L. (1999) Changes in the components of dry-fermented sausages during ripening. *Critical Reviews in Food Science and Nutrition* 39, 329–367.
O'Sullivan, L., Ross, R.P. and Hill, C. (2002) Potential of bacteriocin-producing lactic acid bacteria for improvements in food safety and quality. *Biochimie* 84, 593–604.
Paludan-Müller, C., Huss, H.H. and Gram, L. (1999) Characterization of lactic acid bacteria isolated from a Thai low-salt fermented fish product and the role of garlic as substrate for fermentation. *International Journal of Food Microbiology* 46, 219–229.
Paludan-Müller, C., Valyasevi, R., Huss, H.H. and Gram, L. (2002) Genotypic and phenotypic characterization of garlic-fermenting lactic acid bacteria isolated from som-fak, a Thai low-salt fermented fish product. *Journal of Applied Microbiology, 92,* 307–314.
Piard, J.C., Muriana, P.M., Desmazeaud, M.J. and Klaenhammer, T.R. (1992) Purification and partial characterization of lacticin 481, a lanthionine-containing bacteriocin produced by *Lactococcus lactis* subsp. *lactis* CNRZ481. *Applied Environmental Microbiology* 58, 279–284.
Pucci, M.J., Vedamuthu, E.R., Kunka, B.S. and Vandenbergh, P.A. (1988) Inhibition of *Listeria monocytogenes* by using bacteriocin PA-1 produced by *Pediococcus acidilactici* PAC 1.0. *Applied Environmental Microbiology* 54, 2349–2353.
Radke-Mitchell, L. and Sandine, W.E. (1984) Associative growth and differential enumeration of *Streptococcus thermophilus* and *Lactobacillus bulgaricus*: a review. *Journal of Food Protection* 47, 245–248.
Rammelsberg, M., Muller, E. and Radler, F. (1990) Caseicin 80: purification and characterization of a new bacteriocin from *Lactobacillus casei. Archives of Microbiology* 154, 249–252.
Ray, B. (1992) Bacteriocins of starter culture bacteria as food biopreservatives: an overview. In: Ray, B. and Daeschel, M. (eds), *Food Biopreservatives of Microbial Origin.* CRC Press, Boca Raton, Florida.
Rayman M.K., Aris, B. and Hurst, A. (1981) Nisin: a possible alternative or adjunct to nitrite in the preservation of meats. *Applied Environmental Microbiology* 41, 375–380.
Rosenquist, H. and Hansen, A. (1998) The antimicrobial effect of organic acids, sour dough and nisin against *Bacillus subtilis* and B. *licheniformis* isolated from wheat bread. *Journal of Applied Microbiology* 85, 621–631.
Ross, R.P., Morgan, S. and Hill, C. (2002) Preservation and fermentation: past, present and future. *International Journal of Food Microbiology* 79, 3–16.
Russell, J.B. (1992) Another explanation for the toxicity of fermentation acids at low pH: anion accumulation versus uncoupling. *Journal of Applied Bacteriology* 73, 363–370.
Ryan, M.P., Rea, M.C., Hill, C. and Ross, R.P. (1996) An application in cheddar cheese manufacture for a strain of *Lactococcus lactis* producing a novel broad spectrum bacteriocin, lacticin 3147. *Applied Environmental Microbiology* 62, 612–619.
Ryan, M.P., Meaney, W.J., Ross, R.P. and Hill, C. (1999) Evaluation of lacticin 3147 and a teat seal containing this bacteriocin for inhibition of mastitis pathogens. *Applied Environmental Microbiology* 64, 2287–2290.
Sahl, H.G. and Bierbaum, G. (1998) Lantibiotics: biosynthesis and biological activities of uniquely modified peptides from gram-positive bacteria. *Annual Reviews in Microbiology* 52, 41–79.
Savard, T., Beaulieu, C., Gardner, N.J. and Champagne, C.P. (2002) Characterization of spoilage yeasts isolated from fermented vegetables and inhibition by lactic, acetic and propionic acids. *Food Microbiology* 19, 63–373.
Schillinger, U. and Lucke, F-K. (1989) Antibacterial activity of *Lactobacillus sake* isolated from meat. *Applied Environmental Microbiology* 55, 1901–1906.
Schillinger, U., Kaya, M. and Lucke, F-K. (1991) Behaviour of *Listeria monocytogenes* in meat and its control by a bacteriocin-producing strain of *Lactobacillus sake. Journal of Applied Bacteriology* 70, 473–478.
Schved, F., Lalazar, A., Henis, Y. and Juven, B.J. (1993) Purification, partial characterization and plasmid-linkage of pediocin SJ-1, a bacteriocin produced by *Pediococcus acidilactici. Journal of Applied Microbiology* 74, 67–77.
Sreeramulu, G., Zhu, Y. and Knol, W. (2000) Kombucha fermentation and its antimicrobial activity. *Journal of Agricultural and Food Chemistry* 48, 2589–2594.

Sreeramulu, G., Zhu, Y. and Knol, W. (2001) Characterization of antimicrobial activity in Kombucha fermentation. *Acta Biotechnologica,* 21, 49–56.
Steffen, C., Eberhard, P., Bosset, P.O. and Ruegg, M. (1993) Swiss-type varieties. In: P.F. Fox (ed.), *Cheese, Chemistry, Physics and Microbiology*, Vol. 2. Chapman & Hall, London, pp. 83–110.
Steinkraus, K.H. (1983) Lactic acid fermentation in the production of foods from vegetables, cereals and legumes. *Antonie van Leeuwenhoek,* 49, 337–348.
Steinkraus, K.H., Shapiro, K.B., Hotchkiss, J.H. and Mortlock, R.P. (1996) Investigations into the antibiotic activity of tea fungus/Kombucha beverage. *Acta Biotechnologica*, 16, 291–296.
Stevens, K.A., Sheldon, B.W., Klapes, N.A. and Klaenhammer, T.R. (1992) Effective treatment conditions on nisin inactivation of Gram negative bacteria. *Journal of Food Protection* 55, 763–766.
Sugihara, T.F., Kline, L. and Miller, M.W. (1971) Microorganisms of the San Francisco sour dough bread process. I. Yeasts responsible for the leavening action. *Applied Microbiology* 21, 456–458.
Sung, H-W., Chen, C-N., Liang, H-F. and Hong, M-H. (2003) A natural compound (reuterin) produced by *Lactobacillus reuteri* for biological-tissue fixation. *Biomaterials* 24, 1335–1347.
Suomalainen, H. and Oura, E. (1971) Yeast nutrition and solute uptake. In: Rose, A.H. and Harrison, J.S. (eds), *The Yeasts,* Vol. 2, *Physiology and Biochemistry of Yeasts,* Academic Press, London, pp. 3–74.
Suomalainen, T.H. and Mäyrä-Mäkinen, A.M. (1999) Propionic acid bacteria as protective cultures in fermented milks and breads. *Lait* 79, 165–174.
Tang, J.S. and Gillevet, P.M. (2003) Reclassification of ATCC 9341 from *Micrococcus luteus* to *Kocuria rhizophila. International Journal of Systematic and Evolutionary Microbiology* 53, 995–997.
Teoh, A.L., Heard, G. and Cox, J. (2004) Yeast ecology of Kombucha fermentation. *International Journal of Food Microbiology* 95, 119–126.
Valdez, G.F., de Giori, G.S., Garro, M., Mozzi, F. and Oliver, G. (1990) Lactic acid bacteria from naturally fermented vegetable. *Microbiologie Aliments Nutrition* 8, 175–179.
Waites, M.J., Morgan, N.L., Rockey, J.S. and Higton, G. (2001) *Industrial Microbiology: An Introduction.* Blackwell Science Ltd, Oxford.
Wu, S.C., Yen, G.C., Wang, B.S., Chiu, C.K., Yen, W.J. and Chang, L.W. (2007) Antimutagenic and antimicrobial activities of pu-erh tea. *LWT-Food Science and Technology* 40, 506–512.
Xu, X.Q., Mo, H.Z., Yan, M.C. and Zhu, Y. (2007) Analysis of characteristic aroma of fungal fermented Fuzhuan brick-tea by gas chromatography/mass spectrophotometry. *Journal of the Science of Food and Agriculture.* 87, 1502–1504.
Yang, Z., Suomalainen, T., Mäyrä-Mäkinen, A. and Huttunen, E. (1997) Antimicrobial activity of 2-pyrrolidone-5-carboxylic acid produced by lactic acid bacteria. *Journal of Food Protection* 60, 786–790.

7 Antimicrobials from Marine Algae

Mohamed Faid

Introduction

Natural antimicrobials are acquiring more attention in the field of food preservation and food safety. They are mainly extracted from plants, but some are being produced by bacteria, such as bacteriocins. Seaweeds are a potential source of many compounds for food safety and preservation but not much data are available and more investigations are required in this field. The nutritional benefits of marine algae have been studied in depth and have been widely developed, while the extraction of some antimicrobials is now at a relative premium. Little is known about the antimicrobials and/or some preserving agents from marine algae.

The natural preservation of living materials against spoilage had been carried out for thousands of years, long before the discovery of heat treatment, by discovering means and procedures for keeping processes as natural as possible. Foodstuffs both animal and vegetable, including sea products, are subject to spoilage and decaying phenomena. These occur by the action of various microorganisms including bacteria, moulds and yeasts. Such phenomena can lead to a heavy loss when they occur in food products, especially raw materials, and can also lead to some health problems engendered by the occurrence of food poisoning and infections.

The removal of microorganisms from foods is not only for food safety but also for keeping the quality of the food, which may include the preservation of the organoleptic characteristics and the nutritional components. The most relevant factor in food safety is the microbial hazard. Microorganisms may impact on food safety in many ways, which can be summarized into two categories:

- Direct danger by the ingestion of pathogens: *Salmonella, Yersinia, Listeria, Campylobacter, E. coli,* etc.
- Indirect danger by the ingestion of the microbial toxins released in foods by microorganisms during handling before processing. This involves two types of toxins: (i) bacterial toxins from *Staphylococcus, Clostridium* and *Bacillus* and (ii) mycotoxins from moulds (*Aspergillus, Penicillium, Monilia,* etc.).

Unfortunately the use of chemical antimicrobials potentially dangerous for health was authorized by medical authorities throughout the world. The synthesized antimicrobial chemicals are now being rejected by the consumer because most of them were shown to cause side effects that in many cases led to some hard-to-treat diseases. The search for natural preservatives and/or quality-improving factors is now more and more encouraged.

Although less studied than terrestrial plants, marine algae are now recognized as potential materials for natural antimicrobials (Salvador *et al.*, 2007; Thillairajasekar *et al.*, 2009; Plaza *et al.*, 2010). Several studies of macroalgae-derived compounds report on various biological activities such as antibacterial, antifungal and antiviral, as well as other properties including enzyme inhibition, free-radical scavenging, antioxidation, anti-inflammatory, anti-mitotic and anti-neoplastic activities (Naqvi *et al.*, 1980; Hodgson, 1984; Fenical and Paul, 1984; Ballesteros *et al.*, 1992; Kamat *et al.*, 1992; Bhosale *et al.*, 2002; Souhaili *et al.*, 2008). Research concerning the screening of antimicrobial activities began in the 1970s (Glombitza, 1970; Horsney and Hide, 1974; Henriquez *et al.*, 1977).

Marine Algae: a Survey

More than 150,000 algae and seaweed species exist in oceans and seas but many of them have not been identified. Because of the wide distribution of algae throughout the coastal lands around many continents, studies concerning marine algae are not yet deeply and widely performed to know all the aspects related to these vital materials. Aspects dealing with the nutrition were mostly investigated, as well as some pharmaceutics, but seaweeds may contain natural antimicrobials which could be used in food safety and preservation.

Several chemical compounds extracted from marine algae are widely used, especially in the food industry and pharmacology. The most well-known compounds are:

- Carrageenan, extracted from red algae species and used as stabilizing and/or gelling agents in many food products such as chocolate and instant milk.
- Agar, a colloidal agent, extracted from algae (*Gelidium*) to substitute gelatin, anti-drying agents in bread and pastry, as well as thickening and gelling many kinds of foods such as frozen dairy products, processed cheeses, mayonnaise, puddings, cream and jellies.
- Alginates, extracted from the brown algae, which thicken water products and also make them creamier and more stable over wide ranges of temperature, pH and time. They are used to prevent crystal formation in ice cream.
- Green algae pigments, such as beta carotene, which is used as a natural food colourant, and phycocyanin, another derived colourant from spirulina, a blue green algae.

Marine algae are now being promoted as a food of the future. Many ingredients are being used in pills, drinks, snacks, soups and brews. It is also noteworthy that algae and seaweeds do not undergo any decaying phenomenon because of the potent antimicrobials and antifungals they contain.

The Potential Antimicrobials in Marine Algae

Studies of the antimicrobials from plants are hindered by the unknown chemical nature of these compounds because many factors make them hard to monitor. The wide range of algal varieties throughout the world, the tiny concentration of the active component in these plants, the seasonal variation and their distribution are the most relevant factors that could handicap the research in this field.

Compounds in the seaweed may depend on the availability of some lab equipment and materials in coastal developing countries where scientific research is not yet developed. Almost all the work carried out on the algal antimicrobial compounds is controversial; their chemical structures have not been accurately defined. In fact, some authors have published only antimicrobial activities of the broad extract from some algal species. These may also vary with the nature of the species studied, with the test microorganisms used and with the solvent used to extract the active compound.

In the early 1960s, some authors studied antimicrobial activities in algae (Burkholder *et al.*, 1960). The authors called these compounds antibiotics. Horsney and Hide

(1974) reported, in a wide study, the production of antimicrobials by algae from the British coast and called them antibiotics. These authors used acetone for the extraction and *Staphylococcus aureus* (a Gram-positive bacterium) as a test microorganism. It is well known that Gram-positive bacteria are more sensitive than Gram-negative bacteria. The latter may include *Salmonella* and *Pseudomonas,* which are the bacteria most involved in food poisoning and food spoilage, respectively. The same authors also studied a seasonal variation among 11 algae species to check the season for maximum activity production (Horsney and Hide, 1974). According to the same authors, the algae species were classified into four groups or types: (i) species uniformly active throughout the year; (ii) species showing a peak of activity in winter; (iii) species showing a peak of activity in summer; and (iv) species showing a spring peak of activity. It is very important to know the exact period of maximum activity production in algae for an ideal extraction of the active principles by chemical methods, so the yield is higher. The extraction yield should be as high as possible for economic reasons to encourage the industrial production of these compounds.

Salvador *et al.* (2007) investigated a huge number of marine algae and microalgae species (82 taxa) from the Mediterranean and Atlantic coast of the Iberian Peninsula for their bioactivity, and found that Gram-positive bacteria were more sensitive to the algae than Gram-negative bacteria. These authors also stated that nevertheless red algae had both the highest values and the broadest spectrum of bioactivity. This conclusion seems tangible since red algae may have an antifungal activity as well as an antibacterial one (Souhaili *et al.*, 2004).

Crasta *et al.* (1997) used acetone and ethanol as solvents for the extraction of the antimicrobials from algae. The authors studied five species from the south-western coast of India by screening the extracts on five bacterial species and four fungal species for antimicrobial activities. This study showed the inhibition of *Bacillus subtilis* (a Gram-positive spore-forming bacterium) and the yeast species *Candida albicans,* while *Pseudomonas aeruginosa* and moulds (*Aspergillus* and *Fusarium*) were not inhibited by any of the extracts. Gram-positive bacteria are, in general, the most sensitive to the antimicrobials from plants. The same pattern is observed with conventional antibiotics such as penicillin. The authors could not demonstrate the inhibition against fungi, whereas yeasts were inhibited. This is most probably due to the method used to study the antimicrobial activities, as well as the sensitivity of the strains used as test microorganisms. Similar results were reported by Thillairajasekar *et al.* (2009) who used hexane and ethyl acetate for extracting the active principles from the algae *Trichodesmium erythraeum.* The authors showed antimicrobial effects on Gram-positive and Gram-negative bacteria, moulds and yeasts. According to the same authors the inhibitory effect was due to the presence of some fatty acids, namely myristic acid, palmitic acid, linoleic acid and oleic acid, in the extract.

In other studies, some authors (Tuney *et al.* 2006) used four solvents (acetone, ethanol, methanol and ether) for extracting antimicrobials in algae species collected from the coast of Usla (Izmir Turkey). These authors studied 11 species of algae and used five bacterial species (three Gram positive and two Gram negative) as test microorganisms. The authors showed that ethanol and ether were the most effective solvents for the extraction of natural antimicrobials from algae. According to the same authors, both Gram-positive and Gram-negative bacteria were inhibited, but the former were more sensitive than the latter. This finding about the sensitiveness of Gram-positive bacteria is in accordance with the data acquired in the field of natural antimicrobials and their activities on bacteria. Indeed Salvador *et al.* (2007) reported that *Bacillus cereus* (Gram positive) was the most sensitive to the extracts from algae, while *Pseudomonas* was the most resistant (Gram negative).

Latigan *et al.* (2009) reported that the broadest spectrum of the antimicrobial activity was exhibited by the aqueous fraction of some algae species. These authors studied the crude extract from 19 species of marine algae collected from the coast of South Africa for their antimicrobial effect on Gram-positive and Gram-negative bacteria. Plaza *et al.* (2010) demonstrated the antioxidant and antimicrobial activities of marine algae. The authors used pressurized liquid extract and identified some compounds such as phytol, fucosterol, neophytadiene, palmitic, palmitoleic and oleic acids using a GC/MS system.

The antimicrobal activities in marine algae follows the same pattern as the synthesized antimicrobial compounds, in such a way that Gram-positive bacteria are more inhibited than Gram-negative ones, and also that moulds are sensitive to antimicrobials from algae. These activities would depend on the method used to evaluate them. It should be mentioned here that the antimicrobial compounds in algae may differ from each other by their chemical structure and properties. Some are extractable using acetone or ethanol whereas some others are more soluble in methanol or ether, so the concentration is usually lower because of their fractionation. Together these compounds would have higher antimicrobial activities.

Extraction and fractionation were applied in studying the antimicrobials from four species of algae (Orandy *et al.*, 2004). The authors used methanol, ether, ethanol, acetone and water for extracting the antimicrobials from algae and the most-inhibitory extract was fractioned and tested on *Staphylococcus aureus* and *Candida albicans*. The authors confirmed the higher inhibition obtained by the ether extract and they also stated that acetone and methanol extracts showed no difference. Recently, acetone was shown to be the most effective solvent for extracting antimicrobials from algae (Kolanjinathan and Stella, 2009). The authors used acetone, ethanol and methanol but they did not use ether in their study. The confirmation of acetone as the most effective solvent is therefore by comparison to ethanol and methanol only. The ether extract of the active antimicrobials from algae was more effective than the other solvent but not for all the species (Tuney *et al.*, 2006).

Lima-Filho *et al.* (2002) showed the cell extracts of various algae from the north-eastern Brazilian coast had antibacterial activity against Gram-positive and Gram-negative bacteria. Similarly Chiheb *et al.* (2009) screened 32 macroalgae from the coast of Morocco for their antimicrobial activities on both Gram-positive and Gram-negative bacteria. The authors used methanol as the only solvent for the extraction by the method of Soxhlet and showed the algae had a higher inhibitory activity on Gram-positive bacteria than on Gram-negative bacteria.

It is difficult to draw conclusions about a solvent or a species since the method is crucial in studies concerning the antimicrobials from plants including in marine algae. The sensitivity of the method is relative to the experimental conditions used in the study. It is not only the solvent used for extracting the antimicrobial principles, but also the extraction parameters (time and temperature) and factors related to the plant (season and species), as well as the microbiological procedure used for evaluating the antimicrobial activity in the lab. Solid media (agar diffusion method) may lower the inhibition more than liquid media. The diffusion of the compounds into the solid medium is also a factor to take into account when studying the antimicrobials.

The species to be used as test microorganisms should be from a standard collection to avoid an acquired resistance. The strains should be sensitive and the evaluation should be compared to antibiotics or disinfectants.

Antifungal Activities

Moulds and yeasts are more resistant to antibiotics than bacteria, but they are not resistant to natural antimicrobials from plants, including algae. When studying the antimicrobial activities of natural compounds one may check first if they have antibacterial or antifungal activities. Both are interesting in the field of food preservation and safety.

When starting work on antimicrobial activities in marine algae, we noticed that there was no growth inhibition of bacteria but all the moulds and yeasts species we worked on were inhibited. We then focused our research on the antifungal activities (Souhaili *et al.*, 2004). Four solvents were used to extract the antimicrobials from the species *Cystoseira tamariscifolia* (Table 7.1). The unexpected observation that all the extracts other than the ethanol extract had not shown an effect on the microbial growth can be tentatively explained by the nature of the compounds that are not extractable by the ethanol or by the low concentration of the active principles. Bennamara *et al.* (1999) had isolated and purified an antimicrobial compounds from the species *C. tamariscifolia* called methoxybifurcarenone. This is a meroditerpenoid which may characterize the brown algae. It is known that diterpenoids are used as chemotaxonomic markers in the Cystoseiraceae.

Among the four solvents, ethanol was the most suitable for extracting antifungal compounds from the algae (Souhaili *et al.* 2004). Both yeasts and moulds were inhibited and only weak activities were observed in the methanolic extract. Some work concerning the antimicrobial activities of marine algae showed some activities on bacteria (Hellio *et al.*, 2000).

Some investigations to identify the active compounds in algae were carried out by several authors. Culioli *et al.* (2000) reported the identification of geranylgeraniol-derived diterpenes. The same authors in another work (Culioli *et al.*, 2001) demonstrated the presence of four novel diterpenes from the brown algae collected from the Moroccan Atlantic coast. Daoudi *et al.* (2001) have also identified acyclic diterpenes and sterols from the genera *Bifurcaria* and *Bifurcariopsis*. Furthermore, Bennamara *et al.* (1999) isolated a meroditerpenoid from the brown algae *C. tamariscifolia* which was identified as methoxybifurcarenone. These authors showed the antifungal activity of these compounds on *Botrytis cinerea, Fusarium oxysporum* and *Verticillium albo-atrum*. These species are not all involved in food hazards. In other investigations, authors identified sesquiterpenes from red algae: pacifenol (Sims *et al.*, 1971) and its precursor prepacifenol (Sims *et al.*, 1973). These compounds and other chemically pure structures from red algae were tested by Sims *et al.* (1975) on four bacterial species, namely *Staphylococcus aureus, Salmonella choleraesuis, Mycobacterium smegmatis* and *Escherichia coli* and a yeast species *Candida albicans*. These compounds included chondriol, cyclo-eudesmol, pre-pacifenol, laurinterol and debromolaurinterol. The authors showed that cycloeudesmol, laurinterol and debromo-laurinterol exhibited activity at effective

Table 7.1. Antimicrobial activity test of the algae extract on yeasts (Souhaili *et al.*, 2004).

	Extracts (10%)							
	Methanolic		Hexanic		Ethanolic		Water	
	T	C	T	C	T	C	T	C
Saccharomyces cerevisiae 1	±	+	+	+	–	+	+	+
Saccharomyces cerevisiae 2	±	+	+	+	–	+	+	+
Saccharomyces cerevisiae 3	±	+	+	+	–	+	+	+
Saccharomyces cerevisiae 4	±	+	+	+	–	+	+	+
Kluyveromyces	±	+	+	+	–	+	+	+
Debaryomyces	±+	+	+	+	–	+	+	+
Pichia	±	+	+	+	–	+	+	+
Rhodotorula	±	+	+	+	–	+	+	+

+, growth; ± , low growth; – , inhibition; T, Trial; C, Control.

concentrations approaching that of streptomycin. Glombitza and Pauli (2003) isolated fucols and phlorethols from the brown algae *Scytothamnus australis* which are known by their antimicrobial activities.

Mycotoxin inhibition

Almost all the work carried out on the antimicrobials from marine algae has focused on the microbial growth. It should be emphasized that mould growth in foods is not as interesting in food safety as mycotoxin formation. Growth and myctoxin formation in moulds are separate phenomena; growth is not usually accompanied by mycotoxin formation. Antifungal compounds should be studied for both growth and mycotoxin formation because the latter is more interesting for food preservation and safety.

The inhibition of mycotoxin formation in moulds by the algal compounds has not been investigated as thoroughly as has microbial growth. Mabrouk *et al.* (1985) showed the ability of five marine algae species, namely *Sargassum despiense, Turbinaria decurrense, Dilophus ligulatus, Cystoseira myrica* and *Padina pavonia,* to inhibit aflatoxin formation in *Aspergillus flavus*. A preliminary promoting effect of marine algae on mycotoxin formation in toxigenic mould species was stated by the authors, but more research is needed in the field. Aflatoxin is by far the most dangerous mycotoxin and its formation in food may occur during storage and/or handling of raw or improperly processed foods.

Souhaili *et al.* (2004) studied the effect of antimicrobials from marine algae on mycotoxin formation in *Penicillium* and *Aspergillus,* which are the most involved in food hazards. The authors demonstrated a very interesting phenomenon related to the inhibition of mycotoxin formation by the crude ethanolic extract from the brown algae *C. tamariscifolia*. The mycotoxin formation from *A. flavus* was strongly inhibited by the crude ethanolic extract as shown in Fig. 7.1, whereas a mild inhibition was observed with the methanolic extract (Fig. 7.2). The assay was duplicated and the mycotoxin spot was potently reduced compared to the two controls.

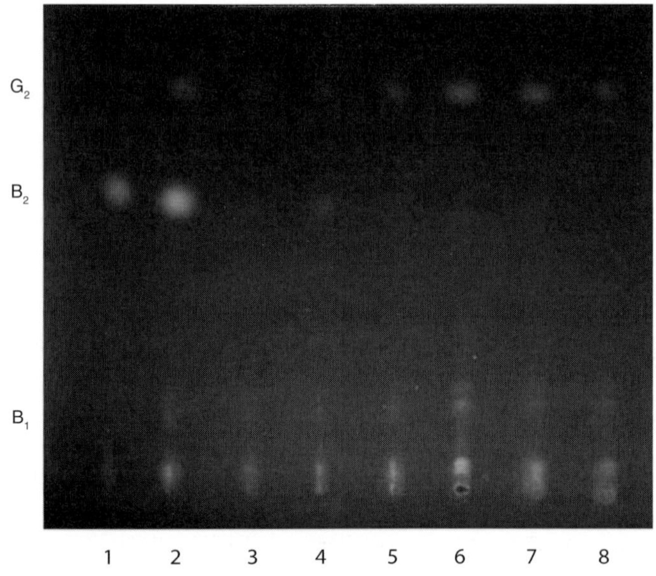

Fig. 7.1. Mycotoxins B1, B2 and G2 surveyed from cultures of *Aspergillus flavus* in a medium containing the ethanolic algae extracts (Souhaili *et al.*, 2004). 1: control; 2: control ethanol; 3: 1% extract; 4: 2.5% extract; 5: 5% extract; 6, 7 and 8 duplicates of 7.5% extract.

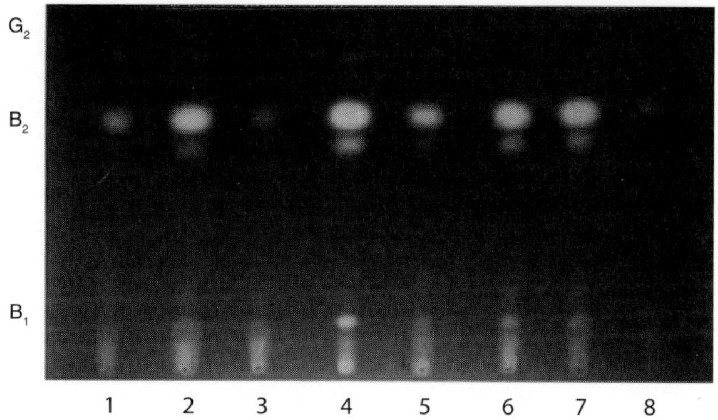

Fig. 7.2. Mycotoxins B1, B2 and G2 surveyed from cultures of *Aspergillus flavus* in a medium containing the methanolic algae extracts (Souhaili *et al.*, 2004). 1: control; 2: control methanol; 3: 1% extract; 4: 2.5% extract; 5: 5% extract; 6, 7 and 8 duplicates of 7.5% extract.

Bennamara *et al.* (1999) demonstrated the antifungal effect of meroditerpenoids from the brown algae *C. tamariscifolia* on *B. cinerea*, *F. oxysporum* and *V. albo-atrum*. The antifungal effect on these agricultural moulds led us to study the effect on species such as *Aspergillus* and *Penicillium* that are more involved in food safety and quality (Souhaili *et al.* 2004). The inhibition of mycotoxin formation in moulds could be applied in the field of food safety and food preservation. The mycotoxin formation in some dried food products is now a relevant problem for food processing. The inhibition of mycotoxin formation in foods is a securing procedure for the food industry and in food packaging.

Enzyme inhibition

Chemical deterioration of food containing fats or oils is mainly due to lipolysis and oxydation of the free unsaturated fatty acids. The inhibition of lipase by natural inhibitors is a very interesting approach for the food industry to avoid using the chemical anti-oxidants tertbuty-l4-hydroxytoluene (BHT) and BHA, which are hazardous to consumers. Bitou *et al.* (1999) demonstrated the inhibitory effect of marine algae extracts on pancreatic lipase (triacylglycerol acyl-hydrolase EC 3.1.1.3). The authors screened 54 marine algae and purified an inhibitor, caulerpenyne, from one species *Caulerpa taxifolia* which showed an inhibitory activity on pancreatic lipase. Lipases are inducing some food deteriorations in fatty materials, which may occur in two stages: an enzymatic lipolysis breakdown followed by a chemical oxidation of the free fatty acids. Lipolysis inhibition by natural compounds is preferred to chemical preservatives.

Barwell *et al.* (1989) isolated some polyphenolic compounds from brown algae (*Ascophyllum nodosum*) with a high inhibiting activity of amylase and trypsin. The same polyphenols were also isolated from *Fucus vesiculosis* and identified as phlorotannins by Koivikko *et al.* (2007) who reported detailed chemical structures of the compounds. Phlorotannins (brown algal polyphenols) are a class of natural products with potential uses in pharmacology and food preservation, such as the prevention of fat oxidation. Xiaojun *et al.* (1996) showed that phlorotannins from *Sargassum kjellmanianum* can prevent fish oil rancidity. The activity was about 2.6 times higher than that of 0.02% BHT. The enzyme inhibition activities of phlorotannins from brown algae (*Ecklonia stolonifera*) were also demonstrated by Jung *et al.* (2006) on angiotensin-converting enzyme.

In a study of structural diversity in the total pool of phlorotannins in brown algae, Cérantola *et al.* (2006) have shown that *Fucus spiralis* may produce two types of polymeric phlorotannins: fucol and fucophlorethol. These components exhibited a higher antioxidant activity than that of ascorbic acid.

Conclusion

Like terrestrial organisms, marine algae are also an important source of biologically active metabolites which may have potential applications in food preservation and also in pharmaceutical preparations. Research carried out on antimicrobials from marine algae showed the presence of bioactive compounds in these organisms. Many antimicrobial compounds including antibacterials, antifungals and antivirals were reported. A few discrepancies among the studies were logically due to some naturally occurring variants including algae species, time of harvesting (season), the region and the method used by the authors to study the antimicrobial activities. The inhibitory activities in marine algae are confirmed and well assessed for their possible applications in the field of food preservation.

References

Ballesteros, E., Martin, D. and Uriz, M.J. (1992) Biological activities of extracts from some Mediterranean macrophytes. *Botanica Marina* 35, 481–485.

Barwell, C.J., Blunden, G. and Maundhar, P.D. (1989) Isomation and characterization of brown algal polyphenols as inhibitors of alpha amylase, lipase, and trypsin. *Journal of Applied Phycology* 1, 319–323.

Bennamara, A., Abouricha, A., Berrada, M., Charrouf, M., Chaib, N., Boudouma, M. and Gameau, F.X. (1999) Methoxybifurcarinone an antifungal and antibacterial merodíterpenoid from the brown algae *Cystoseira tamariscifolia. Phytochemistry* 52, 37–40.

Bhosale, S.H., Nagle, V.I. and Jagtap, T.G. (2002) Antifouling potential of some marine organisms from Indian species of *Bacillus* and *Pseudomonas. Marine Biotechnology* 4, 111–118.

Bitou, N., Ninomiya, A., Tsujita, T. and Okuda, H. (1999) Inhibitors from marine algae *Lipids* 34, 441–445.

Burkholder, P.R., Burkholder, L.M. and Almondovar, L.R. (1960) Antibiotic activity of some marine algae of Puerto Rico. *Botanica Marina* 2, 149–156.

Cérantola, S., Breton, F., ArGall, E. and Deslandes, E. (2006) Co-occurrence and antioxidant activities of fucol and fucophlorethol classes of polymeric phenols in *Fucus spiralis. Botanica Marina* 49, 347–351.

Chiheb, I., Riadi, H., Martinez-Lopez, J., Dominguez Seglar, J.F., Gomez, Vidal, J.A., Bouziane, H. and Kadiri, M. (2009) Screening of antibacterial activity in marine green and brown macroalgae from the coast of Morocco. *African Journal of Biotechnology* 8, 1258–1262.

Crasta, P.J., Raviraja, N.S. and Sridhar, K.R. (1997) Antimicrobial activity of some marine algae of southwest coast of India. *Indian Journal of Marine Sciences* 26, 201–205.

Culioli, G., Di Guardia, S., Valls, R. and Piovetti, L. (2000) Geranylgeraniol-derived diterpenes from the brown algae *Bifurca bifurcata.* Comparison with two others cystoseiraceae species. *Biochemical Systematics and Ecology* 28, 185–187.

Culioli, G., Daoudi, M., Ortalo-Magne, A., Valls, R. and Piovetti, L. (2001) S-(12)-hydroxygeranylgeraniol-derived diterpenes from the brown algae *Bifurca bifurcata. Phytochemstry* 57, 529–535.

Daoudi, M., Bakkas, S., Culioli, G., Ortalo-Magne, A., Piovetti, L. and Guiry, M., D. (2001) Acyclic diterpenes and sterols from the genera *Bifurcaria* and *Bifurcariopsis* (*Cystoseiracea, Phaeophyceae*). *Biochemical Systematics and Ecology* 29, 973–978.

Fenical, W. and Paul, X.J. (1984) Antibiotic and cytotoxic terpenoids from tropical green algae of the family Udoteacea. *Hidrobiologica* 116/117, 137–140.

Glombitza, K.W. (1970) Antimicrobial constituents in algae. Quantitative determination of acrylic acid in sea algae. *Planta Medica* 18, 210–221.

Glombitza, K.W. and Pauli, K. (2003) Fucols and phlorethols from the brown algae *Scytothamnus australis* Hook. et Harv. *Botanica Marina* 46, 315–320.

Hellio, G.B., Bremer, A.M., Pons, G., Cottenceau, Y. and LeGal, Y. (2000) Borgougnon antibactrial and antifungal activities of extracts of marine algae from Brittany France. Use as antifouling agents. *Applied Microbiology and Biotechnology* 54, 543–549.

Henriquez, P., Zemelman, R.H., Moncada, M.A. and Benoit, I.L. (1977) Proppiedades antimicrobioticas de algas marinas. *Bol. Soc. Biolo. Concepcion* 11, 119–122.

Hodgson, L.M. (1984) Antimicrobial and anti neoplastic activity in some south Florida seaweeds. *Botanica Marina* 27, 387–390.

Horsney, I.S. and Hide, D. (1974) The production of antimicrobial compounds by British marine algae. II Seasonal variation production of antibiotics. *British Phycology Journal* 11, 61–67.

Jung, H.A., Hyun, S.K., Kim, H.R. and Choi, J.S. (2006) Angiotensin-converting enzyme. Inhibitory activity of phlorotannins from *Ecklonia stolonifera. Fisheries Science* 72, 1292–1299.

Kamat, S.Y., Wahidulla, S., D'Souza, L., Naik, C.G., Ambiye, V., Bhakuni, D.S., Goel, A.K. and Garg, H.S. (1992) Bioactivity of marine organisms.VI. Antiviral evaluation of marine algal extracts from the Indian coast. *Botanica Marina* 35, 161–164.

Koivikko, R., Loponen, J., Pihlaja, K. and Jormalainen, V. (2007) Screening of marine algae for potential tyrosinase inhibitor: Those inhibitors reduced tyrosinase activity and melanin synthesis in zebrafish. *Phytochemical Analysis* 18, 4, 326–332.

Kolanjinathan, K. and Stella, D. (2009) Antibacterial activity of marine macro algae against human pathogens. *Recent Research in Science and Technology* 1, 20–22.

Lategan, C., Kellerman, T., Afolayan, A..F., Mann, M.G., Antunes, E.M., Smith P.J., Bolton, J.J. and Beukes, D.R. (2009) Antiplasmodial and antimicrobial activities of South African marine algal extracts. *Pharmaceutical Biology* 47, 408–413.

Lima-Filho, J.V.M., Carvalho, A.F.F.U., Freitas, S.M. and Melo, V.M.M. (2002) Antibacterial activity of extracts of six macroalgae from the northeastern Brazilian coast. *Brazilian Journal of Microbiology* 33, 311–313.

Mabrouk, S.S., El-Shayeb, N.M.A., El-Refai, A.H., Sallam, I.A.R. and Hamdy, A.A. (1985) Inhibitory activities of some marine algae on Aflatoxin accumulation. *Applied Microbiology and Biotechnology* 22, 152–155.

Naqvi, S.W., A., Kamat, S.Y., Fernandes, L. and Reddy, C.V.G. (1980) Screening of some marine plants from the indian coast for biological activity. *Botanica Marina* 24, 51–55.

Orandy, M.A., Verde, M.J, Martinez-Lozano, S.J. and Waksman, N.H. (2004) Active fractions from four species of marine alage. *Phyton* (B.Aires) 73, 165–170.

Plaza, M., Santoyo, S., Jaime, L., García-Blairsy Reina, G., Herrero, M., Señoráns, M.J. and Ibáñez, E. (2010) Screening for bioactive compounds from algae. *Journal of Pharmaceutical and Biomedical Analysis.* 51, 450–455.

Salvador, N., Garreta, A.G., Lavelli, L. and Ribera, M.A. (2007) Antimicrobial activity of Iberian macroalgae. *Scientia Marina* 71, 101–113.

Sims, J.J., Fenical, W., Wing, R.M. and Radlick, P. (1971) Marine natural products I Pacifenol, a rare sesquiterpene containing bromine and chlorine from the red algae, *Lawrencia pacifica. Journal of the American Chemical Society* 93, 3774–3775.

Sims, J.J., Fenical, W., Wing, R.M. and Radlick, P. (1973) Marine natural products IV. Prepacifenol a halogenated epoxy sesquiterpene and precursor to pacifenol from the red algae, *Lawrencia pacifica. Journal of the American Chemical Society* 95, 972–974.

Sims, J.J., Donnell, M.S., Leary, J.V. and Lacy, G.H. (1975) Antimicrobial agents from marine algae. *Antimicrobial Agents and Chemotherapy* 7, 320–321.

Souhaili, Z., Lagzouli, M., Faid, M. and Fellat-Zerrouk, K. (2004) Inhibition of growth and mycotoxin formation in moulds by marine algae *Cystoseira tamariscifolia. African Journal of Biotechnology* 3, 71–75.

Souhaili, Z., Mohammadi, H., Habti, N. and Faid, M. (2008) Effet letal de l'extrait aqueux de l'algue brune marine *Cystoseira tamariscifolia* sur la souris et sur les cellules tumorales du myélome murin. *Afrique Science* 4, 580–590.

Thillairajasekar, K., Duraipandiyan, V., Perumal, P. and Ignacimuthu, S. (2010) Antimicrobial activity of *Trichodesmium erythraeum* (Ehr) (microalga) from east coast of Tamil Nadu; India. *International Journal of Integrative Biology* 5, 167–170.

Tuney, I., Cadirci, U.D. and Sukatar, A. (2006) Antimicrobial activities of the extracts of marine algae from the coast of Urla (Izmir Turkey). *Turkish Journal of Biology* 30 171–175.

Xiaojun, Y., Xiancui, L., Chengxu, Z. and Fan Xia, (1996) Prevention of fish oil rancidity by phlorotannins from *Sargassum kjellmanianum. Journal of Applied Phycology* 8, 201–203.

8 Antimicrobial Secondary Metabolites from Fungi for Food Safety

Maira Peres de Carvalho and Wolf-Rainer Abraham*

Introduction

Fungi have been involved in food processing by humans during the entire period of human existence. First, fungi were only a danger for humans because of the threat to spoil our food. However, several thousand years ago, humans learned to use fungi for food processing. Debatably, the oldest biotechnological application of fungi was in the conservation of food by fermentation. Ethanol and citric acid act as antibiotically active agents that have the additional advantages of improving taste or acting as a mild drug in the case of ethanol. The fermentation of various food ingredients became an art in many cultures, leading to unique-tasting food compounds, and, simply by experience, people discovered that many of these fermented food products were beneficial for health and could treat or even prevent certain diseases. Only two centuries ago chemistry and especially analytical chemistry reached a level of maturity that the underlying chemistry for these medicinal effects could be tackled. In the past 70 years more and more compounds produced by fungi during the fermentation processes have been identified and biological activities have been assigned to many of them. Also in the past few decades our knowledge increased considerably on the biologically active compounds from medicinal fungi. Some of these fungi are explicitly used for the treatment of diseases but many others are ingredients of food in certain regions, helping as mild antibiotics to conserve food. Due to globalization and international cuisine these fungi are now used in many widespread countries. From these medicinal fungi, a huge number of compounds have been isolated and their biological activity elucidated. Only in the past few decades has it been discovered that a multitude of compounds produced by fungi during food processing have beneficial effects on human health. Fermentation products of rice or soya consumed for many centuries in Asia recently came into the focus of scientists and physicians because it was shown that several of them could prevent modern diseases such as Alzheimer's disease, obesity, inflammatory bowel disease (especially Crohn's disease) or hypercholesterolemia. They therefore become more and more popular worldwide as prevention therapy or 'healthy' food. Food products containing these compounds are usually referred to as functional food and these compounds are also discussed here.

A significant proportion of the secondary metabolites produced by moulds

* Corresponding author.

must be regarded as mycotoxins. Some of them are severe toxins and should be avoided by any means during food production. Several food poisonings known for centuries can nowadays be attributed to mycotoxins, e.g. ergot alkaloids produced by *Claviceps*, *Aspergillus* and *Penicillium* species causing St Anthony's fire after consuming spoiled wheat or the hepatoxic aflatoxins produced by many *Aspergillus* species growing on grain or nuts. Today strict regulations in food production exclude these compounds from food in most countries. Mycotoxins are normally defined as those secondary metabolites that in small concentrations are toxic to vertebrates and other animals (Samson *et al.*, 1995). According to this definition mycotoxins are not antibiotics and hence are not discussed here. However, some secondary metabolites known for decades as mycotoxins have been recently shown to have antimicrobial activities as well. These compounds are also included in this chapter.

The antibiotic activities of fungi connected with our food can be divided into two areas: protection of our food against spoiling microbes; and the influence on our health by antimicrobial metabolites produced by fungi in our food.

Food Protection by Fungi

One of the oldest applications of fungi by humans is the processing of food by fermentation. The alcohol produced in this process also acts as antimicrobial agent against pathogens and food spoilage. Other fermentations produce large amounts of acids, protecting the food against pathogens by this shift in the pH. One well-known application of this process is the biotechnological production of citric acid by *Aspergillus niger*. In the fermentation processes of food a huge number of further metabolites are produced, both by the fermenting fungus but also by biotranformations of organic compounds. Nearly all of the compounds contribute to the taste of fermented food; however, only recently has it become clear that many of these compounds have biological activities beyond their tastes or odours.

The characteristic smell of several fungi is caused by the formation of 1-octen-3-ol (**1**) (Fig. 8.1). This compound is fungicidal against many fungi (Okull *et al.*, 2003). Furthermore, a number of volatile terpenes from fungi contribute to the specific smell of the fruiting bodies or the fermented food. Many of these terpenes do not only smell but have other biological activities as well. Several fungi produce terpenes that have antibiotic activities, e.g. linalool (**2**) and its oxides (Breheret *et al.*, 1997) and bisabolol (**3**), which are also involved in the flavour of our food, including wine. Farnesol (**4**) known from many fungi acts as an autoinducer for biofilm formation (Hornby *et al.*, 2001). Because of its wide occurrence in fungi it is found in several fermentation products, including wine. Interestingly, it is also an antibiotic against *Staphylococcus epidermidis* and it has been reported to possess comparable or even better activity than that of vancomycin or tetracycline (Gomes *et al.*, 2009).

The use of moulds on sausage surfaces can lead to both desirable and undesirable effects. The pursued effects are: the typical flavour and taste, protection against spontaneous microbial colonization, the delay of rancidity and stabilization of colour, reduced water loss, and easy skin peeling. Inoculations of sausages with moulds were traditionally done with the indigenous flora of the processing plants, the so-called 'house flora', which was mainly composed of penicillia and aspergilli (Sunesena and Stahnke, 2003). Some of the fungi most frequently isolated from fermented and cured meat products such as *Penicillium chrysogenum* and *P. nalgiovense* are known penicillin producers. *P. nalgiovense* produces penicillin (**5**) on the surface of a Spanish fermented sausage (fuet) and the presence of the antibiotic can be detected in the outer layers of the sausages. The occurrence of penicillin in food must be avoided, since it can lead to allergic reactions and the arising of penicillin resistance in human-pathogenic bacteria. From the *Penicillium* spp. growing on food products or used as starters for those

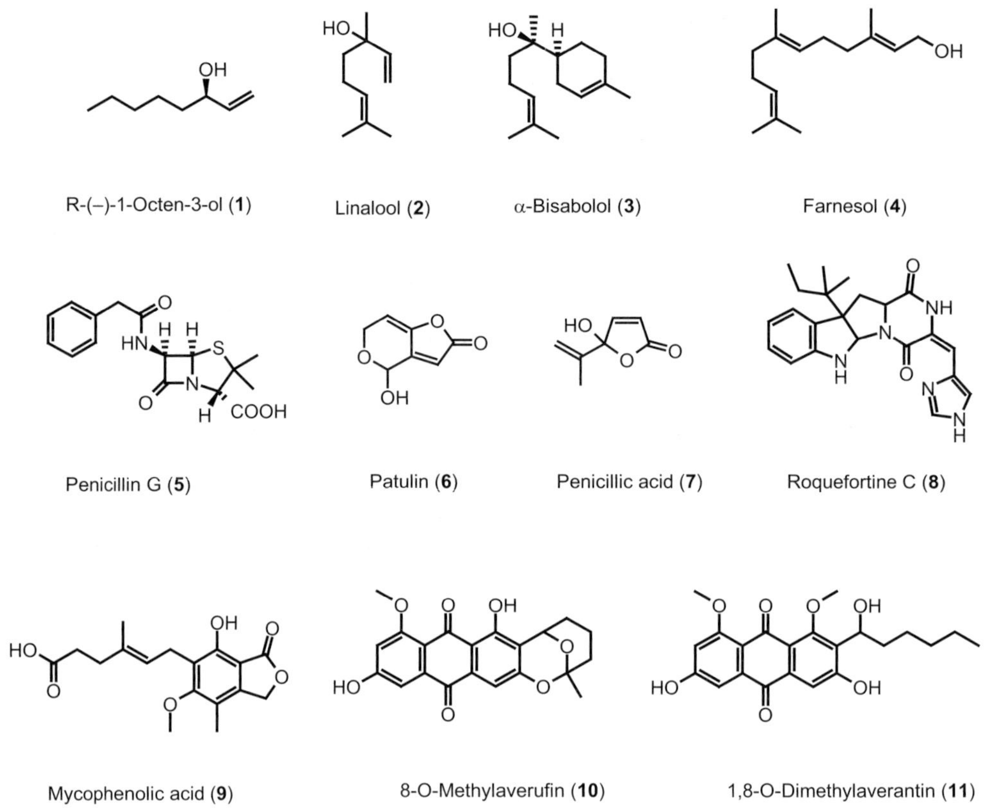

Fig. 8.1. Bioactive metabolites from fungi showing antimicrobial activity.

products, no antibacterial activity was observed after five bioassays for *P. roqueforti, P. camembertii, P. brevicompactum, P. commune, P. solitum, P. expansum, P. implicatum, P. hirsutum, P. aurantiogriseum, P. viridicatum, P. echinulatum, P. purpurogenum* and *Paecilomyces variotii*. It has been demonstrated that *P. chrysogenum, P. nalgiovense, P. griseofulvum, P. verrucosum* and *P. crustosum* had antibacterial activities. While for *P. chrysogenum, P. nalgiovense* and some *P. griseofulvum* strains this antibiotic activity could be destroyed by β-lactamase, the remaining *P. griseofulvum* strains, *P. verrucosum* and *P. crustosum* maintained their antibiotic activity indicating the presence of antibiotics lacking the β-lactam moiety (Laich *et al.*, 2002). The antibacterial activity observed with *P. verrucosum* may be due to the presence of patulin (**6**) or penicillic acid (**7**), reported to be produced by strains of this fungus (Young *et al.*, 1998). Only recently has it been shown that patulin and penicillic acid also block the communication of bacteria in biofilms, a process known as quorum-quenching (Rasmussen *et al.*, 2005).

Furthermore, in 85% of 123 isolates of *Penicillium chrysogenum*, roquefortine C (**8**) was found (El-Banna *et al.*, 1987). Roquefortine C was long regarded as only a mycotoxin but in 1979 it was demonstrated that the growth of Gram-positive organisms containing hemins was inhibited, whereas Lactobacteria and Clostridia were only impaired and the growth of Gram-negative bacteria was not affected (Knopp and Rehm, 1979).

Several species of *Penicillium*, including *P. brevicompactum, P. stoloniferum, P. scabrum, P. nagemi, P. szaferi, P. patris-mei, P. grisco-brunneum* and *P. viridicatum* were reported to produce mycophenolic acid (**9**) (Clutterbuck *et al.*, 1932). This metabolite has diverse biological properties such as antiviral, antifungal, antibacterial, anti-tumour, immunosuppressive and anti-psoriasis activities. The biosynthesis of this interesting compound has been elucidated (Muth and Nash III, 1975). The bioactive components 8-O-methylaverufin (**10**) and 1,8-O-dimethylaverantin (**11**) were isolated from the culture broth of *Penicillium chrysogenum*. Both compounds have moderate antifungal activity (Maskey *et al.*, 2003).

Fungi Involved in Functional Food

Fungi also contribute to functional food. Functional food is understood here to be potentially healthful products including any modified food or ingredient that may provide a health benefit beyond the traditional nutrients it contains. From *Penicillium camembertii,* used for the production of the famous Camembert cheese, brefeldin A (**12**) (Fig. 8.2) has been isolated (Abraham and Arfmann, 1992). Brefeldin A was first discovered from cultures of *Eupenicillium brefeldianum* as an antiviral compound but was later also detected in *Curvularia subulata, Nectria radicicola, Phoma medicaginis* and several *Penicillium* species. Later antibiotic activity and anticancer activity was shown. The finding of brefeldin A in *P. camembertii* is remarkable because this fungus is still used for cheese production. It would be quite interesting to explore the effect of a continuous uptake of brefeldin A through regular food and its outcome on

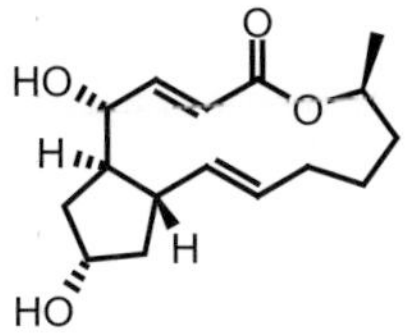

Brefeldin A (**12**)

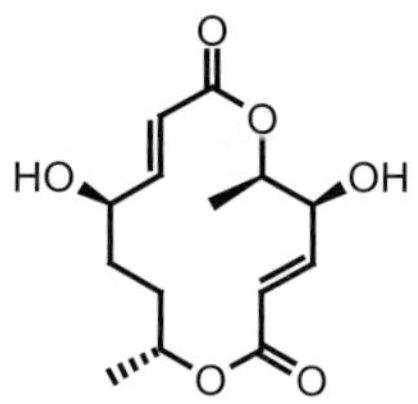

Clonostachydiol (**13**)

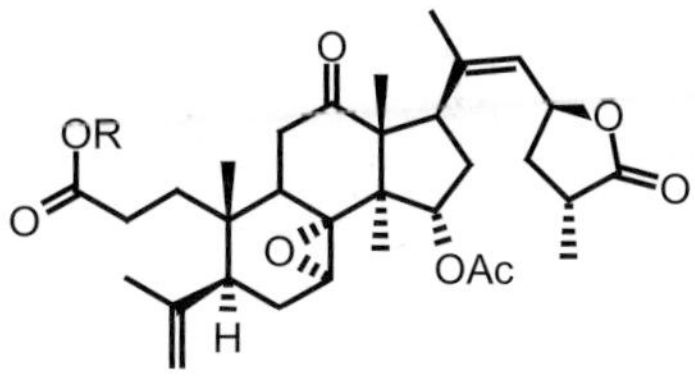

Australic acid: R = H (**14**)
Methyl australate: R = CH_3 (**15**)

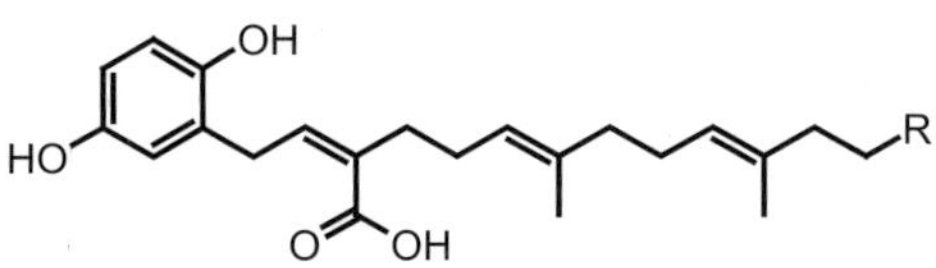

Ganomycin A: R = OH (**16**)
Ganomycin B: R = H (**17**)

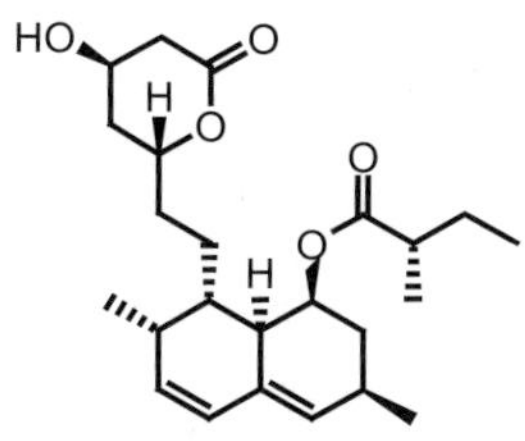

Monacoline K (**18**)

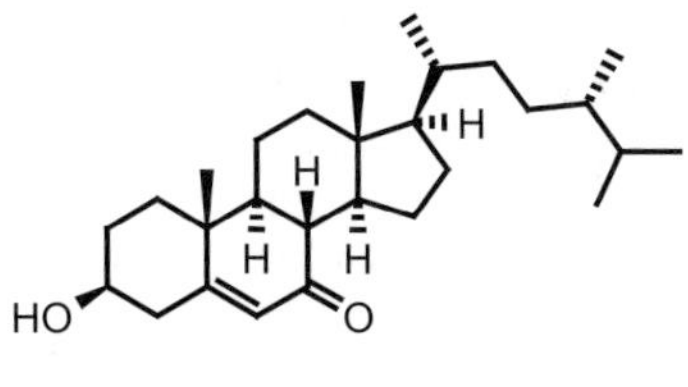

(24S)-3-Hydroxyergost-5-en-7-one (**19**)

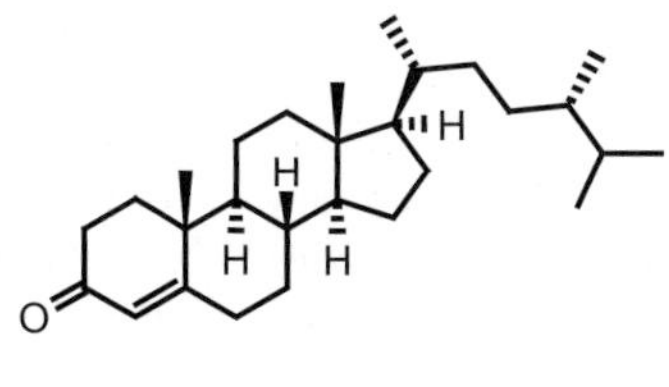

(24S)-Ergost-4-en-3-one (**20**)

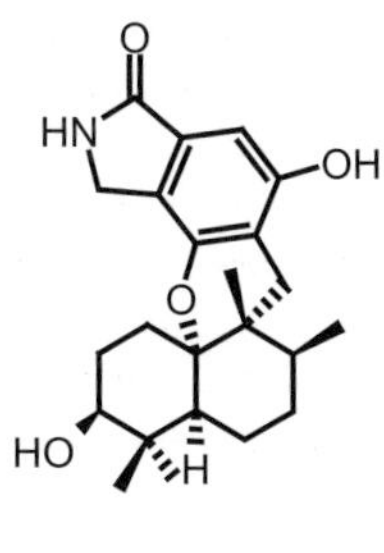

Stachyflin (**21**)

Fig. 8.2. Bioactive metabolites from fungi showing beneficial activity in our food.

virus infection and cancer diseases of the consumers. The same arguments are valid for clonostachydiol (**13**), found in *Xylaria obovata*. Clonostachydiol is a potent anthelminthic and has been tested in field studies for the control of worm infections of sheep (Abate *et al.*, 1997).

In Asia, a variety of dietary products have been used for centuries as popular medicines to prevent or treat different diseases, including fruiting bodies of mushrooms. Mushrooms such as *Ganoderma lucidum* (Reishi), *Lentinus edodes* (Shiitake), *Grifola frondosa* (Maitake), *Hericium erinaceum* (Yamabushitake) and *Inonotus obliquus* (Chaga) have been collected and consumed in China, Korea and Japan for centuries. These mushrooms contain a large variety of bioactive substances, including triterpenes, proteins, polysaccharides, lipids and phenols, displaying a wide variety of biological activities in humans. Arguably, the best known of these mushrooms is the genus *Ganoderma*, also referred as Lingzhi. The dried powder is currently used worldwide in the form of dietary supplements. It has been used for the treatment of migraine, hypertension, arthritis, bronchitis, asthma, gastritis, haemorrhoids, diabetes, hypercholesterolaemia, hepatitis and cardiovascular problems. Usually the reports are on *Ganoderma lucidum* but it remains questionable whether the taxonomy is always correct and these reports really are only on *G. lucidum* (Russell and Paterson, 2006). Antibacterial activity has been observed against Gram-positive bacteria from the basidiocarp extracts of *G. lucidum* (Kim *et al.*, 1993). Interesting is the additive effect on the activity of an aqueous extract of *G. lucidum* with four known antibiotics leading to an increase of the antibacterial activity (Yoon *et al.*, 1994). Steroidal compounds from the basidiocarps of *G. applanatum* were found to have broad spectrum activities and bactericidal effects. These are steroidal compounds (Smania *et al.*, 1999) and australate (**14**) and its methyl ester (**15**) (Smania *et al.*, 2007). Ganomycins A (**16**) and B (**17**), from *G. pfeifferi* exhibited antibacterial activity against Gram-negative and Gram-positive bacteria (Mothana *et al.*, 2000).

Secondary metabolites inhibiting the biosynthesis of cholesterol are known from a number of fungi. Their ecological function is probably that of antifungal compounds and control of the growth of fungi trying to invade the niche occupied by the producing fungi. Inhibitors of cholesterol biosynthesis are usually not applied as fungicides but to lower cholesterol levels in patients struggling with high blood pressure. Therefore, it is rather convenient to consume these compounds not as pills but in functional food. The production of red yeast rice, also known as red Koji or Hongqu, used as foodstuff was recorded in China for more than a thousand years. It is produced by fermenting the steamed rice with a *Monascus* sp., usually *Monascus purpureus*. It is now used to increase the colour and delicacy of meat fish and soybean. Clinical observations have clearly shown that that red yeast rice lowers the blood-lipid level in humans. This dietary effect is caused by monacolines, mainly monacoline K (**18**) and dehydromonacoline K, produced by the fungus (Ma *et al.*, 2000). Monacolines have also been reported from some *Penicillium* species and from the oyster mushroom *Pleurotus ostreatus* (Bobek *et al.*, 1991).

Cyttaria species grow specifically on *Nothofagus* trees so they are found only in Chile and Argentina. The fruiting bodies of these Ascomycota are traditionally consumed by Indians. The analysis of fruiting bodies of *Cyttaria johowii* revealed several ergostane derivatives (Abraham and Schmeda-Hirschmann, 1994). Among these triterpenes are the cytotoxic (24S)-3β-hydroxyergost-5-en-7-one (**19**) and (24S)-ergost-4-en-3-one (**20**), claimed in a Japanese patent to be a hair-growth promoter. The activities of these compounds are not very impressive but, as for many compounds present in food, repeated consumption even at low levels may have a beneficial effect for the consumers. From a *Stachybotrys* species, novel substances inhibiting the influenza A virus, especially the H1N1 type, have been isolated and described as stachyflin (**21**) and its acetyl derivative (Minagawa *et al.*, 2002a). They are highly active compounds: it has been reported that stachyflin (**21**) has 1760-

fold higher anti-influenza A virus activity than that of amantadine (IC_{50} = 5.3 μM) and 250-fold higher activity than that of zanamivir (IC_{50} = 0.75 μM) (Minagawa *et al.*, 2002b).

Malaria is a major public health problem, mainly due to the development of resistance by the most lethal causative parasitic species, *Plasmodium falciparum*, to important drugs such as chloroquine. New drugs with novel structures and mechanism of action are urgently required to treat drug-resistant strains of malaria. In recent years it became more and more evident that secondary metabolites known for a specific activity such as phytotoxins or mycotoxins often have more complex biological activities. The eremophilane sesquiterpenes (+)-phaseolinone (**22**) (Fig. 8.3) and (+)-phomenone (**23**) produced by a number of fungi including *Fusarium* spp. or *Xylaria* spp. are known as phytotoxins (Abate *et al.*, 1997). They also exhibit promising antimalarial activity (EC_{50} = 0.50 and 0.32 μg ml^{-1}, respectively) (Isaka *et al.*, 2000). Benzoquinone metabolite (**24**) and xylariaquinone A (**25**) from the endophytic fungus *Xylaria* sp. have been shown to possess antimalarial activity as well (IC_{50} = 1.84 and 6.68 μM, respectively) (Tansuwan *et al.*, 2007). From another fungal endophyte, *Drechslera dematioidea*, the antiplasmodial merosesquiterpene isocochlioquinone A (**26**, IC_{50} = 1.41 μg ml^{-1}) was reported (Osterhage *et al.*, 2002). The spirodihydrobenzofuran terpenes Mer-NF5003F (**27**) and (**28**) isolated from the fungus *Stachybotrys nephrospora* also displayed antimalarial activity (Sawadjoon *et al.*, 2004). Both possessed antiplasmodial activity (IC_{50} = 0.85 and 0.15 μg ml^{-1}, respectively) and were not toxic in cell lines. Many endophytes cause their host plants to produce flavonoids acting as phytoalexins but some of these flavonoids display antimalarial activities as well. The exact mechanism of antimalarial action of flavonoids is unclear but some flavonoids are shown to inhibit the influx of L-glutamine and myoinositol into infected erythrocytes. The cyclic tetrapeptide apicidin (**29**), isolated from the cultures of *Fusarium pallidoroseum* inhibits protozoal histone deacetylase (HAD) at low nanomolar concentration and is orally active against *P. berghei* in mice (Darkin-Rattray *et al.*, 1996). HAD is a key nuclear enzyme involved in transcriptional control. The continuous acetylation/deacetylation of the ε-amino group of specific histone lysine residues is required for this process, and the inhibition of histone deacetylation interferes with transcriptional control and thus cell proliferation. The 2-amino-8-oxo-decanoic moiety of apicidin presumably mimics the ε-amino acetylated lysine residues of histone substrates, resulting in potent reversible inhibition of HAD. Two cyclodepsipeptides, beauvericin (**30**) and beauvericin A (**31**), isolated from the insect pathogenic fungus *Paecilomyces tenuipes*, exhibited moderate antiplasmodial activities (EC_{50} = 1.60 and 12.0 μg ml^{-1}, respectively) (Nilanonta *et al.*, 2000). From an *Acremonium* species, KS-501a (**32**), a phenolic compound possessing a rather simple structure has been reported. It exhibited considerable antimalarial activity against *Plasmodium falciparum* (IC_{50} = 9.9 μM) while its mono- and digalactopyranosides which were also found in this fungus did not display such an activity (Bunyapaiboonsri *et al.*, 2008). These examples show that many endophytes connected to our crops produce antimalarial compounds and it can be assumed that these compounds are also in our food, turning it into functional food for some sort of antimalarial prophylaxis.

In the past few decades several attempts were made to influence the microbial community of the gut to suppress pathogens. One approach is to add living microorganisms, known as probiotics, as additives to food. While many of these probiotics failed to stand clinical validation, some of them turned out to be valuable for the control of a number of diseases otherwise difficult to control. The non-pathogenic yeast *Saccharomyces boulardii* has been prescribed in the past 30 years for prophylaxis and treatment of diarrhoeal diseases caused by bacteria. *Escherichia coli* EHEC is an important bacterial pathogen that causes gastroenteritis and colitis that in many cases is associated with serious morbidity and mortality. While *S. boulardii*

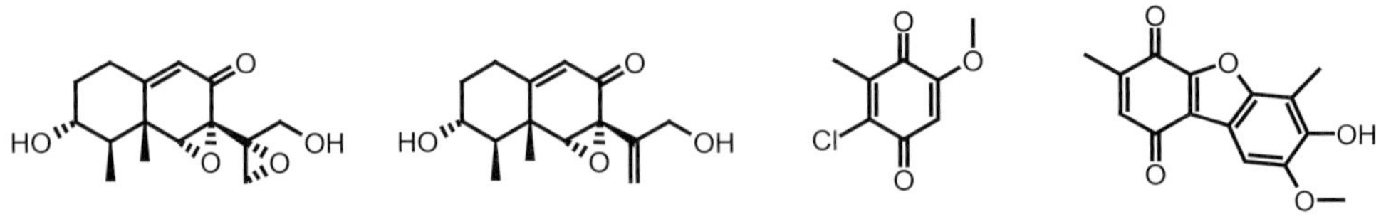

Phaseolinone (**22**) Phomenone (**23**) (**24**) Xylariaquinone A (**25**)

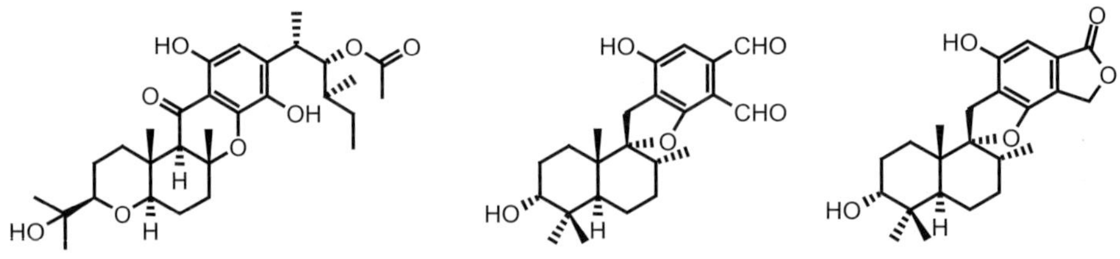

Isocochlioquinone A (**26**) MER-NF5003F (**27**) (**28**)

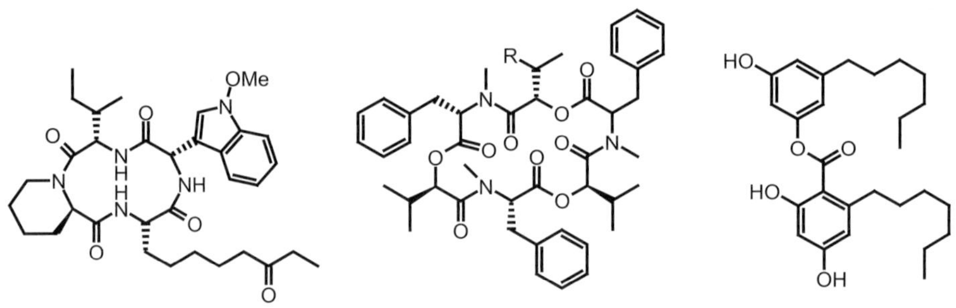

Apicidin (**29**) Beauvericin: R = Me (**30**) Beauvericin A: R = Et (**31**) KS-501a (**32**)

Fig. 8.3. Secondary metabolites from fungi possessing anti-malaria activities.

does not have a significant effect on EHEC growth or EHEC adhesion, it modifies important host signalling pathways that are activated by bacterial invasion with EHEC and reduces the severity of inflammation by EHEC. Lipopolysaccharide (LPS) is an endotoxin released from pathogenic *E. coli* strains that stimulates proinflammatory responses in the intestine and other organs. It has been demonstrated that a protein phosphatase released by *S. boulardii* is able to dephosphorylate LPS from *E. coli* O55B5 at two phosphorylation sites important for expression of its proinflammatory activity (Buts *et al.*, 2006). The most prominent causative agent of antibiotic-associated diarrhoea and colitis, *C. difficile*, mediates intestinal inflammation and mucosal damage by releasing two potent exotoxins, toxin A and toxin B. *S. boulardii* protects against *C. difficile* infections by release of a 54 kDa protease that digests both toxin A and its receptor-binding sites (Castagliuolo *et al.*, 1996). *S. boulardii* also secretes factors able to alter inflammatory diarrhoea by modifying important inflammatory signalling pathways relevant not only to *C. difficile* toxin-induced intestinal inflammation, but to other forms of gut inflammation as well. Finally, animal studies even showed that *S. boulardii* should be also advantageous to treat bowel disease (Pothoulakis, 2009). However, treating patients with *S. boulardii* should be done with high hygienic standards as an outbreak of fungemia in patients neighbouring those treated with this probiotic has been reported (Cassone *et al.*, 2003).

Conclusions

It is well known that fungi are heavily involved in our food production. Some fungi are severe plagues, such as the potato-infecting *Phytophthora infestans* which caused severe famine in Ireland in the mid-19th century, but others protect plants including our crops against infections of phytopathogenic fungi, bacteria or even viruses. Another area is the involvement of fungi in the modification and conservation of food. Arguable the best known example is here the fermentation with *Saccharomyces cerevisiae* or the preparation of cheese involving moulds such as *Penicillium camembertii.* Almost all of these fungi produce secondary metabolites, most of them are welcome because of their pleasant taste, some of them are also antimicrobial, e.g. linalool, but others are mycotoxins, e.g. aflatoxins, brefeldin A or the ergot alkaloids. Modern analytical methods detected and identified a multitude of low-molecular compounds produced during the fermentation of wine, beer, cheese or meat. Only in the past few decades did it become obvious that a number of these secondary metabolites possess antibacterial, antifungal or antiviral activities. A relatively new field of application of fungi in food production is the formation of compounds acting as drugs and promoting or protecting our health. Due to the urgent demand for potent compounds to treat malaria, several antiplasmodial metabolites have been isolated from fungi. Some of these fungi are also involved in food production and it can be speculated that their spectrum of metabolites may influence the frequency and the severity of malaria infections. A fascinating example of such a 'functionalized food' is the red mould rice produced by fermentation with *Monascus purpureus* and traditionally used in Asia. In addition to its effect of lowering cholesterol due to the secondary metabolite lovastatin it has recently been shown that red mould rice extract protects brain cells against amyloid β-peptide, a risk of Alzheimer's disease. Other functional food applications and probiotics involve antimicrobials as well. The application of the yeast *S. boulardii* against diarrhoea or inflammatory bowel disease is only one of several examples. Our growing understanding of microbial ecology and fungus–host interactions will inevitably lead to many more applications of fungi in our food in the near future.

Acknowledgements

M.P.C. acknowledges support of a PhD stipend from the German Academic Exchange Service (DAAD).

References

Abate, D., Abraham, W.-R. and Meyer, H. (1997) Cytochalasins and phytotoxins from the fungus *Xylaria obovata*. *Phytochemistry* 44, 1443–1448.

Abraham, W.-R. and Arfmann, H.-A. (1992) *Penicillium camembertii* Thom – A new source of brefeldin A. *Planta Medica* 58, 484.

Abraham, W.-R. and Schmeda-Hirschmann, G. (1994) (24S)-3β-Hydroxy-ergost-5-en-6-one from *Cyttaria johowii*. *Phytochemistry* 36, 459–461.

Bobek, P., Ginter, E., Jurcovicová, M. and Kuniak, L. (1991) Cholesterol-lowering effect of the mushroom *Pleurotus ostreatus* in hereditary hypercholesterolemic rats. *Annals of Nutrition and Metabolism* 35, 191–195.

Breheret, S., Talou, T., Rapior, S. and Bessiere, J.-M. (1997) Monoterpenes in the aromas of fresh wild mushrooms (Basidiomycetes). *Journal of Agricultural Food Chemistry* 45, 831–836.

Bunyapaiboonsri, T., Yoiprommarat, S., Khonsanit, A. and Komwijit, S. (2008) Phenolic glycosides from the filamentous fungus *Acremonium* sp. BCC 14080. *Journal of Natural Products* 71, 891–894.

Buts, J.P., Dekeyser, N., Stilmant, C., Delem, E., Smets, F. and Sokal, E. (2006) *Saccharomyces boulardii* produces in rat small intestine a novel protein phosphatase that inhibits *Escherichia coli* endotoxin by dephosphorylation. *Pediatric Research* 60, 24–29.

Cassone, M., Serra, P., Mondello, F., Girolamo, A., Scafetti, S., Pistella, E. and Venditti, M. (2003) Outbreak of *Saccharomyces cerevisiae* subtype *boulardii* fungemia in patients neighboring those treated with a probiotic preparation of the organism. *Journal of Clinical Microbiology* 41, 5340–5343.

Castagliuolo, I., LaMont, J.T., Nikulasson, S.T. and Pothoulakis, C. (1996) *Saccharomyces boulardii* protease inhibits *Clostridium difficile* toxin A effects in the rat ileum. *Infection and Immunity* 64, 5225–5232.

Clutterbuck, P.W., Percival, W., Oxford, A.E., Raistrick, H. and Smith, G. (1932) The metabolic products of the *Penicillium brevi-compactum* series. *Biochemical Journal* 26, 1441–1458.

Darkin-Rattray, S.J., Gurnett, A.M., Myers, R.W., Dulski, P.M., Crumley, T.M., Allocco, J.J., Cannova, C., Meinke, P.T., Colletti, S.L., Bednarek, M.A., Singh, S.B., Goetz, M.A., Dombrowski, A.W., Polishook, J.D. and Schmatz, D.M. (1996) Apicidin: a novel antiprotozoal agent that inhibits parasite histone deacetylase. *Proceedings of the National Academy of Sciences USA* 93, 13143–13147.

El-Banna, A.A., Pitt, J.I. and Leistner, L. (1987) Production of mycotoxins by *Penicillium* species. *Systematic and Applied Microbiology* 10, 42–46.

Gomes, F., Cerca, N., Texeira, P., Azeredo, J. and Oliveira, R. (2009) The comparative effect of farnesol and antibiotics against *Staphylococcus epidermidis.* Abstract ASM Conference Biofilms 2009, November 15–19, 2009, Cancun, Mexico.

Hornby, J.M., Jensen, E.C., Lisec, A.D., Tasto, J.J., Jahnke, B., Shoemaker, R., Dussault, P. and Nickerson, K.W. (2001) Quorum sensing in the dimorphic fungus *Candida albicans* is mediated by farnesol. *Applied and Environmental Microbiology* 67, 2982–2992.

Isaka, M., Jaturapat, A., Kladwang, W., Punya, J., Lertwerawat, Y., Tanticharoen, M. and Thebtaranonth, Y. (2000) Antiplasmodial compounds from the wood-decayed fungus *Xylaria* sp. BCC 1067. *Planta Medica* 66, 473–475.

Kim, B.K., Cho, H.Y., Kim, J.S., Kim, H.W. and Choi, E.C. (1993) Studies on constituents of higher fungi of Korea (LXVIII). Antitumor components of the cultured mycelia of *Ganoderma lucidum. Korean Journal of Pharmacology* 24, 203–212.

Knopp, B. and Rehm, H.J. (1979) Antimicrobial action of roquefortine. *Applied Microbiology and Biotechnology* 6, 397–401.

Laich, F., Fierro, F. and Martin, J.F. (2002) Production of penicillin by fungi growing on food products: Identification of a complete penicillin gene cluster in *Penicillium griseofulvum* and a truncated cluster in *Penicillium verrucosum. Applied and Environmental Microbiology* 68, 1211–1219.

Ma, J., Li, Y., Ye, Q., Li, J., Hua, Y., Ju, D., Zhang, D., Cooper, R. and Chang, M. (2000) Constituents of red yeast rice, a traditional Chinese food and medicine. *Journal of Agricultural and Food Chemistry* 48, 5220–5225.

Maskey, R.P., Grün-Wollny, I. and Laatsch, H. (2003) Isolation, structure elucidation and biological activity of 8-O-methylaverufin and 1,8-O-dimethylaverantin as new antifungal agents from *Penicillium chrysogenum. Journal of Antibiotics* (Tokyo) 56, 459–463.

Minagawa, K., Kouzuki, S. and Kamigauchi, T. (2002a) Stachyflin and acetylstachyflin, novel anti-influenza A virus substances, produced by *Stachybotrys* sp. RF-7260. II. Synthesis and preliminary structure-activity relationships of stachyflin derivatives. *Journal of Antibiotics* (Tokyo) 55, 165–171.

Minagawa, K., Kouzuki, S., Yoshimoto, J., Kawamura, Y., Tani, H., Iwata, T., Terui, Y., Nakai, H., Yagi, S., Hattori, N., Fujiwara, T. and Kamigauchi, T. (2002b) Stachyflin and acetylstachyflin, novel anti-influenza A virus substances, produced by *Stachybotrys* sp. RF-7260. I. Isolation, structure elucidation and biological activities. *Journal of Antibiotics (Tokyo)* 55, 155–164.

Mothana, R.A., Jansen, R., Julich, W.D. and Lindequist, U. (2000) Ganomycins A and B, new antimicrobial farnesyl hydroquinones from the basidiomycete *Ganoderma pfeifferi. Journal of Natural Products* 63, 416–418.

Muth, W.L. and Nash III, C.H. (1975) Biosynthesis of mycophenolic acid: Purification and characterization of *S*-adenosyl-L-methionine: Demethylmycophenolic acid *O*-methyltransferase. *Antimicrobial Agents Chemotherapy* 8, 321–327.

Nilanonta, C., Isaka, M., Kittakoop, P., Palittapongarnpim, P., Kamchonwongpaisan, S., Pittayakhajonwut, D., Tanticharoen, M. and Thebtaranonth, Y. (2000) Antimycobacterial and antiplasmodial

cyclodepsipeptides from the insect pathogenic fungus *Paecilomyces tenuipes* BCC 1614. *Planta Medica* 66, 756–758.
Okull, D.O., Beelman, R.B. and Gourama, H. (2003) Antifungal activity of 10-oxo-trans-8-decenoic acid and 1-octen-3-ol against *Penicillium expansum* in potato dextrose agar medium. *Journal of Food Protection* 66, 1503–1505.
Osterhage, C., König, G.M., Höller, U. and Wright, A.D.J. (2002) Rare sesquiterpenes from the algicolous fungus *Drechslera dematioidea. Journal of Natural Products* 65, 306–313.
Pothoulakis, C. (2009) Anti-inflammatory mechanisms of action of *Saccharomyces boulardii. Alimentary Pharmacology and Therapeutics* 30, 826–833.
Rasmussen, T.B., Skindersoe, M.E., Bjarnsholt, T., Phipps, R.K., Christensen, K.B., Jensen, P.O., Andersen, J.B., Koch, B., Larsen, T.O., Hentzer, M., Eberl, L., Hoiby, N. and Givskov, M. (2005) Identity and effects of quorum-sensing inhibitors produced by *Penicillium* species. *Microbiology* 151, 1325–1340.
Russell, R. and Paterson, M. (2006) *Ganoderma* – A therapeutic fungal biofactory. *Phytochemistry* 67, 1985–2001.
Samson, R.A., Hoekstra, E.S., Frisvad, J.C. and Filtenborg, O. (1995) *Introduction to Food-borne Fungi.* Centraalbureau voor Schimmelcultures, Baarn, The Netherlands.
Sawadjoon, S., Kittakoop, P., Isaka, M., Kirtikara, K., Madla, S. and Thebtaranonth, Y. (2004) Antiviral and antiplasmodial spirodihydrobenzofuran terpenes from the fungus *Stachybotrys nephrospora. Planta Medica* 70, 1085–1086.
Smania Jr., A., Delle Monache, F., Smania, E.F.A. and Cuneo, R.S. (1999) Antibacterial activity of steroidal compounds isolated from *Ganoderma applanatum* (Pers.) Pat. (Aphyllophoromycetideae) fruit body. *International Journal for Medicinal Mushrooms* 1, 325–330.
Smania, E.F.A., Delle Monache, F., Yunes, R.A., Paulert, R. and Smania Jr., A. (2007) Antimicrobial activity of methyl australate from *Ganoderma australe. Revista Brasileira de Farmacognosia* 17, 14–16.
Sunesena, L.O. and Stahnke, L.H. (2003) Mould starter cultures for dry sausages – selection, application and effects. *Meat Science* 65, 935–948.
Tansuwan, S., Pornpakakul, S., Roengsumran, S., Petsom, A., Muangsin, N., Sihanonta, P. and Chaichit, N.J. (2007) Antimalarial benzoquinones from an endophytic fungus, *Xylaria* sp. *Journal of Natural Products* 70, 1620–1623.
Yoon, S.Y., Eo, S.K., Kim, Y.S., Lee, C.K. and Han, S.S. (1994) Antimicrobial activity of *Ganoderma lucidum* extract alone and in combination with some antibiotics. *Archive in Pharmaceutical Research* 17, 438–442.
Young, O.S., Min, C.I., Bong, P.S., Hun, Y.S., Oh, S.Y., Chung, I.M., Paik, S.B. and Yu, S.H. (1998) Survey and control of the occurrence of mycotoxins from postharvest cereals. 1. Mycotoxins produced by *Penicillium* isolates from corn and wheat. *Korean Journal of Plant Pathology* 14, 700–704.

9 Antimicrobial Films and Coatings from Milk Proteins

Khaoula Khwaldia

Introduction

Growing demands for petroleum-based plastics, which have been of concern in terms of pollution and sustainability in combination with continuous consumer interest in high-quality food and food safety, have encouraged exploration for new bio-based packaging materials, such as edible and biodegradable films from renewable resources (Tharanathan, 2003). Edible and/or biodegradable films can be defined as a thin layer of natural biopolymers, which can be formed on a food as coating or pre-formed as a film that can be placed between food components, used as a food wrap or formed into a pouch to contain foods (Hernandez-Izquierdo and Krochta, 2008). Because edible films are considered a packaging as well as a food component, they should fulfil a number of requirements, such as good sensory qualities, high barrier and mechanical efficiencies, biochemical, physico-chemical and microbial stability, non-toxic, simple, non-polluting and low cost (Debeaufort *et al.*, 1998).

Naturally renewable biopolymers have been the focus of much research in recent years because of interest in their potential use as edible and biodegradable films and coatings for food packaging. The properties, technology, functionalities and potential uses of biopolymer films and coatings have been extensively reviewed by Kester and Fennema (1986), Gennadios *et al.* (1994), Gontard and Guilbert (1994), Krochta *et al.* (1994), Anker (1996), Guilbert *et al.* (1997), Krochta and De Mulder-Johnston (1997), and Khwaldia *et al.* (2004). These properties, such as barrier, thermal, mechanical and sensory properties, depend mainly on the nature of the components and film and their structure.

Edible films can improve shelf life and food quality by serving as selective barriers to moisture transfer, oxygen uptake, lipid oxidation, and production of volatile aromas and flavours (Kester and Fennema, 1986; Bravin *et al.*, 2006; Jagannath *et al.*, 2006). By functioning as barriers, such edible films and coatings can feasibly reduce the complexity and thus improve the recyclability of packaging materials, compared to the more traditional non-environmentally friendly packaging materials, and may be able to substitute such synthetic polymer films. In addition to enhanced barrier properties, edible films and coatings control adhesion, cohesion and durability, and improve the appearance of coated foods (Krochta, 1997).

Biopolymer-based packaging materials originate from naturally renewable resources such as polysaccharides, proteins and lipids or combinations of those components. They can serve as carriers for a wide range of beneficial food additives such as antioxidants, antimicrobials, colouring agents, flavours, fortifying nutrients and spices (Petersen *et al.*, 1999; Ozdemir and Floros, 2001; Han and Gennadios, 2005; Pranoto *et al.*, 2005; Atarés *et al.*, 2010; Siripatrawan and Harte, 2010).

Most recently, the food industry showed an increasing interest in antimicrobial edible films to enhance food safety and product shelf life. These films could prolong the shelf life and safety of foods by preventing growth of pathogenic and spoilage microorganisms as a result of their lag-phase extension and/ or their growth rate reduction (Quintavalla and Vicini, 2002).

Several studies have previously reported on the incorporation of natural antimicrobial compounds into hydrocolloid-based matrices, including milk proteins. Seydim and Saricus (2006) developed whey-protein-based antimicrobial coatings by adding oregano, rosemary and garlic essential oils. Kristo *et al.* (2008) reported antimicrobial, mechanical and moisture-barrier properties of sodium caseinate films containing potassium sorbate, sodium lactate or nisin.

The main objectives of this chapter are to review the properties of important milk-protein-based films and coatings, to discuss existing and potential applications of milk-protein films in antimicrobial food packaging, and finally to summarize the barrier, mechanical and other properties possessed by milk-protein-based packaging carrying antimicrobial agents.

Milk-Protein Packaging

Milk proteins have become excellent materials for forming edible films because of their high nutritional value and numerous functional properties that are important for film formation. Milk-protein-based edible films and coatings provide the potential to control transfer of moisture, oxygen, aroma, oil and flavour compounds in food systems. Food applications of milk-protein packaging are summarized in Table 9.1.

Several studies have been published on the use of films made from milk proteins for the preservation of fruits and vegetables during storage. Milk-protein packaging can preserve the quality of fruits and vegetables by reducing moisture and solute migration, gas exchange, respiration and oxidative reaction rates, as well as by reducing or even suppressing physiological disorders (Park, 1999). However, limited research has been conducted on the application of milk-protein packaging on muscle foods due to the expected susceptibility of proteins to proteolytic enzymes present in these foods (Gennadios *et al.*, 1997).

Milk-protein-based films have also been claimed as effective carriers of many functional ingredients, such as antimicrobial agents to improve safety and stability of foods, antioxidants to prevent lipid oxidation, and flavourings and pigments to improve quality of food. The concepts of incorporating nutraceuticals into edible coatings and films to enhance the nutritional value of foods have also been reported (Mei and Zhao, 2003).

Whey protein has received much attention for its potential use as an edible film and coating because it has been shown to make transparent films and coatings that can act as excellent oxygen, oil and aroma barriers, as well as gloss enhancers (McHugh and Krochta, 1994; Miller and Krochta, 1997; Perez-Gago and Krochta, 1999; Lee *et al.*, 2002). However, the hydrophilic nature of whey-protein coatings causes them to be less effective as moisture barriers. The oxygen permeability of whey-protein films has been reported to be very low and comparable to that of ethylene vinyl alcohol (EVOH) polymer at low or intermediate relative humidity conditions (McHugh and Krochta, 1994). In addition, their film-forming capability is favoured in more alkaline film solutions since thiol (SH) reactivity increases at $pH > 8$ (Banerjee and Chen, 1995).

Casein and caseinates can readily form edible films from aqueous solutions, are more permeable to water vapour than plastic films, and are capable of retarding water transfer to some degree (Schou *et al.*, 2005). Many authors have studied physical properties of sodium or calcium caseinate based films (Khwaldia *et al.*, 2004; Kristo *et al.*, 2007; Fabra *et al.*, 2008b, 2009; Caprioli *et*

Table 9.1. Milk protein films and coatings for food applications.

Milk protein	Application	Function	Reference
WPI[a]	Apples	Reduce respiration	Cisneros-Zevallos and Krochta (2003)
Calcium caseinate	Potato slices	Prevent oxidative browning	Le Tien *et al.* (2001)
Caseinate–AM[b]	Celery sticks	Moisture barrier	Avena-Bustillos *et al.* (1997)
Casein–stearic acid	Peeled carrots	Moisture retention	Avena-Bustillos *et al.* (1993)
WPI	Asparagus	Reduce weight loss	Tzoumaki *et al.* (2009)
Sodium caseinate	Cherries	Reduce water loss	Certel *et al.* (2004)
Calcium caseinate	Cut potatoes Cut carrots	Reduce oxidative browning Reduce moisture loss	Shon and Haque (2007)
WPC[c]–BW[d]	Cut persimmons	Reduce browning Reduce weight loss	Perez-Gago *et al.* (2005)
WPC–BW	Cut apple slices	Reduce enzymatic browning	Perez-Gago *et al.* (2006)
WPI	Strawberry pieces	Reduce rehydration	Huang *et al.* (2009)
WPI	Roasted peanuts	Reduce rancidity	Maté and Krochta (1996) Lee *et al.* (2002)
Calcium caseinate	Peanut	Reduce oil migration	Han *et al.* (2009)
WPI	Walnut	Reduce rancidity	Maté *et al.* (1996)
WPI	Breakfast cereal, raisins	Moisture barrier	Chen (1995)
WPI	Milk fat	Microencapsulation	Rosenberg and Young (1993)
WPI	Flavour (d-limonene)	Aroma barrier	Miller and Krochta (1997)
WPI	Pastry mix	Reduce fat uptake during frying	Albert and Mittal (2002)
WPI–AM	Frozen salmon	Reduce rancidity Reduce moisture loss	Stuchel and Krochta (1995)
WPI	Dried chicken dice	Reduce mechanical loss	Alcantra and Krochta (1996)
Casein–AM	Frozen fish	Reduce moisture loss	Hirasa (1991)
Potassium caseinate–rennet casein	Frozen fish fillets	Improve sensorial properties	Kilincceker *et al.* (2009)

[a]WPI, whey protein isolate.
[b]AM, acetylated monoglyceride.
[c]WPC, whey protein concentrate.
[d]BW, beeswax.

al., 2009). Calcium caseinate films have more rigid structure with better barrier properties than sodium caseinate (NaCAS) films. This could be attributed to divalent calcium cations which promote cross-linking between protein chains. Nevertheless, NaCAS films have better tensile and optical properties.

To improve functional properties of milk-protein films numerous physical, chemical or enzymatic treatments have been used. For example, the addition of plasticizers such as glycerol or sorbitol is a successful way to obtain a flexible proteinous material by weakening the hydrogen bonding. The hydrophilic nature of the plasticizer significantly affects the moisture-barrier ability of protein films. Indeed, Bodnar *et al.* (2007) reported that the water vapour permeability values of films manufactured from whey protein isolate (WPI) increased with increasing glycerol concentration. Cross-linking may be performed enzymatically or using bifunctional cross-linkers (Ghosh *et al.*, 2009).

An alternative promising strategy to improve the properties of milk-protein-based edible films and coatings is through combining milk proteins with polysaccharides such as chitosan (Pereda *et al.*, 2008; Ferreira *et al.*, 2009), methylcellulose (Erdohan and Turhan, 2005), pullulan (Gounga *et al.*, 2007), starch and alginate (Ciesla *et al.*, 2006). Gounga *et al.* (2007) reported that the addition of pullulan to a whey-protein film resulted in reduction of water vapour and oxygen permeabilities. However, increasing the amount of polysaccharide led to a decrease in these barrier properties. On the other hand, Ferreira *et al.* (2009) produced blend chitosan–whey protein films at acidic pH, carrying a high amount of protein (up to 75% mass). Although some of the film functionality might be compromised due to the incompatibility between the polysaccharide and protein components within the film matrix, the blended films, especially those with intermediated protein amount, may have useful applications on those food systems where the edible films should break up during the cooking or mastication process. Further investigation is still needed to verify the antimicrobial properties of the blend chitosan–whey protein films.

Ouattara *et al.* (2001; 2002) reported that milk-protein-based (casein and/ or whey protein) edible coatings can inhibit aerobic microbial growth by acting as barriers against oxygen transfer to the food surface. They also investigated the combined effect of this antimicrobial coating and gamma irradiation on shelf-life extension of pre-cooked shrimp. Combined use of casein–whey protein based coatings and irradiation may have a synergistic inhibitory effect on microbial growth as a result of improvements in the barrier properties of the coatings by irradiation induced intermolecular protein cross-linking. Conversely, Nortjé *et al.* (2006) found that casein–whey protein edible coatings are not effective to inhibit aerobic microbial growth on moist beef biltong. Moreover, the use of irradiation at 4 kGy together with the coating also does not cause synergistic inhibition of microbes on this very moist meat product.

Antimicrobial Milk Protein Films and Coatings

Antimicrobial packaging is a form of active packaging that could extend the shelf life of food products while providing microbial safety for consumers (Rooney, 1995). The rationale for incorporating antimicrobials into the packaging is to prevent surface growth in foods where a large portion of spoilage and contamination occurs (Appendini and Hotchkiss, 2002; LaCoste *et al.*, 2005). The direct applications of antimicrobial agents onto the surface of the food by dipping, dusting or spraying may result in rapidly lost antimicrobial activity due to inactivation of the antimicrobials by food constituents or dilution below active concentration due to migration into the bulk food matrix (Halek and Garg, 1989; Han and Floros, 1997; Ming *et al.*, 1997; Padgett *et al.*, 1998; Vermeiren *et al.*, 2002). In this sense, antimicrobial edible films and coatings may provide increased inhibitory effects against spoilage and pathogenic bacteria by maintaining effective concentrations of the active compounds on the food surfaces (Gennadios *et al.*, 1997). This approach can reduce the addition of larger quantities of antimicrobials required to achieve the target shelf life (Min and Krochta, 2005). Several compounds have been proposed for antimicrobial activity in food packaging, including organic acids such as sorbic, propionic and benzoic, or their respective acid anhydrides, fatty acid esters (glyceryl monolaurate), polypeptides (lysozyme, peroxidase, lactoferrin and nisin), plant essential oils (cinnamon, oregano and lemongrass), nitrites and sulfites, among others (Franssen and Krochta, 2003).

The choice of antimicrobial components is related to the compatibility of the component with the packaging material and its heat stability. Thus it is important to choose a proper coating matrix, antimicrobial agents and plasticizers. The ability of edible films and coatings to serve as carriers for incorporating antimicrobials is mainly related to their good film-forming properties, high-retention ability and release ability.

The use of milk-protein-based films and coatings to carry antimicrobial agents to a variety of microbial media or food surfaces has been the aim of many studies. They have the ability to carry and release antimicrobials such as potassium sorbate (K sorbate), natamycin (Franssen *et al.*, 2004), *p*-aminobenzoic acid (PABA), sorbic acid (SA) (Cagri *et al.*, 2001), sodium lactate (Na lactate), ε-polylysine (ε-PL) (Zinoviadou *et al.*, 2010), nisin (Kristo *et al.*, 2008), lysozyme (Han, 2000; Min *et al.*, 2008), lactoperoxidase system (LPOS) (Min *et al.*, 2005), and essential oils (Oussallah *et al.*, 2004; Zinoviadou *et al.*, 2009). Table 9.2 reviews some typical applications of milk proteins in antimicrobial food packaging.

Diffusion of antimicrobials through an edible film is influenced by the properties of the polymer film (nature, composition and manufacturing procedure), food (pH and water activity) and storage conditions (temperature, relative humidity and duration) (Cagri *et al.*, 2004). To assess the ability of a polymer film to act as an antimicrobial carrier, diffusion coefficients for the selected antimicrobials should be determined. In this sense, Guilbert (1988) investigated sorbic acid (SA) retention by casein films treated with lactic acid, and placed over an aqueous model food system with a water activity (a_w) of 0.95. Casein films retained 30% of the original SA after 35 days at 25°C. They also conducted microbiological tests to determine the effectiveness of SA incorporated in casein films and showed improved microbial stability. Ozdemir and Floros (2001) produced antimicrobial films from commercial WPI and investigated the release mechanism of potassium sorbate from the films. They found that Non-Fickian diffusion of potassium sorbate was the predominant mechanism of release from sorbitol-plasticized whey-protein films. Potassium sorbate diffusion coefficients in the films ranged from 5.38 to 9.76 $\times 10^{-11}$ m^2/s and are similar to those of WPI films plasticized with glycerol (Franssen *et al.*, 2004). These values were approximately one order of magnitude smaller than those determined in intermediate moisture food systems. This indicates that whey protein films carrying potassium sorbate can be used on food surfaces as active edible preservative releasing systems. Compared with sorbate diffusion in WPI films, natamycin diffusion was found to be slower, probably because of its larger size and bulky shape (Franssen *et al.*, 2004). In another study, Ozdemir and Floros (2003) studied the effect of film composition on potassium sorbate diffusion in whey-protein films using mixture response surface methodology. Increasing the relative amounts of protein and beeswax in the films decreased potassium sorbate diffusivity, while increasing the relative amounts of plasticizer and initial potassium sorbate in the films increased the diffusion of potassium sorbate. Further research is needed to gain more knowledge regarding the effect of film formulation on antimicrobial diffusion to evaluate the performance of edible films and coatings as antimicrobial carriers and to optimize the controlled release of antimicrobials for specific food applications.

Weak organic acids, which are the most common classical preservative agents, inhibit the outgrowth of both bacterial and fungal cells. The inhibitory action of organic acids is believed to be due to the fact that protonated acids are membrane soluble, and can enter the cytoplasm by simple diffusion (Ricke, 2003). Cagri *et al.* (2001) developed low pH (5.2) WPI-based edible films containing 0.5 to 1.5% p-aminobenzoic acid (PABA) or SA and tested their growth inhibition effects against *Listeria monocytogenes*, *Escherichia coli* O157:H7, and *Salmonella typhimurium* DT104 in a disc-diffusion assay. Average inhibition zone diameters were 21.8, 14.6, 13.9 using 1.5% PABA, and 26.7, 10.5, 9.7 mm using 1.5% SA, for *L. monocytogenes*, *E. coli* O157:H7 and *S. typhimurium* DT104, respectively. Three strains of *S. typhimurium* DT104 were resistant to 0.5% SA. These films, which

Table 9.2. Applications of milk proteins in antimicrobial food packaging.

Protein	Antimicrobial agents	Medium/food	Microorganisms	References
Casein	Sorbic acid	Papaya cubes	*Staphylococcus rouxii, Aspergillus niger*	Guilbert (1988)
WPI[a]	Potassium sorbate	Water-glycerol	*Saccharomyces cerevisiae, A. niger, Penicillium roqueforti*	Ozdemir (1999)
WPI	Nisin	Phosphate buffer	*Listeria monocytogenes*	Ko *et al.* (2001)
WPI	Sodium lactate ε-polylysine	Fresh beef cut portions	Total flora, pseudomonas Lactic acid bacteria	Zinoviadou *et al.* (2010)
NaCAS[b]	Nisin, potassium sorbate	Model solid food	*L. monocytogenes*	Kristo *et al.* (2008)
WPI	Propionic/sorbic acid	Frankfurters	*L. monocytogenes*	McDade *et al.* (1999)
WPI	Nisin with EDTA, lysozyme with EDTA	Culture medium	*Brochothrix thermosphacta, Staphylococcus aureus, Salmonella typhimurium, L. monocytogenes, Escherichia coli*	Rodrigues and Han (2000)
WPI	Lysozyme	Culture medium	*Bochothrix thermosphacta*	Han (2000)
WPI	*p*-aminobenzoic acid, sorbic acid	Culture medium, sliced bologna and summer sausage	*L. monocytogenes, E. coli, S. typhimurium*	Cagri *et al.* (2001)
WPI	Malic acid/natamycin Malic acid/ nisin/ natamycin	Culture medium	*Yarrowia lipolytica, Penicillium* spp.	Pintado *et al.* (2010)
WPI	Lactoperoxidase system	Culture medium	*Salmonella enterica, E. coli*	Min *et al.* (2005)
WPI	Oregano essential oil	Culture medium	*S. aureus, Salmonella enteritidis, L. monocytogenes, E.coli, Lactobacillus plantarum*	Seydim and Saricus (2006)
Calcium caseinate–WPI–carboxymethyl cellulose	Oregano essential oil	Beef muscle pieces	*E. coli, Pseudomonas* spp.	Oussallah *et al.* (2004)
WPI	Oregano essential oil	Beef cuts	Total flora, pseudomonads, lactic acid bacteria	Zinoviadou *et al.* (2009)

[a]WPI, whey protein isolate.
[b]NaCAS, sodium caseinate.

retained their antimicrobial activity for 21 days, also showed considerable promise in extending the shelf life of sliced bologna and summer sausage. Indeed, WPI films containing SA or PABA decreased *Listeria, E. coli* O157:H7 and *S. typhimurium* DT104 populations by 3.4 to 4.1, 3.1 to 3.6, and 3.1 to 4.1 logs, respectively, on both products after 21 days of aerobic storage at 4°C (Cagri *et al.*, 2002). Background flora was also inhibited with WPI-based antimicrobial films compared with antimicrobial-free control films. Moreover, the authors claimed that SA and PABA were more effective in inhibiting the three test pathogens when used together rather than alone. On the other hand, McDade and others (1999) reported that growth of *L. monocytogenes* was inhibited on frankfurters during 2 to 3 weeks of storage at 4°C by coating the frankfurters with a whey-protein-film-forming solution that contained propionic acid/SA (pH 5.2). However, non-uniformity of the antimicrobial coating on frankfurters after dipping, draining and drying would probably produce a less effective antimicrobial barrier than using pre-casted films.

While a considerable amount of research has been conducted on the use of various antimicrobial films for controlling meat pathogens, there are limited studies on their effect on the spoilage microflora of these products. In this sense, Zinoviadou *et al.* (2010) evaluated the ability of WPI films containing Na lactate or ε-PL to control beef's spoilage flora on fresh beef cuts during storage at 5°C. According to these authors, the use of films made from 2% w/w Na lactate in film-forming solutions resulted in a significant inhibition of growth of the total flora and pseudomonas population during the entire storage. On the other hand, the addition of ε-polylysine (ε-PL) at a concentration of 0.75% w/w induced complete inhibition of the Gram-positive lactic acid bacteria (LAB) and a significant increase of the lag phase by a factor of 2 for the total viable count and pseudomonads population, whereas sodium lactate did not seem to suppress the growth of LAB.

Ozdemir (1999) determined the antimicrobial activity of the WPI films containing potassium sorbate against *Saccharomyces cerevisiae, Aspergillus niger* and *Penicillium roqueforti*. Active WPI films were shown to suppress the growth of yeasts and moulds by extending the lag period before the growth became apparent. Potassium-sorbate-incorporated whey-protein films help provide improved food safety and stability, thereby increasing shelf life. On the other hand, Kristo *et al.* (2008) found no additional effect on *L. monocytogenes* growth in NaCAS-containing potassium sorbate films when the concentration of the antimicrobial exceeded 10% (w/w, film dry basis). However, the incorporation of nisin (0.05 and 0.5 mg nisin per film) in NaCAS films substantially reduced the levels of *L. monocytogenes* on Tryptic Soy Agar (TSA) NaCl medium for 10 days at 10°C, regardless of the nisin concentration. The nisin-containing NaCAS films were more effective in reducing growth of *L. monocytogenes*, followed by potassium-sorbate-impregnated NaCAS films, whereas films containing sodium lactate were only slightly effective and only at the higher concentration (40% w/w film dry bases) employed. This may be attributed to the high retention of nisin in NaCAS films. Being a bulkier molecule than the two salts of organic acids, nisin may be released more slowly from the NaCAS matrix and also slowly diffuse through the agar medium, maintaining thus, an effective concentration on the agar surface. These findings are in accordance with previous investigations on the antimicrobial effectiveness of nisin as a component of edible films, particularly against pathogens like *L. monocytogenes* (Padgett *et al.*, 1998; Ko *et al.*, 2001; Lungu and Johnson, 2005). Nisin, a hydrophobic and cationic polypeptide, is the most widely studied and commercially available bacteriocin. It remains the only natural antimicrobial peptide approved by the FDA for use as a food preservative. The mode of action proposed for nisin suggests that the molecule interacts with the phospholipids in the cytoplasmic membrane of bacteria, thus disrupting membrane function and preventing outgrowth of spores by inhibiting the swelling process of germination (Abee *et al.*, 1995).

Ko *et al.* (2001) also studied the effect of nisin added into several films (WPI, soy protein isolate, egg albumin and wheat gluten) at varying pH values (2.0 to 8.0) on the reduction of *L. monocytogenes* counts. All tested films exhibited greatest inhibitory effects against *L. monocytogenes* under an acidic environment. Furthermore, WPI films containing nisin were the most effective in reducing *L. monocytogenes* counts, whereas wheat gluten films with nisin showed the lowest antimicrobial activity against *L. monocytogenes*. These results are due to nisin being more active against *L. monocytogenes* at acidic and hydrophobic conditions (Klaenhammer, 1993). Indeed, the WPI film had the highest hydrophobicity because it is water insoluble due to the formation of covalent disulfide bonds. Pintado *et al.* (2006), however, showed that WPI (7.0% w/v)-based films with glycerol (3.0% w/v) as plasticizer, malic acid (3.0% w/v) as antimicrobial and acidifying agent, and nisin (50 IU/ml) as antimicrobial agent are effective to control food-borne pathogenic *L. monocytogenes* strains isolated from cheese samples when tested in a disc-diffusion assay. In another study, Pintado *et al.* (2010) found that the introduction of the antimycotic agent natamycin in the WPI–glycerol–malic acid films inhibited *Yarrowia lipolytica* and *Penicillium* spp. strains. Moreover, the inhibitory activity of nisin and malic acid against *L. monocytogenes* was improved with the addition of sucrose esters, Tween80 or EDTA. Welscher *et al.* (2008) revealed that natamycin blocks fungal growth by binding specifically to ergosterol, present almost exclusively in the fungi plasma membranes and not present in bacteria.

Rodrigues and Han (2000) investigated the growth inhibition effects of antimicrobial WPI films against *L. monocytogenes, E. coli* O157:H7, *Brochothrix thermosphacta, Staphylococcus aureus* and *S. typhimurium*. Films carrying lysozyme or nisin inhibited *B. thermosphacta* and *S. aureus*. When incorporated into WPI films, EDTA effectively inhibited *L. monocytogenes* and *S. typhimurium*. Likewise, Han (2000) demonstrated the effectiveness of lysozyme incorporated into WPI films against *B. thermosphacta*. The lysozyme, slowly released from the film, effectively inhibited the growth of the microorganism. This film may have great potential as a microbial hurdle against Gram-positive spoilage and pathogenic bacteria. Furthermore, Mendes de Souza *et al.* (2010) reported on the effective incorporation of lysozyme to pH- or glyoxal-modified caseinate films. The release of lysozyme from NaCAS films has been observed to be decelerated after cross-linking with glyoxal or immersion in solutions at pH ~4.6 for 120 min.

Min *et al.* (2005) reported the antimicrobial effect of whey-protein edible films incorporating the LPOS against *Salmonella enterica* and *E. coli* O157:H7. WPI films incorporating 0.15 g LPOS/g film (dry basis) completely inhibited both pathogens (4 log CFU/cm^2), whether inoculation was on an agar medium before placement of the film or the inoculation was on the film itself. Thus, LPOS–WPI films have potential for inhibiting those microorganisms already present on food products, as well as for the control of their growth from contamination after wrapping or coating of the food products.

Essential oils and their components, which are naturally occurring antimicrobial agents, are well known for their potency against pathogenic and spoilage microorganisms (Benkeblia, 2004). The hydrophobicity of essential oils is an important characteristic, which makes them able to pass through cell membranes and enter mitochondria, disturbing the internal structures and rendering the membranes more permeable (Burt, 2004). Seydim and Saricus (2006) demonstrated efficacy of some essential oils in WPI films based on the zone of inhibition assay. Oregano and garlic extracts added to films exhibited larger inhibitory zones on *S. aureus, S. enteritidis, L. monocytogenes, E. coli* and *L. plantarum* as compared to films incorporated with rosemary essential oil. Notably, rosemary essential oil did not maintain its known antimicrobial activity in WPI-based edible films. These researchers suggested that the different inhibitory effects of essential oils may be attributed to the differences in the biological properties of the main compounds in the essential oils. Although these films showed their efficacy *in vitro*, additional studies are required to test them in real food systems.

Oussallah *et al.* (2004) studied the effectiveness of milk-protein-based edible films containing 1% essential oils of oregano, pimento or an oregano–pimento mixture in a ratio 1:1 (w/w) against *Pseudomonas* spp. and *E. coli* O157:H7 on the surface of beef muscle pieces. The incorporation of essential oils into the cross-linked coating film formulation significantly decreased the microorganism level in meat samples. They also found that the most effective films against the growth of both pathogenic bacteria are those containing oregano extracts. The antimicrobial potential of oregano essential oils is related to their high phenolic contents, particularly carvacrol and thymol (Sivropoulou *et al.*, 1996). These results support the hypothesis that the inhibition of pathogenic bacteria growth by essential oils depends on the nature of the phenolic compounds and are in agreement with those of Zinoviadou *et al.* (2009), who demonstrated the effectiveness of oregano-oil-containing whey-protein films to increase the shelf life of fresh beef. According to these authors, the growth of lactic acid bacteria was completely inhibited, while the maximum specific growth rate of total flora and pseudomonads were significantly reduced by a factor of two with the use of antimicrobial films (1.5% w/w oil in the film-forming solution).

In contrast to the large amount of information on the effectiveness of films and coatings containing essential oils against a wide spectrum of microorganisms, little is known about their possible impact on organoleptic food properties. Moreover, the variability of the composition of essential oils and their variable activity in foods due to interactions with food components can also limit their use in food products. Thus, it is recommended to study the influence of the incorporation of antimicrobial compounds into edible films and coatings on sensory characteristics of food products and to better understand the interactions between the product matrix and antimicrobials.

Despite the good results achieved so far with the incorporation of essential oils into edible films and coatings, the high drying temperatures usually employed to form edible films and coatings may result in a loss of a high percentage of the aromatic components, negatively affecting food quality. Spice oleoresins, which contain volatile as well as non-volatile components, are convenient substitutes for essential oils in the food-processing industry. Ponce *et al.* (2008) reported that edible films prepared from NaCAS with olive, rosemary, onion, capsicum, cranberry, garlic and oreganum oleoresins at 1% concentrations had no inhibitory activity against either the native microflora of butternut or against *L. monocytogenes*. Similar behaviour was observed for carboxymethylcellulose- and chitosan-film-forming solutions containing 1% of different oleoresins. Indeed, chemical interactions between amino groups in edible films and carboxyl groups in oleoresins could block the active antibacterial sites (Pranoto *et al.*, 2005). Moreover, no antimicrobial activity was detected when these films enriched with oleoresins were applied to butternut slices in *in vivo* assays.

Properties of Antimicrobial Milk Protein Films

When antimicrobial agents are added to edible films and coatings, mechanical, sensory and even functional properties can be dramatically affected. However, there is a lack of available information about the possible impact of the antimicrobial agents on the mechanical and physical properties of the films. Thus, studies on the interactions between antimicrobials and film-forming materials are necessary since the overall performance of the films, which depends strongly on their physicochemical properties, must be maintained when antimicrobial agents are added. The influence of an antimicrobial may depend on their chemical structure, concentration, degree of dispersion in the film and degree of interaction with the polymer (Kester and Fennema, 1986). Combined analyses of antimicrobial, tensile and physical properties are crucial for predicting the behaviour of antimicrobial edible films.

Barrier properties

Ozdemir and Floros (2008a) investigated the effect of protein, sorbitol, beeswax and potassium sorbate concentrations in whey-protein films on their water vapour permeability (WVP) and water solubility properties using mixture response surface methods. Potassium sorbate significantly increased WVP and water solubility. The mixture proportions of protein = 0.53, sorbitol = 0.38, beeswax = 0.08 and potassium sorbate = 0.01 would yield an edible film with good WVP, water resistance and sensory properties. According to Cagri *et al.* (2001), addition of SA and PABA to the WPI film-forming solutions increased WVP. Indeed, both antimicrobials are hydrophilic compounds and addition of polar additives may increase the hydrophilic character and the solubility coefficient of the film (McHugh and Krochta, 1994). Moreover, the main effect of additives such as SA or PABA may be based on disrupting intermolecular protein–protein hydrogen interactions in protein films or coatings, hence reducing the protein packing density and increasing water mobility (Guilbert, 1986). In agreement with these results, Ozdemir (1999) reported an increase of WVP of whey-protein films with increasing potassium sorbate concentration. Similarly, Kristo *et al.* (2008) pointed out that the addition of sodium lactate and potassium sorbate significantly increased film WVP, with sodium lactate inducing a greater increase than potassium sorbate at similar concentrations. These results demonstrated that both antimicrobials may function as plasticizers for NaCAS films. In contrast, the addition of nisin did not cause significant changes in the WVP of NaCAS films.

Zinoviadou *et al.* (2010) demonstrated that addition of ε-PL into the WPI film matrix did not alter the WVP and the water sorption properties of the films. Being a cationic peptide, ε-PL increases the pH of the film-forming solutions, which increases the SH reactivity, promotes a better protein network structure and thereby a lower WVP. On the contrary, addition of sodium lactate assisted a higher moisture uptake that resulted in a higher WVP. On the other hand, Min *et al.* (2005) reported that the oxygen barrier property of WPI films was improved with the incorporation of LPOS at 0.15 to 0.25 g/g (dry basis). This may be explained by the formation of protein aggregates in LPOS–WPI films due to presence of gluconolactone. Aggregated protein domains would be more strongly associated than the amorphous protein domains of the film, providing greater resistance to the diffusion of oxygen.

The WVP of antimicrobial whey-protein films containing malic acid, nisin and natamycin was not affected by the incorporation of sucrose esters (S970, SP30 or SP50), EDTA or Tween80 (Pintado *et al.*, 2010). However, the increase of plasticizer concentration resulted in a linear relation between the amount of plasticizer and WVP. Several authors reported the same effect when hygroscopic films were studied (Irissin-Mangata *et al.*, 2001; Vanin *et al.*, 2005).

Theoretically, addition of nisin would be expected to improve the water barrier property of films by hydrophobic interaction with protein constituents because nisin itself is a hydrophobic protein (Klaenhammer, 1993). However, nisin at the amounts added (4.0– 160 IU/film disc) had no significant effect on the WVP values of WPI films (Ko *et al.*, 2001). Likewise, Zinoviadou *et al.* (2009) pointed out that the WVP of WPI films was not affected by the addition of oregano oil at any of the concentrations used (0.5%, 1% and 1.5% w/w). In contrast, other studies have indicated that the incorporation of fats or lipids into edible film formulations can enhance the efficiency of water vapour barrier properties (Perez-Gago and Krochta, 1999, Fabra *et al.*, 2008a).

Mechanical properties

Mechanical properties of edible films are important to ensure that the film has adequate mechanical strength and integrity during transportation, handling and storage of foods coated with edible films. Tensile strength (TS), elongation (E) and Young's modulus (YM) are the most commonly reported responses to describe mechanical properties of edible films and coatings. TS is a measure of the ability of a film to resist breaking under tension. E shows the ability of a film to stretch before it breaks. YM is the ratio of stress to strain over the linear part of stress–strain curve, and it is a measure of film stiffness (Banker *et al.*, 1966).

Chen (1995) and Hotchkiss (1995) pointed out that incorporation of additives such as antimicrobial agents into edible film formulations could affect the mechanical and optical properties of the film. On the other hand, Han (2000) showed that the incorporation of lysozyme in edible WPI films produced clear films and maintained the TS at concentrations of up to 100 mg of lysozyme/g of dried film. Likewise, the mechanical properties of caseinate films were not remarkably modified after the addition of lysozyme at 1% w/w (Mendes de Souza *et al.*, 2010). However, Park *et al.* (2004) found that lysozyme has a low film-forming capacity and usually weakens film structure and integrity.

Rodriguez and Han (2000) reported that the physical properties of WPI films were not affected by adding nisin, EDTA or propyl-*p*-benzoic acid. In contrast, Kristo *et al.* (2008) pointed out that increasing the amount of nisin led to a decrease in TS and to an increase in E of the resulting NaCAS-based antimicrobial films at water contents higher than 8% (w/w).

Ozdemir and Floros (2008b) investigated the effect of protein, sorbitol, beeswax and potassium sorbate concentrations in whey-protein films on their ultimate TS, YM, E and transparency using mixture response surface methods. Potassium sorbate adversely affected TS and YM, and showed significant interactions with some other components in the mixture. The optimum mixture proportions of protein = 0.58, sorbitol = 0.38, beeswax = 0 and potassium sorbate = 0.04 would be recommended to obtain mechanically good and transparent films. According to Cagri *et al.* (2001), increasing PABA and SA concentrations led to a decrease in TS and to an increase in E of WPI films. The E of antimicrobial-free WPI films was about 11 times and 6 times lower than that of antimicrobial films containing 1.5% SA and PABA, respectively. Moreover, films containing SA exhibited lower TS and higher E as compared to films containing PABA. This may be attributed to the straight chain of SA, leading to easier penetration into WPI chains, higher mobility between protein chains, and greater flexibility. These results suggested that SA and PABA antimicrobials acted as plasticizers for the WPI films. These results are in agreement with those of Kristo *et al.* (2008), who showed that an increase in sodium lactate and potassium sorbate concentration resulted in reduction of Young modulus and maximum TS and increasing of E suggesting that both antimicrobials acted as plasticizers for the NaCAS films. Similar results have been also reported for WPI–based edible films containing 1.0% and 1.5% Na lactate (w/w) (Zinoviadou *et al.*, 2010). The same authors have pointed out that addition of ε-PL into the WPI films at concentrations of 0.50% and 0.75% (w/w) in the film-forming solution resulted in significantly smaller TS and improved extensibility compared to the antimicrobial-free control film. Indeed, increasing the amount of additives other than cross-linking agents generally produced films with lower TS and greater E, since these molecules insert between protein chains to form hydrogen bonds with amide groups of proteins (Kester and Fennema, 1986). Reduced interactions between these protein chains lead to increased flexibility and movement.

Cagri *et al.* (2002) found that the TS of antimicrobial films containing PABA and/or SA decreased while E remained unchanged following 72 h of contact with bologna and summer sausage slices at 4°C. Moisture absorption by the antimicrobial films in contact with meat slices probably led to decreased TS and increased flexibility. On the other hand, Min *et al.* (2005) pointed out that the tensile properties of WPI films were affected by the incorporation of the LPOS. Incorporation of 0.25 g/g of LPOS decreased YM, TS and E. This may be related to changes in the WPI film structure induced by gluconolactone, which promotes protein aggregation. Discontinuities between aggregated protein domains and amorphous protein domains of the film could result in reduced tensile properties.

According to Zinoviadou *et al.* (2009), increasing the amount of oregano oil in the sorbitol-plasticized WPI films led to a decrease in TS and YM. However, addition of oregano oil at a concentration up to 1% in the film-forming solution resulted in an increase in E properties. This effect can be attributed to the development of discontinuities in the polymer network induced by lipid addition. Indeed, oregano oil, which is liquid at room temperature, forms oil droplets in the film, enhancing its stretching ability and extensibility. This coincides with the results reported by Rojas-Graü *et al.* (2007) when adding plant essential oils and oil compounds to an alginate–apple puree matrix and is in agreement with the effect of the structural discontinuities provoked by the incorporation of the oil on the mechanical behaviour.

Ko *et al.* (2001) showed that nisin addition increased the TS of WPI films, whereas the TS of soy protein isolate films were not affected by nisin addition. This result may be due to the lower hydrophobicity values of soy protein isolate films compared to that of WPI films. Being a hydrophobic protein, nisin incorporation into WPI-film-forming solutions may have caused rearrangement of disulfide and hydrophobic bonds or more interactions between nisin molecules and protein molecules forming the film network.

Conclusions

Milk-protein-based edible films and coatings can act as suitable carriers for delivering effective antimicrobials. More studies are necessary to gain more knowledge regarding the interactions between the film matrix, antimicrobial compounds and target microorganisms to evaluate the materials' performance and to optimize the compositions of active packaging. Mechanical, sensory and functional properties of antimicrobial edible films are of great merit for future research. Studies on this subject are rather limited, and more information is required in order to develop new packaging applications with improved functionality and high sensory performance.

Although most of antimicrobial milk-protein films have shown their efficacy *in vitro*, further practical tests are needed to determine the ability of such films to deliver antimicrobial compounds in real food systems. Moreover, further studies on food applications should be focused on a commercial scale with the purpose of providing more realistic information that can be used to commercialize food products coated with milk-protein-based edible films or coatings. Therefore, cost, organoleptic, consumer preference, toxicological, safety and regulatory considerations should be addressed if these types of packaging are to be used by the food industry.

References

Abee, T., Krockel, L. and Hill, C. (1995) Bacteriocins: modes of action and potentials in food preservation and control of food poisoning. *International Journal of Food Microbiology* 28, 169–185.

Albert, S. and Mittal, G.S. (2002) Comparative evaluation of edible coatings to reduce fat uptake in a deep-fried cereal product. *Food Research International* 35, 445–458.

Alcantra, C.R. and Krochta, J.M. (1996) Mechanical properties of whey-protein-isolate-coated freeze-dried chicken white meat dice. In: Book of Abstracts, IFT Annual Meeting, 22–26 June 1996. Institute of Food Technologists, Chicago.

Anker, M. (1996) *Edible and biodegradable films and coatings for food packaging – A Literature Review.* SIC, Goteborg, Sweden.

Appendini, P. and Hotchkiss, J.H. (2002) Review of antimicrobial food packaging. *Innovative Food Science and Emerging Technologies* 3, 113–126.

Atarés, L., Bonilla, J. and Chiralt, A. (2010) Characterization of sodium caseinate-based edible films incorporated with cinnamon or ginger essential oils. *Journal of Food Engineering* 100, 678–687.

Avena Bustillos, R.D.J., Cisneros-Zevallos, L.A., Krochta, J.M. and Saltveit, M.E. (1993) Optimization of edible coatings on minimally processed carrots using response surface methodology. *Transactions of the ASAE* 36, 801–805.

Avena-Bustillos, R.D.J., Krochta, J.M. and Saltveit, M.E. (1997) Water vapour resistance of red delicious apples and celery sticks coated with edible caseinate-acetylated monoglyceride films. *Journal of Food Science* 62, 351–354.

Banerjee, R. and Chen, H. (1995) Functional properties of edible films using whey protein concentrate. *Journal of Dairy Science* 78, 1673–1683.

Banker, G.S., Gore, A.Y. and Swarbrick, J. (1966) Water vapour transmission properties of free polymer films. *Journal of Pharmacy and Pharmacology* 18, 457–466.

Benkeblia, N. (2004) Antimicrobial activity of essential oil extracts of various onions (*Allium cepa*) and garlic (*Allium sativum*). *Lebensmittel Wissenschaft und Technologie* 37, 263–268.

Bodnar, I., Alting, A.C. and Verschueren, M. (2007) Structural effects on the permeability of whey protein films in an aqueous environment. *Food Hydrocolloids* 21, 889–895.

Bravin, B., Peressini, D. and Sensidoni, A. (2006) Development and application of polysaccharide-lipid edible coating to extend shelf-life of dry bakery products. *Journal of Food Engineering* 76, 280–290.

Burt, S. (2004) Essential oils: their antibacterial properties and potential applications in foods: a review. *International Journal of Food Microbiology* 94, 223–253.

Cagri, A., Ustunol, Z. and Ryser, E. T. (2001) Antimicrobial, mechanical, and moisture barrier properties of low pH whey protein-based edible films containing p-aminobenzoic or sorbic acids. *Journal of Food Science* 66, 865–870.

Cagri, A., Ustunol, Z. and Ryser, E.T. (2002) Inhibition of three pathogens on bologna and summer sausage slices using antimicrobial edible films. *Journal of Food Science* 67, 2317–2324.

Cagri, A., Ustunol, Z. and Ryser, E.T. (2004) Antimicrobial edible films and coatings. *Journal of Food Protection* 67, 833–848.

Caprioli, I., O'Sullivan, M. and Monahan, F.J. (2009) Use of sodium caseinate/glycerol edible films to reduce lipid oxidation in sliced turkey meat. *European Food Research and Technology* 228, 433–440.

Certel, M., Uslu, M.K. and Ozedemir, F. (2004) Effects of sodium caseinate- and milk protein concentrate-based edible coatings on the postharvest quality of bing cherries. *Journal of the Science of Food and Agriculture* 84, 1229–1234.

Chen, H. (1995) Functional properties and applications of edible films made of milk proteins. *Journal of Dairy Science* 78, 2563–2583.

Ciesla, K., Salmieri, S. and Lacroix, M. (2006) Modification of the properties of milk protein films by gamma radiation and polysaccharide addition. *Journal of the Science of Food and Agriculture* 86, 908–914.

Cisneros-Zevallos, L. and Krochta, J.M. (2003) Whey protein coatings for fresh fruits and relative humidity effects. *Journal of Food Science* 68, 176–181.

Debeaufort, F., Quezada-Gallo, J.A. and Voilley, A. (1998) Edible films and coatings: tomorrow's packagings: a review. *Critical Reviews in Food Science and Nutrition* 38, 299–313.

Erdohan, Z.O. and Turhan, K.N. (2005) Barrier and mechanical properties of methylcellulose–whey protein films. *Packaging Technology and Science* 18, 295–302.

Fabra, M. J., Talens, P. and Chiralt, A. (2008a) Tensile properties and water vapour permeability of sodium caseinate films containing oleic acid–beeswax mixtures. *Journal of Food Engineering* 85, 393–400.

Fabra, M.J., Hambleton, A., Talens, P., Debeaufort, F., Chiralt, A. and Voilley, A. (2008b) Aroma barrier properties of sodium caseinate-based edible films. *Biomacromolecules* 9, 1406–1410.

Fabra, M.J., Talens, P. and Chiralt, A. (2009) Microstructure and optical properties of sodium caseinate films containing oleic acid-beeswax mixtures. *Food Hydrocolloids* 23, 676–683.

Ferreira, C.O., Carla, A., Nunes, C.A., Ivonne Delgadillo, I. and Lopes-da-Silva, J.A. (2009) Characterization of chitosan–whey protein films at acid pH. *Food Research International* 42, 807–813.

Franssen, L. R. and Krochta, J. M. (2003) Edible coatings containing natural antimicrobials for processed foods. In: Roller, S. (ed.) *Natural Antimicrobials for Minimal Processing of Foods.* CRC Press, Boca Raton, pp. 250–262.

Franssen, L.R., Rumsey, T.R. and Krochta, J.M. (2004) Whey protein film composition effects on potassium sorbate and natamycin diffusion. *Journal of Food Science* 69, C347–C350.

Gennadios, A., McHugh, T., Weller, C.L. and Krochta, J.M. (1994) Edible coatings and films based on proteins. In: Krotcha, J.M., Baldwin, E.A. and Nisperos-Carriedo, M.O. (eds) *Edible Coatings and Films to Improve Food Quality.* Technomic Publishing Company, Inc., Lancaster, PA.

Gennadios, A., Hanna, M.A. and Kurth, L.B. (1997) Application of edible coatings on meats, poultry and seafoods: a review. *Lebensmittel Wissenschaft und Technologie* 30, 337–350.

Ghosh, A., Ali, M.A. and Dias, G.J. (2009) Effect of cross-linking on microstructure and physical performance of casein protein. *Biomacromolecules* 10, 1681–1688.

Gontard, N. and Guilbert, S. (1994) Biopackaging: technology and properties of edible and/or biodegradable materials of agricultural origin. In: Mathlouthi, M. (ed.) *Food Packaging and Preservation.* Blackie Academic & Professional, London, pp. 159–181.

Gounga, M.E., Xu, S.Y. and Wang, Z. (2007) Whey protein isolate-based edible films as affected by protein concentration, glycerol ratio and pullulan addition in film formation. *Journal of Food Engineering* 83, 521–530.

Guilbert, S. (1986) Technology and application of edible protective films. In: Mathlouthi, M. (ed.) *Food Packaging and Preservation: Theory and Practice.* Elsevier Applied Science Publishing Co., London, pp. 371– 394.

Guilbert, S. (1988) Use of superficial edible layer to protect intermediate moisture foods: application to the protection of tropical fruit dehydrated by osmosis. In: Seow, C.C., Teng, T.T. and Quah, C.H. (eds) *Food Preservation by Moisture Control.* Elsevier Applied Science, London, pp. 199–219.

Guilbert, S., Cuq, B. and Gontard, N. (1997) Recent innovations in edible and/or biodegradable packaging materials. *Food Additives and Contaminants* 14, 741–751.

Halek, W. and Garg, A. (1989) Fungal inhibition by a fungicide coupled to an ionomeric film. *Journal of Food Safety* 9, 215–222.

Han, J.H. (2000) Antimicrobial food packaging. *Food Technology* 54, 56–65.

Han, J.H. and Floros, J.D. (1997) Casting antimicrobial packaging films and measuring their physical properties and antimicrobial activity. *Journal of Plastic Film and Sheeting* 13, 287–298.

Han, J.H. and Gennadios, A. (2005) Edible films and coatings: A review. In: Han, J.H. (ed.) *Innovations in Food Packaging.* Elsevier Academic Press, London, pp. 239–262.

Han, J., Bourgeois, S. and Lacroix, M. (2009) Protein-based coatings on peanut to minimise oil migration. *Food Chemistry* 115, 462–468.

Hernandez-Izquierdo, V.M. and Krochta, J.M. (2008) Thermoplastic processing of proteins for film formation – a review. *Journal of Food Science* 73, R30–R39.

Hirasa, K. (1991) Moisture loss and lipid oxidation in frozen fish – effect of a casein-acetylated monoglyceride edible coating. MSc thesis, University of California, Davis.

Hotchkiss, J.S. (1995) Safety considerations in active packaging. In: Rooney, M.L. (ed.) *Active Food Packaging.* Blackie Academic & Professional, Glasgow, pp. 238–253.

Huang, L.L., Zhang, M., Yan, W.Q., Mujumdar, A.S. and Sun, D.F. (2009) Effect of coating on post-drying of freeze-dried strawberry pieces. *Journal of Food Engineering* 92, 107–111.

Irissin-Mangata, J., Bauduin, G., Boutevin, B. and Gontard, N. (2001) New plasticizers for wheat gluten films. *European Polymer Journal* 37, 1533–1541.

Jagannath, J.H., Nanjappa, C., Gupta, D.D. and Bawa, A.S. (2006) Studies on the stability of an edible film and its use for the preservation of carrot (*Daucus carota*). *International Journal of Food Science and Technology* 41, 498–506.

Kester, J.J. and Fennema, O.R. (1986) Edible films and coatings: a review. *Food Technology* 40, 47–59.

Khwaldia, K., Perez, C., Banon, S., Desobry, S. and Hardy, J. (2004) Milk proteins for edible films and coatings. *Critical Reviews in Food Science and Nutrition* 44, 239–251.

Kilincceker, O., Dogan, I.S. and Kucukoner, E. (2009) Effect of edible coatings on the quality of frozen fish fillets. *LWT – Food Science and Technology* 42, 868–873.

Klaenhammer, T.R. (1993) Genetics of bacteriocins produced by lactic acid bacteria. *FEMS Microbiology Reviews* 12, 39–86.

Ko, S., Janes, M.E., Hettiarachchy, N.S. and Johnson, M.G. (2001) Physical and chemical properties of edible films containing nisin and their action against *Listeria monocytogenes*. *Journal of Food Science* 66, 1006–1011.

Kristo, E., Biliaderis, C.G. and Zampraka, A. (2007) Water vapour barrier and tensile properties of composite caseinate–pullulan films: biopolymer composition effects and impact of beeswax lamination. *Food Chemistry* 101, 753–764.

Kristo, E., Koutsoumanis, K.P. and Biliaderis, C.G. (2008) Thermal, mechanical and water vapor barrier properties of sodium caseinate films containing antimicrobials and their inhibitory action on *Listeria monocytogenes*. *Food Hydrocolloids* 22, 373–386.

Krochta, J.M. (1997) Edible protein films and coatings. In: Damodaran, S. and A. Paraf, A. (eds) *Food Proteins and Their Applications*. Marcel Dekker, Inc., New York, pp. 529– 549.

Krochta, J.M. and De Mulder-Johnston, C.D. (1997) Edible and biodegradable polymer films: challenges and opportunities. *Food Technology* 51, 61–74.

Krochta, J.M., Baldwin, E.A. and Nisperos-Carriedo, M.O. (1994) *Edible Coatings and Films to Improve Food Quality*. Technomic Publishing Company, Inc., Lancaster, PA.

LaCoste, A., Schaich, K. M., Zumbrunnen, D. and Yam, K. L. (2005) Advancing controlled release packaging through smart blending. *Packaging Technology and Science* 18, 77–87.

Le Tien, C.L., Vachon, C., Mateescu, M.A. and Lacroix, M. (2001) Milk protein coatings prevent oxidative browning of apples and potatoes. *Journal of Food Science* 66, 512–516.

Lee, S.Y., Trezza, T. A., Guinard, J.X. and Krochta, J. M. (2002) Whey protein-coated peanuts assessed by sensory evaluation and static headspace gas chromatography. *Journal of Food Science* 67, 1212–1218.

Lungu, B. and Johnson, M.G. (2005) Fate of *Listeria monocytogenes* inoculated onto the surface of model turkey frankfurter pieces treated with zein coatings containing nisin, sodium diacetate, and sodium lactate at 4°C. *Journal of Food Protection* 68, 855–859.

Maté, J.I. and Krochta, J.M. (1996) Whey protein coating effect on the oxygen uptake of dry roasted peanuts. *Journal of Food Science* 61, 1202–1206, 1210.

Maté, J.I., Saltveit, M.E and Krochta, J.M. (1996) Peanut and walnut rancidity: effects of oxygen concentration and relative humidity. *Journal of Food Science* 61, 465–469.

McDade C.R., Zutara, S.M., Ryser, E., Donnelly, C.W. and Chen, H. (1999) Use of whey-based edible film containing antimicrobial agents to inhibit *L. monocytogenes* in frankfurters [abstract]. In: Program and Abstract Book, Annual Meeting of the International Association for Food Protection, Aug. 1–4, Dearborn, Mich. Abstract nr T10.

McHugh, T.H. and Krochta, J.M. (1994) Sorbitol vs glycerol-plasticized whey protein edible films: integrated oxygen permeability and tensile property evaluation. *Journal of Agricultural and Food Chemistry* 42, 841–845.

Mei, Y. and Zhao, Y. (2003) Barrier and mechanical properties of milk protein-based edible films containing nutraceuticals. *Journal of Agricultural and Food Chemistry* 51, 1914–1918.

Mendes de Souza, P., Fernández, A., López-Carballo, G., Gavara, R. and Hernández-Muñoz, P. (2010) Modified sodium caseinate films as releasing carriers of lysozyme. *Food Hydrocolloids* 24, 300–306.

Miller, K.S. and Krochta J.M. (1997) Oxygen and aroma barrier properties of edible films: a review. *Trends in Food Science and Technology* 8, 228–237.

Min, S. and Krochta, J.M. (2005) Inhibition of *Penicillium* commune by edible whey protein films incorporating lactoferrin, lacto-ferrin hydrolysate, and lactoperoxidase systems. *Journal of Food Science* 70, 87–94.

Min, S., Harris, L.J. and Krochta, J.M. (2005) Antimicrobial effects of lactoferrin, lysozyme, and the lactoperoxidase system and edible whey protein films incorporating the lactoperoxidase system against *Salmonella enterica* and *Escherichia coli* O157:H7. *Journal of Food Science* 70, M332–M338.

Min, S., Rumsey, T.R. and Krochta, J.M. (2008) Diffusion of the antimicrobial lysozyme from a whey protein coating on smoked salmon. *Journal of Food Engineering* 84, 39–47.

Ming, X., Weber, G.H., Ayres, J.W. and Sandine, W.E. (1997) Bacteriocins applied to food packaging materials to inhibit *Listeria monocytogenes* on meats. *Journal of Food Science* 62, 413–415.

Nortjé, K., Buys, E.M. and Minnaar, A. (2006) Use of g-irradiation to reduce high levels of *Staphylococcus aureus* on casein–whey protein coated moist beef biltong. *Food Microbiology* 23, 729–737.

Ouattara, B., Sabato, S.F. and Lacroix, M. (2001) Combined effect of antimicrobial coating and gamma irradiation on shelf life extension of pre-cooked shrimp (*Penaeus spp.*). *International Journal of Food Microbiology* 68, 1–9.

Ouattara, B., Sabato, S.F. and Lacroix, M. (2002) Use of gamma-irradiation technology in combination with edible coating to produce shelf-stable foods. *Radiation Physics and Chemistry* 63, 305–310.

Oussalah, L., Caillet, S., Salmieri, S., Saucier, L. and Lacroix, M. (2004) Antimicrobial and antioxidant effects of milk protein-based film containing essential oils for the preservation of whole beef muscle. *Journal of Agricultural and Food Chemistry* 52, 5598–5605.

Ozdemir, M. (1999) Antimicrobial releasing edible whey protein films and coatings. PhD dissertation, Purdue University, West Lafayette, Indiana, USA.

Ozdemir, M. and Floros, J. D. (2001) Analysis and modeling of potassium sorbate diffusion through edible whey protein films. *Journal of Food Engineering* 47, 149–155.
Ozdemir, M. and Floros, J.D. (2003) Film composition effects on diffusion of potassium sorbate through whey protein films. *Journal of Food Science* 68, 511–516.
Ozdemir, M. and Floros, J.D. (2008a) Optimization of edible whey protein films containing preservatives for water vapour permeability, water solubility, and sensory characteristics. *Journal of Food Engineering* 86, 215–224.
Ozdemir, M. and Floros, J.D. (2008b) Optimization of edible whey protein films containing preservatives for mechanical and optical properties. *Journal of Food Engineering* 84, 116–123.
Padgett, T., Han, I.Y. and Dawson, P.L. (1998) Incorporation of food-grade antimicrobial compounds into biodegradable packaging films. *Journal of Food Protection* 61, 1330–1335.
Park, H. J. (1999) Development of advanced edible coatings for fruits. *Trends in Food Science and Technology* 10, 254–260.
Park, S.I., Daeschel, M.A. and Zhao, Y. (2004) Functional properties of antimicrobial lysozyme–chitosan composite films. *Journal of Food Science* 69, 215–221.
Pereda, M., Aranguren, M.I. and Marcovich, N.E. (2008) Characterization of chitosan/caseinate films. *Journal of Applied Polymer Science* 107, 1080–1090.
Perez-Gago, M.B. and Krochta, J.M. (1999) Water vapour permeability of whey protein emulsion films as affected by pH. *Journal of Food Science* 64, 695–698.
Perez-Gago, M.B., Del Rio, M.A. and Serra, M. (2005) Effect of whey protein-beeswax edible composite coating on color change of fresh-cut persimmons cv. 'Rojo Brillante'. *Acta Horticulturae* 682, 1917–1923.
Perez-Gago, M.B., Serra, M. and Del Rio, M.A. (2006) Colour change of fresh-cut apples coated with whey protein concentrate-based edible coatings. *Postharvest Biology and Technology* 39, 84–92.
Petersen, K., Nielsen, P.V., Bertelsen, G., Lawther, M., Olsen, M.B., Nilsson, N.H. and Mortensen, G. (1999) Potential of biobased materials for food packaging. *Trends in Food Science and Technology* 10, 52–68.
Pintado, C.M.B.S., Sousa, I. and Ferreira, M.A.A.S. (2006) Edible films with bioactive characteristics against *Listeria monocytogenes*. In: Book of proceedings of the 2006 EFFoST Annual Meeting/Total Food 2006 – Sustainability of the Agri-Food Chain, The Hague, The Netherlands.
Pintado, C.M.B.S., Ferreira, M.A.A.S. and Sousa, I. (2010) Control of pathogenic and spoilage microorganisms from cheese surface by whey protein films containing malic acid, nisin and natamycin. *Food Control* 21, 240–246.
Ponce, A., Roura, S., del Valle, C. and Moreira, M. (2008) Antimicrobial and antioxidant activities of edible coatings enriched with natural plant extracts: *In vitro* and *in vivo* studies. *Postharvest Biology and Technology* 49, 294–300.
Pranoto, Y., Rakshit, S. and Salokhe, V. (2005) Enhancing antimicrobial activity of chitosan films by incorporating garlic oil, potassium sorbate and nisin. *LWT – Food Science and Technology* 38, 859–865.
Quintavalla, S. and Vicini, L. (2002) Antimicrobial food packaging in meat industry. *Meat Science* 62, 373–380.
Ricke, S.C. (2003) Perspectives on the use of organic acids and short chain fatty acids as antimicrobials. *Poultry Science* 82, 632–639.
Rodrigues, E. T. and Han, J.H. (2000) Antimicrobial whey-protein films against spoilage and pathogenic bacteria. In: Book of Abstracts, IFT Annual Meeting. Institute of Food Technologists, Chicago, pp. 30, 78E.
Rojas-Graü, M.A., Avena-Bustillos, R.J., Olsen, C., Friedman, M., Henika, P.R., Martín-Belloso, O., Pan, Z. and McHugh, T.H. (2007) Effects of plant essential oils and oil compounds on mechanical, barrier and antimicrobial properties of alginate–apple puree edible films. *Journal of Food Engineering* 81, 634–641.
Rooney M.L. (1995) Active packaging in polymer films. In: Rooney, M.L. (ed.) *Active Food Packaging*. Blackie Academic and Professional, Glasgow, pp. 74–110.
Rosenberg, M. and Young, S.L. (1993) Whey proteins as microencapsulating agents. Microencapsulation of anhydrous milkfat — Structure evaluation. *Food Structure* 12, 31–41.
Schou, M., Longares, A., Montesinos-Herrero, C., Monahan, F., O'Riordan, D. and O'Sullivan, M. (2005) Properties of edible sodium caseinate films and their application as food wrapping. *LWT – Food Science and Technology* 38, 605–610.

Seydim, A.C. and Saricus, G. (2006) Antimicrobial activity of whey protein based edible films incorporated with oregano, rosemary, and garlic essential oils. *Food Research International* 39, 639–644.

Shon, J. and Haque, Z.U. (2007) Efficacy of sour whey as a shelf-life enhancer: use in antioxidative edible coatings of cut vegetables and fruit. *Journal of Food Quality* 30, 581–593.

Siripatrawan, U. and Harte, B.R. (2010) Physical properties and antioxidant activity of an active film from chitosan incorporated with green tea extract. *Food Hydrocolloids* 24, 770–775.

Sivropoulou, A., Papanikolaou, E., Nikolanou, C., Kokkini, S., Lanaras, T. and Arsenakis, M. (1996) Antimicrobial and cytotoxic activities of *Origanum* essential oils. *Journal of Agricultural and Food Chemistry* 44, 1202–1205.

Stuchel, Y.M. and Krochta, J.M. (1995) Edible coatings on frozen king salmon: Effect of whey protein isolate and acetylated monoglycerides on moisture loss and lipid oxidation. *Journal of Food Science* 60, 28–31.

Tharanathan, R. N. (2003) Biodegradable films and composite coatings: past, present and future. *Trends in Food Science and Technology* 14, 71–78.

Tzoumaki, M.V., Biliaderis, C.G. and Vasilakakis, M. (2009) Impact of edible coatings and packaging on quality of white asparagus (*Asparagus officinalis, L.*) during cold storage. *Food Chemistry* 117, 55–63.

Vanin, F.M., Sobral, P.J.A., Menegalli, F.C., Carvalho, R.A. and Habitante, A.M.Q.B. (2005) Effects of plasticizers and their concentrations on thermal and functional properties of gelatin-based films. *Food Hydrocolloids* 19, 899–907.

Vermeiren, L., Devlieghere, F. and Debevere, J. (2002) Effectiveness of some recent antimicrobial packaging concepts. *Food Additives and Contaminants* 19, 163–171.

Welscher, Y.M., Napel, H.H., Balagué, M.M., Souza, C.M., Riezman, H. and Kruijff, B. (2008) Natamycin blocks fungal growth by binding specifically to ergosterol without permeabilizing the membrane. *Journal of Biological Chemistry* 283, 6393–6401.

Zinoviadou, K.G., Koutsoumanis, K.P. and Biliaderis, C.G. (2009) Physico-chemical properties of whey protein isolate films containing oregano oil and their antimicrobial action against spoilage flora of fresh beef. *Meat Science* 82, 338–345.

Zinoviadou, K.G., Koutsoumanis, K.P. and Biliaderis, C.G. (2010) Physical and thermo-mechanical properties of whey protein isolate films containing antimicrobials, and their effect against spoilage flora of fresh beef. *Food Hydrocolloids* 24, 49–59.

10 Antimicrobial and other Beneficial Applications of Chitosans

Mendel Friedman* and Vijay K. Juneja

Introduction

Chitin, a component of the exoskeletons of insects and crustaceae including crab and shrimp, consists of N-acetylglucosamine residues joined by β(1-4) glycosidic links. It is the second most abundant biopolymer in the world after cellulose. Structurally, chitin (poly-N-acetylglucosamine) resembles cellulose, except that the substituent at the carbon-2 atom is an acetylated amino (-NH-CO-CH_3) instead of a hydroxyl (OH) group. Deacetylation is achieved by exposing chitin to strong NaOH solutions or to the enzyme chitosinase (Fig. 10.1). Interest in chitin resides in the fact that its deacetylated product called chitosan exhibits desirable functional and biological traits, including antimicrobial and antioxidative properties, and it appears to be non-toxic.

Chitosan exhibits strong antimicrobial effects against a variety of pathogenic and spoilage organisms (reviewed in Rabea *et al.* 2003; Roller, 2003; Shahidi and Abuzaytoun, 2005; No *et al.*, 2007; Dutta *et al.*, 2009). Numerous studies have been carried out on the antimicrobial, antifungal and antiviral effects of chitosan and chitosan derivatives (Jung *et al.*, 1998; You-Jin and Kim, 2001; Avadi *et al.*, 2004; Liu *et al.*, 2004; Park *et al.*, 2004; Qin *et al.*, 2004; Kenawy *et al.*, 2005; Je and Kim, 2006; Liu *et al.*, 2007; Kanatt *et al.*, 2008b; Másson *et al.*, 2008; Zhong *et al.*, 2008; Aranaz *et al.*, 2009; Liu *et al.*, 2009; Sousa *et al.*, 2009) as well as chitosan coatings and films (Ouattar *et al.*, 2000; Möller *et al.*, 2004; Kulkarni *et al.*, 2005; Pranoto *et al.*, 2005a; 2005b; Dutta *et al.*, 2009; Pelissari *et al.*, 2009; Portes *et al.*, 2009; Vásconez *et al.*, 2009). The molecular size of chitosan, which may range from about 2000 to more than 100,000 Da, as well as particle size (Qi *et al.*, 2004; Chen *et al.*, 2009) also influence biological activity (Rhoades and Roller, 2000; Savard *et al.*, 2002; Kim *et al.*, 2003; Omura *et al.*, 2003; Tsai *et al.*, 2004; Zivanovic *et al.*, 2004).

Generally, chitosan is more effective against Gram-negative than against Gram-positive bacteria (Devlieghere *et al.*, 2004). The biopolymer also protected mice against infection by *Listeria monocytogenes* (Okawa *et al.*, 2003), suggesting that the antimicrobial action may also occur in humans.

The objective of this chapter is largely limited to outlining and summarizing reported studies on the beneficial effects of chitosan on microbial safety and quality of cereal and dairy products, fruits, fruit juices, meat and poultry products including eggs, seafood and vegetables. Also covered are

* Corresponding author.

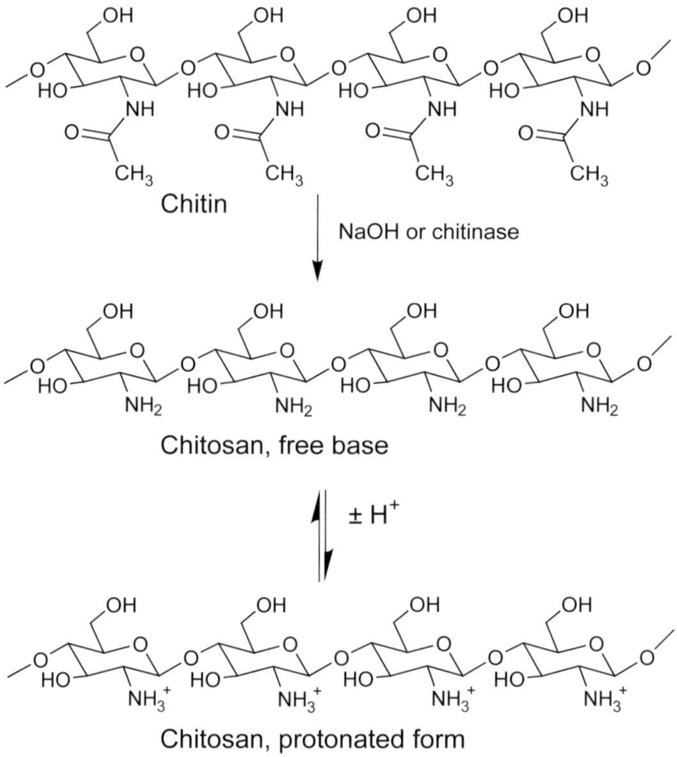

Fig. 10.1. Deacetylation of chitin to chitosan and acid–base equilibrium of chitosan. The antimicrobial effect of the free base is postulated to involve chelating to trace elements and metalloenzymes, and of the protonated form to disruption of cell membranes.

antimicrobial fabrics, possible mechanisms of beneficial antimicrobial and antioxidative effects, suggested medicinal uses and food-related research needs.

Applications of Chitosan in Food

Cereal and legume products

Grains and legumes are occasionally contaminated with two major environmental hazards: toxic weed seeds (Dugan *et al.*, 1989; Friedman *et al.*, 1989; Friedman and Levin, 1989; Crawford *et al.*, 1990; Friedman and Dao, 1990; Friedman and Henika, 1991; Friedman, 2004) and food-borne pathogens and spoilage organisms (Stewart, 2009). Here we mention efforts to overcome microbial contaminants with the aid of chitosan in two cereal products and one legume product.

Pasta

The combined effect of chitosan and modified atmosphere packaging improved the microbiological safety of amaranth-based homemade fresh pasta (Del Nobile *et al.*, 2009a; 2009b). The treatment inhibited the growth of mesophilic bacteria, *Staphylococcus* spp. yeasts, moulds and total coliforms during storage at 4°C for 2 months. The cited effects and the compatible sensorial properties of the chitosan-added pasta suggest the potential of this non-thermal preservation for large-scale use.

Rice

Treatment with chitosan extended the shelf life of rice cakes and rice noodles (Lee *et al.*, 2000). Black rice cultivars contain strong antioxidative compounds that merit study for potential antimicrobial properties in food (Nam *et al.*, 2005a,b, 2006).

Chickpeas

Chitosan at low pH inhibited the growth of spoilage organisms (*Candida* sp., *Zygosaccharomyces bailii*) in a chickpea (hummus) dip (Rhoades and Roller, 2000).

Produce

Protecting fruit against pathogenic and spoilage organisms is a challenging problem. The following observations suggest that the use of chitosan in various forms may help ameliorate this problem.

Fresh fruit

Studies with cantaloupe and pineapple showed that (i) coating of fresh-cut cantaloupes with chitosan-methyl cellulose films reduced the growth of mesophilic aerobes, psychotrophs, lactic acid bacteria, yeasts and moulds, and prevented the multiplication of *Escherichia coli* and *Salmonella* spp. organisms (Krasaekoopt and Mabumrung, 2008); and (ii) chitosan/methyl cellulose vanillin films inhibited the growth of *E. coli* and *Saccharomyces cerevisiae* yeast on fresh-cut cantaloupe and pineapple, while maintaining quality attributes of the fruit (Sangsuwan *et al.*, 2008, 2009).

Chitosan and chitosan coatings increase postharvest quality, reduce postharvest rotting, and extend the shelf life of citrus (*Murcott tangor*) fruit (Chien *et al.*, 2007b; Canale Rappussi *et al.*, 2009). Studies with grapes showed that (i) chitosan and grapefruit seed extract appear to act synergistically in reducing postharvest fungal rot of table grapes caused by *Botrytis cinerea* (Xu *et al.*, 2007); and (ii) chitosan acetate effectively controlled postharvest grey mould of table grapes at cold and ambient storage temperatures, without any apparent injury to the grapes (Meng *et al.*, 2008; Romanazzi *et al.*, 2009). Several studies showed that (i) application of chitosan coatings maintained quality parameters and extended the shelf life of peeled lychee (*Litchi chinensis*) fruit (Dong *et al.*, 2004); (ii) chitosan coatings enhanced the microbial safety of cold-stored lychee fruit at ambient temperatures (Jiang *et al.*, 2005); and (iii) chitosan improved the quality of lychee cultivars (De Reuck *et al.*, 2009). Chitosan and chitosan coatings also delayed the ripening, improved the quality, reduced decay, and extended the shelf life of mangoes (*Mangifera indica*) fruit (Chien *et al.*, 2007c; Jitareerat *et al.*, 2007; Wang *et al.*, 2007; Zhu *et al.*, 2008).

Studies with berries showed that (a) a chitosan coating significantly reduced the decay of fresh strawberries and raspberries and had beneficial effects on firmness and anthocyanin and vitamin C content of the berries (Zhang and Quantick, 1998); (b) a chitosan–lactic acid/sodium lactate dip solution inhibited the growth of pathogenic and spoilage organisms on strawberries (*Fragaria x ananassa*) and lettuce (Devlieghere *et al.*, 2004); (c) chitosan-based coating exhibited antifungal properties against *Cladosporium* sp. and *Rhizopus* sp. on strawberries. The treatment also reduced total aerobic count, coliforms, and weight loss of the strawberries during storage (Park *et al.*, 2005); (d) chitosan coatings did not affect consumer acceptability of flavour, sweetness or firmness of strawberries (Han *et al.*, 2005); and (e) treatment with 1% chitosan solutions reduced the microflora, and improved the quality and prolonged the shelf of fresh-cut strawberries (Campaniello *et al.*, 2008).

Fresh vegetables

Chitosan inhibits the late-blight-causing fungus *Phytophthora infestans* as well as *Colletotrichum* in tomatoes and grapes (Atia *et al.*, 2005; Muñoz *et al.*, 2009). Chitosan-coated squash slices showed significant log reductions of mesophilic aerobic micro-organisms (Ponce *et al.*, 2008; Moreira *et al.*, 2009). Those dried at 50°C for 30 min showed the highest reductions (5.02 log CFU/g). Chitosan also enhanced the quality characteristics of sweet potatoes during 17-day refrigerated storage (Waimaleongora-Ek *et al.*, 2008).

Fruit juices

Studies with apple juice indicated that (a) chitosan exhibited antifungal properties in juice (Roller and Covill, 1999); (b) Low molecular weight chitosan exhibited higher antioxidative and free radical scavenging effects than did high molecular weight chitosans (Chien *et al.*, 2007a); and (c) chitosan and pressure at 193 MPa at 20°C exerted synergistic effects against *E. coli*, *Staphylococcus aureus*, psychrophiles, psychrotrophs and yeast during storage of the juice (Malinowska-Panczyk *et al.*, 2009). Added chitosan extended the shelf-life of fresh orange juice. The preservative action of chitosan merits further study as a replacement of heat pasteurization (Martín-Diana *et al.*, 2009). Possible beneficial effects of chitosan in vegetable juices have apparently not been studied.

Dairy products

Here we briefly outline reported studies on the application of chitosan-based films and coatings and free chitosan to improve the shelf life, quality and microbial safety of cheese, milk and yogurt. Selected studies in this area include the following findings.

Cheese

- Chitosan-based edible coating adjusted to pH 5 inhibited the growth of Gram-positive *L. monocytogenes* and *S. aureus*, but not Gram-negative *Pseudomonas aeruginosa*, on a cheese food product (Coma *et al.*, 2003). Because chitosan increased the microbial lag phase and decreased microbial density, the chitosan films and coatings have the potential to be used to preserve dairy products.
- Chitosan–lysozyme films and coatings inhibited the growth of *E. coli*, *L. monocytogenes*, *Pseudomonas fluorescens* as well as moulds and yeast in mozzarella cheese (Duan *et al.*, 2007). The authors suggest that these films could be used as cheese packaging to control post-processing microbial contaminants. Chitosan also prolonged the shelf life of mozzarella cheese (Altieri *et al.*, 2005).
- A combination chitosan coating and modified atmosphere packaging inhibited the growth of coliform and *Pseudomonas* spp. bacteria resulting in improved microbial and sensory quality as well as longer shelf life of stored Fior di latte cheese (Del Nobile *et al.*, 2009c).
- Addition of chitosan increased the encapsulation efficiency of the enzyme flavourzyme used to control cheese ripening (Anjani *et al.*, 2007).

Milk

Chitosan-containing antimicrobial paper board suppressed the growth of aerobic bacteria in milk and yeast in orange juice at 3°C and 10°C, but not at 20°C (Lee *et al.*, 2004).

Yogurt

- Addition of probiotics encapsulated in chitosan-coated chitosan beads increased survival of the probiotic bacteria (*Lactobacillus acidophilus*) in yogurt during storage (Krasaekoopt *et al.*, 2006).
- Addition of chitosan to yogurt reduced the *in vitro* availability of nutrients such as glucose and calcium, suggesting that added chitosan behaved as a dietary fibre (Rodríguez *et al.*, 2008).

Meat products

Chitosan has been extensively evaluated for its antibiotic and antioxidative (preservative) properties in a variety of meat products. Here we briefly review selected reported observations for different meat categories, in alphabetical order.

Bacon

Low molecular weight irradiated chitosan exhibited enhanced antioxidative activity without affecting antimicrobial potency of bacon and mutton seek kababs (Rao *et al.*, 2005).

Beef

Edible chitosan films dissolved in acetic or lactic acids reduced *L. monocytogenes* pathogens on the surface of ready-to-eat (RTE) roast beef by 2–3 logs on day 14 (Beverlya *et al.*, 2008). The acetic acid chitosan coatings were more effective in controlling pathogens than the lactic acid coatings.

Chitosan alone and in combination with either rosemary extract or α-tocopherol had better antioxidative properties during frozen storage of beef burgers than either rosemary or α-tocopherol alone (Georgantelis *et al.*, 2007b). The best antioxidative effects were observed with the combination of chitosan and rosemary extract.

The shelf life of minced meat containing irradiated chitosan and lysozyme was extended up to 15 days at chilled temperatures (Rao *et al.*, 2008). This beneficial effect, associated with the synergistic action of chitosan and lysozyme, was accompanied by complete elimination of *Bacillus cereus, E. coli,* and *P. fluorescens,* and a reduced load of *S. aureus* cells as well as enhanced resistance to oxidative spoilage.

The inhibition of *Clostridium perfringens* in ground beef (Juneja *et al.*, 2006) is described in the section on turkey.

Bologna

The application of chitosan films enriched with oregano essential oil on bologna followed by sensory evaluation showed that 45 ppm oregano oil in bologna originating from chitosan–oregano films would be acceptable to consumers (Chi *et al.*, 2006). Related studies on the effectiveness of antimicrobial chitosan films are described in Ouattar *et al.* (2000) and Zivanovic *et al.* (2005).

Ham

Antimicrobial films prepared from chitosan inhibited *L. monocytogenes* and surface-spoilage bacteria in processed hams (Ouattar *et al.*, 2000; Ye *et al.*, 2008a).

Lamb

Irradiated chitosan reduced the rancidity of radiation-processed lamb meat (Kanatt *et al.*, 2004). Chito-oligosaccharides produced by γ-irradiation of chitosan and the antimicrobial lysozyme exhibited a synergistic effect (they were more effective than the individual compounds) against pathogens on meat (Rao *et al.*, 2008). The combined treatment induced complete elimination of *B. cereus, E. coli* and *P. fluorescens*. The shelf life of treated lamb meat was extended up to 15 days at chilled temperatures.

Pork

Sagoo *et al.* (2002) found that (i) chitosan at 0.05% in 0.9% saline at pH 6.2 inhibited the growth of the spoilage microorganism *Saccharomycodes ludwigii;* (ii) a higher concentration (0.25 to 0.5%) inactivated *Lactobacillus viridescens* and *Listeria innocua;* and (iii) dipping of skinless pork sausages in chitosan solutions (1.0%) reduced the native microflora by 1–3 logs during 18 days at 7°C and increased shelf life from 7 to 15 days.

The combination of chitosan and a rosemary extract showed intense antioxidative and antimicrobial effects in fresh pork sausages stored at 4°C, similar to those mentioned above for beef burgers (Georgantelis *et al.*, 2007a).

Because meat is susceptible to both microbial and oxidative spoilage, Kannat *et al.* (2008a) evaluated the combination of chitosan and an antioxidant mint-plant extract as a preservative for meat products. The shelf life of pork cocktail salami was enhanced following exposure to the combination. The combination also induced a reduction in the following pathogenic and spoilage organisms: *B. cereus, E. coli, Salmonella typhimurium* and *P. fluorescens*. These observations suggest that combinations of chitosan–mint mixture are a potent antimicrobial and antioxidative agent that can be used for preservation and shelf-life extension of meats and meat products.

The combination of chitosan (0.5% and 1%) with nitrites (150 ppm) reduced the viable counts of lactic acid bacteria, *Pseudomonas* spp., *Brochothrix thermosphacta, Enterobacteriaceae,* yeast and moulds in fresh pork sausages (Soultos *et al.*, 2008). The rate of lipid oxidation was also significantly decreased. The cited data indicate that chitosan can be used to extend shelf life of stored pork products.

Poultry products

Here we outline the use of chitosan to reduce pathogens in chicken, turkey and eggs.

Chicken

- Chitosan coatings reduced *Salmonella* levels in modified-atmosphere-packaged fresh chicken breasts (Cooksey, 2005).
- Combinations of chitosan (1.5%) and thyme oil (0.2%) inhibited the growth of spoilage lactic acid bacteria, *Pseudomonas* and *B. thermosphacta* as well as *Enterobacteriaceae* in ready-to-cook (RTC) chicken–pepper kebab stored under anaerobic condition at 4°C for 12 days (Giatrakou *et al.*, 2010). The treatment extended the shelf life of the organoleptically acceptable products by 4 to 6 days.

Turkey

- We investigated the inhibition of *C. perfringens* spore germination and outgrowth by the biopolymer chitosan during abusive chilling of cooked ground beef and turkey obtained from a retail store (Juneja *et al.*, 2006). We found that chilling of ground beef resulted in germination and outgrowth of *C. perfringens* spores and that added chitosan reduced the outgrowth of the pathogen. Our data show that, in the control samples without chitosan, cooling from 54.4°C to 7.2°C in 12, 15, 18 or 21 h resulted in 3.10, 4.51, 5.03 and 4.70 $\log_{10}$ CFU/g increases, respectively, in *C. perfringens* populations of the ground meat. The corresponding increases for ground turkey are 5.27, 4.52, 5.11, and 5.38 $\log_{10}$ CFU/g. The results suggest that incorporation of 3% chitosan into ground beef or turkey may reduce the potential risk of *C. perfringens* spore germination during abusive cooling from 54.4°C to 7.2°C in 12, 15 or 18 h.
- A chitosan-based film induced 1.7 log reduction of *L. monocytogenes* on turkey breast after 10 days and 1.2 log reduction after 15 days of storage at 4°C (Joerger *et al.*, 2009).

Eggs

- Hard-boiled eggs coated with a chitosan–lysozyme composite coating inhibited the growth of injected *L. monocytogenes* and *S. enterica* as well as multiplication of coliforms, moulds and yeasts (Kim *et al.*, 2008, 2009). The treatment also retarded moisture loss and pH changes during storage at 10°C. The authors conclude that chitosan–lysozyme based coatings can enhance extend shelf life of hard-boiled eggs by controlling post-processing contamination and delaying undesirable changes in egg quality.
- A related study (Kim *et al.*, 2009) showed that chitosan coating extended the shelf life and quality of eggs. The total amino acid content of albumen and fatty acid composition of non-coated and chitosan-coated eggs was the same after storage for 5 weeks. Similar observations are reported by Su *et al.*(2007).
- Added chitosan inhibited the growth of spoilage organisms *S. liquefaciens* and *Z. bailii* in egg-containing mayonnaise stored at 25°C (Oh *et al.*, 2001).

Seafood

Although sea-derived chitosan alone possesses antimicrobial properties, the use of chitosan in combination with other antimicrobials enhances its activity in seafood. The following examples illustrate the use of chitosan formulations to improve shelf life and the microbial safety of seafood.

Fish

COD. Low levels (50 to 200 ppm) of chitosan prepared from snowcrab with different molecular weights and viscosities were effective in controlling the oxidation of lipids in comminuted cod (*Gadus morhua*) during cooking (Shahidi *et al*., 2002). The inhibition of lipid oxidation was concentration dependent. The mechanism of the protective effect may involve formation of chitosan–iron complexes, thus reducing or eliminating the pro-oxidant effect of this metal ion in the cod. A related study further demonstrated the ability of chitosan to act as a preservative in cod (Jeon *et al*., 2002).

Chitosan-fish oil coatings increased total lipid and omega-3 fatty acid content threefold, reduced lipid oxidation, and inhibited growth of total and psychotropic microorganisms in fresh lingcod (*Ophidian elongates*). The coatings did not affect the colour of the fish fillets but lowered the pH and moisture content of the samples. These observations suggest that chitosan coatings containing fish oils have the potential to be applied in fish packaging to increase omega-3 fatty acid content and extend shelf life of seafood (Duan *et al*., 2010).

FISH FILLETS. Pretreatment of fish fillets (*Oncorhynchus nerka*) with a 1% chitosan solution for 3 h retarded the increase in the growth of coliforms, *Vibrio* spp., *Aeromonas* spp., mesophiles and psychotrophs and extended the shelf life from 5 to 9 days (Tsai *et al*., 2002).

HERRING. Chitosan effectively protected cooked comminuted herring (*Clupea harengus*) samples against oxidation (Jeon *et al*., 2002; Kamil *et al*., 2002; Shahidi *et al*., 2002). The formation of hydroperoxides and 2-thiobarbituric acid reactive substances (TBARS) were reduced in herring after 8 days of storage by 61 and 52%, respectively. Because chitosan has the potential to prevent lipid oxidation, the growth of spoilage bacteria and moisture loss, chitosan extracted from crab-processing wastes appears to be a natural antioxidant and antimicrobial for stabilizing lipid-containing foods.

SALMON. Chitosan coatings were more effective than edible films prepared from the solution against *Lactobacilli* spp. and *Z. bailii* on salmon slices (Vásconez *et al*., 2009). Chitosan-coated plastic films incorporating the antimicrobials nisin, sodium lactate, sodium benzoate, sodium diacetate and potassium sorbate inhibited the growth of a five-strain cocktail of *L. monocytogenes* on cold-smoked salmon samples (Ye *et al*., 2008b). The film incorporating nisin plus sodium lactate completely inhibiting the growth of the bacteria during 10 days of storage. Storage of the test samples in a refrigerator continued to inhibit the bacteria for up to 6 weeks. Chitosan coatings improved the quality and shelf life of pink salmon (Sathivel, 2005; Sathivel *et al*., 2007).

TROUT. Immersion of contaminated rainbow trout fingerlings in a chitosan solution of 125 μg/ml caused 100% mortality of *Saprolegnia parasitica* zoospores (Yuasa and Hatai, 1996). A chitosan–cinnamon oil coating improved the quality of refrigerated rainbow trout (Ojagh *et al*., 2010).

Shellfish

OYSTERS. Oysters are highly susceptible to contamination by pathogens (Ravishankar *et al*., 2010). A water-soluble sulfbenzoyl chitosan derivative inhibited the growth of strong of *Aeromonas hydrophila*, *B. cereus*, *S. typhimurium* and *Shigella dysenteriae* (Chen *et al*., 1998). Solutions containing 1000 and 2000 ppm of the chitosan derivative also retarded the growth of coliforms, *Aeromonas*, *Pseudomonas* and *Vibrio* species and extended the shelf life of oysters.

- Chitosan inhibited the growth of *S. enterica*, *S. aureus*, and *Vibrio vulnificus* in oysters (Chhabra *et al*., 2006).
- Treatment of raw Pacific oysters (*Crassostrea gigas*), the most abundant harvested shellfish, with chitosan solutions (5 g/l) reduced the microbial flora (*Pseudomonas* and *Vibrionaceae*) and extended the shelf life of the oysters stored at 5°C from 8–9 days to 14–15 days (Cao *et al*., 2009).
- Evaluation of water-soluble chitosan oligosaccharides *in vitro* and *in vivo* against the Gram-negative pathogen *V. vulnificus* that contaminates oysters and other shellfish showed that the

oligosaccharide with a molecular weight of 10,000 at concentrations of 0.5 to 10 mg/ml suppressed the growth of the pathogen. The same oligosaccharide at concentrations of 0.1–0.5 mg/mouse also increased the survival of infected mice (Lee *et al.*, 2009a). The number of viable pathogens in the blood, liver, small intestine and spleen was significantly lower in the treated mice compared to controls. These results suggest that chitosan has the potential to prevent and treat humans infected with *V. vulnificus* originating from raw oysters.

SHRIMP. Chitosan coatings (9 mg/g of shrimp) inhibited the growth of spoilage flora in raw shrimp from 8 log CFU in the controls to 4 log CFU during four weeks (Roller and Covill, 2000). Additional studies indicate that chitosan can also be used for the preservation of shrimp salad and other seafood (Simpson *et al.*, 1997; Tsai *et al.*, 2002). These results suggest chitosan has the potential for fish and shellfish preservation.

Antimicrobial textiles

Chitosan-treated fabrics could potentially be used to protect food against pathogens, similar to the proposed uses of other wraps. Here we briefly mention reported studies with cotton and wool.

Cotton

Cellulose fibres cross-linked to chitosan can be used to prepare antimicrobial cotton fabrics (Alonso et al., 2009). These fabrics inhibited the growth of *E. coli* and *Penicillium chrysogenum*. Another study showed that chitosan is irreversibly bound to the cotton fabric (Cakara *et al.*, 2009).

Wool

Wool cross-linked by chitosan biguanidine exhibited antimicrobial properties that were retained after washing (Zhao *et al.*, 2009). These observations suggest that positively charged pyridinium side chains in wool (Friedman and Noma, 1970; Friedman, 2001) may also exhibit antimicrobial properties (Hsieh *et al.*, 2004; Ammayappan and Jeyakodi Moses, 2009).

Antiviral activities

Because pathogenic viruses can also contaminate food, we briefly summarize the following observed antiviral activities of chitosans.

- Treatment of tobacco plants with 0.1% chitosan solutions suppressed the growth of the tobacco necrosis necrovirus (TNV) (Iriti *et al.*, 2006).
- Although the monomeric chitosan molecules glucosamine and N-acetylglucosamine exhibited no antiviral activity, low-molecular-weight chitosans at concentrations of 10 or 100 μg/ml prevented infection of beans (*Phaseolus vulgaris*) by a mosaic virus (Kulikov *et al.*, 2006).
- Sulfated chitosan at a concentration as low as 0.29 μg/ml completely inhibited the infection of the AIDS virus (HIV-I) in blood cells (T-lymphocytes) (Nishimura *et al.*, 1998).
- Chitosan facilitates the absorption and biological utilization of the antiviral drug acyclovir (Dhaliwal *et al.*, 2008; Shah *et al.*, 2008) and of peptide antibiotics (Van Der Merwe *et al.*, 2004).
- Possible mechanisms of viral inhibition by polycationic chitosan and derivatives are described in Chirkov (2002), Rabea *et al.* (2003) and Carlescu *et al.* (2009).

The observed cytotoxicity of chitosan against viruses suggests that chitosans can be used to manage viral-induced plant and human diseases. There is a need to find out whether the biopolymer will inhibit food-borne viruses such as the hepatitis A virus on spinach (Shieh *et al.*, 2009) and noroviruses in other foods (Mokhtari and Jaykus, 2009; Sala Farré *et al.*, 2009).

Mechanisms of Beneficial Effects

Understanding the molecular basis of the beneficial effects of chitosans should minimize adverse microbial and maximize beneficial sensory, nutritional and health effects of treated foods in the diet. Such efforts should lead to better and safer foods and improved human health.

Antimicrobial mechanisms

The main mechanism that seems to govern the bacteriostatic and bactericidal effects of chitosan appears to involve binding of its positively charged amino ($-NH_3^+$) groups to negatively charged carboxylate ($-COO^-$) groups located on the surface of the bacterial cell membranes (Rabea *et al.*, 2003). Such electrochemical binding can alter the distribution of negative and positive charges on the surfaces of the cell membranes, leading to weakening and/or disruption of the membranes followed by leakage of cell components. This mechanism is supported by electron microscopy studies that showed that the polymer binds to and weakens the outer membrane of bacteria (Helander *et al.*, 2001) as well as by atomic force microscopy studies which indicate that chitosan nanoparticles induced disruption of cell membranes and leakage of cytoplasm of *Salmonella choleraesuis* organisms (Qi *et al.*, 2004).

The pH of the microenvironment in which chitosan operates determines the relative ratios of unprotonated and protonated amino groups, which are governed by the equilibrium:

$$\text{Chitosan-NH}_2\text{ (unprotonated)} + H^+ \leftrightarrow \text{Chitosan-NH}_3^+\text{ (protonated); pKa} = 6.5 \quad (1)$$

At a pH = pKa, 50% of amino group are protonated. At pH 5.5, the positively charged amino group contribute 90% and at pH 4.5, 99% to the equilibrium shown in equation (1).

The antimicrobial effectiveness of chitosan seems to be highest below pH 6 where the protonated form predominates and where chitosan is most soluble. By contrast, only the unprotonated form can chelate essential metal ions. These considerations suggest that, depending on pH, different mechanisms may operate in different food categories and that lowering the internal pH of meat may enhance the antimicrobial activity of chitosan. The internal pH of the ground meat and turkey were 6.25 and 6.46, respectively (Juneja *et al.*, 2006).

The following experimental findings contribute to our understanding of possible mechanisms of bactericidal and antifungal effects of chitosans and chitosan metallo-complexes. These effects seem to be largely, but not exclusively, due to their specific perturbations of the ordered structure of phosphatidylcholine and phosphatidylethanolamine bilayers constituting bacterial cell wall membranes via electrochemical and hydrogen-bond interactions, as described in detail elsewhere for tea catechins (Friedman, 2007; Sirk *et al.*, 2008).

- Among six pathogens tested, *Campylobacter* spp. was most susceptible to inactivation by chitosan with minimum inhibitory concentration (MIC) values ranging from 0.005 to 0.05% (Ganan *et al.*, 2009). Inhibition of *Campylobacter* was accompanied by loss of membrane integrity. There was a change in cell-membrane resistance toward a loss of integrity as the cells entered the stationary phases. This finding confirms our previous observations that *Campylobacter* strains are much more susceptible to inactivation than other food-borne pathogens (Friedman *et al.*, 2002, 2003).
- The mechanism of antibacterial action of chitosan may involve cross-linking or association between positively charged amino groups and negatively charged anions on the bacterial surface. This results in changes in the membrane permeability, and leakage of cell components followed by cell death (Jeon *et al.*, 2002).

- Chitosan caused leakage of glucose and lactate dehydrogenase from *E. coli* cells (Tsai and Su, 1999).
- Based on the observed relative antibacterial lethality rates induced by chitosan powder against *E. coli*, *Enterococcus faecalis*, *P. aeruginosa* and *Staphylococcus saprophyticus*, Andres *et al.* (2007) suggest that the antibacterial mechanism involves cell-wall disruption induced by chitosan amino groups.
- The following MICs (in ppm) of chitosan against fungi show a 500-fold difference in resistance to inhibition among seven fungi that contaminate food plants and food: *Botrytis cinerea*, *Drechstera sorokiana*, *Micronectriella nivalis* (10; least resistant to inhibition), *Fusarium oxysporum* (100), *Rhizoctonia solani* (1000), *Trichophyton equinum* (2500) and *Piricularia oryzae* (5000; most resistant) (Rabea *et al.*, 2003).
- Chitosan inhibits RNA synthesis by binding to RNA of microorganisms (Sudarshan *et al.*, 1992; Rabea *et al.*, 2003).

Antioxidative mechanisms

Because the antioxidative effect of chitosan also contributes to preservation and shelf-life extension of foods, there is a need to better understand the molecular basis of the reported antioxidative free-radical-scavenging activity. An antioxidative effect can, in principle, occur during mixing of chitosan with foods containing peroxide free radicals that results in transfer of the free electron from the peroxide to the electron sink of chitosan. It is not, however, immediately apparent why chitosan behaves as an antioxidant in foods because it does not contain carbonyl or phenolic groups that are reported to stabilize the free electron abstracted from other food ingredients, as discussed in detail elsewhere (Friedman, 1997; Nam *et al.*, 2005b; Choe and Min, 2006; Choi *et al.*, 2007). The following observations suggest that chitosan does exhibit antioxidative effects in foods and *in vivo*.

- Added chitosans of different molecular weights (30, 89 and 120 kDa) at 0.2% and 0.5% (w/v) exhibited antioxidative abilities in salmon (Ye *et al.*, 2008b; Vásconez *et al.*, 2009). Scavenging of free radicals increased with concentration and decreased with increased molecular weights of chitosans. Antioxidative activities may be due to bonding of chitosan amino groups to iron in ferritin, hemoglobin and myoglobin present in salmon. The iron in these metalloproteins is known to be released during storage. The released ferrous ions can then activate oxygen and initiate lipid oxidation (Darmadji and Izumimoto, 1994; Xue *et al.*, 1998; Kamil *et al.*, 2002).
- Grafted eugenol and carvacrol enhanced both antimicrobial and antioxidative activities of chitosan nanoparticles (Chen *et al.*, 2009). Modification with flavonoids also increased both antimicrobial and antioxidative effects of chitosan (Sousa *et al.*, 2009).
- Tomida *et al.* (2009) measured the ability of several chitosans with a range of molecular weights to protect plasma proteins of human volunteers against oxidation by peroxyl radicals. They observed a linear correlation between antioxidant activity and the molecular weights of the chitosans *in vitro*. Low-molecular-weight chitosans (20–30 kDa) were most effective in protecting the proteins against the formation of carbonyl oxidation products. Differences in intramolecular hydrogen bonding between amino and hydroxyl groups may govern the observed relative antioxidative effects.
- A 1% chitosan solution exhibited higher radical-scavenging activity than did the known food preservative butylated hydroxytoluene (BHT) (Kim and Thomas, 2007).

Metal-ion-chelating mechanisms

Friedman and colleagues (Friedman and Waiss, 1972; Masri *et al.*, 1974) found that chitosan and other natural biopolymers have a strong affinity for toxic (cadmium, cobalt, copper, gold, iron, lead and mercury) and for nutritionally essential (copper, iron, manganese and zinc) metal ions. Based on this observation, they suggested that the natural biopolymers may be useful for the removal of toxic metal salts from contaminated water supplies. This suggestion was later realized by numerous studies designed to demonstrate this possibility (Piron and Domard, 1998; Jeon and Park, 2005; Lv *et al.*, 2009; Miretzky and Cirelli, 2009; Zhou *et al.*, 2009).

The inhibitory effect of chitosan–metal complexes with Cu (II), Zn (II) and Fe (II) ions against two Gram-positive bacteria (*S. aureus* and *S. epidermidis*), two Gram-negative bacteria (*E. coli* and *P. aeruginosa*) and two fungi (*Candida albicans* and *Candida parapsilosis*) was dependent on the metal ion, the molecular weight, the degree of deacetylation of chitosan and the pH of the environment (Wang *et al.*, 2005). Electron microscopy studies indicated that the antimicrobial action of the chitosan–Cu (II) complex against *S. aureus* resulted in the disruption of the bacterial cell envelope (Wang *et al.*, 2005). Binding of bacterial trace metals by chitosan inhibited both microbial growth and production of bacterial toxins (Knowles and Roller, 2001). It is also likely that binding of chitosan to trace elements, such as ferric and zinc ions that the bacteria need for growth, may contribute to its antimicrobial action (Rabea *et al.*, 2003). The use of chitosan to eliminate toxic metal ions from contaminated liquid foods (milk, fruit and vegetable juices, etc.) merits study.

Medicinal Properties of Chitosans

In addition to antimicrobial activities, chitosans are reported to have numerous medicinal properties, reviewed in Yin *et al.* (2009). These include anti-carcinogenic, anti-cholesterol, anti-obesity, anti-hypertensive, improved drug delivery, haemostatic, immune-stimulating and neuroprotective effects. Food scientists, microbiologists, physicians, veterinarians and the general public who may be interested in anti-microbial effects of chitosans should also be aware of these potential health benefits of chitosans. Here, we briefly discuss selected recent studies on potential applications of chitosans in human and veterinary medicine.

Anti-carcinogenic activities

The following observations suggest that chitosan has the potential to be used in chemotherapy of cancers with minimal side effects. (i) Succinyl-chitosan nanoparticles were found to inhibit K562 cancer cells (Luo *et al.*, 2009). Cell death by apoptosis and necrosis was accompanied by a decrease in intracellular reactive oxygen species (ROS). (ii) The apparent effectiveness of chitosan–alginate–DNA nanoparticles and ultrasound to control gene transfection of HeLa and 293 T cancer cells suggests that this formulation shows promise for use in gene therapy of cancers (Yang *et al.*, 2010).

Anti-cholesterol effects

Positively charged chitosan amino groups can, in principle, bind negatively charged lipids, including long-chain fatty acids, cholesterol and bile acids forming complexes, which can then be eliminated in the faeces (Wydro *et al.*, 2007). A meta-analysis of six controlled studies (n = 416 patients) through May 2008 revealed that chitosan significantly lowered total cholesterol but not low-density lipoprotein (LDL) cholesterol, high-density lipoprotein (HDL) cholesterol or plasma triglyceride levels (Baker *et al.*, 2009). We found that diets containing the tomato glycoalkaloid tomatine also reduced plasma cholesterol levels in mice by a similar mechanism (Friedman *et al.*, 2000a,b).

Anti-obesity effects

The administration of six 500 mg chitosan capsules per day to 134 human overweight volunteers facilitated the depletion of excess body fat with minimal loss of fat-free or lean body mass (Kaats *et al.*, 2006). Chitosan appears to act as a potent down-regulator of obesity-related gene expression and is an effective agent in controlling food intake, body weight gain, blood glucose and lipid profile in genetically modified mice (Kumar *et al.*, 2009). Chitosan also induced hypolipidemia in obese rats (Khairunnuur *et al.*, 2010). The possible benefits of chitosan in the treatment of childhood obesity, the most common paediatric disorder in children (Rogovik *et al.*, 2010), merits study.

Facilitation of drug delivery

The following studies report that chitosan-containing adhesives, gels and capsules facilitate controlled delivery of several medicines to target tissues, reviewed in Amidi *et al.* (2010). These include (i) amoxicillin for the treatment of *Helicobacter pylori* infections of the digestive tract (Chang *et al.*, 2010); (ii) ellagic acid for the treatment of brain cancer (Kim *et al.*, 2010); (iii) 5-fluorouracil for the treatment of multi-organ cancers (Huang *et al.*, 2010); (iv) oral insulin for the treatment of diabetes (Sonaje *et al.*, 2010); (v) nimodipine for the treatment of high blood pressure (Hassan *et al.*, 2010); and (vi) timolol for the treatment of glaucoma (Gupta *et al.*, 2010).

Facilitation of haemostasis

Several studies report that chitosan-containing dressings and sponges have the potential to diminish blood flow in wounded tissues during surgery, thus limiting post-operative bleeding. These include nasal surgeries (Dailey *et al.*, 2009; Valentine *et al.*, 2010) and liver injuries (Bochicchio *et al.*, 2010; Gu *et al.*, 2010). Because chitosans appear to be compatible with blood and cells, they merit use as absorbable, implantable agent for promoting post-operative hemostasis (Gu *et al.*, 2010).

Human antifungal effects

High-molecular-weight chitosans inhibited the growth of clinical strains of *Candida* spp., suggesting that these compounds could potentially be used in vulvovaginal candidiasis (Tapia *et al.*, 2009). Related studies showed that chitosans have the potential to be used in the prophylaxis of medical devices infected with fungal biofilms (Martinez *et al.*, 2010) and of phytopathogenic fungi in pear fruit (Meng *et al.*, 2010).

Neuroprotection

Topical application of chitosan to the spinal cord of guinea pigs sealed compromised nerve-cell membranes, thus serving as a potent neuroprotector of spinal chord trauma and brain injury (Cho *et al.*, 2010). The protective effect was accompanied by suppression of the generation of reactive oxygen species and the resultant lipid peroxidation of membranes.

Vaccine adjuvants

Chitosans enhanced the protective immunity effect in mice of a vaccine against the influenza A virus (Sui *et al.*, 2010) and were found to be a promising adjuvant for the delivery of live vaccine against Newcastle viral disease in poultry (Rauw *et al.*, 2010). The mechanism of immune stimulation appears to involve enhancement of the antigen-specific cell-mediated immune response in the spleen.

The cited observations show that numerous beneficial effects of free chitosans complement antimicrobial activities. An unanswered question is whether chitosan-treated food will exhibit similar medicinal properties.

Research Needs

To further enhance the potential of chitosans to help assure food quality and safety, future studies need to address the following food-related aspects.

1. Determine whether the potent antibiotic preservative effects of chitosan *in vitro* can be duplicated *in vivo*, especially in humans. Are inhibitory effects by chitosan clinically significant, i.e. would human consumption of chitosan-containing food be expected to reduce symptoms of infectious diseases?
2. Define additive and/or synergistic activities of mixtures of chitosans with other plant-derived antimicrobials such as oregano oil, sodium lactate, and polyphenol-rich apple, grape, olive and tea extracts (Juneja *et al.*, 2007, 2010; Wong *et al.*, 2008) and with medical antibiotics such as methicillin and vancomycin. Combinations of natural antimicrobials that act synergistically will lessen amounts needed to design effective antimicrobial food formulations. They will be safer and will affect flavour and taste less compared to the use of individual compounds.
3. Determine whether chitosan will inhibit both pathogens and carcinogenic heterocyclic amines during baking and grilling of meat, poultry and seafood products (Friedman *et al.*, 2009).
4. Determine whether human consumption of chitosan-treated food will contribute to reported beneficial effects of chitosan against cancers (Lee *et al.*, 2009b; Das *et al.*, 2010), cholesterol (Alsarra, 2009; Baker *et al.*, 2009; Bangoura *et al.*, 2009), obesity (Jull *et al.*, 2008; Rogovik and Goldman, 2008) and wound healing (Alsarra, 2009; Ribeiro *et al.*, 2009).
5. Evaluate effectiveness of chitosan against antibiotic-resistant pathogens (Friedman *et al.*, 2004; Friedman, 2006; Ravishankar *et al.*, 2008).
6. Determine whether chitosan added to animal feed can replace standard antibiotics, whose use is being discontinued (Alali *et al.*, 2009; Apata, 2009).
7. Determine the fate of chitosan during processing food by heat, microwaves and radiation.
8. Determine long-term stabilities of added natural antimicrobials in chitosan films and microspheres (Altiok *et al.*, 2009; Ravishankar *et al.*, 2009).
9. Determine whether chitosan continues inhibiting the growth of bacteria during post-thermal processing and storage of ground beef and poultry products (Juneja *et al.*, 2010).
10. Determine whether molecular modelling of chitosan structure–cell membrane interactions can be used to predict antibiotic activities of chitosan and derivatives, as seems to be the case for antimicrobial tea catechins (Sirk *et al.*, 2008, 2009).
11. Determine whether chitosan can concurrently inactivate pathogenic bacteria such as *Clostridium botulinum, E. coli, S. aureus* and the botulinum, Shiga, and staphylococcus toxins produced by these bacteria (Friedman, 2007; Quiñones *et al.*, 2009; Rasooly *et al.*, 2010a,b,c).
12. Determine whether the mechanisms of protection against food-borne pathogens are similar to those that may govern the inhibition of phytopathogenic bacteria, including strains of *Agrobacterium, Clavibacter, Pseudomonas, Erwinia* and *Xanthomonas* that contaminate cabbage, aubergines, grapes, lettuce, onions, potatoes and tomatoes (Richards and Beuchat, 2005).
13. Determine whether chitosan can be used to both inactivate pathogenic organisms and eliminate toxic and radioactive metals (cadmium, lead, mercury and uranium) from liquid and solid foods (Miretzky and Cirelli, 2009; Rana *et al.*, 2009; Wang *et al.*, 2009).
14. Determine whether chitosan-treated cotton and wool fabrics can be used as antimicrobial wraps for food.
15. Determine whether added chitosans will inactivate the hepatitis A virus and the norovirus in food (Butot *et al.*, 2009; Grove *et al.*, 2009; Schmid *et al.*, 2009).

Acknowledgements

We thank Carol E. Levin for assistance with the preparation of the manuscript.

References

Alali, W.Q., Scott, H.M., Christian, K.L., Fajt, V.R., Harvey, R.B. and Lawhorn, D.B. (2009) Relationship between level of antibiotic use and resistance among *Escherichia coli* isolates from integrated multi-site cohorts of humans and swine. *Preventive Veterinary Medicine* 90, 160–167.

Alonso, D., Gimeno, M., Olayo, R., Vázquez-Torres, H., Sepúlveda-Sánchez, J.D. and Shirai, K. (2009) Cross-linking chitosan into UV-irradiated cellulose fibers for the preparation of antimicrobial-finished textiles. *Carbohydrate Polymers* 77, 536–543.

Alsarra, I.A. (2009) Chitosan topical gel formulation in the management of burn wounds. *International Journal of Biological Macromolecules* 45, 16–21.

Altieri, C., Scrocco, C., Sinigaglia, M. and Del Nobile, M.A. (2005) Use of chitosan to prolong mozzarella cheese shelf life. *Journal of Dairy Science* 88, 2683–2688.

Altiok, D., Altiok, E., Bayraktar, O. and Tihminlioglu, F. (2009) Stability of trans-resveratrol incorporated in chitosan microspheres. In: Proceedings of 2009 14th National Biomedical Engineering Meeting, BIYOMUT 2009.

Amidi, M., Mastrobattista, E., Jiskoot, W. and Hennink, W.E. (2010) Chitosan-based delivery systems for protein therapeutics and antigens. *Advanced Drug Delivery Reviews* 62, 59–82.

Ammayappan, L. and Jeyakodi Moses, J. (2009) Study of antimicrobial activity of aloevera, chitosan, and curcumin on cotton, wool, and rabbit hair. *Fibers and Polymers* 10, 161–166.

Andres, Y., Giraud, L., Gerente, C. and Le Cloirec, P. (2007) Antibacterial effects of Chitosan powder: Mechanisms of action. *Environmental Technology* 28, 1357–1363.

Anjani, K., Kailasapathy, K. and Phillips, M. (2007) Microencapsulation of enzymes for potential application in acceleration of cheese ripening. *International Dairy Journal* 17, 79–86.

Apata, D. F. (2009) Antibiotic resistance in poultry. *International Journal of Poultry Science* 8, 404–408.

Aranaz, I., Mengíbar, M., Harris, R., Paños, I., Miralles, B., Acosta, N., Galed, G. and Heras, Á. (2009) Functional characterization of chitin and chitosan. *Current Chemical Biology* 3, 203–230.

Atia, M.M.M., Buchenauer, H., Aly, A.Z. and Abou-Zaid, M.I. (2005) Antifungal activity of chitosan against *Phytophthora infestans* and activation of defence mechanisms in tomato to late blight. *Biological Agriculture and Horticulture* 23, 175–197.

Avadi, M.R., Sadeghi, A.M.M., Tahzibi, A., Bayati, K., Pouladzadeh, M., Zohuriaan-Mehr, M.J. and Rafiee-Tehrani, M. (2004) Diethylmethyl chitosan as an antimicrobial agent: Synthesis, characterization and antibacterial effects. *European Polymer Journal* 40, 1355–1361.

Baker, W.L., Tercius, A., Anglade, M., White, C.M. and Coleman, C.I. (2009) A meta-analysis evaluating the impact of chitosan on serum lipids in hypercholesterolemic patients. *Annals of Nutrition and Metabolism* 55, 368–374.

Bangoura, M., Wenshui, X. and Jiali, Z. (2009) In vitro binding capacity of cholesterol and bile salts by partially depolymerized chitosans. *American Journal of Food Technology* 4, 126–135.

Beverlya, R.L., Janes, M.E., Prinyawiwatkula, W. and No, H.K. (2008) Edible chitosan films on ready-to-eat roast beef for the control of *Listeria monocytogenes*. *Food Microbiology* 25, 534–537.

Bochicchio, G.V., Kilbourne, M.J., Keledjian, K., Hess, J. and Scalea, T. (2010) Evaluation of a new hemostatic agent in a porcine grade V liver injury model. *American Surgeon* 76, 317–320.

Butot, S., Putallaz, T., Amoroso, R. and Sanchez, G. (2009) Inactivation of enteric viruses in minimally processed berries and herbs. *Applied and Environmental Microbiology* 75, 4155–4161.

Cakara, D., Fras, L., Bracic, M. and Kleinschek, K.S. (2009) Protonation behavior of cotton fabric with irreversibly adsorbed chitosan: A potentiometric titration study. *Carbohydrate Polymers* 78, 36–40.

Campaniello, D., Bevilacqua, A., Sinigaglia, M. and Corbo, M.R. (2008) Chitosan: antimicrobial activity and potential applications for preserving minimally processed strawberries. *Food Microbiology* 25, 992–1000.

Canale Rappussi, M.C., Pascholati, S.F., Benato, E.A. and Cia, P. (2009) Chitosan reduces infection by *Guignardia citricarpa* in postharvest 'Valencia' oranges. *Brazilian Archives of Biology and Technology* 52, 513–521.

Cao, R., Xue, C.-H. and Liu, Q. (2009) Changes in microbial flora of Pacific oysters (*Crassostrea gigas*) during refrigerated storage and its shelf-life extension by chitosan. *International Journal of Food Microbiology* 131, 272–276.

Carlescu, I., Scutaru, D., Popa, M. and Uglea, C.V. (2009) Synthetic sialic-acid-containing polyvalent antiviral inhibitors. *Medicinal Chemistry Research* 18, 477–494.

Chang, C.H., Lin, Y.H., Yeh, C.L., Chen, Y.C., Chiou, S.F., Hsu, Y.M., Chen, Y.S. and Wang, C.C. (2010) Nanoparticles incorporated in pH-sensitive hydrogels as amoxicillin delivery for eradication of *Helicobacter pylori*. *Biomacromolecules* 11, 133–142.
Chen, C.-S., Liau, W.-Y. and Tsai, G.-J. (1998) Antibacterial effects of N-sulfonated and N-sulfobenzoyl chitosan and application to oyster preservation. *Journal of Food Protection* 61, 1124–1128.
Chen, F., Shi, Z., Neoh, K.G. and Kang, E.T. (2009) Antioxidant and antibacterial activities of eugenol and carvacrol-grafted chitosan nanoparticles. *Biotechnology and Bioengineering* 104, 30–39.
Chhabra, P., Huang, Y.-W., Frank, J.F., Chmielewski, R. and Gates, K. (2006) Fate of *Staphylococcus aureus*, *Salmonella enterica* serovar Typhimurium, and *Vibrio vulnificus* in raw oysters treated with chitosan. *Journal of Food Protection* 69, 1600–1604.
Chi, S., Zivanovic, S. and Penfield, M.P. (2006) Application of chitosan films enriched with oregano essential oil on bologna – active compounds and sensory attributes. *Food Science and Technology International* 12, 111–117.
Chien, P.-J., Sheu, F., Huang, W.-T. and Su, M.-S. (2007a) Effect of molecular weight of chitosans on their antioxidative activities in apple juice. *Food Chemistry* 102, 1192–1198.
Chien, P.-J., Sheu, F. and Lin, H.-R. (2007b) Coating citrus (*Murcott tangor*) fruit with low molecular weight chitosan increases postharvest quality and shelf life. *Food Chemistry* 100, 1160–1164.
Chien, P.-J., Sheu, F. and Yang, F.-H. (2007c) Effects of edible chitosan coating on quality and shelf life of sliced mango fruit. *Journal of Food Engineering* 78, 225–229.
Chirkov, S.N. (2002) The antiviral activity of chitosan (review). *Applied Biochemistry and Microbiology* 38, 1–8.
Cho, Y., Shi, R. and Borgens, R.B. (2010) Chitosan nanoparticle-based neuronal membrane sealing and neuroprotection following acrolein-induced cell injury. *Journal of Biological Engineering*, in press, 2.
Choe, E. and Min, D. B. (2006) Chemistry and reactions of reactive oxygen species in foods. *Journal of Food Science* 70, R142–R159.
Choi, S.P., Kang, M.Y., Koh, H.J., Nam, S.H. and Friedman, M. (2007) Antiallergic activities of pigmented rice bran extracts in cell assays. *Journal of Food Science* 72, S719–S726.
Coma, V., Deschamps, A. and Martial-Gros, A. (2003) Bioactive packaging materials from edible chitosan polymer – antimicrobial activity assessment on dairy-related contaminants. *Journal of Food Science* 68, 2788–2792.
Cooksey, K. (2005) Effectiveness of antimicrobial food packaging materials. *Food Additives and Contaminants* 22, 980–987.
Crawford, L., McDonald, G.M. and Friedman, M. (1990) Composition of sicklepod (*Cassia obtusifolia*) toxic weed seeds. *Journal of Agricultural and Food Chemistry* 38, 2169–2175.
Dailey, R.A., Chavez, M.R. and Choi, D. (2009) Use of a chitosan-based hemostatic dressing in dacryocystorhinostomy. *Ophthalmic Plastic and Reconstructive Surgery* 25, 350–353.
Darmadji, P. and Izumimoto, M. (1994) Effect of chitosan in meat preservation. *Meat Science* 38, 243–254.
Das, R.K., Kasoju, N. and Bora, U. (2010) Encapsulation of curcumin in alginate-chitosan-pluronic composite nanoparticles for delivery to cancer cells. *Nanomedicine: Nanotechnology, Biology, and Medicine* 6, e153–e160.
De Reuck, K., Sivakumar, D. and Korsten, L. (2009) Effect of integrated application of chitosan coating and modified atmosphere packaging on overall quality retention in litchi cultivars. *Journal of the Science of Food and Agriculture* 89, 915–920.
Del Nobile, M.A., Di Benedetto, N., Suriano, N., Conte, A., Corbo, M.R. and Sinigaglia, M. (2009a) Combined effects of chitosan and MAP to improve the microbial quality of amaranth homemade fresh pasta. *Food Microbiology* 26, 587–591.
Del Nobile, M.A., Di Benedetto, N., Suriano, N., Conte, A., Lamacchia, C., Corbo, M.R. and Sinigaglia, M. (2009b) Use of natural compounds to improve the microbial stability of Amaranth-based homemade fresh pasta. *Food Microbiology* 26, 151–156.
Del Nobile, M.A., Gammariello, D., Conte, A. and Attanasio, M. (2009c) A combination of chitosan, coating and modified atmosphere packaging for prolonging Fior di latte cheese shelf life. *Carbohydrate Polymers* 78, 151–156.
Devlieghere, F., Vermeulen, A. and Debevere, J. (2004) Chitosan: antimicrobial activity, interactions with food components and applicability as a coating on fruit and vegetables. *Food Microbiology (London)* 21, 703–714.
Dhaliwal, S., Jain, S., Singh, H.P. and Tiwary, A.K. (2008) Mucoadhesive microspheres for gastroretentive delivery of acyclovir: In vitro and in vivo evaluation. *AAPS Journal* 10, 322–330.

Dong, H., Cheng, L., Tan, J., Zheng, K. and Jiang, Y. (2004) Effects of chitosan coating on quality and shelf life of peeled litchi fruit. *Journal of Food Engineering* 64, 355–358.
Duan, J., Park, S.-I., Daeschel, M.A. and Zhao, Y. (2007) Antimicrobial chitosan-lysozyme (CL) films and coatings for enhancing microbial safety of mozzarella cheese. *Journal of Food Science* 72, M355–M362.
Duan, J., Cherian, G. and Zhao, Y. (2010) Quality enhancement in fresh and frozen lingcod (*Ophiodon elongates*) fillets by employment of fish oil incorporated chitosan coatings. *Food Chemistry* 119, 524–532.
Dugan, G.M., Gumbmann, M.R. and Friedman, M. (1989) Toxicological evaluation of jimson weed (*Datura stramonium*) seed. *Food and Chemical Toxicology* 27, 501–510.
Dutta, P. K., Tripathi, S., Mehrotra, G.K. and Dutta, J. (2009) Perspectives for chitosan based antimicrobial films in food applications. *Food Chemistry* 114, 1173–1182.
Friedman, M. (1997) Chemistry, biochemistry, and dietary role of potato polyphenols. A review. *Journal of Agricultural and Food Chemistry* 45, 1523–1540.
Friedman, M. (2001) Application of the S-pyridylethylation reaction to the elucidation of the structures and functions of proteins. *Journal of Protein Chemistry* 20, 431–453.
Friedman, M. (2004) Analysis of biologically active compounds in potatoes (*Solanum tuberosum*), tomatoes (*Lycopersicon esculentum*), and jimson weed (*Datura stramonium*) seeds. *Journal of Chromatography A* 1054, 143–155.
Friedman, M. (2006) Structure-antibiotic activity relationships of plant compounds against nonresistant and antibiotic-resistant foodborne pathogens. In: Juneja, V.K., Cherry, J.P. and Tunick, M.H. (eds) *Advances in Microbial Food Safety, ACS Symposium Series*. American Chemical Society, Washington, DC, pp. 167–183.
Friedman, M. (2007) Overview of antibacterial, antitoxin, antiviral, and antifungal activities of tea compounds. *Molecular Nutrition and Food Research* 51, 116–134.
Friedman, M. and Dao, L. (1990) Effect of autoclaving and conventional and microwave baking on the ergot alkaloid and chlorogenic acid contents of morning glory (*Ipomoea tricolor* Cav. cv.) heavenly blue seeds. *Journal of Agricultural and Food Chemistry* 38, 805–808.
Friedman, M. and Henika, P.R. (1991) Mutagenicity of toxic weed seeds in the Ames test: Jimson weed (*Datura stramonium*), velvetleaf (*Abutilon theophrasti*), morning glory (*Ipomoea* spp.), and sicklepod (*Cassia obtusifolia*). Journal of Agricultural and Food Chemistry 39, 494–501.
Friedman, M. and Levin, C.E. (1989) Composition of jimson weed (*Datura stramonium*) seeds. *Journal of Agricultural and Food Chemistry* 37, 998–1005.
Friedman, M. and Noma, A.T. (1970) Cystine content of wool. *Textile Research Journal* 40, 1073–1078.
Friedman, M. and Waiss, S.J., Jr. (1972) Mercury uptake by selected agricultural products and byproducts. *Environmental Science and Technology* 6, 457–458.
Friedman, M., Dao, L. and Gumbmann, M.R. (1989) Ergot alkaloid and chlorogenic acid content in different varieties of morning-glory (*Ipomoea* spp.) seeds. *Journal of Agricultural and Food Chemistry* 37, 708–712.
Friedman, M., Fitch, T.E., Levin, C.E. and Yokoyama, W.H. (2000a) Feeding tomatoes to hamsters reduces their plasma low-density lipoprotein cholesterol and triglycerides. *Journal of Food Science* 65, 897–900.
Friedman, M., Fitch, T.E. and Yokoyama, W.E. (2000b) Lowering of plasma LDL cholesterol in hamsters by the tomato glycoalkaloid tomatine. *Food and Chemical Toxicology* 38, 549–553.
Friedman, M., Henika, P.R. and Mandrell, R.E. (2002) Bactericidal activities of plant essential oils and some of their isolated constituents against *Campylobacter jejuni*, *Escherichia coli*, *Listeria monocytogenes*, and *Salmonella enterica*. *Journal of Food Protection* 65, 1545–1560.
Friedman, M., Henika, P.R. and Mandrell, R.E. (2003) Antibacterial activities of phenolic benzaldehydes and benzoic acids against *Campylobacter jejuni, Escherichia coli, Listeria monocytogenes*, and *Salmonella enterica*. *Journal of Food Protection* 66, 1811–1821.
Friedman, M., Buick, R. and Elliott, C.T. (2004) Antibacterial activities of naturally occurring compounds against antibiotic-resistant *Bacillus cereus* vegetative cells and spores, *Escherichia coli*, and *Staphylococcus aureus*. *Journal of Food Protection* 67, 1774–1778.
Friedman, M., Zhu, L., Feinstein, Y. and Ravishankar, S. (2009) Carvacrol facilitates heat-induced inactivation of *Escherichia coli* O157:H7 and inhibits formation of heterocyclic amines in grilled ground beef patties. *Journal of Agricultural and Food Chemistry* 57, 1848–1853.
Ganan, M., Carrascosa, A.V. and Martínez-Rodríguez, A.J. (2009) Antimicrobial activity of chitosan against

Campylobacter spp. and other microorganisms and its mechanism of action. *Journal of Food Protection* 72, 1735–1738.

Georgantelis, D., Ambrosiadis, I., Katikou, P., Blekas, G. and Georgakis, S.A. (2007a) Effect of rosemary extract, chitosan and a-tocopherol on microbiological parameters and lipid oxidation of fresh pork sausages stored at 4 °C. *Meat Science* 76, 172–181.

Georgantelis, D., Blekas, G., Katikou, P., Ambrosiadis, I. and Fletouris, D.J. (2007b) Effect of rosemary extract, chitosan and a-tocopherol on lipid oxidation and colour stability during frozen storage of beef burgers. *Meat Science* 75, 266–274.

Giatrakou, V., Ntzimani, A. and Savvaidis, I.N. (2010) Effect of chitosan and thyme oil on a ready to cook chicken product. *Food Microbiology* 27, 132–136.

Grove, S.F., Lee, A., Stewart, C.M. and Ross, T. (2009) Development of a high pressure processing inactivation model for hepatitis A virus. *Journal of Food Protection* 72, 1434–1442.

Gu, R., Sun, W., Zhou, H., Wu, Z., Meng, Z., Zhu, X., Tang, Q., Dong, J. and Dou, G. (2010) The performance of a fly-larva shell-derived chitosan sponge as an absorbable surgical hemostatic agent. *Biomaterials* 31, 1270–1277.

Gupta, H., Velpandian, T. and Jain, S. (2010) Ion- and pH-activated novel in-situ gel system for sustained ocular drug delivery. *Journal of Drug Targeting* 18, 499–505.

Han, C., Lederer, C., McDaniel, M. and Zhao, Y. (2005) Sensory evaluation of fresh strawberries (*Fragaria ananassa*) coated with chitosan-based edible coatings. *Journal of Food Science* 70, S172–S178.

Hassan, N., Ali, M. and Ali, J. (2010) Novel buccal adhesive system for anti-hypertensive agent Nimodipine. *Pharmaceutical Development and Technology* 15, 124–130.

Helander, I.M., Nurmiaho-Lassila, E.L., Ahvenainen, R., Rhoades, J. and Roller, S. (2001) Chitosan disrupts the barrier properties of the outer membrane of gram-negative bacteria. *International Journal of Food Microbiology* 71, 235–244.

Hsieh, S.-H., Huang, Z.K., Huang, Z.Z. and Tseng, Z.S. (2004) Antimicrobial and physical properties of woolen fabrics cured with citric acid and chitosan. *Journal of Applied Polymer Science* 94, 1999–2007.

Huang, L., Sui, W., Wang, Y. and Jiao, Q. (2010) Preparation of chitosan/chondroitin sulfate complex microcapsules and application in controlled release of 5-fluorouracil. *Carbohydrate Polymers* 80, 168–173.

Iriti, M., Sironi, M., Gomarasca, S., Casazza, A.P., Soave, C. and Faoro, F. (2006) Cell death-mediated antiviral effect of chitosan in tobacco. *Plant Physiology and Biochemistry* 44, 893–900.

Je, J.-Y. and Kim, S.-K. (2006) Chitosan derivatives killed bacteria by disrupting the outer and inner membrane. *Journal of Agricultural and Food Chemistry* 54, 6629–6633.

Jeon, C. and Park, K.H. (2005) Adsorption and desorption characteristics of mercury(II) ions using aminated chitosan bead. *Water Research* 39, 3938–3944.

Jeon, Y.J., Kamil, J.Y.V.A. and Shahidi, F. (2002) Chitosan as an edible invisible film for quality preservation of herring and Atlantic cod. *Journal of Agricultural and Food Chemistry* 50, 5167–5178.

Jiang, Y., Li, J. and Jiang, W. (2005) Effects of chitosan coating on shelf life of cold-stored litchi fruit at ambient temperature. *Lebensmittel-Wissenschaft und-Technologie* 38, 757–761.

Jitareerat, P., Paumchai, S., Kanlayanarat, S. and Sangchote, S. (2007) Effect of chitosan on ripening, enzymatic activity, and disease development in mango (*Mangifera indica*) fruit. *New Zealand Journal of Crop and Horticultural Science* 35, 211–218.

Joerger, R.D., Sabesan, S., Visioli, D., Urian, D. and Joerger, M.C. (2009) Antimicrobial activity of chitosan attached to ethylene copolymer films. *Packaging Technology and Science* 22, 125–138.

Jull, A.B., Ni Mhurchu, C., Bennett, D.A., Dunshea-Mooij, C.A. and Rodgers, A. (2008) Chitosan for overweight or obesity. *Cochrane Database of Systematic Reviews (Online)* 3, CD003892.

Juneja, V.K., Thippareddi, H., Bari, L., Inatsu, Y., Kawamoto, S. and Friedman, M. (2006) Chitosan protects cooked ground beef and turkey against *Clostridium perfringens* spores during chilling. *Journal of Food Science* 71, M236–M240.

Juneja, V.K., Bari, M.L., Inatsu, Y., Kawamoto, S. and Friedman, M. (2007) Control of *Clostridium perfringens* spores by green tea leaf extracts during cooling of cooked ground beef, chicken, and pork. *Journal of Food Protection* 70, 1429–1433.

Juneja, V.K., Hwang, C.-A. and Friedman, M. (2010) Thermal inactivation and post-thermal treatment growth during storage of multiple *Salmonella* serotypes in ground beef as affected by sodium lactate and oregano oil. *Journal of Food Science* 75, M1–M6.

Jung, B.O., Kang, S.T. and Chung, S.J. (1998) The anti-microbial activity of modified chitosan. *Korean Journal of Applied Microbiology and Biotechnology* 26, 338–344.
Kaats, G.R., Michalek, J.E. and Preuss, H.G. (2006) Evaluating efficacy of a chitosan product using a double-blinded, placebo-controlled protocol. *Journal of the American College of Nutrition* 25, 389–394.
Kamil, J.Y.V.A., Jeon, Y.-J. and Shahidi, F. (2002) Antioxidative activity of chitosans of different viscosity in cooked comminuted flesh of herring (*Clupea harengus*). *Food Chemistry* 79, 69–77.
Kanatt, S.R., Chander, R. and Sharma, A. (2004) Effect of irradiated chitosan on the rancidity of radiation-processed lamb meat. *International Journal of Food Science and Technology* 39, 997–1003.
Kanatt, S.R., Chander, R. and Sharma, A. (2008a) Chitosan and mint mixture: A new preservative for meat and meat products. *Food Chemistry* 107, 845–852.
Kanatt, S.R., Chander, R. and Sharma, A. (2008b) Chitosan glucose complex – a novel food preservative. *Food Chemistry* 106, 521–528.
Kenawy, E.-R., Abdel-Hay, F.I., El-Magd, A.A. and Mahmoud, Y. (2005) Biologically active polymers: Modification and anti-microbial activity of chitosan derivatives. *Journal of Bioactive and Compatible Polymers* 20, 95–111.
Khairunnuur, F.A., Zulkhairi, A., Hairuszah, I., Azrina, A., Nursakinah, I., Fazali, F., Kamal, M.N.H., Zamree, M.S. and Kamilah, K.A.K. (2010) Hypolipemic and weight reducing properties from *Tamarindus indica* L. pulp extract in diet-induced obese rats. *International Journal of Pharmacology* 6, 216–223.
Kim, K.W. and Thomas, R.L. (2007) Antioxidative activity of chitosans with varying molecular weights. *Food Chemistry* 101, 308–313.
Kim, K.W., Thomas, R.L., Lee, C. and Park, H.J. (2003) Antimicrobial activity of native chitosan, degraded chitosan, and O-carboxymethylated chitosan. *Journal of Food Protection* 66, 1495–1498.
Kim, K.W., Daeschel, M. and Zhao, Y. (2008) Edible coatings for enhancing microbial safety and extending shelf life of hard-boiled eggs. *Journal of Food Science* 73, M227–M235.
Kim, S., Nishimoto, S.K., Bumgardner, J.D., Haggard, W.O., Gaber, M.W. and Yang, Y. (2010) A chitosan/b-glycerophosphate thermo-sensitive gel for the delivery of ellagic acid for the treatment of brain cancer. *Biomaterials* 31, 4157–4166.
Kim, S.H., Youn, D.K., No, H.K., Choi, S.W. and Prinyawiwatkul, W. (2009) Effects of chitosan coating and storage position on quality and shelf life of eggs. *International Journal of Food Science and Technology* 44, 1351–1359.
Knowles, J. and Roller, S. (2001) Efficacy of chitosan, carvacrol, and a hydrogen peroxide-based biocide against foodborne microorganisms in suspension and adhered to stainless steel. *Journal of Food Protection* 64, 1542–1548.
Krasaekoopt, W. and Mabumrung, J. (2008) Microbiological evaluation of edible coated fresh-cut cantaloupe. *Kasetsart Journal – Natural Science* 42, 552–557.
Krasaekoopt, W., Bhandari, B. and Deeth, H.C. (2006) Survival of probiotics encapsulated in chitosan-coated alginate beads in yoghurt from UHT- and conventionally treated milk during storage. *Lebensmittel-Wissenschaft und-Technologie* 39, 177–183.
Kulikov, S.N., Chirkov, S.N., Il'ina, A.V., Lopatin, S.A. and Varlamov, V.P. (2006) Effect of the molecular weight of chitosan on its antiviral activity in plants. *Prikladnaia Biokhimiia i Mikrobiologiia* 42, 224–228.
Kulkarni, A.R., Kulkarni, V.H., Keshavayya, J., Hukkeri, V.I. and Sung, H.-W. (2005) Anti-microbial activity and film characterization of thiazolidinone derivatives of chitosan. *Macromolecular Bioscience* 5, 490–493.
Kumar, S.G., Md. Rahman, A., Lee, S.H., Hwang, H.S., Kim, H.A. and Yun, J.W. (2009) Plasma proteome analysis for anti-obesity and anti-diabetic potentials of chitosan oligosaccharides in ob/ob mice. *Proteomics* 9, 2149–2162.
Lee, B.C., Kim, M.S., Choi, S.H., Kim, K.Y. and Kim, T.S. (2009a) In vitro and in vivo antimicrobial activity of water-soluble chitosan oligosaccharides against *Vibrio vulnificus*. *International Journal of Molecular Medicine* 24, 327–333.
Lee, C.H., Park, H.J. and Lee, D.S. (2004) Influence of antimicrobial packaging on kinetics of spoilage microbial growth in milk and orange juice. *Journal of Food Engineering* 65, 527–531.
Lee, E., Kim, H., Lee, I.-H. and Jon, S. (2009b) In vivo antitumor effects of chitosan-conjugated docetaxel after oral administration. *Journal of Controlled Release* 140, 79–85.
Lee, J.W., Lee, H.H. and Rhim, J.M. (2000) Shelf life extension of white rice cake and wet noodles by treatment with chitosan. *Korean Journal of Food Science and Technology* 32, 828–833.

Liu, H., Du, Y., Yang, J. and Zhu, H. (2004) Structural characterization and antimicrobial activity of chitosan/ betaine derivative complex. *Carbohydrate Polymers* 55, 291–297.
Liu, X., Song, L., Li, L., Li, S. and Yao, K. (2007) Antibacterial effects of chitosan and its water-soluble derivatives on *E. coli*, plasmids DNA, and mRNA. *Journal of Applied Polymer Science* 103, 3521–3528.
Liu, X., Chen, H.-X. and Chen, W.-G. (2009) Preparation of modified chitosan with quaternary ammonium salt and its antibacterial property. *Fang Zhi Gao Xiao Ji Chu Ke Xue Xue Bao* 22, 85–89.
Luo, H., Li, J. and Chen, X. (2009) Antitumor effect of N-succinyl-chitosan nanoparticles on K562 cells. *Biomedicine and Pharmacotherapy* 64, 521–526.
Lv, K.L., Du, Y.L. and Wang, C.M. (2009) Synthesis of carboxylated chitosan and its adsorption properties for cadmium (II), lead (II) and copper (II) from aqueous solutions. *Water Science and Technology* 60, 467–474.
Malinowska-Panczyk, E., Kolodziejska, I., Murawska, D. and Wolosewicz, G. (2009) The combined effect of moderate pressure and chitosan on *Escherichia coli* and *Staphylococcus aureus* cells suspended in a buffer and on natural microflora of apple juice and minced pork. *Food Technology and Biotechnology* 47, 202–209.
Martín-Diana, A.B., Rico, D., Barat, J.M. and Barry-Ryan, C. (2009) Orange juices enriched with chitosan: Optimisation for extending the shelf-life. *Innovative Food Science and Emerging Technologies* 10, 590–600.
Martinez, L.R., Mihu, M.R., Tar, M., Cordero, R.J., Han, G., Friedman, A.J., Friedman, J.M. and Nosanchuk, J.D. (2010) Demonstration of antibiofilm and antifungal efficacy of chitosan against candidal biofilms, using an in vivo central venous catheter model. *Journal of Infectious Disease* 201, 1436–1440.
Masri, M.S., Reuter, F.W. and Friedman, M. (1974) Binding of metal cations by natural substances. *Journal of Applied Polymer Science* 18, 675–681.
Másson, M., Holappa, J., Hjálmarsdóttir, M., Rúnarsson, O.V., Nevalainen, T. and Järvinen, T. (2008) Antimicrobial activity of piperazine derivatives of chitosan. *Carbohydrate Polymers* 74, 566–571.
Meng, X., Li, B., Liu, J. and Tian, S. (2008) Physiological responses and quality attributes of table grape fruit to chitosan preharvest spray and postharvest coating during storage. *Food Chemistry* 106, 501–508.
Meng, X., Yang, L., Kennedy, J.F. and Tian, S. (2010) Effects of chitosan and oligochitosan on growth of two fungal pathogens and physiological properties in pear fruit. *Carbohydrate Polymers* 81, 70–75.
Miretzky, P. and Cirelli, A.F. (2009) Hg(II) removal from water by chitosan and chitosan derivatives: A review. *Journal of Hazardous Materials* 167, 10–23.
Mokhtari, A. and Jaykus, L.-A. (2009) Quantitative exposure model for the transmission of norovirus in retail food preparation. *International Journal of Food Microbiology* 133, 38–47.
Möller, H., Grelier, S., Pardon, P. and Coma, V. (2004) Antimicrobial and physicochemical properties of chitosan-HPMC-based films. *Journal of Agricultural and Food Chemistry* 52, 6585–6591.
Moreira, M.D.R., Ponce, A., Del Valle, C.E. and Roura, S.I. (2009) Edible coatings on fresh squash slices: Effect of film drying temperature on the nutritional and microbiological quality. *Journal of Food Processing and Preservation* 33, 226–236.
Muñoz, Z., Moret, A. and Garcés, S. (2009) Assessment of chitosan for inhibition of *Colletotrichum* sp. on tomatoes and grapes. *Crop Protection* 28, 36–40.
Nam, S.H., Choi, S.P., Kang, M.Y., Koh, H.J., Kozukue, N. and Friedman, M. (2005a) Bran extracts from pigmented rice seeds inhibit tumor promotion in lymphoblastoid B cells by phorbol ester. *Food and Chemical Toxicology* 43, 741–745.
Nam, S.H., Choi, S.P., Kang, M.Y., Kozukue, N. and Friedman, M. (2005b) Antioxidative, antimutagenic, and anticarcinogenic activities of rice bran extracts in chemical and cell assays. *Journal of Agricultural and Food Chemistry* 53, 816–822.
Nam, S.H., Choi, S.P., Kang, M.Y., Koh, H.J., Kozukue, N. and Friedman, M. (2006) Antioxidative activities of bran extracts from twenty one pigmented rice cultivars. *Food Chemistry* 94, 613–620.
Nishimura, S.-I., Kai, H., Shinada, K., Yoshida, T., Tokura, S., Kurita, K., Nakashima, H., Yamamoto, N. and Uryu, T. (1998) Regioselective syntheses of sulfated polysaccharides: Specific anti-HIV-1 activity of novel chitin sulfates. *Carbohydrate Research* 306, 427–433.
No, H.K., Meyers, S.P., Prinyawiwatkul, W. and Xu, Z. (2007) Applications of chitosan for improvement of quality and shelf life of foods: A review. *Journal of Food Science* 72, R87–R100.
Oh, H.I., Kim, Y.J., Chang, E.J. and Kim, J.Y. (2001) Antimicrobial characteristics of chitosans against food spoilage microorganisms in liquid media and mayonnaise. *Bioscience Biotechnology and Biochemistry* 65, 2378–2383.

Ojagh, S.M., Rezaei, M., Razavi, S.H. and Hosseini, S.M.H. (2010) Effect of chitosan coatings enriched with cinnamon oil on the quality of refrigerated rainbow trout. *Food Chemistry* 120, 193–198.
Okawa, Y., Kobayashi, M., Suzuki, S. and Suzuki, M. (2003) Comparative study of protective effects of chitin, chitosan, and N-acetyl chitohexaose against *Pseudomonas aeruginosa* and *Listeria monocytogenes* infections in mice. *Biological and Pharmaceutical Bulletin* 26, 902–904.
Omura, Y., Shigemoto, M., Akiyama, T., Saimoto, H., Shigemasa, Y., Nakamura, I. and Tsuchido, T. (2003) Antimicrobial activity of chitosan with different degrees of acetylation and molecular weights. *Biocontrol Science* 8, 25–30.
Ouattar, B., Simard, R.E., Piett, G., Bégin, A. and Holley, R.A. (2000) Inhibition of surface spoilage bacteria in processed meats by application of antimicrobial films prepared with chitosan. *International Journal of Food Microbiology* 62, 139–148.
Park, P.-J., Je, J.-Y., Byun, H.-G., Moon, S.-H. and Kim, S.-K. (2004) Antimicrobial activity of hetero-chitosans and their oligosaccharides with different molecular weights. *Journal of Microbiology and Biotechnology* 14, 317–323.
Park, S.-I., Stan, S.D., Daeschel, M.A. and Zhao, Y. (2005) Antifungal coatings on fresh strawberries (*Fragaria* x *ananassa*) to control mold growth during cold storage. *Journal of Food Science* 70, M202–M207.
Pelissari, F.M., Grossmann, M.V.E., Yamashita, F. and Pined, E.A.G. (2009) Antimicrobial, mechanical, and barrier properties of cassava starch-chitosan films incorporated with oregano essential oil. *Journal of Agricultural and Food Chemistry* 57, 7499–7504.
Piron, E. and Domard, A. (1998) Interaction between chitosan and uranyl ions. Part 2. Mechanism of interaction. *International Journal of Biological Macromolecules* 22, 33–40.
Ponce, A.G., Roura, S.I., del Valle, C.E. and Moreira, M.R. (2008) Antimicrobial and antioxidant activities of edible coatings enriched with natural plant extracts: *in vitro* and *in vivo* studies. *Postharvest Biology and Technology* 49, 294–300.
Portes, E., Gardrat, C., Castellan, A. and Coma, V. (2009) Environmentally friendly films based on chitosan and tetrahydrocurcuminoid derivatives exhibiting antibacterial and antioxidative properties. *Carbohydrate Polymers* 76, 578–584.
Pranoto, Y., Rakshit, S.K. and Salokhe, V.M. (2005a) Enhancing antimicrobial activity of chitosan films by incorporating garlic oil, potassium sorbate and nisin. *Lebensmittel-Wissenschaft und -Technologie* 8, 859–865.
Pranoto, Y., Salokhe, V.M. and Rakshit, S.K. (2005b) Physical and antibacterial properties of alginate-based edible film incorporated with garlic oil. *Food Research International* 38, 267–272.
Qi, L., Xu, Z., Jiang, X., Hu, C. and Zou, X. (2004) Preparation and antibacterial activity of chitosan nanoparticles. *Carbohydrate Research* 339, 2693–2700.
Qin, C., Xiao, Q., Li, H., Fang, M., Liu, Y., Chen, X. and Li, Q. (2004) Calorimetric studies of the action of chitosan-N-2-hydroxypropyl trimethyl ammonium chloride on the growth of microorganisms. *International Journal of Biological Macromolecules* 34, 121–126.
Quiñones, B., Massey, S., Friedman, M., Swimley, M.S. and Teter, K. (2009) Novel cell-based method to detect Shiga toxin 2 from *Escherichia coli* O157:H7 and inhibitors of toxin activity. *Applied and Environmental Microbiology* 75, 1410–1416.
Rabea, E.I., Badawy, M.E.-T., Stevens, C.V., Smagghe, G. and Steurbaut, W. (2003) Chitosan as antimicrobial agent: applications and mode of action. *Biomacromolecules* 4, 1457–1465.
Rana, M.S., Halim, M.A., Safiullah, S., Mamun Mollah, M., Azam, M.S., Goni, M.A., Kamal Hossain, M. and Rana, M.M. (2009) Removal of heavy metal from contaminated water by biopolymer crab shell chitosan. *Journal of Applied Sciences* 9, 2762–2769.
Rao, M.S., Chander, R. and Sharma, A. (2005) Development of shelf-stable intermediate-moisture meat products using active edible chitosan coating and irradiation. *Journal of Food Science* 70, M325–M331.
Rao, M.S., Chander, R. and Sharma, A. (2008) Synergistic effect of chitooligosaccharides and lysozyme for meat preservation. *Lebensmittel-Wissenschaft und-Technologie* 41, 1995–2001.
Rasooly, R., Do, P.M. and Friedman, M. (2010a) Inhibition of biological activity of Staphylococcal Enterotoxin A (SEA) by apple juice and apple polyphenols. *Journal of Agricultural and Food Chemistry* 58, 5421–5426.
Rasooly, R., Do, P.M., Griffey, S.M., Vilches-Moure, J.G. and Friedman, M. (2010b) Ingestion of Shiga toxin 2 (Stx2) causes histopathological changes in kidney, spleen and thymus tissues and mortality in mice. *Journal of Agricultural and Food Chemistry* 58, 9281–9286.

Rasooly, R., Do, P.M., Levin, C.E. and Friedman, M. (2010c) Inhibition of Shiga toxin 2 (Stx2) in apple juices and its resistance to pasteurization. *Journal of Food Science* 75, M296–M301.
Rauw, F., Gardin, Y., Palya, V., Anbari, S., Gonze, M., Lemaire, S., van den Berg, T. and Lambrecht, B. (2010) The positive adjuvant effect of chitosan on antigen-specific cell-mediated immunity after chickens vaccination with live Newcastle disease vaccine. *Veterinary Immunology and Immunopathology* 134, 249–258.
Ravishankar, S., Zhu, L., Law, B., Joens, L. and Friedman, M. (2008) Plant-derived compounds inactivate antibiotic-resistant *Campylobacter jejuni* strains. *Journal of Food Protection* 71, 1145–1149.
Ravishankar, S., Zhu, L., Olsen, C.W., McHugh, T. H. and Friedman, M. (2009) Edible apple film wraps containing plant antimicrobials inactivate foodborne pathogens on meat and poultry products. *Journal of Food Science* 74, M440–M445.
Ravishankar, S., Zhu, L., Reyna-Granados, J., Law, B., Joens, L. and Friedman, M. (2010) Carvacrol and cinnamaldehyde inactivate antibiotic-resistant *Salmonella enterica* in buffer and on celery and oysters. *Journal of Food Protection* 73, 234–240.
Rhoades, J. and Roller, S. (2000) Antimicrobial actions of degraded and native chitosan against spoilage organisms in laboratory media and foods. *Applied and Environmental Microbiology* 66, 80–86.
Ribeiro, M.P., Espiga, A., Silva, D., Baptista, P., Henriques, J., Ferreira, C., Silva, J.C., Borges, J.P., Pires, E., Chaves, P. and Correia, I.J. (2009) Development of a new chitosan hydrogel for wound dressing. *Wound Repair and Regeneration* 17, 817–824.
Richards, G.M. and Beuchat, L.R. (2005) Infection of cantaloupe rind with *Cladosporium cladosporioides* and *Penicillium expansum*, and associated migration of *Salmonella poona* into edible tissues. *International Journal of Food Microbiology* 103, 1–10.
Rodríguez, M.S., Montero, M., Staffolo, M.D., Martino, M., Bevilacqua, A. and Albertengo, L. (2008) Chitosan influence on glucose and calcium availability from yogurt: In vitro comparative study with plants fibre. *Carbohydrate Polymers* 74, 797–801.
Rogovik, A.L. and Goldman, R.D. (2008) Pharmacological treatment of pediatric obesity. *International Pediatrics* 23, 99–102.
Rogovik, A.L., Chanoine, J.-P. and Goldman, R.D. (2010) Pharmacotherapy and weight-loss supplements for treatment of paediatric obesity. *Drugs* 70, 335–346.
Roller, S. (2003) Chitosan: New food preservative of laboratory curiosity. In: Roller, S. (ed.) *Natural Antimicrobials for the Minimal Processing of Foods*. Woodhead Publishing, Cambridge, England, pp. 158–175.
Roller, S. and Covill, N. (1999) The antifungal properties of chitosan in laboratory media and apple juice. *International Journal of Food Microbiology* 47, 67–77.
Roller, S. and Covill, N. (2000) The antimicrobial properties of chitosan in mayonnaise and mayonnaise-based shrimp salads. *Journal of Food Protection* 63, 202–209.
Romanazzi, G., Gabler, F. M., Margosan, D., MacKey, B.E. and Smilanick, J. L. (2009) Effect of chitosan dissolved in different acids on its ability to control postharvest gray mold of table grape. *Phytopathology* 99, 1028–1036.
Sagoo, S., Board, R. and Roller, S. (2002) Chitosan inhibits growth of spoilage micro-organisms in chilled pork products. *Food Microbiology* 19, 175–182.
Sala Farré, M.R., Arias, C., Domínguez, A., Bartolomé, R. and Muntada, J.M. (2009) Foodborne outbreak of gastroenteritis due to Norovirus and *Vibrio parahaemolyticus*. *Epidemiology and Infection* 137, 626–629.
Sangsuwan, J., Rattanapanone, N. and Rachtanapun, P. (2008) Effect of chitosan/methyl cellulose films on microbial and quality characteristics of fresh-cut cantaloupe and pineapple. *Postharvest Biology and Technology* 49, 403–410.
Sangsuwan, J., Rattanapanone, N., Auras, R.A., Harte, B.R. and Rachtanapun, P. (2009) Factors affecting migration of vanillin from chitosan/methyl cellulose films. *Journal of Food Science* 74, C549–C555.
Sathivel, S. (2005) Chitosan and protein coatings affect yield, moisture loss, and lipid oxidation of pink salmon (*Oncorhynchus gorbuscha*) fillets during frozen storage. *Journal of Food Science* 70, E455–E459.
Sathivel, S., Liu, Q., Huang, J. and Prinyawiwatkul, W. (2007) The influence of chitosan glazing on the quality of skinless pink salmon (*Oncorhynchus gorbuscha*) fillets during frozen storage. *Journal of Food Engineering* 83, 366–373.
Savard, T., Beaulieu, C., Boucher, I. and Champagne, C.P. (2002) Antimicrobial action of hydrolyzed chitosan against spoilage yeasts and lactic acid bacteria of fermented vegetables. *Journal of Food Protection* 65, 828–833.

Schmid, D., Fretz, R., Buchner, G., Kanig, C., Perner, H., Sollak, R., Tratter, A., Hell, M., Maass, M., Strasser, M. and Allerberger, F. (2009) Foodborne outbreak of hepatitis A, November 2007–January 2008, Austria. *European Journal of Clinical Microbiology and Infectious Diseases* 28, 385–391.

Shah, P., Jogani, V., Mishra, P., Mishra, A.K., Bagchi, T. and Misra, A. (2008) In vitro assessment of acyclovir permeation across cell monolayers in the presence of absorption enhancers. *Drug Development and Industrial Pharmacy* 34, 279–288.

Shahidi, F. and Abuzaytoun, R. (2005) Chitin, chitosan, and co-products: Chemistry, production, applications, and health effects. *Advances in Food and Nutrition Research* 49, 93–135.

Shahidi, F., Kamil, J., Jeon, Y.-J. and Kim, S.-K. (2002) Antioxidant role of chitosan in a cooked cod (*Gadus morhua*) model system. *Journal of Food Lipids* 9, 57–64.

Shieh, Y.C., Stewart, D.S. and Laird, D.T. (2009) Survival of hepatitis A virus in spinach during low temperature storage. *Journal of Food Protection* 72, 2390–2393.

Simpson, B.K., Gagné, N., Ashie, I.N.A. and Noroozi, E. (1997) Utilization of chitosan for preservation of raw shrimp (*Pandalus borealis*). *Food Biotechnology* 11, 25–44.

Sirk, T.W., Brown, E.F., Sum, A.K. and Friedman, M. (2008) Molecular dynamics study on the biophysical interactions of seven green tea catechins with lipid bilayers of cell membranes. *Journal of Agricultural and Food Chemistry* 56, 7750–7758.

Sirk, T.W., Brown, E.F., Friedman, M. and Sum, A.K. (2009) Molecular binding of catechins to biomembranes: Relationship to biological activity. *Journal of Agricultural and Food Chemistry* 57, 6720–6728.

Sonaje, K., Chen, Y.-J., Chen, H.-L., Wey, S.-P., Juang, J.-H., Nguyen, H.-N., Hsu, C.-W., Lin, K.-J. and Sung, H.-W. (2010) Enteric-coated capsules filled with freeze-dried chitosan/poly(g-glutamic acid) nanoparticles for oral insulin delivery. *Biomaterials* 31, 3384–3394.

Soultos, N., Tzikas, Z., Abrahim, A., Georgantelis, D. and Ambrosiadis, I. (2008) Chitosan effects on quality properties of Greek style fresh pork sausages. *Meat Science* 80, 1150–1156.

Sousa, F., Guebitz, G.M. and Kokol, V. (2009) Antimicrobial and antioxidant properties of chitosan enzymatically functionalized with flavonoids. *Process Biochemistry* 44, 749–756.

Stewart, E.M. (2009) Food irradiation. In: Stadler, R.H. and Lineback, D.R. (eds) *Process-Induced Food Toxicants – Occurrence, Formation, Mitigation, and Health Risks*. Wiley, New York, pp. 387–412.

Su, H.K., Hong, K.N. and Prinyawiwatkul, W. (2007) Effect of molecular weight, type of chitosan, and chitosan solution pH on the shelf-life and quality of coated eggs. *Journal of Food Science* 72, S044–S048.

Sudarshan, N.R., Hoover, D.G. and Knorr, D. (1992) Antibacterial action of chitosan. *Food Biotechnology* 6, 257–272.

Sui, Z., Chen, Q., Wu, R., Zhang, H., Zheng, M., Wang, H. and Chen, Z. (2010) Cross-protection against influenza virus infection by intranasal administration of M2-based vaccine with chitosan as an adjuvant. *Archives of Virology*, 1–10.

Tapia, P.C., Soto, M.D., Vergara, G.L., Alburquerque, O.C., Maccioni, R.A., Matamala, C.A.M., Hermosilla, D.G. and Bucarey, V.S. (2009) Antifungal effect of high molecular weight chitosan on *Candida* spp isolated from clinical samples. *Revista Chilena de Infectologia* 26, 515–519.

Tomida, H., Fujii, T., Furutani, N., Michihara, A., Yasufuku, T., Akasaki, K., Maruyama, T., Otagiri, M., Gebicki, J.M. and Anraku, M. (2009) Antioxidant properties of some different molecular weight chitosans. *Carbohydrate Research* 344, 1690–1696.

Tsai, G.-J. and Su, W.-H. (1999) Antibacterial activity of shrimp chitosan against *Escherichia coli*. *Journal of Food Protection* 62, 239–243.

Tsai, G.-J., Su, W.-H., Chen, H.-C. and Pan, C.-L. (2002) Antimicrobial activity of shrimp chitin and chitosan from different treatments and applications of fish preservation. *Fisheries Science* 68, 170–177.

Tsai, G.-J., Zhang, S.-L. and Shieh, P.-L. (2004) Antimicrobial activity of a low-molecular-weight chitosan obtained from cellulase digestion of chitosan. *Journal of Food Protection* 67, 396–398.

Valentine, R., Athanasiadis, T., Moratti, S., Hanton, L., Robinson, S. and Wormald, P.-J. (2010) The efficacy of a novel chitosan gel on hemostasis and wound healing after endoscopic sinus surgery. *American Journal of Rhinology and Allergy* 24, 70–75.

Van Der Merwe, S.M., Verhoef, J.C., Verheijden, J.H.M., Kotzé, A.F. and Junginger, H.E. (2004) Trimethylated chitosan as polymeric absorption enhancer for improved peroral delivery of peptide drugs. *European Journal of Pharmaceutics and Biopharmaceutics* 58, 225–235.

Vásconez, M.B., Flores, S.K., Campos, C.A., Alvarado, J. and Gerschenson, L.N. (2009) Antimicrobial activity and physical properties of chitosan-tapioca starch based edible films and coatings. *Food Research International* 42, 762–769.

Waimaleongora-Ek, P., Corredor, A.J.H., No, H.K., Prinyawiwatkul, W., King, J.M., Janes, M.E. and Sathivel, S. (2008) Selected quality characteristics of fresh-cut sweet potatoes coated with chitosan during 17-day refrigerated storage. *Journal of Food Science* 73, S418–S423.

Wang, G., Liu, J., Wang, X., Xie, Z. and Deng, N. (2009) Adsorption of uranium (VI) from aqueous solution onto cross-linked chitosan. *Journal of Hazardous Materials* 168, 1053–1058.

Wang, J., Wang, B., Jiang, W. and Zhao, Y. (2007) Quality and shelf life of mango (*Mangifera indica* L. cv. 'Tainong') coated by using chitosan and polyphenols. *Food Science and Technology International* 13, 317–322.

Wang, X., Du, Y., Fan, L., Liu, H. and Hu, Y. (2005) Chitosan-metal complexes as antimicrobial agent: synthesis, characterization and structure-activity study. *Polymer Bulletin* 55, 105–113.

Wong, S.Y.Y., Grant, I.R., Friedman, M., Elliott, C.T. and Situ, C. (2008) Antibacterial activities of naturally occurring compounds against *Mycobacterium avium* subsp. *paratuberculosis*. *Applied and Environmental Microbiology* 74, 5986–5990.

Wydro, P., Krajewska, B. and Hac-Wydro, K. (2007) Chitosan as a lipid binder: A langmuir monolayer study of chitosan–lipid interactions. *Biomacromolecules* 8, 2611–2617.

Xu, W.-T., Huang, K.-L., Guo, F., Qu, W., Yang, J.-J., Liang, Z.-H. and Luo, Y.-B. (2007) Postharvest grapefruit seed extract and chitosan treatments of table grapes to control *Botrytis cinerea*. *Postharvest Biology and Technology* 46, 86–94.

Xue, C., Yu, G., Hirata, T., Terao, J. and Lin, H. (1998) Antioxidative activities of several marine polysaccharides evaluated in a phosphatidylcholine-liposomal suspension and organic solvents. *Bioscience Biotechnology and Biochemistry* 62, 206–209.

Yang, S.-J., Chang, S.-M., Tsai, K.-C., Chen, W.-S., Lin, F.-H. and Shieh, M.-J. (2010) Effect of chitosan-alginate nanoparticles and ultrasound on the efficiency of gene transfection of human cancer cells. *Journal of Gene Medicine* 12, 168–179.

Ye, M., Neetoo, H. and Chen, H. (2008a) Control of *Listeria monocytogenes* on ham steaks by antimicrobials incorporated into chitosan-coated plastic films. *Food Microbiology* 25, 260–268.

Ye, M., Neetoo, H. and Chen, H. (2008b) Effectiveness of chitosan-coated plastic films incorporating antimicrobials in inhibition of *Listeria monocytogenes* on cold-smoked salmon. *International Journal of Food Microbiology* 127, 235–240.

Yin, H., Du, Y. and Zhang, J. (2009) Low molecular weight and oligomeric chitosans and their bioactivities. *Current Topics in Medicinal Chemistry* 9, 1546–1559.

You-Jin, J. and Kim, S.-K. (2001) Effect of antimicrobial activity by chitosan oligosaccharide N-conjugated with asparagine. *Journal of Microbiology and Biotechnology* 11, 281–286.

Yuasa, K. and Hatai, K. (1996) Investigation of effective chemicals for treatment of saprolegniasis caused by *Saprolegnia parasitica*. *Journal of Antibacterial and Antifungal Agents, Japan* 24, 27–31.

Zhang, D. and Quantick, P.C. (1998) Antifungal effects of chitosan coating on fresh strawberries and raspberries during storage. *Journal of Horticultural Science and Biotechnology* 73, 763–767.

Zhao, X., Qiao, Z.-Z., He, J.-X. and Zhan, Y.-Z. (2009) Effects of chitosan biguanidine hydrochloride antimicrobial property on wool fabric. *Wool Textile Journal* 37, 23–26.

Zhong, Z., Xing, R., Liu, S., Wang, L., Cai, S. and Li, P. (2008) Synthesis of acyl thiourea derivatives of chitosan and their antimicrobial activities in vitro. *Carbohydrate Research* 343, 566–570.

Zhong, Z., Li, P., Xing, R. and Liu, S. (2009) Antimicrobial activity of hydroxylbenzenesulfonailides derivatives of chitosan, chitosan sulfates and carboxymethyl chitosan. *International Journal of Biological Macromolecules* 45, 163–168.

Zhou, Y.-T., Branford-White, C., Nie, H.-L. and Zhu, L.-M. (2009) Adsorption mechanism of Cu2+ from aqueous solution by chitosan-coated magnetic nanoparticles modified with α-ketoglutaric acid. *Colloids and Surfaces B: Biointerfaces* 74, 244–252.

Zhu, X., Wang, Q., Cao, J. and Jiang, W. (2008) Effects of chitosan coating on postharvest quality of mango (*Mangifera indica* L. cv. Tainong) fruits. *Journal of Food Processing and Preservation* 32, 770–784.

Zivanovic, S., Basurto, C.C., Chi, S., Davidson, P.M. and Weiss, J. (2004) Molecular weight of chitosan influences antimicrobial activity in oil-in-water emulsions. *Journal of Food Protection* 67, 952–959.

Zivanovic, S., Chi, S. and Draughon, A.F. (2005) Antimicrobial activity of chitosan films enriched with essential oils. *Journal of Food Science* 70, M45–M51.

11 Reduction of Biogenic Amine Levels in Meat and Meat Products

Claudia Ruiz-Capillas,* Ana Maria Herrero and Francisco Jimenez-Colmenero

Introduction

Biogenic amines (BAs) are present in a wide range of foods, including meat and meat products. BAs are non-volatile amines that can be formed and degraded as a result of normal metabolic activity in animals, plants and microorganisms. Chemically, BAs are basic, low molecular weight, nitrogenous compounds and can be classified as biogenic amines (e.g. putrescine, cadaverine, agmatine, tyramine, histamine, phenylethylamine and tryptamine) and natural polyamines (agmatine spermidine, spermine and putrescine). Biogenic amines are mainly formed by decarboxylation of free amino acids (FAAs) by the action of amino acid decarboxylase, an enzyme of microbial origin, with formation being conditioned by numerous factors (raw material, FAAs, microbiological considerations, additives, processing, storage, etc.) (Fig. 11.1).

The presence of biogenic amines in foodstuffs is doubly important. First, consumption of products having high concentrations of BAs may represent a potential public health concern on account of their physiological and toxicological effects (Halász *et al.*, 1994; Bardócz, 1995; Shalaby, 1996). Second, they may act as quality and/or acceptability indicators in some foods (Mietz and Karmas, 1977; Hernández-Jover *et al.*, 1997a; Ruiz-Capillas and Moral, 2001; Ruiz-Capillas and Jiménez-Colmenero, 2004a).

As a consequence of these health and quality concerns, there is clear interest in reducing the amounts of biogenic amines in foods, including meat and meat products. Meat products have a wide range of BA contents depending on product type, and certain factors may also cause concentrations to vary widely within one and the same product (Ruiz-Capillas and Jiménez Colmenero, 2004a,b).

Meat is an important component of the diet in developed countries, supplying around 30% of the protein and 20% of the fat in the diet. The presence of biogenic amines in meat products could lead to rejection by consumers because of potential health repercussions, and for this reason procedures for lowering BA contents have begun to hold great interest as a field of study. In addition, certain polyamines are involved in biological reactions, and hence their presence in foods is also important. As a consequence, it has become necessary to try to reduce or regulate biogenic amine levels in meat and meat products and in foodstuffs generally with a view to meeting daily requirements of

* Corresponding author.

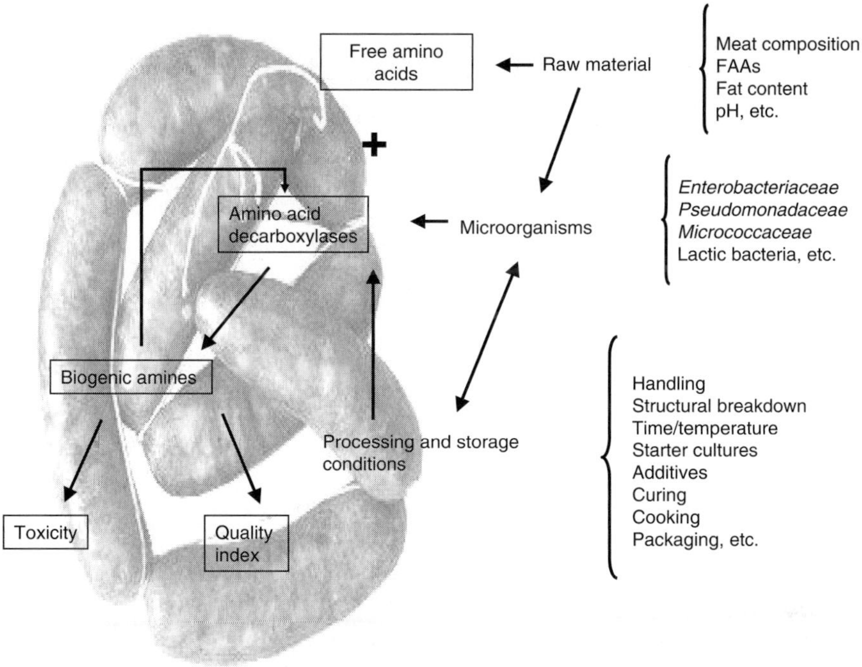

Fig. 11.1. Formation of biogenic amines and factors affecting their formation in meat and meat products (adapted from Ruiz-Capillas and Jiménez-Colmenero, 2004a).

these amines while at the same time preventing overconsumption.

Biological Importance of Biogenic Amines

Polyamines have been implicated in a wide range of biological reactions affecting important bodily functions (Bardócz, 1995). All of the body's organs require polyamines for growth, renewal and metabolism. Thanks to their involvement in signal transduction and in nearly every step of DNA, RNA and protein synthesis, natural polyamines are essential to growth and cell proliferation. Spermidine and spermine have important functions in reproduction (Bardócz, 1995; Kalač and Krausová, 2005).

In the past the polyamines necessary to bodily functions were thought to be formed *in situ*, but the source of polyamines building up in the small bowel and other organs has been shown to be *de novo* biosynthesis, the diet and bacteria present in the gut lumen (Bardócz, 1995). The intestinal tract, in its turn, plays a vital role in nutrition and maintaining human health. While every cell can synthesize polyamines, the body also appears to rely on a continuous supply of polyamines from food. This should not be overlooked when severely limiting the polyamine levels present in foods.

Toxicological Importance of Biogenic Amines

In addition to the vital functions that have been described for certain amines, high amine consumption has also been related to nausea, respiratory distress, hot flushes, sweating, heart palpitations, strong migraines, headaches, bright red rash, burning mouth, hyper or hypotension, stomach and bowel problems, and pseudoallergic reactions. Tyramine, which is vasoactive, and histamine, which is vasoactive and psychoactive, are the most biologically active biogenic amines. Histamine is the main biogenic amine in fish and fish products and is the main factor in 'scombroid poisoning' or 'histamine poisoning', caused by eating fish containing high levels of histamine and/or other biogenic amines (Taylor, 1986; Ruiz-Capillas and Moral, 2003, 2005). This amine is not relevant in meat products, in which tyramine is present in higher concentrations, as it is in cheese. Typical symptoms of tyramine poisoning are migraine, headache and raised blood pressure (ten Brink *et al.*, 1990). Besides tyramine being toxic in itself, recent studies have shown it to promote adhesion of pathogens such as *Escherichia coli* O157:H7 to the gastric mucosa (Lyte, 2004). Other amines like spermidine and spermine have also been associated with food allergies (Bardócz, 1995; Kalač and Krausová, 2005). While putrescine and cadaverine are not themselves toxic, they are also implicated in these illnesses, in that they enhance the toxicity of histamine and tyramine (Rice *et al.*, 1976; Sattler *et al.*, 1988).

The toxicity of these amines in the body depends on the efficiency of the body's detoxification system (Halász *et al.*, 1994; Bardócz, 1995). In normal circumstances, the human body is able to quickly detoxify the histamine and tyramine absorbed from foods by means of the enzymes monoamine oxidase (MAO; EC 1.4.3.4), diamine oxidase (DAO; EC 1.4.3.6) and polyamine oxidase (PAO; EC 1.5.3.11) (Rice *et al.*, 1976; Bardócz, 1995). However, these detoxification mechanisms may be altered by genetic factors, either because an individual is allergic or because he or she ingests or is undergoing treatment with oxidase enzymes (e.g. monoamine oxidase inhibitor; MAOI), which inhibit aminooxidases or cause aminooxidase deficiency (McCabe, 1986; Halász *et al.*, 1994). The amount of tyramine needed to be toxic in a normal person is 125 mg/kg. In contrast, 6 mg/kg of tyramine would be toxic if ingested with MAOIs (McCabe, 1986). This is particularly important in today's society in view of the high consumption of MAOIs as antidepressants (Sattler *et al.*, 1988). Studies of biogenic amine concentrations in commercially processed Spanish meat products reported that 63 % of 'salchichon' sausage samples and 64 % of 'chorizo' sausage samples (both ripened meat products) contained enough tyramine to poison consumers taking MAOIs (Vidal-Carou *et al.*, 1990). Similar levels of biogenic amines were also observed in various meat products (cooked, fermented, etc.) (Ruiz-Capillas and Jiménez-Colmenero, 2004b).

The toxicity of histamine and tyramine in foods may be heightened by high levels of such other amines as agmatine, cadaverine and putrescine. These last-mentioned amines are not toxic in themselves, but where present they can boost the toxicity of histamine and tyramine by interacting with aminooxidases, thereby interfering with the detoxification mechanism (Rice *et al.*, 1976; Taylor, 1986; Sattler *et al.*, 1988). BA toxicity is also conditioned by other factors such as alcohol and acetaldehyde, which can increase the toxic potential of biogenic amines, by helping to promote transport of BAs through the intestinal wall. Eating acidic foods could also inhibit the enzymes responsible for metabolizing histamine, thereby heightening its toxicity. Accordingly, BA toxicity will depend on both quantitative and qualitative factors associated with the foods that are eaten and on consumer-related factors (individual susceptibility and state of health). All this makes it extremely difficult to set BA levels for foodstuffs and determining the amine concentrations in each case does not suffice to assess the toxic potential of a given food.

For its part, the European Union only sets legal limits on histamine (Directive 91/493/EEC) at 100 mg/kg for scombrid and cupleoid fish only, measurement values to be calculated from nine representative samples from each batch, with no individual reading to be in excess of 200 mg/kg, except for products that have undergone enzyme ripening treatment in brine, which may have values up to as much as 400 mg/kg. The US Food and Drug Administration (FDA, 1990) sets the limit for histamine at 50 mg/kg, lower than the EU. There are, however, no official limits for other amines or for other food products, e.g. meat, in which high levels of such amines as tyramine have been observed, even in conjunction with histamine, putrescine and cadaverine, particularly in fermented products. It should also be borne in mind that this legislation sets limits for one foodstuff, whereas the quantity of BAs an individual ingests is the sum total of all the amines present in all the different foods and beverages taken together at a meal (meat, fish, cheese, wine etc.), which contribute to the total amount of BAs present in the body.

Nitrosamines

BAs may be involved in the formation of potent carcinogens such as nitrosamines (dimethylnitrosamine, diethylnitrosamine, nitrosopiperidine, nitrosopyrrolidine, methylethylnitrosamine and nitrosomorpholine), compounds with teratogenic, mutagenic and carcinogenic effects that are highly dangerous to human health (Cassens, 1997). These N-nitroso compounds have been associated with human cancers, including brain tumours and oesophageal and gastric cancers (Warthesen *et al.*, 1975).

Nitrosamines are formed by reactions between amines (spermidine, spermine, tyramine, putrescine and cadaverine) and nitrites under certain conditions (Warthesen *et al.*, 1975). This takes on particular importance in certain meat products with high BA levels to which nitrates and nitrites are added during processing. Domanska-Blicharz *et al.* (2005) observed that nitrosodimethylamine and nitrosopiperidine increased significantly in nitrite-treated meat products during storage at 4–8°C for 72 h. The USA government has set 10 μg/kg nitrosopyrrolidine as the maximum allowable limit in bacon (Domanska-Blicharz *et al.*, 2005)

The food industry has tried reducing or even avoiding the use of nitrites and nitrates in order to effectively limit nitrosamine formation, and other additives that block the chemical mechanism underlying nitrosamine formation have also been used. These include ascorbic acid (E-330) and its derivatives and tocopherols (E-306 ff.), the former particularly effective in aqueous systems, the latter in fatty systems. These additives have been shown to significantly decrease nitrosamine formation in cured meat products (Cassens, 1997). Nitrites are used in association with ascorbic acid, and indeed use in tandem is mandatory in the USA.

Biogenic Amines as Indicators of Hygienic Quality

Some research has aimed at establishing relationships between biogenic amines and the most widely used spoilage or quality indices in different foods, including meat and meat products (Wortberg and Woller, 1982; Edwards *et al.*, 1985, 1987; Hernández-Jover *et al.* 1996; Ruiz-Capillas and Moral, 2001; Silva and Gloria, 2002; Vinci and Antonelli 2002; Ruiz-Capillas and Jiménez Colmenero, 2004a; Ruiz-Capillas *et al.*, 2007a). Biogenic amines can be used as quality control indices because they undergo changes during meat processing and storage. They are present in fresh meat at very low levels but tend to form progressively during storage in association with bacterial spoilage (Edwards *et al.*, 1985; Ruiz-Capillas and Jiménez-Colmenero, 2004a). In the case of meat, tyramine, putrescine and cadaverine are the main amines implicated in the deterioration of food quality. Because of their endogenous origin, levels of BAs like spermidine and spermine decrease or remain constant during storage; hence they are not usually employed as quality indicators in meat (ten Brink *et al.*, 1990; Halász *et al.*, 1994; Bardócz, 1995). However, the ratio of spermidine to spermine has been proposed as a quality index in chicken meat during refrigerated storage (Silva and Gloria, 2002).

BAs have been used as quality indicators both singly and in combination. Wortberg and Woller (1982) reported that high cadaverine concentrations were clearly indicative of spoilage in meat and meat products. Tyramine too has been used as a quality indicator for vacuum-packaged beef and cooked ham subjected to high-pressure processing (Edwards *et al.*, 1985, 1987; Ruiz-

Capillas *et al.*, 2007a). Using putrescine and cadaverine together has been suggested as an acceptability index for fresh meat, because their levels increase before spoilage and correlate well with the microbial load (Edwards *et al.*, 1985). Similarly cadaverine and tyramine levels during storage could be used to monitor spoilage in red (adult bovine) meat and white (chicken) meat (Vinci and Antonelli, 2002). The presence of still acceptable levels of tyramine and putrescine has been suggested as indicative of the onset of undesired fermentation during ham processing (Virgili *et al.*, 2007). Most of these indices combining various amines have been based on the biogenic amine index of Mietz and Karmas (1977), the most widely used measure of spoilage in fish based on the histamine, cadaverine, putrescine, spermidine and spermine contents. The index has subsequently been applied to assess meat quality and how it is related to microbial growth (Maijala *et al.*, 1995a). A biogenic amine index (BAI) comprising the sum of the amines putrescine, cadaverine, histamine and tyramine has been developed for bologna sausage and minced beef and pork (Wortberg and Woller, 1982; Hernández-Jover *et al.*, 1996).

However, the usefulness of BAs as a quality index will depend on numerous factors: product attributes (pH, water activity (a_w), FAAs, etc.), type and level of contamination of the raw material, degree of structural breakdown, manufacturing practices and processing stages employed, the starters used, storage conditions or technology, etc.

BAs are non-volatile, heat-stable compounds that do not degrade on cooking. That makes them useful, and they have been used to evaluate the sanitary condition of the raw materials used to manufacture heat-treated products (Hernández-Jover *et al.*, 1996). Their usefulness as quality indicators in fermented products is less clear, in that the various processing factors associated with this type of product (fermentation and/or ripening, a_w, starter, proteolysis, additives, etc.) are potentially conducive or unconducive to BA formation.

Procedures for Reducing Biogenic Amine Formation in Meat and Meat Products

Meat and meat products present highly variable quantities of BAs, and this is even true for different samples of the same product. The BA content will depend on a number of interrelated factors, such as the raw material (meat composition, pH, handling and hygienic conditions, etc.), additives (salt, sugar, nitrites, etc.) that affect FAA availability, microbiological aspects (bacterial species and strain, bacterial growth, etc.), the technical processing undergone by the meat or meat products (e.g., steaks, roasts, hams, ground, restructured, comminuted, fresh, cooked, smoked and fermented meats, etc.), and storage conditions (time/temperature, packaging, temperature abuse, etc.). The combined action of all these factors together will mostly determine final BA concentrations by directly or indirectly determining the presence and activity of substrate and enzyme (Ruiz-Capillas and Jiménez-Colmenero 2004a). Implementation of procedures intended to limit BA formation in meat and meat products should therefore be aimed at the raw materials, technical aspects of the manufacturing processes and storage. The specific procedures employed should address the various factors that condition BA formation (Fig. 11.1). This means that steps taken should mainly target levels of the FAAs that are the precursors of BAs; the enzyme amino acid decarboxylase, whose action results in decarboxylation of the FAAs; and final BA formation.

Meat Raw Materials

Meat is an excellent protein-rich food and a major natural source of FAAs. It has high water activity (a_w) and is extremely perishable and hence is a good substrate for potential BA-producing microorganisms to grow in (Ruiz-Capillas and Jiménez-Colmenero, 2004a). BA formation has been shown to depend on the FAA content of the raw material (e.g. meat), but still in many cases no direct relationship has been observed between the FAA content of a meat or fish food product and the BAs formed (Fig. 11.1) (Eerola *et al.*, 1996; Ruiz-Capillas and Moral 2002; 2003; Ruiz-Capillas *et al.*, 2007a). Modifying FAA levels is complicated, and in any event FAA concentrations in meat as a raw material are so high that it is hard to see how this factor could be used to limit BA formation in these products.

Both quantitatively and qualitatively the BAs present in meat will depend on meat type, which conditions pH and a_w, and on the presence of high fat levels (Smith 1980; Halász *et al.*, 1994; Eerola *et al.*, 1996; Hernández-Jover *et al.*, 1997a; Ruiz-Capillas and Jiménez-Colmenero, 2004a; Virgili *et al.*, 2007). Meat quality (microbial load) is one of the factors that has the greatest effect on BA production. Several studies have focused on lowering BA levels in meat products on the basis of meat quality (Bover-Cid *et al.*, 2001a; Maijala *et al.*, 1995a; Suzzi and Gardini, 2003; Ruiz-Capillas and Jiménez-Colmenero, 2004a). BA formation has been reported not to take place in sterile meat, while BA concentrations in meat have been observed to increase along with rising microorganism levels (Slemr and Beyermann, 1985). In general, BA production in meat products has been attributed to the action of a variety of microorganisms, both Gram positive and Gram negative, which have expressed amino acid decarboxylase activity. These include bacteria in the families *Enterobacteriaceae* and *Micrococcaceae*, and lactic acid bacteria, and such diverse genera as *Pseudomonas*, *Citrobacter*, *Klebsiella*, *Proteus*, *Salmonella*, *Shigella*, *Staphylococcus*, *Micrococcus*, *Morganella*, *Vibrio*, *Lactobacillus*, *Enterococcus*, *Carnobacterium*, *Pediococcus* and *Lactococcus* (Halász *et al.*, 1994; Silla-Santos, 1996; Roig-Sagués *et al.*, 2009). However, decarboxylation activity, it should be noted, is strain specific, and hence any given species may include both high-BA-producing strains and other strains that express only limited decarboxylase activity. The differing BA levels in the various food products will thus be dependent on the different microorganisms and strains present. In addition, the negative (or positive) responses recorded in screening media do not necessarily reflect similar behaviour in food products (Bover-Cid and Holzapfel, 1999; de la Rivas *et al.*, 2008; Roig-Sagués *et al.*, 2009). Behaviour has, though, been described as being similar in the case of slightly fermented sausages, in which the enzymes released early in the preservation process are responsible for BA accumulation and may remain active even in the absence of viable cells of *Enterobacteriaceae* (Bover-Cid *et al.*, 2001b; Roig-Sagués *et al.*, 2009). These bacteria are chiefly responsible for cadaverine and putrescine. Histidine decarboxylase activity is related, in particular, to certain species belonging to the genera *Micrococcus* and *Staphylococcus* (Silla-Santos, 1996). *Staphylococcus carnosus* strains analysed in pressurized dry-cured Spanish chorizo sausage produced phenylethyamine or both phenylethyamine and tyramine together (de la Rivas *et al.*, 2008). Lactic acid bacteria are the main producers of BAs in fermented meat and are associated with tyramine production. Some strains of *Lactococcus* and *Leuconostoc* have been described as tyramine producers. BA production by *Carnobacterium* has been observed, in particular considerable tyramine production by *C. divergens*, *C. piscicola* and *C. gallinarum* (Masson *et al.*, 1996).

With this in mind, employing any procedure or technology effective at preserving the hygienic quality of the raw material by curbing microbial growth will be a suitable procedure for reducing BA formation in food products. Frozen storage of the raw material inhibits microbial growth and the activity of BA-producing enzymes. Biogenic amine levels in frozen foodstuffs are related to low raw material quality or previous proliferation of microorganisms, as has been reported in cooked products. Nevertheless, proper control of the thawing factors (time and temperature) is very important, because improper thawing can affect BA levels in the product in which the raw material is then used (Bover-Cid *et al.*, 2001a). Depending on the temperature, refrigeration can limit microorganism growth to a great extent (Maijala *et al.*, 1995b).

The starter cultures used to manufacture fermented meat products are a source of microorganisms that initiate rapid acidification of the raw meat batter and, depending on their natures, ultimately yield the desirable sensory qualities of the finished product. Many starter culture components are the primary BA producers in the products in which they are used. For this reason, one of the most effective procedures for lowering BA concentrations in these types of products would be to employ different starter cultures with lower amino acid decarboxylase activity that are nonetheless still capable of conferring suitable sensory attributes on the products. To this end, a range of starter cultures comprising different strains of lactobacilli, pediococci, staphylococci or micrococci have been tested, with very promising results. Decreases of from 25 to 95 % have been attained, depending on the starter culture and the individual amine concerned (Maijala *et al.*, 1995a; Hernández-Jover *et al.*, 1997b; Roig-Sagués and Eerola, 1997; Bover-Cid *et al.*, 2000a; Bover-Cid *et al.*, 2001a,c; Suzzi and Gardini, 2003; Roig-Sagués *et al.*, 2009; Latorre-Moratella *et al.*, 2010). *Lactobacillus sakei* CTC 494 is one of the most effective starters at reducing BA formation, lowering the tyramine, putrescine and cadaverine contents and inhibiting β-phenylethylamine, tryptamine and histamine formation in Spanish fuet sausage. However, its effectiveness is reduced or counteracted where the raw meat has high levels of microbial contamination (*Enterobacteriaceae* counts are not to exceed 10^3 cfu/g) (Bover-Cid *et al.*, 2001a; Latorre-Moratella *et al.*, 2010). Different combinations of starters have also been studied. *L. sakei*, along with *S. carnosus* and *Staphylococcus xylosus*, lowered the total BA content by 80–90 % in the manufacture of Spanish fuet and chorizo sausages (Bover-Cid *et al.*, 2000c). Reductions in BA contents of 50% have also been observed in sausages fermented using *Lactobacillus curvatus* CTC 371 in association with a proteolytic strain of *S. xylosus* (Bover-Cid *et al.*, 2001c). In contrast, single-strain starter cultures of *Pediococcus cerevisiae* or *Lactobacillus plantarum* did not decrease either the tyramine content or the total BA content compared with spontaneous fermentation (Rice *et al.*, 1976).

Thus, another strategy for reducing BAs entails rapid detection of the bacteria responsible for BA formation so as to eliminate certain amine-producing bacterial strains, replacing them with other non-producing strains or with strains having low decarboxylase activity, thereby lowering BA levels in the finished product. Studies carried out to date have focused on isolating strains and hydrolysing the amino acids, but this is a long and tedious method. Work on employing newer and faster methods, e.g. polymerase chain reaction (PCR) and spectroscopy, is therefore under way. Along these lines, a multiplex PCR assay aimed at simultaneously detecting lactic acid bacteria (LAB) strains that are potential producers of histamine, tyramine and putrescine in fermented foods has been developed (Marcobal *et al.*, 2005). As an example of a spectroscopic method, Fourier transform infrared (FT–IR) has been shown to be a useful tool for rapidly characterizing bacterial spoilage in beef as a result of proteolysis, resulting in changes in amide and amine levels (Ellis *et al.*, 2004) and has also been used to identify closely related lactobacilli (*L. sakei, L. plantarum, L. curvatus* and *L. paracasei*) naturally present in meat and used as starter cultures in meat products (Oust *et al.*, 2004).

Some ingredients and additives are used to enhance technical properties and/or organoleptic attributes during manufacturing or in the finished product, though in many cases their main function is to inhibit microbial growth, thereby conditioning BA formation. Consequently, management of these ingredients and additives may be a strategy for lowering BA levels in meat products. Ingredients such as salt in meat products will condition BA levels because of their effect in inhibiting microbial growth, on the one hand, and in interfering with endogenous and exogenous proteolytic enzymatic activity, on the other, thereby potentially limiting FAA formation (Virgili *et al.*, 2007). The salt content in fermented products significantly affects BA production, mainly through its role in reducing a_w. Experiments carried out on restructured steaks revealed BA levels to be lower in this products that contained salt than in restructured steaks that did not, whereas the converse held true for microbial growth (Ruiz-Capillas, unpublished data). This suggests that the salt content might select a specific strain of microorganism less capable of producing BAs than the strains that grow in the unsalted product. Sodium chloride can enhance the activity of histidine decarboxylases in halotolerant and halophilic *Staphylococcus* spp. (Hernández-Herrero *et al.*, 1999). Straub *et al.* (1994) reported that 30 g of NaCl/kg in fermented sausages enhanced tyramine production by promoting the growth of *L. curvatus*.

Food additives such as preservatives (sulfite, potassium sorbate, nitrate and nitrite, etc.) also limit microbial growth, and hence BAs, in meat products (Straub *et al.*, 1994; Shalaby, 1996; Ruiz-Capillas *et al.*, 2006; Lorenzo *et al.*, 2007). Nitrates and nitrites are commonly employed as curing salts in meat products. No clear relationship between the use of these additives and lower BA levels has been found. The presence of these additives in other meat products [soy hamburger texturizers, meat batter, lacon (a form of cured pork picnic shoulder)] resulted in appreciable concentrations of tyramine (Straub *et al.*, 1994; Ruiz-Capillas *et al.*, 2006; Lorenzo *et al.*, 2007).

Technical Food Manufacturing Processes

Careful control of the processes used to prepare meat raw materials and to manufacture meat products is essential if BA levels in the finished product are to be lowered. Supervision of technical processing methods starts at the processing plant and extends to product preservation and storage. Such factors as meat handling, temperatures, processing times and the like will exert a direct influence on the microbial population present.

Structural breakdown (grinding, chopping, sectioning, slicing, etc.) is a critical stage in product manufacture, because if it is not carried out properly it may result in microbial contamination. Raised BA levels have been observed in highly comminuted meat products such as burgers (Ruiz-Capillas and Jiménez-Colmenero, 2004a). Proper handling of the raw materials and proper use and maintenance of equipment to ensure sanitary conditions with processing temperatures at 2–4ºC are required to prevent contamination and thus lower BA formation. The cold chain must be kept unbroken during carcass handling and product manufacture up to production of the finished product (Maijala *et al.*, 1995b).

Additionally, it should be borne in mind that each meat product entails a specific type of processing that will need to be properly implemented in each case in order to be able to effectively reduce BA concentrations in the finished product. Indeed, heat-treated meat products (meat batter, frankfurter sausages, cooked ham, etc.) have very low BA levels (except spermidine and spermine) if the entire process is carried out in the proper conditions.

However, a series of circumstances (contamination of the raw material, pH, a_w, starter, temperature, etc.) associated with processing methods make fermentation of meat products one of the technical processes most conducive to BA formation (Vidal-Carou *et al.*, 1990; Maijala *et al.*, 1995a; Bover-Cid *et al.*, 2000b, 2001b, Roig-Xagués *et al.*, 2009). Reducing amine levels in products of this type involves careful control of the factors discussed above. Apart from contaminating microorganisms, the microorganisms responsible for BA production in these types of products include the bacterial components of the starter culture and the secondary microbiota. Consequently, lowering BA levels in these products will require, on the one hand, controlling the contaminating flora by means of proper handling, processing and storage and, on the other, choosing starter cultures composed of microorganisms that have no or very low levels of amino acid decarboxylase activity. Beyond that, a_w, temperature, relative humidity and ripening time also need to be carefully controlled. High temperatures (higher than 22–24ºC) during ripening increase BA levels (Maijala *et al.*, 1995b;

Eerola *et al.*, 1998). An appreciable decrease in a_w could also inhibit BA formation (Eerola *et al.*, 1996; Treviño *et al.*, 1997).

The development of new products, including functional foods, also needs to be monitored to avoid increased BA contents in the finished product. Qualitative and/or quantitative changes in composition are intended either to enhance the presence of components that have displayed beneficial effects (n-3 polyunsaturated fatty acids (PUFAs), dietetic fibre, probiotics, antioxidants, etc.) or act to reduce other components that have adverse consequences for health (saturated fat, sodium, etc.) (Jiménez-Colmenero, 2007). Obviously, changes in composition and/or processing may also alter factors that affect BA formation, for instance, product type and nature and the microbial flora, which may have potential repercussions on the presence and activity of both the substrate and enzymes (Ruiz-Capillas and Jiménez-Colmenero, 2004a). Walnut-enriched restructured meats have been shown to have a quantitative and qualitative effect on BAs content (Ruiz-Capillas *et al.*, 2004).

Packaging and Storage Conditions

Product packaging upon manufacture presents a major barrier to recontamination and BA formation. Recontamination of meat products after manufacture is an important source of microorganisms that acts to limit product shelf life and contributes to BA formation. The use of protective atmospheres and vacuums has proven to be advantageous in lowering BA levels or altering the BA profile of the finished product. The gas mixture used to prepare modified atmospheres for meat and meat products will have a direct effect on microorganism growth and on amino acid decarboxylase enzymatic activity levels (Wortberg and Woller, 1982; Edwards *et al.*, 1987; Ruiz-Capillas and Moral, 2002, 2003; Ruiz-Capillas and Jiménez-Colmenero, 2004a). Modifying O_2 levels in the package may condition the growth of one flora in favour of another anaerobic or facultatively aerobic flora with a different level of BA production. High amounts of CO_2 in the gas blend inhibit microorganism growth and hence BA production (Nadon *et al.*, 2001). Changes in CO_2/O_2 levels in modified atmospheres have also been observed to result in a rise in amine concentrations by selecting for high BA-producing strains. For instance, tyramine was observed to increase in dry fermented Spanish chorizo sausage and cadaverine was observed to increase in PAP-stored sliced cooked ham (Ruiz-Capillas and Jiménez-Colmenero, 2004b).

Employing other minimal processing technologies designed to minimize alterations in product quality attributes and prolong shelf life have also been found to be beneficial for controlling the BA content of the finished product. Putrescine, tyramine, spermidine and spermine have been observed to be most sensitive to irradiation (Kim *et al.*, 2005; Min *et al.*, 2007). Subjecting meat products (cooked ham, chorizo, meat butter, etc.) to high hydrostatic pressures has also been reported to decrease BA levels. High-pressure treatment of sliced chorizo sausage also brought about a significant decrease in tyramine, putrescine, cadaverine and spermine along with a significant increase in spermidine (Ruiz-Capillas *et al.*, 2007b) but, to achieve this, strict control of product temperature during storage was required, since BAs in meat products are affected not just by processing but to an even larger extent by storage conditions (Ruiz-Capillas *et al.*, 2007a; Ruiz-Capillas *et al.*, 2007c). Fluctuations in storage temperature between 2 and 12ºC affected final BA production (Ruiz-Capillas *et al.*, 2007a).

Conclusion

Lowering BA content in foods generally and in meat and meat products in particular has implications for researchers, consumers, food manufacturers and health authorities because of the potential roles of BAs in health and food quality. Meat and meat products are highly perishable protein-rich foods which are good substrates for the growth of microorganisms that are potential

BA producers. Consequently, employing a variety of procedures to lower BA levels in the finished products is vitally important. Procedures for lowering BA levels in meat and meat products need to regulate all the factors involved in BA formation in each product as a whole. Comprehensive control of the raw materials, additives, treatments and processing employed, preservation treatments, packaging and storage is required. Most such factors (temperature, time, etc.) are mainly directed at inhibiting the growth of BA-producing micro-organisms or at promoting the growth of microbial communities with low levels of amino acid decarboxylase activity using suitable starters which also help develop the requisite sensory attributes of each product. It should additionally be noted that these procedures are all interrelated and contingent upon complex interactions among different factors operating concurrently in meat and meat products.

Acknowledgements

This research was also supported under projects AGL2003-00454, AGL2007-61038/ALI of the Plan Nacional de Investigación Científica, Desarrollo e Innovación Tecnológica (I+D+I), the Consolider CSD2007-00016, Ministerio de Ciencia y Tecnología, and the Intramural Special Project 200970I104.

References

Bardócz, S. (1995) Polyamines in food and their consequences for food quality and human health. *Trends in Food Science and Technology* 6, 341–346.

Bover-Cid, S. and Holzapfel, W.H. (1999) Improved screening procedure for biogenic amine production by lactic acid bacteria. *International Journal of Food Microbiology* 53, 33–41.

Bover-Cid, S., Izquierdo-Pulido, M. and Vidal-Carou, M.C. (2000a) Mixed starter cultures to control biogenic amine production in dry fermented sausages. *Journal Food Protection* 63, 1556–1562.

Bover-Cid, S., Izquierdo-Pulido, M. and Vidal-Carou, M.C. (2000b) Influence of hygienic quality of raw material son biogenic amine production during ripening and storage of dry fermented sausages. *Journal Food Protection* 63, 1544–1550.

Bover-Cid, S., Izquierdo-Pulido, M. and Vidal-Carou, M.C. (2001a) Effectiveness of a *Lactobacillus sakei* starter culture in reduction of biogenic amine accumulation as a function of the raw material quality. *Journal Food Protection* 64, 367–373.

Bover-Cid, S., Izquierdo-Pulido, M. and Vidal-Carou, M.C. (2001b) Changes in biogenic amine and polyamine contents in slightly fermented sausages manufactured with and without sugar. *Meat Science* 57, 215–221.

Bover-Cid, S., Izquierdo-Pulido, M. and Vidal-Carou, M.C. (2001c) Effect of the interaction between a low tyramine-producing Lactobacillus and proteolytic staphylococci on biogenic amine production during ripening and storage of dry sausages. *International Journal of Food Microbiology* 65, 113–123.

Cassens, R.G. (1997) Residual nitrite in cured meat. *Food Technology* 51, 53–55.

de las Rivas, B., Ruiz-Capillas, C., Carrascosa, A.V., Curiel, J.A., Jiménez-Colmenero, F. and Muñoz, R. (2008) Biogenic amine production by Gram-positive bacteria isolated from Spanish dry-cured "chorizo" sausage treated with high pressure and kept in chilled storage. *Meat Science* 80, 272–277.

Directive 91/493/CEE (1991). Directive of July 22, 1991, concerning the normative for production and merchandising of fishery products (91/493/CEE). Diario Oficial de la Comunidades Europeas, L286, 15–34.

Domanska-Blicharz, K., Rachubik, J. and Kowalski, B. (2005) Occurrence of volatile N-Nitrosamines in polish tinned food. *Bulletin of the Veterinary Institute in Pulawy* 49, 319–322.

Edwards, R.A., Dainty, R.H. and Hibard, C.M. (1985) Putrescine and cadaverine formation in vacuum packed beef. *Journal of Applied Bacteriology* 58, 13–19.

Edwards, R.A., Dainty, R.H., Hibard, C.M. and Ramantanis, S.V. (1987) Amines in fresh beef of normal pH and the role of bacteria in changes in concentration observed during storage in vacuum packs at chill temperatures. *Journal of Applied Bacteriology* 63, 427–434.

Eerola, S., Maijala, R., Roig Sagués, A.X., Salminen, M. and Hirvi, T. (1996) Biogenic amines in dry sausages as affected by starter culture and contaminant amine-positive *Lactobacillus*. *Journal of Food Science* 61, 1243–1246.

Eerola, H.S., Roig-Sagués, A.X. and Hirvi, T.K. (1998) Biogenic amines in Finnish dry sausages. *Journal of Food Safety*. 18, 127–138.

Ellis, D.I., Broadhurst, D. and Goodacre, R. (2004) Rapid and quantitative detection of the microbial spoilage of beef by Fourier transform infrared spectroscopy and machine learning. *Analytica Chimica Acta* 514, 193–201.

FDA (Food and Drug Administration) (1990) Decomposition of histamines; raw, frozen tuna and malu malu, canned tuna, and related species. Revised compliance policy guide, Availability-Federal Register, 60, (149), 39574–39756.

Halász, A., Simon-Sarkadi, L. and Holzapfel, W. (1994) Biogenic amines and their production by microorganisms in food. *Trends in Food Science and Technology* 5, 42–49.

Hernández-Herrero, M.M., Roig-Sagués, A.X., Rodríguez, J.J. and Mora-Ventura, M.T. (1999) Halotolerant and halophilic histamine-forming bacteria isolated during the ripening of salted anchovies (Engraulis encrasicholus). Journal of Food Protection 62, 509–514.

Hernández-Jover, T., Izquierdo-Pulido, M., Veciana-Nogues, M.T. and Vidal-Carou, M.C. (1996) Biogenic amine sources in cooked cured shoulder pork. *Journal of Agricultural and Food Chemistry* 44, 3097–3101.

Hernández-Jover, T., Izquierdo-Pulido, M., Veciana-Nogues, M.T., Mariné-Font, A. and Vidal-Carou, M.C. (1997a) Biogenic amine and polyamine contents in meat and meat products. Journal of Agricultural and *Food Chemistry* 45, 2098–2102.

Hernández-Jover, T., Izquierdo-Pulido, M., Veciana-Nogues, M.T., Mariné-Font, A. and Vidal-Carou, M.C. (1997b) Effect of starter cultures on biogenic amine formation during fermented sausage production. *Journal of Food Protection* 60, 825–830.

Jiménez-Colmenero, F. (2007) Healthier lipid formulation approaches in meat-based functional foods. Technological options for replacement of meat fats by non-meat fats, *Trends in Food Science and Technology* 18, 567–578.

Kalač, P. and Krausová, P. (2005) A review of dietary polyamines: Formation, implications for growth and health and occurrence in foods. *Food Chemistry* 90, 219–230.

Kim, J.H., Kim, D.H., Ahn, H.J., Park, H.J. and Byun, M.W. (2005) Reduction of the biogenic amine contents in low salt-fermented soybean paste by gamma irradiation. *Food Control* 16, 43–49.

Latorre-Moratalla, M.L., Bover-Cid, S., Talon, R., Garriga, M., Zanardi, E., Ianieri, A., Fraqueza, M.J., Elias, M., Drosinos, E.H. and Vidal-Carou, M.C. (2010) Strategies to reduce biogenic amine accumulation in traditional sausage manufacturing. *LWT - Food Science and Technology* 43, 20–25.

Lorenzo, J.M., Martínez, S., Franco, I. and Carballo, J. (2007) Biogenic amine content during the manufacture of dry-cured lacón, a Spanish traditional meat product: Effect of some additives. *Meat Science* 77, 287–293.

Lyte, M. (2004) The biogenic amine tyramine modulates the adherence of *Escherichia coli* O157:H7 to intestinal mucosa. *Journal of Food Protection* 6, 878–883.

Maijala, R., Eerola, S., Lievonen, S., Hill P. and Hirvi, T. (1995a) Formation of biogenic amines during ripening of dry sausages as affected by starter culture and thawing time of raw materials. *Journal of Food Science* 60, 1187–1190.

Maijala, R., Nurmi, F. and Fischer, A. (1995b) Influence of processing temperature on the formation of biogenic amines in dry sausages. *Meat Science* 39, 19–22.

Marcobal, A., de las Rivas, B., Moreno-Arribas, M.V. and Muñoz, R. (2005) Multiplex-PCR method for the simultaneous detection of acid lactic bacteria producing histamine, tyramine and putrescine, three major biogenic amines. *Journal of Food Protection* 68, 874–878.

Masson, F., Talon, R. and Montel, M.C. (1996) Histamine and tyramine production by bacteria from meat products. *International Journal of Food Microbiology* 32, 199–207.

McCabe, B.J. (1986) Dietary tyramine and other precursors amines in MAOI regimens: A review. *Journal of the American Dietetic Association* 86, 1059–1064.

Mietz, J.L. and Karmas, E. (1977) Polyamine and histamine content of rockfish, salmon, lobster and shrimp as an indicator of decomposition. *Journal of the Association of Official Analytical Chemists* 61, 139–145.

Min, J.S., Lee, S.O., Jang, A., Jo, Ch. and Lee, M. (2007) Irradiation and organic acid treatment for microbial control and the production of biogenic amines in beef and pork. *Food Chemistry*, 104, 791–799.

Nadon, C.A., Ismond, M.A. and Holley, R. (2001) Biogenic amines in vacuum-packaged and carbon dioxide-controlled atmosphere-packaged fresh pork stored at −1.5 degrees C. *Journal of Food Protection* 64, 220–227.

Oust, A., Møretrø, T., Kirschner, C., Narvhus, JA. and Kohler A. (2004) FT-IR spectroscopy for identification of closely related lactobacilli. *Journal of Microbiological Methods* 59, 149–162.

Rice, S.L., Eitenmiller, R.R. and Koehler, P.E. (1976) Biologically active amines in food: a review. *Journal of Milk and. Food Technology* 39, 353–358.

Roig-Sagués, A.X. and Eerola, S. (1997) Biogenic amines in meat inoculated with Lactobacillus sake starter strains and an amine-positive lactic acid bacterium. *Zeitschrift für Lebensmitteluntersuchung und -Forschung A*, 205, 227–231.

Roig-Sagués, A.X., Ruiz-Capillas, C., Espinosa, D. and Hernández, M. (2009) The decarboxylating bacteria present in foodstuffs and the effect of emerging technologies on their formation. In: Dandrifosse, G (ed.) *Biological Aspects of Biogenic Amines, Polyamines and Conjugates.* Transworld Research Network, Kerala, India pp. 201–230.

Ruiz-Capillas, C. and Moral, A. (2001) Production of biogenic amines and their potential use as quality control indices for hake (*Merluccius merluccius L.*) stored in ice. *Journal of Food Science* 66, 1030–1032.

Ruiz-Capillas, C. and Moral, A. (2002) Effect of controlled and modified atmospheres on the production of biogenic amines and free amino acids during storage of hake. *European Food Research and Technology* 214, 476-481.

Ruiz-Capillas, C. and Moral, A. (2003) Free amino acids and biogenic amines in red and white muscle of tuna stored in controlled atmospheres. *Amino acids* 26, 125–132.

Ruiz-Capillas, C. and Jiménez-Colmenero, F. (2004a) Biogenic amines in meat and meat products. *Critical Reviews in Food Science and Nutrition* 44, 489–599.

Ruiz-Capillas, C. and Jiménez-Colmenero, F. (2004b) Biogenic amine content in Spanish retail market meat products treated with protective atmosphere and high pressure. *European Food Research and Technology* 218, 237–241.

Ruiz-Capillas, C. and Moral, A. (2005) Sensory and biochemical aspects of quality of whole bigeye tuna (*Thunnus obesus*) during bulk storage in controlled atmospheres. *Food Chemistry* 89, 347–354.

Ruiz-Capillas, C., Cofrades, S., Serrano, A. and Jiménez-Colmenero, F. (2004) Biogenic amines in restructured beef steaks as affected by added walnuts and chilling storage. *Journal of Food Protection* 67, 607–609.

Ruiz-Capillas, C., Aller-Guiote, P., Carballo, J. and Jiménez-Colmenero, F. (2006) Biogenic amine formation and nitrite reactions in meat batter as affected by high-pressure processing and chilled storage. *Journal of Agricultural and Food Chemistry* 54, 9959–9965.

Ruiz-Capillas, C., Carballo, J. and Jiménez-Colmenero, F. (2007a) Biogenic amines in pressurized vacuum-packaged cooked sliced ham under different chilled storage conditions. *Meat Science* 75, 397–405.

Ruiz-Capillas, C., Jiménez Colmenero, F. Carrascosa, A.V. and Muñoz, R. (2007b) Biogenic amine production in Spanish dry-cured “chorizo” sausage treated with high-pressure and kept in chilled storage. *Meat Science* 77, 365–371.

Ruiz-Capillas, C., Carballo, J. and Jiménez-Colmenero, F. (2007c) Consequences of high-pressure processing of vacuum-packaged frankfurters on the formation of biogenic amines: effect of chilled storage. *Food Chemistry* 104, 202–208.

Sattler, J., Hafner, D., Klotter, H.J., Lorenz, W. and Wagner, P.K. (1988) Food-induced histaminosis as an epidemiological problem: plasma histamine elevation and haemodynamic alterations after oral histamine administration and blockade of diamine oxidase (DAO). *Agents and Actions* 23, 361–365.

Shalaby, A.R. (1996) Significance of biogenic amines to food safety and human health. *Food Research International* 29, 675–690.

Silla-Santos, M.H. (1996) Biogenic amines: their importance in foods. *International Journal of Food Microbiology* 29, 213–231.

Silva, C.M.G. and Gloria, M.B.A. (2002) Bioactive amines in chicken breast and thigh after slaughter and during storage at 4±1 °C and in chicken-based meat products. *Food Chemistry* 78, 241–248.

Slemr, J. and Beyermann, K. (1985) Concentration profiles of diamines in fresh and aerobically stored pork and beef. *Journal of Agricultural and Food Chemistry* 33, 336–339.

Smith, T.A. (1980) Amines in food. *Food Chemistry* 6, 169–200.

Straub, B.W., Tichaczek, P.S., Kicherer, M. and Hammes, W.P. (1994) Formation of tyramine by *Lactobacillus curvatus* LTH 972. *Zeitschrift für Lebensmitteluntersuchung und -Forschung A.* 1, 9–12.

Suzzi, G. and Gardini, F. (2003) Biogenic amines in dry fermented sausages: A review. *International Journal of Food Microbiology* 88, 41–54.
Taylor, S.L. (1986) Histamine food poisoning: toxicology and clinical aspects. *Critical Reviews in Food Science and Nutrition* 17, 91–128.
ten Brink, B., Damink, C., Joosten, H.M.L.J. and Huis in´t Veld, J.H.J. (1990) Occurrence and formation of biologically active amines in foods. *International Journal of Food Microbiology* 11, 73–84.
Treviño, E., Beil, D. and Steinhart, H. (1997) Formation of biogenic amines during the maturity process of raw meat products, for example of cervelat sausage. *Food Chemistry* 60, 521–526.
Vidal-Carou, M.C., Izquierdo, M. L., Matín, M.C. and Mariné, A. (1990) Histamina y tiramina en derivados cárnico. *Revista de Agroquimica y Tecnologia de Alimentos 30,* 102–108.
Vinci, G. and Antonelli, M.L. (2002) Biogenic amines: quality index of freshness in red and white meat. *Food control* 13, 519–524.
Virgili, R., Saccani, G., Gabba, L., Tanzi, E. and Soresi Bordini, C. (2007) Changes of free amino acids and biogenic amines during extended ageing of Italian dry-cured ham. *LWT - Food Science and Technology* 40, 871–878.
Warthensen, J.J., Scanlan, R.A., Bills, D.D. and Libbely, L.M. (1975) Formation of heterocyclic N-nitrosamines from the reaction of nitrite and selected primary diamines and amino acids. *Journal of Agricultural and Food Chemistry* 23, 898–902.
Wortberg, B. and Woller, R. (1982) Quality and freshness of meat and meat products as related to their content of biogenic amines. *Fleischwirtsschaft* 62, 1457–1463.

12 Biogenic Amines in Wine and Vinegar: Role of Starter Culture in their Inhibition

Isabel M.P.L.V.O. Ferreira* and Olívia Pinho

Introduction

Wine and vinegar are fermented products known by most ancient civilizations. The preservation aspect of fermented foods was obviously important thousands of years ago, when few other preservation techniques existed. However, nowadays, these products are widely studied and appreciated because of their organoleptic properties, uniqueness and increased economic value.

The early manufacturers of fermented products obviously could not have appreciated the actual science involved in their production, since it was only in the last 150 to 200 years that microorganisms and enzymes were 'discovered'. Nowadays, it is well known that the quality of fermented products is closely related to the diversity and composition of the microbial species that develop during the fermentative process (Romano *et al.*, 2003; Hansen, 2002; Pinho *et al.*, 2004). Once the scientific basis of fermentation was established, efforts soon began to identify and cultivate micro-organisms capable of performing specific fermentations, for example, cultures for wine have become the norm for wine producers (Caruso *et al.*, 2002; Hansen, 2002). The primary activity of the culture in wine fermentation is to convert carbohydrates to desired metabolites as alcohol, acetic acid, lactic acid or CO_2. Alcohol and organic acids are good natural preservatives, but also appreciated in the fermented product. A secondary fermentation by lactic acid bacteria is responsible for the reduction of the acidity by converting malic acid to lactic acid. The cultures used in wine fermentations are, however, also contributing by 'secondary' reactions to the formation of flavour and texture. This secondary contribution can often be responsible for the difference between products of different brands, and thereby contribute significantly to the value of the product (Jackson, 2000).

Wine vinegar is mostly produced in countries with enological tradition; its use is as ancient as the use of wine. It can be made from red or white wine, and is the most commonly used vinegar as seasoning or preserving agent in the households of the Mediterranean countries and Central Europe. Although wine vinegar results from a spontaneous process which takes place in wines and musts in contact with air, it is far from being the simple spoilage of wine. Vinegar, as a food by-product from wine, has recently acquired an important role in salad dressings, ketchup and other sauces, etc. (Tesfaye *et al.*, 2002).

* Corresponding author.

Vinegar is produced by a two-stage fermentation process, the first one being the conversion of fermentable sugars to ethanol by yeasts, usually *Saccharomyces* species, and the second the oxidation of ethanol by bacteria, usually *Acetobacter* species (Adams, 1998).

The production of fermented products such as wine and vinegar results in desirable effects, namely, natural preservation, hygienic safety and acquisition of high sensory and nutritive value (Straub *et al.*, 1995). However, the metabolic activity of the microorganisms involved may also give rise to the formation of undesirable compounds, such as biogenic amines (BAs) (Kirschbaum *et al.*, 1997; Ferreira and Pinho, 2006; Costantini *et al.*, 2009). These compounds can be formed and degraded as a result of normal metabolic activity in animals, plants and microorganisms, and are originated mainly from the decarboxylation of amino acids. In fermented products the reaction is catalysed by amino acid decarboxylase enzymes produced by microorganisms (ten Brink *et al.*, 1990; Halasz *et al.*, 1994).

BAs are necessary for several critical functions in man, but high concentrations of BAs can cause undesirable physiological effects in sensitive humans, especially when alcohol is present (Bauza and Teissedre, 1995). The presence of BAs and the origin of these compounds in wine is well documented (Vidal-Carou *et al.*, 1990; Soufleros *et al.*, 1998; Caruso *et al.*, 2002; Ferreira and Pinho, 2006). The main BAs in wine are histamine, tyramine, putrescine, cadaverine and phenylethylamine. Studies on the BA content of wine vinegars are scarce (Kirschbaum *et al.*, 1997).

The purpose of the present chapter is to describe the factors influencing the formation of biogenic amines during wine and wine vinegar making and the role of starter cultures in its reduction.

Process of Wine and Wine Vinegar Making

Winemaking history started way back in 6000 BC. It became popular in ancient Greece, Rome and Egypt, but it was only in 1857 that Pasteur correctly described the science behind fermentation and wine production. Nowadays, winemaking requires a deep scientific knowledge. This process, also named as vinification, is the production of wine, starting with selection of the grapes, followed by the alcoholic fermentation of grape juice, which is stored in a manner to retain its wine-like properties, and ending with bottling the finished wine. Two major types are produced: red wine and white wine (Jackson, 2000).

Red wine is a macerated wine. The extraction of solids from grape (skins, seeds and stems) accompanies the alcoholic fermentation of the juice. The localization of red pigment exclusively in skins permits a slightly tinted or white wine to be made from the colourless juice obtained from a delicate pressing of red grapes (Ribéreau-Gayon *et al.*, 2000).

The classic steps of red winemaking are depicted in Fig. 12.1. Grapes are traditionally destemmed and crushed to break the skin in order to release the pulp and the juice to obtain the must. Grape must is a highly fermentable medium in which yeast find the necessary substances to ensure their vital function. Carbohydrates (glucose and fructose) are used as carbon and energy sources.

The two main organisms involved in vinification are *Saccharomyces cerevisiae* and *Leuconostoc oenos*. They are atypical in withstanding moderately high ethanol concentrations and are somewhat unusual in selectively employing fermentative metabolism. *S. cerevisiae* possesses the ability to respire; however, it predominantly ferments, even in the presence of oxygen. It has the ability to quickly establish itself as the dominant organism in grape must owing to its alcohol tolerance and preference for alcoholic fermentation. Most bacteria that could grow during fermentation are inhibited by *S. cerevisae*. *L. oenos* is less well adapted to growing in grape juice or must than *S. cerevisiae*. It typically grows slowly in wine after *S. cerevisiae* has completed alcoholic fermentation and becomes inactive. It is the main bacterium responsible for malolactic fermentation (MLF), the con-

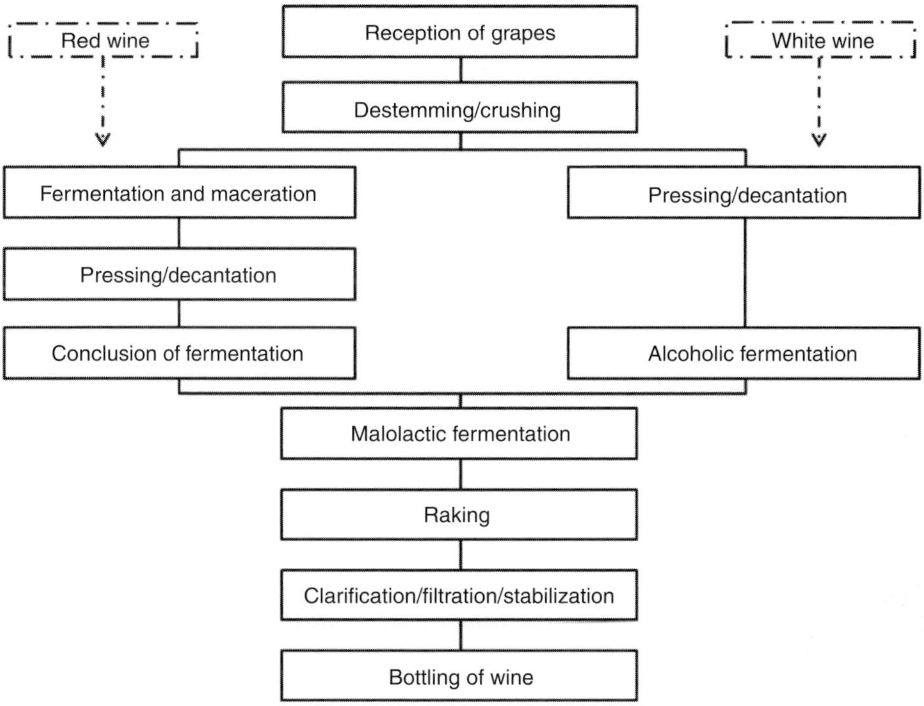

Fig 12.1. Scheme of the red and white wine vinification processes.

version of malic acid to lactic acid, increasing the pH, which is in general a characteristic of red winemaking.

White wines are exclusively produced by the fermentation of grape juice. Thus, in the production of white wines, juice extraction and varying degrees of clarification always precede alcoholic fermentation. It is the absence of skin contact in the alcoholic phase, and not the colour of the grape, that distinguishes white winemaking from red winemaking. This is not to say that white winemaking does not include any maceration. A certain degree of maceration is inevitably associated with white winemaking. It occurs in the absence of alcohol during the pre-fermentation phase, at the time of juice extraction and clarification (Ribéreau-Gayon *et al.*, 2000). MLF has more frequently been involved in red winemaking, but more recently in making white wines as well. The classic steps of white winemaking are depicted in Fig. 12.1.

Until the beginning of the 19th century it was thought that vinegar was derived from the spontaneous acidification of wine ('Vin aigre'). In 1864, Pasteur discovered that vinegar was produced by the action of microorganisms, 'Mycoderma aceti', on the fermented product (wine, in this case). Thus, wine vinegar is obtained by double fermentation: the alcoholic fermentation is performed by yeasts, while the acetic acid bacteria carry out the acetic acid fermentation (oxidative fermentation). In this way, Pasteur suggested the improvement of vinegar production from the original method, known as the Orleans Method or Pasteur's Method (Llaguno, 1991).

In Pasteur's Method, also called surface-culture fermentation or traditional (slow) process, the acetic acid bacterium is placed on the air–liquid interface in a direct contact with atmospheric air (oxygen). The presence of the bacteria is limited to the surface of the acidifying liquid and hence, it is also considered as a static method. Nowadays, this method is employed for the production of traditional and selected vinegars and a very long period of time is required to obtain a high acetic degree (Adams and Moss, 2000; Plessi, 2003; Guizani and Mothershaw, 2006). As a consequence, production time and costs are increased. This process permits simultaneous acetification and ageing (Tesfaye *et al.*, 2002).

The fast and industrial method of producing a lot of vinegar at once for commercial use is by submerged fermentation. This method is used in factories that have large equipment components including an acetator tank and a heating system to control the temperature of the vinegar being created. In submerged fermentation, a quantity of mother of vinegar is used as a catalyst and introduced into a large base in metal tanks known as acetators. These tanks are equipped with a variety of systems that keep the mixture constantly turning, introducing air into the mixture to provide enough oxygen to keep the bacteria working. The mixture is kept at about 80°F to speed up the process.

Although the industrial submerged processes are more efficient, the product is less aromatic than vinegars obtained by slower processes, due to the brief period of contact. The esterases do not have time to perform their function adequately; consequently the characteristic volatile content is low. The product is therefore filtered and, in some cases, put into wooden casks to age. The result is a vinegar of superior quality (reserve).

A schematic representation of wine vinegar elaboration with the two basic methods is presented in Fig. 12.2. Wine vinegar is made from red or white wine. The raw material's quality is crucial in order to obtain good vinegar. However, the production process is also influential in producing a quality product. Ageing the wine vinegar increases its quality, because it is through this maturing process that the vinegar obtains the organoleptic characteristics that the consumer desires (Morales *et al.*, 2002). Vinegar production usually requires lower capital investment, has shorter start-up times, and can generate different types of vinegar when different carbohydrate sources are used (Plessi, 2003; Guizani and Mothershaw, 2006).

Fermentation and Microbiology

Fermentation is an energy-releasing form of metabolism in which both the substrate and by-product are organic compounds using yeasts, bacteria or a combination thereof, under anaerobic conditions. It differs fundamentally from respiration in not

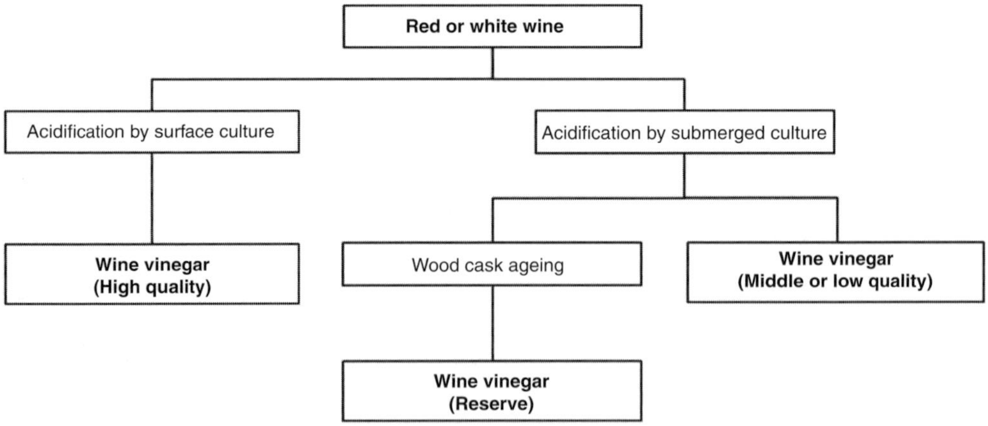

Fig 12.2. Wine vinegar elaboration by the two base methods: surface and submerged culture. Additional wood-cask ageing is an alternative to improve wine-vinegar quality.

requiring the involvement of molecular oxygen. The science of fermentation is known as zymology.

Alcoholic fermentation

Many fermentative pathways exist, but *S. cerevisiae* possesses the most common – alcoholic fermentation. Although this yeast has the ability to respire, it predominantly ferments, even in the presence of oxygen. In its various forms, it may function as the wine yeast, brewer's yeast, distiller's yeast and baker's yeast. Laboratory strains are extensively used in industry and in fundamental studies on genetics, biochemistry and molecular biology.

S. cerevisiae is usually absent or rare on grapes. Mortimer and Polsonelli (1999) estimated about one healthy berry per thousand carries wine yeast. However, on the surface of damaged fruit, the frequency may rise to one in four (~1×10^5 to 1×10^6 yeast cells/berry).

Other *Saccharomyces* species, *S. bayanus* and *S. uvarum*, can also conduct equally effective alcoholic fermentations; they are employed in special winemaking situations (Jackson, 2000).

Glucose and fructose are metabolized to ethanol during alcoholic fermentation primarily via glycolysis. Additional yeast metabolites generate the most common aromatic compounds found in wine. Non-volatile aroma precursors are hydrolysed, liberating aromatic terpenes, phenols and norisoprenoids. Thus, much of the fragrance of wine can be interpreted in terms of the modifications of primary and intermediary yeast-cell metabolism.

Even in healthy grapes, other, non-*Saccharomyces* yeasts may occur at concentrations similar to those of strains of *S. cerevisiae*. Consequently, in spontaneous fermentation there is an early and rapid succession of yeast species. At its commencement, fermentation involves the action of species such as *Kloeckera apiculata, Candida stellata, Toruslaspora delbrueckii* and *Saccharomycodes ludwigii*. In the course of the fermentation process, they are replaced by high-ethanol-tolerant *S. cerevisiae* (Ciani and Picciotti, 1995).

There has been much discussion over the years concerning the relative merits of spontaneous versus induced fermentation. That various strains of *S. cerevisiae* differ markedly in the fragrance they donate is indisputable. Spontaneous fermentations may provide a source of aromatic distinctiveness that can vary from year to year, and location to location. This can contribute an element of uniqueness that is desirable on marketing. It also carries the risk that the indigenous yeast may confer off-odours or possess other undesirable traits.

Malolactic fermentation

The high acidity of must and wine retards or inhibits the growth of most lactic acid bacteria, thus, excludes competitive bacteria. However, in some conditions, malolactic fermentation (MLF) can occur. This fermentation is the conversion of L(–)- malic acid to L(+)-lactic acid, a weaker acid, and it results in an increase in the pH and favours bacterial growth. MLF has three distinct but interrelated effects on wine quality. It reduces acidity, influences microbial stability and may affect the sensory characteristics of the wine. It is more frequent in red wines. The greatest controversy concerning the relative merits and demerits of MLF is related with flavour modifications, because it reduces the incidence of vegetal notes and accentuates fruit flavours.

In common practice, the native malolactic bacteria of grapes accomplish MLF fermentation in wine (Versari *et al.*, 1999). *Lactobacillus, Pediococcus* and *Leuconostoc* spp. are associated with the MLF. Among them, *Leuconostoc oenos*, more recently reclassified as *Oenococcus oeni* (Dicks *et al.*, 1995), is recognized as the bacterium most tolerant to the wine conditions and, in most wines, it is dominant during MLF. In recent years freeze-dried starter cultures using *L. oenos* have become available to initiate this process.

Lactic acid bacteria cannot grow with L-malic acid as a unique carbon source; these microorganisms therefore need an additional energy source, such as residual fermentable sugars or amino acids to allow cell growth (Liu and Pilone, 1998). Substrate cofermentation by *L. oenos* largely depends on the strain used as well as on the environmental conditions (e.g. wine composition, pH and temperature). Wine has a complex composition (carbohydrates, SO_2, ethanol, phenolic compounds, fatty acids, amino acids, micronutrients, etc.), which varies depending on several conditions, such as cultivar, season, and winemaking process. Malolactic bacteria have elaborate nutritional requirements (Buckenhuskes, 1993) and competition for these may inhibit or delay yeast activity during the alcoholic fermentation. Edwards *et al.* (1998) suggested that inoculation of must with starter cultures should take place only after the conclusion of the alcoholic fermentation to avoid the increase of wine volatile acidity due to sugar metabolism by *L. oenos*.

The role of amino acids on MLF is still subject to investigation (González-Marco *et at.*, 2006). Isoleucine, glutamic acid, tryptophan and arginine are essential amino acids for the growth of some *L. oenos* strains in synthetic medium at pH 5.0. Additional amino acids are also required for optimal bacterial growth. On the other hand, the lack of glycine, phenylalanine, proline and tyrosine limits MLF without affecting growth (Fourcassie *et al.*, 1992). Closely related to amino acid composition and MLF is the production of biogenic amines in wine. Microorganisms decarboxylate amino acids in order to provide the cell with energy and to protect the cell against acidic environments by increasing the pH (Landete *et al.*, 2005a).

Oenococcus, Lactobacillus and *Pediococcus* are able to produce biogenic amines. MLF starter cultures of *L. oenos* are selected not to form biogenic amines in wine and *pediococci* are usually associated with wines that have very high levels. An increase in the levels of biogenic amines usually occurs towards the end of MLF or during maturation, when *lactobacilli* and *pediococci* are higher. The levels of biogenic amines in wines from different countries differ according to vinification practices. Nutritive supplements such as yeast autolysates are often added to must in order to activate fermentation and to improve the quality of wine. These, as well as providing nitrogen compounds such as amino acids, also supply fatty acids; however, after MLF it was observed that the concentration of biogenic amines was higher in the wine from the supplemented must (González-Marco *et al.*, 2006).

Acetic acid fermentation

Worldwide, the term vinegar is used to describe acetic acid that is produced by primary microbial metabolism, called 'acetic acid fermentation' or 'vinegar fermentation' (Ebner *et al.*, 1995). It is obtained biologically by oxidative conversion of ethanol-containing solutions by acetic acid bacteria (AAB). These bacteria eat the alcohol and turn it into acetic acid, which forms the most important part of the vinegar along with the original sugar particles still left (Lu *et al.*, 1999). Primary metabolic conversion of ethanol to acetic acid is accompanied by secondary metabolism, which combines producing flavour and typical aroma. Small quantities of volatile substances are formed during secondary metabolism, which include ethane, acetaldehyde, ethyl formate, ethyl acetate, isopentyl acetate, butanol, methylbutanol and 3-hydroxy-2-butanone, which vary from vinegar to vinegar, depending on the starting material, and because of their individual properties produce vinegars with a variety of odour, taste, colour and other properties.

The most important properties of a production strain in the vinegar industry are tolerance to high concentrations of acetic acid, low nutrient requirements and high production rate (Ebner *et al.*, 1996). The bioprocess is usually stopped at a minimum residual ethanol level to avoid over-oxidation. If ethanol concentration falls under this level, acetic acid is oxidized to water and CO_2 (Plessi, 2003). As the bacteria eat away at the alcohol, they form what is known as mother of vinegar, a mass of bacteria and starchy substances that is used as a catalyst for making more vinegar.

A variety of bacteria can produce acetic acid, mostly members of *Acetobacter* (*Gluconacetobacter*) are used commercially, typically the aerobic bacterium *A. aceti* at 27°C to 37°C (Adams and Moss, 2000; O'Toole and Lee, 2003; Plessi, 2003; Guizani and Mothershaw, 2006; Josephsen and Jaspersen, 2006). *Acetobacter europaeus*, now named *Gluconacetobacter europaeus*, was found to be the main producing species of industrial vinegar bioreactors in central Europe (Sievers *et al.*, 1992). Other species frequently isolated from vinegar fermentations include *A. pasterianus*, *A. polyoxogenes*, *G. xylinus*, *G. hansenii*, *G. oboediens*, and *G. intermedius* (Gullo *et al.*, 2006; Baena-Ruano *et al.*, 2006; Raspor and Goranovic, 2008).

The overall theoretical yield of acetic acid produced from glucose is 0.67 g acetic acid per gram glucose. Complete aeration and strict control of the oxygen concentration during bioprocessing are important to keep the bacteria viable and maximize yields (Cheryan, 2000). An interruption in the oxygen supply will result in death of the bacteria. The theoretical amount of air required for 1 l of vinegar containing 6% of acetic acid is about 120 l, whereas, in practice, given the slow rate of liquid–gas exchange, the amount required is much greater (García-García *et al.*, 2009).

The influence of base wines obtained by the fermentation of different yeast species on acetic acid bacteria growth and on the analytical profile of vinegars was investigated (Ciani, 1998). The substrates for wine-vinegar production exerted a strong influence on both acetic acid bacteria growth and analytical profile of vinegars. According to Ciani (1998) the base wine obtained from the alcoholic fermentation of *S. cerevisiae* was not always the best substrate. The *C. stellata* positively influenced the acetic acid bacteria growth and the quality of vinegar, while the wine obtained from the fermentation of *K. apiculata* was a good substrate for acetic acid bacteria growth and acetic acid production and could be used for low-quality vinegar production.

Biogenic Amine Production by Yeast, Lactic Acid Bacteria and Acetic Acid Bacteria from Wine and Vinegar

BAs contain a health risk for sensitive individuals. The presence of high concentrations of histamine, tyramine, and phenylethylamine in wine is related to dietary migraines (Sandler *et al.*, 1974; Rivas-Gonzalo *et al.*, 1983). Symptoms include not only headaches, but also nausea, respiratorial discomfort, hot flushes, cold sweat, palpitations and red rash. Other BAs, namely, putrescine and cadaverine, although not toxic in themselves, intensify adverse effects of the above-mentioned BAs as they interfere with enzymes that metabolize them. Alcohol and acetaldehyde have been found to increase the sensitivity to biogenic amines (Landete *et al.*, 2005a). Secondary amines (spermine, spermidine, dimethylamine, pyrrolidine, etc.) can react with nitrous acid and its salts to form nitrosamines, compounds of known carcinogenic action (Silla-Santos, 1996).

At this stage there are no legal limits, but certain countries have recommended maximum limits with regard to histamine levels (mg/l) that are applicable to imported wines. Upper limits for histamine in wine have been recommended in Germany (2 mg/l), Belgium (5–6 mg/l), and France (8 mg/l) (Lehtonen, 1996).

Several factors affect the content of BAs in wine and wine vinegar, such as the type of soil and nitrogen fertilizer, the degree of maturation of grape, wine- and vinegar-making practices, the presence of precursor amino acids, the yeast strain responsible for fermentation, the number of decarboxylase-positive LAB, spontaneous MLF, residual microbial populations and the pH. It is important to know the ability of microorganisms involved in wine and vinegar making to produce BAs.

The presence of BAs in must and wines is well documented in the literature (Lehtonen, 1996; Soufleros *et al.*, 1998; Gerbaux and Monamy, 2000; Landete *et al.*, 2005b, Millan *et al.*, 2007; Ertan Anlı and Bayram, 2009). However, data are complex and in some cases contradictory. Studies on BAs content of wine vinegars are, however, scarce (Kirschbaum *et al.*, 1997). Table 12.1 summarizes BAs of wines and wine vinegar reported by different researchers. Some biogenic amines are normal constituents of grapes and *Vitis vinifera*, namely, putrescine and spermidine that are usually abundant (20 and 45 mg/kg of fresh fruit, respectively) (Buteau *et al.*,1984; Ough *et al.*, 1987; Radler and Fáth, 1991; Baucom *et al.*, 1996). Other BAs have their origin in the microbial decarboxylation of amino acids (Fig. 12.3). BAs concentration in wines could be decreased, reducing to a minimum the length of the processes that incorporate amino acids to must or wine such as grape skin maceration and the contact with the lees, but this is impossible when aged wines are intended (Landete *et al.*, 2007).

The role of lees in the presence of BAs in wine is crucial, not only as being responsible for amino acids release which can be decarboxylated, but also as a micro-organisms' reservoir (Pérez-Serradilla and Luque de Castro, 2008). The most common microorganisms present in the lees are yeasts, which are responsible for fermentation, but bacteria may also be present in MLF. Thus, the lees can be responsible for the presence in wines of amino acids, decarboxylase-positive microorganisms and decarboxylase enzymes (which can be released during yeast lees autolysis), which, under favourable environmental conditions, can lead to biogenic amines formation in white wines (González-Marco and Ancín-Azpilicueta, 2006) and in red wines (Martín-Álvarez *et al.*, 2006). The main conclusion in both cases was that the overall concentration of BAs in wines matured with lees was higher than in those elaborated without lees.

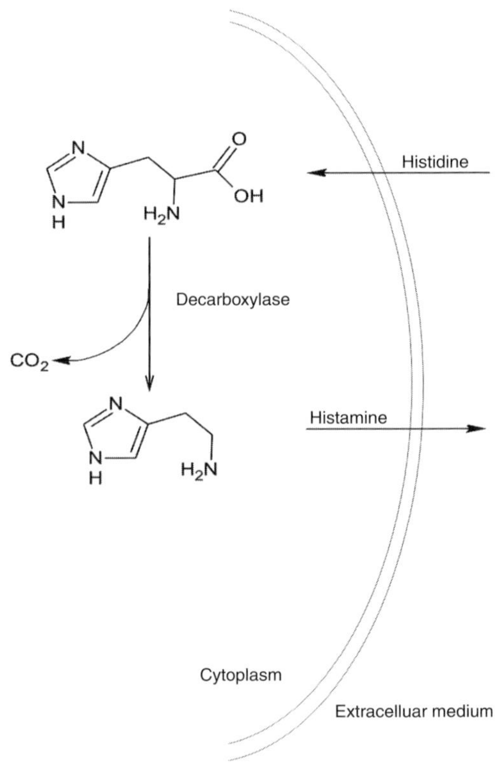

Fig 12.3. The formation of biogenic amines, with an example of histamine production.

Table 12.1. Biogenic amine content (mg/l) of wines and wine vinegars reported by different researchers.

	PUT	HI	CAD	TY	SPD	PEA	SPM	Reference
Red wine	0.94–15.6	0.03–2.8	0.11–4	0.04–3.5	0.01–1	0.05–3.4	0.1–8.6	Anli *et al.*, 2004
Red wine	0.06–13	0.4–8.22	0.07–0.68	0.03–3.2	0.08–1.1	0.09–0.19	–	Millán *et al.*, 2007
Red wine	3.2–8.87	0–0.17	–	0–1.65	–	–	0–2.59	Husnik *et al.*, 2006
Red wine	0–26.54	0–10.77	0–3.15	0–11.32	–	–	0–2.24	Busto *et al.*, 1997
Red wine	0.77–4.33	0–1.73	–	0.33–1.07	0.03–1.63	–	0.2–1.37	Souza *et al.*, 2005
Rioja red wine	32.97	8.72	0.61	4.98	–	–	–	Fernandes and Ferreira, 2000
Azsu wine	4.1	0.31	0.31	1.98	5.34	–	18.4	Hajós *et al.*, 2000
Porto wine	0.14–11.9	0.01–2.35	0.06–0.09	0.006–2.63	–	–	0.01–1.32	Romero *et al.*, 2000
Rioja white wine	3.01	0.84	0.28	0.89	–	–	–	Fernandes and Ferreira, 2000
Sherry wine	7.3	n.d	n.d	n.d	1.6	n.d	0.1	Kirschbaum *et al.*, 1997
Sherry vinegar	37.2	21.9	4.8	15.9	1.9	n.d	n.d	Kirschbaum *et al.*, 1997
Red wine vinegar	20.3	23.8	0.3	13.5	0.2	n.d	0.1	Kirschbaum *et al.*, 1997
Balsamic vinegar	15.0	n.d	2.6	16.8	1.0	n.d	0.2	Kirschbaum *et al.*, 1997
Aromatized vinegar	1.9	n.d	n.d	n.d	n.d	n.d	0.9	Kirschbaum *et al.*, 1997

PUT, putrescine; HI, histamine; CAD, cadaverine; TY, tiramine; SPD, spermidine; PEA, phenylethylamine; SPM, spermine.

The concentration of six BAs was studied in 163 wines from three different regions of Spain to evaluate the influence of grape variety, type of vinification, wine pH, malolactic fermentation, and storage in the bottle (Landete *et al.*, 2005a). Results show important differences in putrescine and histamine concentrations among regions, varieties of grape and type of wine. Differences were less relevant for the remaining BAs studied. Low pH prevented BA formation. MLF and short storage periods in the bottle (3–6 months) showed increases in histamine concentration, whereas longer periods of storage led to a general decrease in histamine.

Histamine and tyramine are, besides putrescine, the most abundant amines in wine. Phenylethylamine is also frequently found. Although tryptamine and cadaverine are also found in wine, they are in much lower concentrations than the others mentioned (Le Jeune *et al.*, 1995; Moreno-Arribas *et al.*, 2003; Landete *et al.*, 2005a,b). The histamine, tyramine and phenylethylamine concentrations found in must are very low or non-existent (Landete *et al.*, 2005b). So, it is normal that the concentrations of these BAs must be the result of the yeast or lactic acid bacteria fermentative activity.

Caruso *et al.* (2002) reported the results of screening 50 yeast strains belonging to five wine genera/species isolated from grapes and wines for BAs production: *S. cerevisiae, K. apiculata, C. stellata, Metschnikowia pulcherrima, Brettanamyces bruxellensis*. Methylamine and agmatine were formed by all the species considered, levels in the range 0.25–0.98 mg/l and 1.12–4.11 mg/l, respectively. The production of other BAs by these five wine yeast species is summarized in Table 12.2. All the species produced very low or non-detectable amounts of histamine. The highest concentration of total BAs was formed by *B. bruxellensis*, followed by *S. cerevisiae*. Two BAs seemed to be species specific, phenylethylamine and ethanolamine, produced in more considerable amounts by *B. bruxellensis* and *S. cerevisiae*, respectively.

Several strains of lactic acid bacteria were isolated from wine, and their ability to form BAs was assayed in synthetic media, grape must and wine (Landete *et al.*, 2005a). *O. oeni* is the bacterium generally responsible for malolactic fermentation; other species such as *Pediococcus parvulus* and *Lactobacillus hilgardii* can develop at the same time. *P. parvulus* and *L. hilgardii* strains exhibited a high histidine decarboxylase activity, whereas those of *O. oeni* showed a much lower activity. A high number of *O. oeni* cells is required to explain the increase in histamine concentration when MLF is performed by this species alone.

Landete *et al.* (2007) investigated BA production (histamine, tyramine, phenylethylamine and putrescine) using cultures of 231 microorganisms representing LAB (155 strains), AAB (40 strains) and yeast (36 strains). BA production was not observed by AAB and yeast; however, production of histamine, tyramine, phenylethylamine and putrescine by LAB was observed. Results are also summarized in Table 12.2. *O. oeni, L. hilgardii, L. mali, Leuconostoc mesenteroides* and *P. parvulus* can contribute to the histamine synthesis in wine, but the main species responsible for high histamine production in wines seem to be *L. hilgardii* and *P. parvulus*.

High levels of BAs correlate fairly well with other wine spoilage components, for example, butyric acid, lactic acid, acetic acid, ethylacetate and diethyl succinate. This is why wines with higher levels usually also have higher levels of volatile acid. Red wines also have higher levels than white wines, mainly because of vinification practices, such as maceration and MLF (Aerny, 1985; Cilliers and van Wyk, 1985); however, other authors have not reported higher BA levels (Ough *et al.*, 1987). Few studies on the production of non-volatile amines by yeast during alcoholic fermentation are described (Buteau *et al.*, 1984).

Role of Starter Culture in Biogenic Amine Inhibition in Wine and Wine Vinegar

The spontaneous fermentation of wine can be inconvenient because the total content of BAs in the final product can derive in part

Table 12.2. Production of BAs by different wine yeast and LAB species, average values (mg/l).

	Strains (number)	PUT	HI	CAD	TRYP	EA	PEA	TY	Reference
S. cerevisiae	10	0.38	0.01	0	0.80	5.98	0.39	–	
K. apiculata	10	0.32	0.15	0.68	0.10	0	2.88	–	
C. stellata	10	0	0	0	0.48	1.94	0	–	Caruso *et al.*, 2002[a]
M. pulcherrima	10	0.83	0	0.19	0	0	6.56	–	
B. bruxellensis	10	1.18	0.20	0.31	0	0	10.07	–	
L. brevis	21	0	0	–	–	–	4.2	7.1	
L. casei	2	0	0	–	–	–	0	0	
L. collinoides	1	0	0	–	–	–	0	0	
L. hilgardii	17	67	24.2	–	–	–	7.8	23.2	
L. mali	3	0	10.1	–	–	–	0	0	
L. paracasei	11	0	0	–	–	–	0	0	Landete *et al.*, 2007[b]
L. plantarum	5	0	0	–	–	–	0	0	
L. vini	1	0	0	–	–	–	0	0	
L. mesenteroides	15	0	17	–	–	–	0	0	
O. ovni	39	42	4.2	–	–	–	0	0	
P. parvulus	37	0	37.3	–	–	–	0	0	
P. pentosaceus	3	0	0	–	–	–	0	0	

PUT, putrescine; HI, histamine; CAD, cadaverine; TRYP, tryptamine; EA, ethylamine; PEA, phenylethylamine; TY, tyramine.
[a]Fermentations were carried out in 125-ml flasks filled with 100 ml of sterilized grape must inoculated with a 5% concentration of 48 h pre-cultures in the same must and incubated at 22°C until complete fermentation.
[b]A pre-culture (0.1 ml) grown with all the microorganisms was inoculated into 10 ml of wine supplemented with 0.5 g/l of histidine, tyrosine, phenylalanine, ornitine, tryptophan and lysine and incubated at 28 °C for 10 days.

also from the activity of unfavourable yeast and LAB strains. Landete *et al.* (2005a) found lower concentrations of histamine, tyramine and phenylethylamine in wines in which MLF was performed with a commercial starter culture.

Commercially available strains of *S. cerevisiae* possess a wide range of characteristics, suitable for most winemaking situations, namely, to enhance varietal flavourants and produce an abundance of fruit esters. Other strains may be selected because of their relative fermentation speed, alcohol tolerance or production of regionally distinctive wines. Others are notable for their production of low levels of compounds such as acetic acid, hydrogen sulfide, urea or BAs. Thus, it is important to choose selected strains of *S. cerevisiae* and LAB not only for the expression of desirable technological traits, but also to avoid potentially negative effects to human health. This last consideration promotes, necessarily, the inclusion in the wine yeast and LAB selection programme of additional strain characteristics, such as low or no BA production activity (Torrea and Ancín, 2002).

Spontaneous MLF has the potential to have higher levels of BAs than when MLF starter cultures are being used, since the starter cultures are selected not to have this characteristic and they also suppress competing LAB. Additionally, control of BA production in wine involves management of pH, since a pH of above 3.5 promotes the growth of *Lactobacilli* and *Pediococci* and the initial numbers on grapes may also be higher.

It is known that commercial yeast starter preparations contain LAB contaminants (Manzano *et al.*, 2005); an important problem can occur if these bacteria are able to produce BAs during alcoholic fermentation. Costantini *et al.* (2009) evaluated 30 commercial starters (14 yeasts, *Saccharomyces cerevisiae*, and 16 bacteria, *Oenococcus oeni*) and demonstrated that the risk of BA production exists, and it depends on the contaminating bacteria species. Additionally, fermentations in grape juice with two yeast commercial preparations containing bacterial contaminants were performed, to check the potential of BA production during winemaking. Authors concluded that BA production was possible in the conditions used.

Future Perspectives

The production of BAs by wine microorganisms continues to be the focus of intensive study because of their potential toxicity. The main goal is to identify the microbial species capable of producing these compounds in order to control their presence and the conditions for its mitigation. However, more studies are needed with respect to BA composition of wine vinegar. It is important to choose commercially available strains of yeasts and bacteria, not only for the expression of desirable technological traits, but also to avoid potentially negative effects to human health. The main advantages of direct inoculation with these commercial strains are reliability, performance and safety, as well as convenience of use.

The culture market can be increased by expanding the application of commercial strains and by increasing the quality and safety of wine and wine vinegar.

References

Adams, M.R. and Moss, M.O. (2000) *Food Microbiology* (2nd edn) Royal Society of Chemistry, Cambridge.

Adams, M.R. (1998) Vinegar. In: Wood, B.J.B. (ed.) *Microbiology of Fermented Foods* (2nd edn) Blackie Academic and Professional, London, UK.

Aerny, J. (1985) Origine de l'histamine des vins. Connaissances actuelles. *Bulletin de l'O.I V.* 656–657, 1016–1019.

Anli, R.E., Vural, N., Yilmaz, S. and Vural, Y.H. (2004) The determination of biogenic amines in Turkish red wines. *Journal of Food Composition and Analysis* 17, 53–62.

Baena-Ruano, S., Jiménez-Ot, C., Santos-Duenas, I.M., Cantero-Moreno, D., Barja, F. and García-García,

I. (2006) Rapid method for total, viable and non-viable acetic acid bacteria determination during acetification process. *Process Biochemistry* 41, 1160–1164.

Baucom, T.L., Tabacchi, M.H., Cottrell, T.H.E. and Richmond, B.S. (1996) Biogenic amine content of New York state wines. *Journal of Food Science* 51, 1376–1377.

Bauza, T. and Teissedre, P.L. (1995) Les amines biogénes du vin. Métabolisme et toxicité. *Bulletin de l'Office International du Vin* 68, 42–67.

Buckenhuskes, H.J. (1993) Selection criteria for lactic acid bacteria to be used as starter cultures for various food commodities. *FEMS Microbiology Reviews* 12, 253–272.

Busto O., Miracle, M., Guasch, J. and Borrull, F. (1997) Determination of biogenic amines in wines by high-performance liquid chromatography with on-column fluorescence derivatization. *Journal of Chromatography A* 757, 311–318.

Buteau, C., Duitschaever, C.L. and Ashton, G.C. (1984) High-performance liquid chromatographic detection and quantitation of amines in must and wine. *Journal of Chromatography* 284, 201–210.

Caruso, M., Fiore, C., Contursi, M., Salzano, G., Paparella, A. and Romano, P. (2002) Formation of biogenic amines as criteria for the selection of wine yeasts. *World Journal of Microbiology and Biotechnology* 18, 159–163.

Cheryan, M. (2000) Acetic acid production. In: Lederberg, J. (ed.), *Encyclopedia of Microbiology,* Academic Press, New York. pp.13–17.

Ciani, M. (1998) Wine vinegar production using base wines made with different yeast species. *Journal of Science Food Agriculture* 78, 290–294.

Ciani, M. and Picciotti, G. (1995) The growth kinetics and fermentation behaviour of some *non-saccharomyces* yeasts associated with wine-making. *Biotechnology Letters* 17, 1247–1250.

Cilliers, J.D. and van Wyk, C.J. (1985) Histamine and tyramine content of South African wine. *South African Journal of Enology and Viticulture* 6, 35–40.

Costantini, A., Vaudano, E., Prete, V.D., Danei, M. and Garcia-Moruno, E. (2009) Biogenic amine production by contaminating bacteria found in starter preparations used in winemaking. *Journal of Agriculture and Food Chemistry* 57, 10664–10669.

Dicks, L.M.T., Dellaglio, F. and Collins, M.D. (1995) Proposal to reclassify *Leuconostoc oenos* as *Oenococcus oeni. International Journal of Systematic Bacteriology* 45, 395–397.

Ebner, H., Follmann, H. and Sellmer, S. (1995) Vinegar. In: Rehm, H. J. and Reed, G. (eds) *Biotechnology: A Multi-Volume Comprehensive Treatise,* VCH, Weinheim, Germany. pp. 579–591.

Ebner, H., Sellmer, S., and Follmann, H. (1996) Acetic acid. In: *Biotechnology: A Multi-Volume Comprehensive Treatise,* (2nd ed.) pp. 381–401.

Edwards, C.G., Haag, K.M. and Collins, M.D. (1998) Identification and characterization of two lactic acid bacteria associated with sluggish/stuck fermentations. *American Journal of Enology and Viticulture* 49, 445–448.

Ertan Anli, R. and Bayram, M. (2009) Biogenic amines in wines. *Food Reviews International* 25, 86–102.

Fernandes, J.O. and Ferreira, M.A. (2000) Combined ion-pair extraction and gas chromatography–mass spectrometry for the simultaneous determination of diamines, polyamines and aromatic amines in port wine and grape juice. *Journal of Chromatography A,* 886, 183–195.

Ferreira, I.M.P.L.V.O. and Pinho, O. (2006) Review: biogenic amines in Portuguese traditional foods and wines. *Journal of Food Protection* 69, 2293–2303.

Fourcassie, P. Makaga-Kabinda-Massard, E., Belarbi, A. and Maujean, A. (1992) Growth, d-glucose utilization and malolactic fermentation by *Leuconostoc oenos* strains in 18 media deficient in one amino acid. *Journal of Applied Bacteriology* 73, 489–496.

García-García, I., Santos-Dueñas I.M., Jiménez-Ot, C., Jiménez-Hornero, J.E. and Bonilla-Venceslada, J.L. (2009) Vinegar engineering. In: Solieri, L. and Giudici, P. (eds) *Vinegars of the World,* Springer-Verlag Italia, pp. 97–119.

Gerbaux, V. and Monamy, C. (2000) Biogenic amines in Burgundy wines. Contents and origin in wines. *Revue Français Oenologie* 183, 25–28.

González-Marco, A. and Ancín-Azpilicueta, C. (2006) Influence of lees contact on evolution of amines in Chardonnay wine. *Journal of Food Science* 71, 544–548.

González-Marco, A., Jiménez-Moreno, N. and Ancín-Azpilicueta, C. (2006) Influence of addition of yeast autolysate on the formation of amines in wine. *Journal of the Science of Food and Agriculture* 86, 2221–2227.

Guizani, N. and Mothershaw, A. (2006) Fermentation: general principles. In: Hui, Y. H. (ed.) Taylor & Francis *Handbook of Food Science, Technology, and Engineering,* London. 63/61–28.

Gullo, M., Caggia, C., De Vero, L., and Giudici, P. (2006) Characterisation of acetic acid bacteria in "traditional balsamic vinegar. *International Journal of Food Microbiology* 106, 209–212.
Hajós, G., Sass Kiss, A., Szerdahelyi, E. and Bardocz, S. (2000) Changes in biogenic amine content of Tokaj grapes, wines, and Aszu-wines. *Journal of Food Science* 65, 1142–1144.
Halasz, A., Barath, A., Simon-Sarkadi, L. and Holzapfel, W. (1994) Biogenic amines and their production by microorganisms in foods. *Trends in Food Science and Technology* 5, 42–49.
Hansen, E.H. (2002) Commercial bacterial starter cultures for fermented foods of the future. *International Journal of Food Microbiology* 78, 119-131.
Husnik, J.I., Volschenk, H., Bauer, J., Colavizza, D., Luo, Z. and van Vuuren, J.J.H. (2006) Metabolic engineering of malolactic wine yeast. *Metabolic Engineering* 8, 315–323.
Jackson, R.S. (2000) *Wine Science, Principles, Practice, Perception* (2nd edn) Academic Press. Elsevier, London, UK.
Josephsen, J. and Jaspersen, L. (2006) Fermented food and starter cultures. In: Hui, Y.H. (ed.) Taylor & Francis *Handbook of Food Science, Technology, and Engineering,* London. 177/171–120.
Kirschbaum, J., Busch, I. and Bruchner, H. (1997) Determination of biogenic amines in food by automated pre-column derivatization with 2-naphthyloxycarbonyl chloride (NOC-Cl). *Chromatographia* 45, 263–268.
Landete, J.M., Ferrer, S. and Pardo, I. (2007) Biogenic amine production by lactic acid bacteria, acetic bacteria and yeast isolated from wine *Food Control* 18, 1569–1574.
Landete, J.M., Ferrer, S., Polo, L. and Pardo, I. (2005a) Biogenic amines in wines from three Spanish regions. *Journal of Agriculture and Food Chemistry* 53, 119–1124.
Landete, J.M., Ferrer, S. and Pardo, I. (2005b) Which are the lactic acid bacteria responsible of histamine production in wine? *Journal of Applied Microbiology,* 99, 580–586.
Le Jeune, C., Lonvaud-Funel, A., ten Brink, B., Hofstra, H. and Van der Vossen, J.M.B.M. (1995) Development of a detection system for histidine decarboxylating lactic acid bacteria on DNA probes, PCR and activity test. *Journal of Applied Bacteriology* 78, 316–326.
Lehtonen, P. (1996) Determination of amines and amino acids in wine. A review. *American Journal of Enology and Viticulture* 47, 127–132.
Liu, S.Q. and Pilone, G.J. (1998) A review: arginine metabolism in wine lactic acid bacteria and its practical significance. *Journal of Applied Microbiology* 84, 315–327.
Llaguno, C. (1991) Antecedentes históricos. In: Llaguno, C. and Polo, M.C. (eds), *El vinagre de vino,* Madrid: CSIC. pp. 1–23.
Lu, S.F., Lee, F.L. and Chen, H.K. (1999) A thermotolerant and high acetic acid-producing bacterium *Acetobacter* sp. I14–2. *Journal of Applied Microbiology* 86, 55–62.
Manzano, M., Giusto, C., Bartolomeoli, I., Buiatti, S. and Comi, G. (2005) Microbiological analyses of dry and slurry yeasts for brewing. *Journal of the Institute of Brewing* 111 (2), 203–208.
Martín-Álvarez, P.J., Marcobal, A., Polo, C. and Moreno-Arribas, M.V. (2006) Influence of technological practices on biogenic amine contents in red wines. *European Food Research Technology* 222, 420–424.
Millan, S., Sampedro, C., Unceta, N., Goicolea, A.M. and Barrio, J.R. (2007) Simple and rapid determination of biogenic amines in wine by liquid chromatography–electrospray ionization ion trap mass spectrometry. *Analytica Chimica Acta* 584, 145–152.
Morales, M.L., Tesfaye, W., García-Parrilla, C.M., Casas, J.A. and Troncoso, A.M. (2002) Evolution of the aroma profile of sherry wine vinegars during an experimental aging in wood. *Journal of Agriculture and Food Chemistry* 50, 3173–3178.
Moreno-Arribas, M.V., Polo, M.C., Jorganes, F. and Muñoz, R. (2003) Screening of biogenic amine production by lactic acid bacteria isolated from grape must and wine. *International Journal of Food Microbiology* 84, 117–123.
Mortimer, R. and Polsonelli, M. (1999) On the origins of wine yeast. *Research Microbiology* 150, 199–204.
O'Toole, D.K. and Lee, Y.K. (2003) Fermented foods In: Lee, Y.K. (ed.) *Microbial Biotechnology: Principles and Applications,* World Scientific, Hackensack, NJ. pp. 201–256.
Ough, C.S., Crowell, E.A., Kunke, R.E., Vilas, M.R.S. and Lagier, S. (1987) A study of histamine production by various wine bacteria in model solution and in wine. *Journal of Food Processing and Preservation* 12, 63–70.
Pérez-Serradilla, J.A. and Luque de Castro, M.D. (2008) Role of lees in wine production: a review. *Food Chemistry* 111, 447–456.
Pinho, O., Pintado, A.I.E., Gomes A.M.P., Pintado, M.M.E., Malcata, F.X. and Ferreira I.M.P.L.V.O. (2004)

Interrelationships among microbiological, physicochemical, and biochemical properties of Terrincho cheese, with emphasis on biogenic amines. *Journal of Food Protection* 67, 2779–2785.

Plessi, M. (2003) Vinegar. In: Caballero, B., Trugo, L.C. and Finglas, P.M. (eds) *Encyclopedia of Food Sciences and Nutrition,* Academic Press, Oxford. pp. 5996–6004.

Radler, F. and Fáth, K.P. (1991) Histamine and other biogenic amines in wines. In: Rantz, J. (ed.), *Proceedings of the International Symposium on Nitrogen in Grapes and Wine.* American Society for Enology and Viticulture, Davis, CA. pp. 5–195.

Raspor, P. and Goranovic, D. (2008) Biotechnological applications of acetic acid bacteria. *Critical Reviews in Biotechnology* 28, 101–124.

Ribéreau-Gayon, P., Dubourdieu, D., Donèche, B. and Lonvaud, A. (2000) *Handbook of Enology. Vol 1 The Microbiology of Wine and Vinifications.* John Wiley & Sons Ltd, Chichester, UK.

Rivas-Gonzalo, J.C., Santos-Hernandez, J.F. and Marine-Font, A. (1983) Study of the evolution of tyramine content during the vinification process. *Journal of Food Science* 48, 417–418.

Romano, P., Fiore, C., Paraggio, M., Caruso, M. and Capece, A. (2003) Function of yeast species and strains in wine flavour. *International Journal of Food Microbiology* 86, 169–180.

Romero, R., Gazquez, D., Bagur, M.G. and Sanchez-Vinas, M. (2000) Optimization of chromatographic parameters for the determination of biogenic amines in wines by reversed-phase high-performance liquid chromatography. *Journal of Chromatography A* 871, 75–83.

Sandler, M., Youdim, M.B.H. and Hanington, E. (1974) A phenylethylamine oxidising defect in migraine. *Nature* 250, 335–337.

Sievers, M., Sellmer, S. and Teuber, M. (1992) *Acetobacter europaeus*, new species, a main component of industrial vinegar fermenters in central Europe. *Systems Applied Microbiology* 15, 386–392.

Silla-Santos, M.H. (1996) Biogenic amines: their importance in foods. *International Journal of Food Microbiology* 29, 213–231.

Soufleros, E., Marie-Lyse, B. and Bertrand, A. (1998) Correlation between the content of biogenic amines and other wine compounds. *American Journal of Enology and Viticulture* 49, 266–277.

Souza, S.C., Theodoro, K.H., Souza, E.R., Motta, S. and Abreu Glória, B.M. (2005) Bioactive amines in Brazilian wines: types, levels and correlation with physico-chemical parameters. *Brazilian Archives of Biology and Technology* 48, 53–62.

Straub, B.W., Kicherer, M. Schilcher, S.M. and Hammes, W. (1995) The formation of biogenic amines by fermentation organisms. *Zeitschrift fur Lebensmittel Untersuchung und Forschung* 201, 79–82.

ten Brink, B., Damink, C., Joosten, H.M.L.J. and Huis in't Veld, J.H.J. (1990) Occurrence and formation of biologically active amines in foods. *International Journal of Food Microbiology* 11, 73–84.

Tesfaye, W., Morales, M.L. and Trancoso, A.M. (2002) Wine vinegar technology, authenticity and quality evaluation. *Trends in Food Science & Technology* 13, 12–21.

Torrea, D. and Ancín, C. (2002) Content of biogenic amines in a Chardonnay wine obtained through spontaneous and inoculated fermentations. *Journal of Agriculture and Food Chemistry* 50, 4895–4899.

Versari, A., Parpinello, G. P. and Cattaneo, M. (1999) *Leuconostoc oenos* and malolactic fermentation in wine: a review. *Journal of Industrial Microbiology & Biotechnology* 23, 447–455.

Vidal-Carou, M.C., Ambatille-Espunyes, A., Ulla-Ulla, M.C. and Marine-Font, A. (1990) Histamine and tyramine in Spanish wines: their formation during the wine making process. *American Journal of Enology and Viticulture* 41, 160–167.

13 Natural Inhibitors of Food-borne Fungi from Plants and Microorganisms

Mehdi Razzaghi-Abyaneh* and Masoomeh Shams-Ghahfarokhi

Introduction

Food-borne fungi (FBF) or food-spoilage fungi are a diverse group of fungal pathogens responsible for food spoilage and crop diseases all over the world. They comprise an expanding list of at least 53 genera and 156 species mainly belonged to the phylum Deuteromycota. The history of food spoilage caused by FBF dates back to 10,000 years ago, when humans started to domesticate plants as a necessity for subsistence. An obvious example is human food poisoning by *Claviceps purpurea*, which was reported more than 200 years ago (Pitt and Hocking, 2009). The real significance of FBF was notified around 50 years ago when the famous 'Turkey X disease' killed 100,000 turkey poults in the UK. It revealed another important aspect of these fungi, i.e. the ability to produce toxic fungal metabolites named 'mycotoxins'. At present, it is estimated that one quarter of the world's food supply is lost through microbial activity, of which approximately 5 to 10% is due to food spoilage by common FBF. Nowadays, the fungal invasion of food and crops is an economic problem that is not yet under adequate control, despite modern food technologies and the range of preservation techniques available. Chemicals are widely used in the food and agricultural industries to control detrimental effects of FBF. Various physicochemical methods have also been employed for food preservation from FBF during storage, including drying, heating and irradiation (Pitt and Hocking, 2009). New synthetic antifungals have been successfully commercialized in recent years as well. Nearly all such methods encounter major problems, in part not only in regard to their adverse side effects on mammalian systems but also for the development of resistance by fungal pathogens, water contamination, residues in food, high costs for safe experiments and undesirable effects on non-target organisms sharing the ecosystem (Ghisalberti, 2000). So, there is a clear tendency towards the search for novel natural antifungals that produce minimal damage to the environment and human health.

In recent years, researchers have focused on finding novel antimicrobials from higher plants, microorganisms, insects, nematodes and vertebrates as safer inhibitors of FBF. Plants and microorganisms are in the first line of investigation because of their much greater diversity than that of any other group of organisms and because they possess valuable pharmaceutically active molecules. After the first description of a natural

* Corresponding author.

pesticide prepared from *Pyrethrum* flowers (containing pyrethrin as effective component) under the commercialized name of 'Caucasian insect powder' in the middle of 19th century (Banerji *et al.*, 1985), considerable efforts have been made to discover novel biocides from natural sources as an urgent necessity.

This chapter highlights the current status of antifungal lead compounds from plants ranging from volatiles of the essential oils to phytoalexins, peptides (defensins), phenylpropanoids, alkaloids, saponins, etc. Next to be addressed will be terrestrial fungi (moulds), bacteria and actinomycetes as microbial factories of novel bioactive molecules inhibitory to FBF.

Food-borne Fungi

The fungal kingdom comprises an estimated number of 1.5 million species on our planet. Among 100,000 known fungal species, less than 10%, are plant pathogens of which around 200 species are known as food contaminants designated under the general name of 'food-borne fungi (FBF)'. They include both moulds and yeasts from different genera and species of which the majority belong to the genera *Aspergillus*, *Penicillium* and *Fusarium* (Samson *et al.*, 2000). FBF are cosmopolitan organisms widely distributed in various environments including soil, air and plant-decaying materials. They are able to contaminate nearly all foods due to their capacity to utilize a wide array of substrates as well as their unique adaptation to undesirable conditions of low pH, water activity and temperature. The growth of these fungi may result in different kinds of food spoilage as a worldwide problem. Besides the economic losses from food spoilage, producing hazardous mycotoxins and potentially allergenic spores are other important aspects of fungal food contamination. Since a large number of FBF are classified in the list of plant pathogens as well, we further discuss how plant- and microbial-derived natural substances can inhibit the growth of all important food and plant pathogenic fungi under the general name of FBF regardless to their taxonomic position.

Plant Products as Antifungal Lead Compounds

The history of using plants as healing agents dates back to around 60,000 years ago when Neanderthals, in present-day Iraq, used hollyhock (*Alcea rosea* L.) as a remedy (Stockwell, 1985). Plants form an important part of the ecosystem as food and also oxygen-delivery systems. It has been estimated that less than 10% of 250,000–500,000 known plant species on our planet have ethno-botanical importance (Borris, 1996).

In addition to essential primary metabolites, a diverse range of low-molecular-weight molecules named 'secondary metabolites' are produced by the plants, which contribute to plant fitness by interacting with the ecosystem. In fact, these molecules play a major role in plant adaptation to the environment besides that of representing an important source of active pharmaceuticals. More than 100,000 secondary metabolites have been discovered from the plant kingdom, of which around 10,000 have been chemically defined (Verpoorte, 1998; Oksman-Caldentey and Inzé, 2004). Flavonoids, phenylpropanoids, cumarins, flavones, phenolics, quinines, tannins, alkaloids, terpenoids and essential oil volatiles are the best-known examples of such compounds.

Plant secondary metabolites are attractive as flavours, fragrances, pesticides, pharmaceuticals and antimicrobials. Although a large number of bioactive plant metabolites have been introduced as antifungal compounds, there are many other lead antifungals waiting to be discovered. Likewise, the biological activity of a large number of secondary metabolites has not been explored toward the FBF. Table 13.1 illustrates detailed data about a large number of best-characterized plant-derived bioactive metabolites with antifungal activity towards a wide array of FBF. We further describe the main categories of plant-derived

Table 13.1 Plants and their bioactive metabolites inhibitory to food-borne fungi.

Family	Species	Common name	Part used	Effective component	Inhibitory concentration	Affected fungi	Reference
Alliaceae	*Allium minutiflorum*	Onion	Bulbs	Saponins: Minutoside A, B and C Alliogenin Neoagigenin	1000 ppm	*A. alternata* *A. porri* *B. cinerea* *F. oxysporum* *F. solani* *F. oxysporum* f. sp. *lycopersici* *R. solani* *T. harzianum* *P. ultimum*	Barile *et al.* (2007)
Apocynaceae	*Alstonia venenata*	Devil tree	Bark tissue	Δ^3-alstovenine (alkaloid)	250–1000 µg/ml	*H. maydis* *A. brassisicola* *C. lunata* *Cercospora* sp. *A. solani* *H. sativum*	Singh *et al.* (1999)
Asteraceae	*Matricaria chamomile*	German chamomile	Flowers	α-bisabolol? (a sesquiterpene alcohol)	Not determined	*A. niger*	Tolouee *et al.* (2010)
	Helianthus annus	Sunflower	Flowers	SAP16 (16 kDa trypsin inhibitor)	5 µg/ml (0.31 M)	*S. sclerotiorum*	Giudici *et al.* (2000)
	Pterocaulon alopecuroides	Quitoco	Aerial parts	5-methoxy-6,7-methylenedioxycoumarin 7-(2´,3´-epoxy-3´-methylbutyloxy)-6-methoxycoumarin	>250 µg/ml (MIC)	*A. flavus* *A. niger*	Stein *et al.* (2006)
	Pterocaulon polystachyum		Aerial parts	5-methoxy-6,7-methylenedioxycoumarin Ayapin Prenyletin Prenyletin-methyl-ether	>250 µg/ml (MIC)	*A. flavus* *A. niger*	Stein *et al.* (2006)
	Pterocaulon balansae		Aerial parts	5-methoxy-6,7-methylenedioxycoumarin	>250 µg/ml (MIC)	*A. flavus* *A. niger*	Stein *et al.* (2006)
Boraginaceae	*Cordia linnaei* Steam.		Roots	Terpenoids: Cordiaquinones B, E, F, G and H	1.5–50 µg/ml (MICs)	*C. cucumerinum*	Ioset *et al.* (1998)

Brassicaceae	*Brassica oleracea* var. *botrytis*	Wild cabbage	Florets	Phytoalexins: Caulilexins A, B and C Brassicanals A and C Arvelexin	5 × 10 $^{-4}$ M	*P. lingam* *S. sclerotiorum* *R. solani*	Pedras *et al.* (2006)
	Brassica alboglabra		Seeds	5907 Da Peptide	0.15 µM 2.1 µM 2.4 µM 4.3 µM	*V. mali* *H. maydis* *M. arachidicola* *F. oxysporum*	Lin and Ng (2008)
	Brassica juncea	Mustard greens Indian mustard	Florets	Chitinase 1 (BjCHI1)	20–50 µg/well	*C. truncatum* *B. cinerea* *C. acutatum* *A. rabiei*	Guan *et al.* (2008)
Caryophyllaceae	*Dianthus caryophyllus*	Clove pink	Leaves	kaempferol 3-O-β-D-glucopyranosyl (1→2)-O-β-D-glucopyranosyl (1→2)-O-[α-L-rhamnopyranosyl-(1→6)]-β-D-glucopyranoside (a flavonoid)	700 µM	*F. oxysporum* f. sp. *dianthi*	Galeotti *et al.* (2008)
Celastraceae	*Euonymus europaeus*	Spindle tree	Bark tissue	Ee-CBP (peptide from bark tissue)	0.2–20 µM	*A. brassisicola* *B. cinerea* *F. culmorum*	Van den Berg *et al.* (2004)
			Leaves	Ee-chitinase (from leaves)	>6.5 µM	*F. oxysporum* *R. solani* *T. hamatum* *P. exigua* *P. ultimum*	
Combretaceae	*Terminalia alata*	Indian laurel	Roots	Glycosides: 5,7,2′-tri-*O*-methylflavanone 4′-*C*-α- L-rhamnopyranosyl-1→4 -β- D-glucopyranoside 2α,3β, 19β,23-tetrahydroxyolean-12-en-28-oic acid 3-*O*-β-D-galactopyranosyl- 1→3 -β-D-glucopyranoside-28-*O*-β-D-glucopyranoside	50 ppm	*A. niger*	Srivastava *et al.* (2001)

Continued

Table 13.1 *Continued*

Family	Species	Common name	Part used	Effective component	Inhibitory concentration	Affected fungi	Reference
Cupressaceae	*Calocedrus macrolepis* var. *formosana*	Chinese incense-cedar	Heartwood	Isopropyl cycloheptatrienolones: β-thujaplicin (hinokitiol) γ-thujaplicin	5–50 μg/ml (MIC)	*F. solani* *T. viride* *P. citrinum* *A. niger* *C. gloeosporioides*	Yen *et al.* (2008)
			Leaves	T-muurolol α-cadinol β-caryophyllene Caryophyllene oxide	15.2–93.0 μg/ml 11.7–51.9 μg/ml 57.9–>200 μg/ml 66.1–>200 μg/ml (IC_{50})	*R. solani* *F. solani* *F. oxysporum* *C. gloeosporioides* *P. funerea* *G. australe*	Chang *et al.* (2008)
Dioscoreaceae	*Dioscorea alata*	Ube Purple yam	Bulb (tubers)	β-sitosterol (a phytosterol)	91.2 μg/ml (ED_{50})	*A. niger*	Aderiye *et al.* (1989)
Fabaceae	*Clitoria ternatea*	Butterfly pea	Seeds	Finotin (a protein)	300 μl/disc	*R. solani* *C. lindemuthianum* *B. oryzae* *P. grisea*	Kelemu *et al.* (2004)
	Geoffroea decorticans	Kumbaro Chanar Chilean palo verde	Leaves Twigs	Flavonoids: Tetrahydroxy-4′-methoxy-5′-prenylisoflavanone	9–18 μg/ml	*A. flavus* *A. parasiticus* *A. nomius*	Quiroga *et al.* (2009)
				Trihydroxy-4′-methoxy-5′-prenylisoflavanone	10–21 μg/ml (MICs)		
	Cicer bijugum	Chickpea	Roots	Cicerfuran [2-(2′-methylenedioxyphenyl)-6-hydroxybenzofuran]	25–400 μg/ml (MIC)	*A. niger* *B. cinerea* *C. herbarum*	Aslam *et al.* (2009)
	Pisum sativum	Pea	Seeds	Lectin (PSL)	0.1–10 mg/ml	*A. flavus* *T. viride* *F. oxysporum*	Sitohy *et al.* (2007)
	Indigofera oblongifolia		Leaves	Peptide 4	15–30 μg/ml 25–50 μg/ml >100 μg/ml (MICs)	*A. niger* *A. flavus* *F. oxysporum*	Dahot (1999) Stein *et al.* (2006)
	Acacia plumose	Unha de Gato	Seeds	Tripsin inhibitors A, B and C	70-140 μg/well	*A. niger* *Colletotrichum* sp. *F. moniliforme*	Lopes *et al.* (2009)
	Lens culinaris	Red lentil Masoor daal	Seeds	11 kDa Peptide	36 μM	*M. arachidicola* *F. oxysporum*	Wang and Ng (2007)

	Cicer arietinum	Chickpea Garbanzo bean Indian pea	Seeds	Ciceratin (a peptide)	8.2 µM 15.3 µM 20.6 µM (IC_{50})	*P. piricola* *M. arachi* *B. cinerea*	Chu *et al.* (2003)
	Cassia garrettiana	Sa-mae-san	Heartwood	3,3′,4,5′-tetrahydroxystilbene	50–100 µg/ml	*A. niger* *C. cladosporioides* *A. terreus* *T. longibranchiatum* *P. thomii*	Inamori *et al.* (1984)
Fagaceae	*Castanopsis chinensis*	Chinquapin Chinkapin	Seeds	30 kDa Protein	0.5 µM	*F. oxysporum* *B. cinerea* *M. arachidicola* *P. piricola*	Chu and Ng (2003)
Fumariaceae	*Corydalis chaerophylla*	Corydalis Fumewort	Roots Leaves	Alkaloids: Berberine (from roots) (±)-bicuculline (from leaves)	 100–1000 ppm	*A. solani* *A. brassicicola* *C. musae* *C. lunata* *C. maculans* *F. udum* *H. spiciferum*	Basha *et al.* (2002)
Lamiaceae	*Satureja hortensis*	Summer savory	Leaves	Phenolics: Thymol Carvacrol	0.86 µM 0.79 µM (IC_{50})	*A. parasiticus*	Razzaghi-Abyaneh *et al.* (2008)
	Rosmarinus officinalis	Rosemary	Leaves	Phenolics: Caffeic acid Rosmarinic acid	3–6 mg/ml	*P. capsici* *P. palmivora* *P. megakarya*	Widmer and Laurent (2006)
Linaceae	*Hugonia castaneifolia*		Root bark	Diterpenes: 3β-hydroxyrosa-1(10),15-dien-2-one 18-hydroxyhugorosenone 18-Hydroxy-3-deoxyhugorosenone [18-hydroxy-2-oxorosa-1(10),15-diene] 12-hydroxy-13-methylpodocarpa-8,11,13-trien-3-one Hugonone B	 12.5–200 µg 6.25–200 µg 50–200 µg 12.5–200 µg 50–200 µg	*C. cucumericum*	Baraza *et al.* (2008)
Magnoliaceae	*Magnolia obovata*	Japanese bigleaf magnolia	Stem bark	Lignans (Phytoesterogens): Honokiol Magnolol	 50 µg/ml 100 µg/ml (MICs)	*A. niger*	Bang *et al.* (2000)

Continued

Table 13.1 *Continued*

Family	Species	Common name	Part used	Effective component	Inhibitory concentration	Affected fungi	Reference
Melanthiaceae	*Veratrum taliense*		Roots	Alkaloids:		*P. capsici*	Zhou *et al.* (2003)
				Neoveratalines A and B	200 μg/ml	*R. cerealis*	
				Veramitaline	80–160 μg/ml		
				Stenophylline B	80–160 μg/ml		
				Jervine	80–160 μg/ml (MICs)		
Myristicaceae	*Virola surinamensis*	Baboonwood	Roots	Methoxy-galbelgin	200 μg	*C. cladosporioides*	Lopes *et al.* (1999)
				Grandicin	200 μg		
				Juruenolide C	100 μg		
				Epi-juruenolide C	100 μg		
				7-hydroxyflavonone	5 μg		
				Biochanin A	5 μg		
				2′-hydroxy-7,4-dimethoxyisoflavone	5 μg		
				7-hydroxy-4′-methoxyisoflavone	1 μg		
				Virolane	10 μg		
				Virolanol C	10 μg		
Phytolaccaceae	*Phytolacca tetramera*	Pokeweed Poke Pokebush	Berries	Phytolaccoside B (a saponin)	50 μg/ml 125 μg/ml (MICs)	*A. flavus* *A. niger*	Escalante *et al.* (2002)
Pinaceae	*Pinus strobus*	Eastern white pine	Knotwood	Phenolics: Dihydroxypinosylvin monomethylether	50 mM	*P. brevicompactum*	Välimaa et al. (2007)
	Pinus sylvestris	Scots pine	Knotwood	Pinosylvin Pinosylvin monomethyl ether			
	Pinus cembra	Swiss pine	Knotwood	Pinocembrin			

Piperaceae	*Piper umbellatum*		Branches	Alkaloids: Piperumbellactam D N-hydroxyaristolam II 4-nerolidylcatechol	200 μg/ml	*A. flavus* *F. solani*	Tabopda *et al.* (2008)
	Piper tuberculatum		Leaves	Fagaramide Methyl 6,7,8-trimethoxydihydrocinnamate Methyl trans-6,7,8-trimethoxycinnamate	10 μg 10 μg 5 μg	*C. cladosporioides*	de Silva *et al.* (2002)
			Seeds	Pellitorine $\Delta^{\alpha\beta}$dihydropiperine Piplartine Dihydropiplartine *cis*-piplartine	5 μg 5 μg 5 μg 5 μg 5 μg		
	Piper arboretum		Leaves	N-[10-(13,14-methylenedioxyphenyl)-7(E),9(Z)-pentadienoyl]-pyrrolidine Arboreumine N-[10-(13,14-methylenedioxyphenyl)-7(E)-pentaenoyl]-pyrrolidine N-[10-(13,14-methylenedioxyphenyl)-7(E),9(E)-pentadienoyl]-pyrrolidine N-[10-(13,14-methylenedioxyphenyl)-pentanoyl]-pyrrolidine	10 μg 5 μg 0.1 μg 0.1μg 5 μg	*C.sphaerospermum* *C. cladosporioides*	
	Piper caldense		Leaves	Caldensinic acid Caldensinic acid derivatives	5–25 μg 5–50 μg	*C. cladosporioides* *C.sphaerospermum*	Freitas *et al.* (2009)
	Piper longum	Long pepper	Fruits	Alkaloids: Piperoctadecalidine Piperine Piperlongumine Pipernonaline	2–20 mM	*A. flavus*	Lee *et al.* (2002)
Polygonaceae	*Polygonum acuminatum*	Knotweed Knotgrass	Aerial parts	Polygodial Isopolygodial Drimenol Confertifolin	>500 μg/ml (MIC)	*A. flavus* *A. niger*	Derita *et al.* (2009)

Continued

Table 13.1 *Continued*

Family	Species	Common name	Part used	Effective component	Inhibitory concentration	Affected fungi	Reference
Rhamnaceae	*Karwinskia parvifolia*		Fruits	-Hydroxyanthracenones: Peroxisomicine A1 Isoperoxisomicine A1	16->32 μg/ml (MICs)	*A. flavus* *A. niger* *R. arrhizus* *A. terreus* *F. solani*	Salazar *et al.* (2006)
Ruscaceae	*Aspidistra elatior*	Cast-iron plant	Leaves	Aspidistrin (a saponin)	10 μg/ml 100 μg/ml >100 μg/ml >100 μg/ml (MICs)	*M. mucedo* *A. niger* *P. chrysogenum* *R. chinensis*	Koketsu *et al.* (1996)
	Aegle marmelos	Bael	Seeds	2-isopropenyl-4-methyl-1-oxa-cyclopenta[b]anthracene-5,10-dione (+)-4-(20-hydroxy-30-methylbut-30-enyloxy)-8H [1,3] dioxolo[4,5-h]chromen-8-one	62.5 μg/ml (MIC)	*T. viride* *A. niger* *A. flavus*	Mishra *et al.* (2010) Stein *et al.* (2006)
	Toddalia asiatica		Leaves	Flindersine	≥ 250 μg/ml (MIC)	*M. grisea* *B. cinerea* *C. lunata* *A. niger* *Scopulariopsis* sp.	Duraipandiyan and Ignacimuthu (2009)
Theaceae	*Gordonia dassanayake*		Stem bark	3-formyl-2,4-dihydroxy-6-methylbenzoic acid 3-hydroxy-4-(methoxycarbonyl)-2,5-dimethylphenyl ester	0.01%	*R. solani* *Curvularia* sp. *Fusarium* sp. *C. gloeosporioides*	Athukoralage *et al.* (2001)
Zingiberaceae	*Aframomum aulacocarpos*		Seeds	Terpenoids: Aframodial Aulacocarpinolide Aulacocarpin A Aulacocarpin B	 12.5 μg/ml 25 μg/ml 6.25 μg/ml 12.5 μg/ml 50 μg/ml (MICs)	 *A. niger* *M. mucedo* *P. crustosum* *R. chinensis* *M. miehei*	Ayafor *et al.* (1994)

For full names of species see the Abbreviations section.

antifungal compounds based on their chemical structures with emphasis on plant–fungus interactions.

Flavonoids, alkaloids and terpenoids

Flavonoids are a class of widely distributed plant secondary metabolites which can be classified into three major groups: flavonoids, isoflavonoids and neoflavonoids. More than 5000 naturally occurring flavonoids have been reported from various plants (Harborne and Williams, 2000). They fulfil many functions such as antioxidant, radical-scavenging, anti-allergic, anti-inflammatory, anti-microbial and anti-cancer activities. The majority of plant flavonoids possessing antifungal activity are isoflavonoids, flavans or flavanones. Alkaloids are nitrogen-containing plant secondary metabolites with low molecular weights and heterocyclic structures. They occur in approximately 20% of all plant species with more than 12,000 identified structures. Besides the plant species, they are produced by a large variety of organisms including bacteria and fungi. Terpenoids are a large and diverse class of plant-derived secondary metabolites with multicyclic structures derived from five-carbon isoprene units. They are extensively used because of their aromatic characteristics and their important role in traditional herbal remedies as antimicrobials and anti-neoplastics. The detailed data of the large number of antifungals reported from these major groups of plant metabolites are summarized in Table 13.1.

Saponins

Saponins are a group of naturally occurring phytoanticipins found in particular abundance in various plant species. They are low-molecular-weight secondary plant metabolites possessing a wide array of biological activities responsible for plant defence against insects, fungi and other hazardous microorganisms. Among three major groups of saponins, i.e. triterpenoids, steroids and steroidal glycoalkaloids, the two latter ones have been studied in detail in relation to bioactivity against FBF (Osbourn, 1996a,b). It is believed that the antifungal property of saponins is a consequence of a direct interaction with fungal membrane sterols causing pore formation and loss of membrane integrity. Avenacins A and B (triterpenoid saponins from young oat roots) and α-tomatine (major steroidal glycoalkaloid saponin of tomato leaves) are the best examples of antifungal saponins inhibitory toward some FBF such as *Gaeumannomyces graminis* var. *tritici* and *Fusarium* species (Osbourn, 1996a,b). Some other plant-originated antifungal saponins are summarized in Table 13.1.

Defensins

Plant defensins are an important family of cationic peptides with multiple functions extending beyond antimicrobial activity against a diverse range of microorganisms. In addition to their activity against a wide array of organisms including bacteria, fungi and viruses, they are known to have inhibitory activity toward cancer cells, microbial key enzymes and ion channels. The first description of antimicrobial activity of defensins is from Terras *et al.* (1992) who described antifungal defensins from the seeds of *Raphanus sativus*. Defensins inhibitory to various FBF have now been identified in many plant species belonging to various genera and families (see details in Carvalho and Gomes, 2009). Although different plant defensins arrest fungal growth by various complicated mechanisms, much evidence supports the hypothesis that these compounds bind to the specific receptors located on fungal plasma membranes, resulting in permeabilization and finally cell death. It has been shown that during plant seed germination, when the seeds become vulnerable to pathogenic fungi, the deposition of defensins in seed tissues provides a good protective mechanism. In fact, plant defensins are natural antimicrobial peptides (AMPs) produced and fortified in an evolutionary process to challenge threats to plant survival in the environment. A peptide from *Lens culinaris* with growth inhibitory activity toward *Fusarium oxysporum* is a good example of an antifungal plant defensin, as described in Table 13.1.

Phytoalexins

The term 'phytoalexins' was first described as 'antifungal substances which are specifically formed when a plant is attacked by a fungus'. Nowadays, they are defined as 'antibiotics formed in plants via a metabolic sequence induced either biologically or in response to chemical or environmental factors' (Grayer and Kokubun, 2001). The interaction of fungi with living plants is often beneficial to plants as in the obvious cases of mycorrhizae and endophytes, but in some cases the fungi result in an imbalance to become plant pathogens. After fungal invasion, the plants respond to a pathogenic fungus by releasing antifungal compounds, generating reactive oxygen species and accumulating newly produced antifungal chemicals named phytoalexins. Phytoalexins are mainly produced by members of Leguminaceae and Rosaceae. Chemically, they have isoflavonoid structures classified mainly as isoflavones, isoflavanones and isoflavans (Harborne and Baxtor, 1999). 6a-Hydroxyisomedicarpine (from *Melilotus alba*), wyerone (from the genera *Lens* and *Vicia*), pisatin (from the genera *Lathyrus* and *Pisum*), lupinisoflavone A (from *Lupinus albus*) and many other plant chemicals produced as a response to FBF attack are known examples of antifungal phytoalexins (Grayer and Kokubun, 2001). Some other important phytoalexins inhibitory to FBF are described in Table 13.1.

Essential oils

Plant essential oils constitute a heterogeneous mixture of chemical compounds with different structures and functions. They occur as monoterpenes, diterpenes, triterpenes and aromatic (aldehydes, phenols and alcohols) compounds which are widely used in pharmaceutical, cosmetic, food and agricultural industries for their antifungal, antibacterial, antioxidant, anti-insecticidal, anti-inflammatory and anticancer properties (Bakkali *et al.*, 2008). Essential oils can be isolated from nearly all plant parts including roots, flowers, buds, leaves, stems, seeds, fruits and bark by the hydro-distillation method. Their composition is influenced by type of extraction, climatic condition (season), and plant characteristics including variety, part, age and growth cycle. Among more than 3000 essential oils known, nearly 300 oils are commercially prepared many of which have antifungal properties (Bakkali *et al.*, 2008). As a part of our ongoing research on natural antimicrobials, the antifungal activity of some essential oils prepared from various Iranian medicinal plants has been described (Razzaghi-Abyaneh *et al.*, 2008, 2009; Razzaghi-Abyaneh *et al.*, 2010 and references therein; Tolouee *et al.*, 2010). In fact, many plant lead compounds inhibitory to FBF (as listed in Table 13.1) originate from the essential oils. So, we are not representing the detailed data on crude essential oils in this chapter. For more information, the biological activity of various plant essential oils toward microorganisms including fungi has been summarized in an excellent review by Bakkali *et al.* (2008).

Miscellaneous

Besides bioactive compounds from major classes of plant secondary metabolites, there are some other unclassified molecules inhibitory toward FBF. Flavours (acetaldehyde, benzaldehyde, cinnamaldehyde, ethanol, hexenel and 2-hexanal), acetic acid, jasmonates (jasmonic acid and methyl jasmonate), glucosinolates (isothiocyanates, nitriles and thiocyanates), cyanigenic glycosides (hydrocyanic acid), fusapyrone and chitosan are the best examples of such compounds that are inhibitory for some members of the fungal genera *Aspergillus, Fusarium, Penicillium* and *Alternaria*, and also *B. cinerea* (Osbourn, 1996b; Tripathi and Dubey, 2004).

Microorganisms as Promising Sources of Antifungal Metabolites

The use of beneficial microorganisms is one of the most promising methods for the development of environmentally friendly alternatives to chemical pesticides in preventing food spoilage and combating crop diseases. It has been reported that, on average, two or three antibiotics derived from microorganisms break into the market each year (Clark, 1996). Likewise, fungal- and bacterial-based biocontrol products have now been commercially developed for the control of fungal spoilage of food products (Sharma *et al.*, 2009). Among beneficial microorganisms, two major groups, i.e. bacteria and fungi, have received major consideration as they are natural factories for a wide array of biologically active antifungals active against FBF. Nowadays, hundreds of compounds have been isolated from a vast array of fungi and bacteria and there are more compounds waiting to be discovered by researchers. Here, we describe a brief history of natural antifungals originating from terrestrial bacteria, actinomycetes and fungi, as summarized in Tables 13.2 and 13.3.

Bacteria and actinomycetes

Many soil bacteria and actinomycetes have now been introduced as potential sources of bioactive molecules, of which the genera *Bacillus, Pseudomonas, Agrobacterium* and *Streptomyces* received major consideration with regard to their production of a diverse range of bioactive metabolites affecting FBF (Ongena and Jacques, 2007). Table 13.2 represents some examples of such metabolites originating from different bacteria and actinomycetes. Antifungal lipopeptides belonging to the iturin, fengycin and surfactin families have been isolated from *Bacillus* species (Ongena and Jacques, 2007). Fengycins A and B, plipastatins A and B, iturin A, mycosubtilin and bacillomycin D are selected examples of *Bacillus*-derived antifungal lipopeptides affecting important FBF such as *A. flavus, B. cinerea*, etc. (Moyne *et al.*, 2001). Lipodepsipeptides (syringomycins, syringostatins and syringotoxins) from *Pseudomonas syringae* are among the most potent antifungal peptides with bacterial origin. They are active against a diverse range of FBF from the genera *Aspergillus* and *Fusarium* (De Lucca and Walsh, 2000). Comprehensive data now exist about antifungal metabolites purified from non-streptomycete actinomycetes (see details in El-Tarabily and Sivasithamparam, 2006). Lactic acid bacteria from the genera *Lactobacillus, Lactococcus, Leuconostoc* and *Pediococcus* are another important group of potential biocontrol agents. They produce a wide range of antifungal metabolites including organic acids, phenolics, hydroxyl fatty acids, hydrogen peroxide, reuterin and peptides which are active against various FBF (see details in Dalié *et al.*, 2010).

Fungi

Fungi are rich sources of novel lead compounds that have not yet been well explored. A large number of antimicrobial substances have now been isolated from a wide array of fungal species. Some good examples of the hundreds of antifungals that are discovered from fungal sources are summarized in Table 13.3. Heptadecenoic acids from *Sporothrix flocculosa*, diatretyne nitrile from the mycorrhizal fungus *Leucopaxillus cerealis*, zaragozic acids from various members of Ascomycota group, 6-pentyl-2H-pyran-2-one, harzianolide, pyridine, trichodermin, harziandione, homothallin II and ketotriol from *Trichoderma* species, [2-(buta-1,3-dienyl)-3-hydroxy-4-(pental,3-dienyl)-tetrahydrofuran] and chaetomin from *Chaetomium globosum*, gliotoxin and gliovirin from *Gliocladium* species, and epicorazines A and B from *Epicoccum purpurescens* are other examples of fungal-based bioactive molecules inhibitory to FBF (for more details, see Ghisalberti, 2000). Screening of fungal species from marine sources and other unusual habitats, selection of appropriate fungal strain, efficient production of fungal biomass, formulation and methods of application are important

Table 13.2. Antifungal substances originated from different groups of bacteria and actinomycetes.

Reference	Affected fungi	Inhibitory concentration	Effective component	Species	Source	Group
Schlingmann *et al.* (1999)	*V. inaequalis* *B. cinerea*	5–25 µg/ml (MICs)	Strevertene A	*Streptoverticillium* sp.	Soil	Actinobacteria
Chakor *et al.* (2008)	*P. capsici* *B. cinerea* *M. mucedo*	10 µg/ml 50 µg/ml 250 µg/ml (MICs)	Thiobutacin (a butanoic acid)	*Lechevalieria aerocolonigenes*	Soil	
Prabavathy *et al.* (2006)	*P. oryzae* *R. solani*	25–100 µg/ml	SPM5C-1	*Streptomyces* sp. PM5	Soil	
Wan *et al.* (2008)	*B. cinerea* *R. solani* *S. sclerotiorum*	Qualitative (0–4 scores for infection of rice leaf)	Volatile compounds?	*Streptomyces platensis*	*Oryza sativa* (seeds)	
Kavitha *et al.* (2010)	*A. flavus* *A. niger* *A. alternata* *C. maculans* *C. lunata* *P. citrinum* *F. oxysporum*	20–200 µg/ml (MICs)	1-Phenylbut-3-ene-2-ol	*Nocardia levis* MK-VL-113	Soil	
Ismet *et al.* (2004)	*P. oryzae*	0.6 mg/disc	Cervinomycins A1 and A2 Phenylacetic acid 2,3-Dihydroxybenzoic acid	*Micromonospora* sp. M39	*Rhizophora* sp. (roots)	
Mukherjee and Sen (2006)	*A. niger* *A. alternata* *H. sativum*	10 µg/disc	Chitinase	*Streptomyces venezuelae* P10	Soil	
Wang *et al.* (2007)	*F. graminearum*	8.0 µg/ml	Fengycin (a lipopeptide)	*Bacillus subtilis* IB	Soil	Bacilli
Wei-Wei *et al.* (2008)	*B. cinerea* *S. sclerotiorum* *A. solani* *A. brassicae* *F. oxysporum* *P. arachnidicola* *F. graminearum* *A. citrullina*	300 µl/plate as volatiles	2,4-Decadienal Diethylphtalate n-Hexadecanoic acid Oleic acid	*Bacillus subtilis* *Bacillus pumilus* *Paenibacillus polymyxa*	Soil	

Moyne et al. 2001	*A. solani* *A. flavus* *B. ribis* *C. gloeosporioides* *F. oxysporum* *H. maydis* *P. gossypii* *S. rolfsii*	6–10 μg 3 μg 3 μg 3–6 μg 6 μg 3 μg 1–3 μg 1–10 μg	Bacillomycin D	*B. subtilis* AU195	AU195	
Kajiyama *et al.* (1998)	*A. candidus*	1.6 μg/ml (MIC)	Nostofungicidine	*Nostoc commune*	Soil	Cyanobacteria
Ojika *et al.* (2004)	*P. capsici*	0.04 μg/disc	Cystothiazole A	*Cystobacter fuscus*	Soil	Myxobacteria
Kaur *et al.* (2006)	*G. graminis var. tritici*	50 mg/ml	Gloconic acid	*Pseudomonas fluorescens*	Soil	Proteobacteria
Moretti *et al.* (2008)	*F. oxysporum f. sp. lycopersici*	1×10^8 CFU/ml	Cell suspension	*Achromobacter xylosoxydans*	Soil	
Walsh *et al.* (2001)	*F. oxysporum* *G. graminis var. tritici* *P. ultimum*	Not determined	2,4-diacetylphloroglucinol Viscosinamide Pyluteorin Pyrrolnitrin	*Pseudomonas* spp.	Soil	
Pedras *et al.* (2003)	*P. lingam* *A. brassicae* *R. solani* *S. sclerotiorum*	0.5 mg/ml	Pseudophomins A and B	*Pseudomonas flourescens* BRG100 (IDAC 141200-1)	International Depository Authority of Canada	
Fernando *et al.* (2005)	*S. sclerotiorum*	100–150 μl/disc	Cyclohexanol Decanal Nonanal Benzothiazole Dimethyltrisulfide	*Pseudomonas* spp.	*Brassica campestris* (seeds)	
Ryazanova *et al.* (2005)	*A. niger* *A. terreus* *A. japonicus* *A. heteromorphus*	0.2 mg/ml	Lysoamidase	*Lysobacter* sp. XL-1	Soil	
De Lucca *et al.* (1999)	*A. flavus* *A. niger* *F. moniliforme* *F. oxysporum*	1.9–7.8 μg/ml (LD_{95})	Syringomycin-E	*Pseudomonas syringae*	Soil	

Table 13.3. Bioactive fungal metabolites inhibitory to food-borne fungi.

Reference	Affected fungi	Inhibitory concentration	Effective component	Species	Source	Group
Ko *et al.* (2010)	*A. brassicicola*	2 µl/drop	Culture filtrate (a 100–1000 Da compound)	*Pseudallescheria boydii*	Soil	Ascomycota
Aneja *et al.* (2005)	*M. roreri* *C. perniciosa*	0.09 µM 0.02 µM	Nonanoic (pelargonic) acid	*Trichoderma harzianum*	Cacao pod	
Brown and Hamilton (1992)	*F. oxysporum* *R. solani* *P. cinnamomi* *P. heterothallicum* *S. sclerotiorum* *P. intermedium*	47–258 µg/ml (ED_{50})	Indol-3-ethanol	*Zygorrhynchus moelleri*	NF87/1	
Wicklow *et al.* (2005)	*A. flavus* *F. verticillioides*	250 µg/disc	Pyrrocidines A and B	*Acremonium zeae*	Soil	
Wicklow *et al.* (2009)	*F. graminearum* *T. viride* *N. oryzae* *A. alternata*	>50 µg/ml (MIC)	Monorden Monicillin I	*Colletotrichum* (*Glomerella*) *graminicola*	Maize kernel	
Skouri-Gargouri and Gargouri (2008)	*F. oxysporum* *F. solani* *B. cinerea* *T. reesei* *A. niger* *A. solani* *S. microbispora*	50–200 µg/ml (MICs)	AcAFP (a peptide)	*Aspergillus clavatus*	CBS-Netherlands	
Yang *et al.* (2007)	*S. sclerotiorum*	10% (v/v) in PDA	Culture filtrate	*Coniothyrium minitans*	Soil	
Yang *et al.* (2008)	*A. alternata* *S. sclerotiorum* *B. cinerea* *G. zeae* *B. maydis* *F. oxysporum* *M. grisea* *R. solani* *S. rolfsii*	Not determined (30–40% zone of inhibition on PDA)	Fungal colony	*Penicillium oxalicum*	Soil	

Barrett (2002)	*A. niger* *A. flavus* *A. terreus* *F. solani*	0.0078 µg/ml 0.0156 µg/ml 0.0156 µg/ml >64 µg/ml (MICs)	FK463 (a lipopeptide)	*Coleophoma empetri*	F-11899	
Macías-Rubalcava *et al.* (2008)	*A. solani* *F. oxysporum* *P. parasitica* *P. capsici*	1.5–58.0 × 10^{-4} M (IC_{50})	Precussomerins EG_1, EG_2 and EG_3 Palmarumycin CP_2	*Edenia gomezpompae*	*Callicarpa acuminata* (leaves)	Ascomycota (endophyte)
You *et al.* (2009)	*Fusarium* sp. *Verticillium* sp. *Rhizoctonia* sp.	0.97 µg/ml	2,6-Dihydroxy-2-methyl-7-(prop-1E-enyl)-1-benzofuran-3 (2H)	*Verticillium* sp.	*Rehmannia glutinosa* (roots)	
Li *et al.* (2008)	*F. culmorum* *G. zeae* *V. albo-atrum*	7.2–236.9 µM (MICs)	Pestalachloride A, B, and C	*Pestalotiopsis adusta*	Unidentified tree (stem)	
Harper *et al.* (2003)	*P. ultimum*	10 µg/ml 40 µg/ml (MICs)	Pestacin (a 1,3-dihydroisobenzofuran) Isopestacin	*Pestalotiopsis microspora*	*Terminalia morobensis* (stem)	
Abdou *et al.* (2009)	*A. terreus* *F. oxysporum*	26.03–238.8 µM (MICs)	Botryorhodines A and B	*Botryosphaeria rhodina*	*Bidens pilosa* (stem)	
Whipps (1997)	*A. flavus*	2.0–8.0 µg/ml (MICs)	Viridiofungins A, B and C	*Trichoderma viride*	MF 5628	
Onishi *et al.* (1997)	*R. solani* *P. ultimum*	Not determined	9Z,12Z-8-hydroxylinoleic acid	*Laetisaria arvalis*	Soil	Basidiomycota
Weber *et al.* (1990)	*P. notatum*	0.3 µg/ml	Strobilurin D	*Cyphellopsis anomala*	Wood	

steps to commercialize active lead antifungals identified from biocontrol fungi.

Concluding Remarks and Future Perspectives

Despite the large amount of data now available on natural antifungal lead compounds from plants and microorganisms, there are thousands of beneficial bioactive molecules waiting to be discovered. Among an estimated number of 250,000–500,000 plant species in the world, less than 10% have been examined for some aspects of biological activity. Indeed, only a small portion (less that 1%) of an estimated number of 1.5 million bacterial and fungal species existing on our planet has yet been identified, of which fewer still are tested for bioactive antifungal metabolites. Recent advances in analytical methods, the design of comprehensive natural products libraries, the optimization of plant-cell-culture methods, cloning, and genetic engineering have provided a unique opportunity for the isolation and structural elucidation of novel bioactive antifungals from natural sources. More efforts should be made to identify lead compounds from newly described plant species, terrestrial bacteria and fungi. Special consideration should be made to unexplored habitats such as marine environments, which contain anaerobic bacteria and fungi, snails, microalgae, cyanobacteria, actinomycetes and plants as unique sources of novel bioactive molecules. Myxobacteria, non-streptomycete and unusual actinomycetes, plant endophytes and unusual moulds and yeasts are good examples of microorganisms from terrestrial environments that have not been thoroughly investigated and thus, they could be considered as potential sources of novel lead antifungals by researchers. Undoubtedly, the investigation of secondary metabolites from plants and microorganisms provides a unique opportunity for the discovery of new antifungal molecules as potential lead structures for developing potent environmentally friendly fungicides. Finally, a better understanding of plant–microbe interactions helps us to manage more powerful strategies in preventing food spoilage and crop diseases due to food-borne fungal pathogens.

Abbreviations

Alternaria (*A. alternata, A. porri, A. solani, A. brassicae, A. brassisicola*), Aspergillus (*A. niger, A. flavus, A. terreus, A. parasiticus, A. nomius, A. candidus, A. japonicus, A. heteromorphus*), Ascochyta (*A. rabiei, A. citrullina*), Bipolaris (*B. maydis, B. oryzae*), Botryosphaeris (*B. ribis*), Botrytis (*B. cinerea*), Cladosporium (*C. cucumerinum, C. cladosporioides, C. sphaerospermum, C. cucumerinum*), Colletotricum (*C. musae, C. truncatum, C. acutatum, C. gloeosporioides, C. lindemuthianum*), Crinipellis (*C. perniciosa*), Curvularia (*C. lunata, C. maculans*), Fusarium (*F. solani, F. oxysporum, F. moniliforme, F. udum, F. culmorum, F. graminearum, F. verticillioides*), Gaeumannomyces (*G. graminis*), Giberella (*G. zeae*), Helminthosporium (*H. spiciferum, H. maydis, H. sativum*), Magnaporthe (*M. grisea*), Moniliophthora (*M. roreri*), Mucor (*M. mucedo, M. miehei*), Mycosphaerella (*M. arachidicola*), Nigrospora (*N. oryzae*), Penicillium (*P. notatum, P. citrinum, P. expansum, P. brevicompactum, P. thomii*), Phoma (*P. arachnidicola, P. lingam, P. exigua*), Phomopsis (*P. gossypii*), Physalospora (*P. piricola*), Phytophtora (*P. cinnamomi, P. parasitica, P. capsici, P. palmivora, P. megakarya*), Pyricularia (*P. orysae, P. grisea*), Pythium (*P. heterothallicum, P. intermedium, P. ultimum*), Rhizopus (*R. arrhizus, R. chinensis*), Rhizoctonia (*R. solani, R. cerealis*), Sclerotinia (*S. sclerotiorum*), Sclerotium (*S. rolfsii*), Stachybotrys (*S. microbispora*), Trichoderma (*T. reesei, T. viride, T. hamatum, T. longibranchiatum*), Valsa (*V. mali*), Venturia (*V. inaequalis*), Verticillium (*V. albo-atrum, V. dahliae*).

References

Abdou, R., Scherlach, K., Dahse, H.-M., Sattler, I. and Hertweck, C. (2009) Botryorhodines A–D, antifungal and cytotoxic depsidones from *Botryosphaeria rhodina*, an endophyte of the medicinal plant *Bidens pilosa*. *Phytochemistry* 71, 110–116.

Aderiye, B.I., Ogundana, S.K., Adesanya, S.A. and Roberts, M.F. (1989) The effect of β-sitosterol on spore germination and germ-tube elongation of *Aspergillus niger* and *Botryodiplodia theobromae*. *International Journal of Food Microbiology* 8, 73–78.

Aneja, M., Gianfagna, T.J. and Hebbar, P.K. (2005) *Trichoderma harzianum* produces nonanoic acid, an inhibitor of spore germination and mycelial growth of two cacao pathogens, *Physiological and Molecular Plant Pathology* 67, 304–307.

Aslam, S.N., Stevenson, P.C., Kokuban, T. and Hall, D.R. (2009) Antibacterial and antifungal activity of cicerfuran and related 2-arylbenzofurans and stilbenes. *Microbiological Research* 164, 191–195.

Athukoralage, P.S., Herath, H.M.T.B., Deraniyagala, S.A., Wijesundera, R.L.C. and Weerasinghe, P.A. (2001) Antifungal constituent from *Gordonia dassanayakei*. *Fitoterapia* 72, 565–567.

Ayafor, F.J., Tchuendem, M.H.K., Nyasse, B., Tillequin, F. and Anke, H. (1994) Novel bioactive diterpenoids from *Aframomum aulacocarpos*. *Journal of Natural Products* 57, 917–923.

Bakkali, F., Averbeck, S., Averbeck, D. and Idaomar, M. (2008) Biological effects of essential oils – a review. *Food and Chemical Toxicology* 46, 446–475.

Banerji, R., Misra, G. and Nigam, S.K. (1985) Role of indigenous plant material in pest control. *Pesticides* 19, 32–38.

Bang K.H., Kim, Y.K., Min, B.S., Na, M.K., Rhee, Y.H., Lee, J.P. and Bae, K.H. (2000) Antifungal activity of magnolol and honokiol. *Archives of Pharmacal Research* 23, 46–49.

Baraza, L.D., Joseph, C.C., Munissi, J.J.E., Nkunya, M.H.H., Arnold, N., Porzel, A. and Wessjohann, L. (2008) Antifungal rosane diterpenes and other constituents of *Hugonia castaneifolia*. *Phytochemistry* 69, 200–205.

Barile, E., Bonanomi, G., Antignani, V., Zolfaghari, B., Ebrahim Sajjadi, S., Scala, F. and Lanzotti, V. (2007) Saponins from *Allium minutiflorum* with antifungal activity. *Phytochemistry* 68, 596–603.

Barrett, D. (2002) From natural products to clinically useful antifungals. *Biochimica et Biophysica Acta* 1587, 224–233.

Basha, S.A., Mishra, R.K., Jha, R.N., Pandey, V.B. and Singh, U.P. (2002) Effect of berberine and (+)-bicuculline isolated from *Corydalis chaerophylla* on spore germination of some fungi. *Folia Microbiologica* 47, 161–165.

Borris, R.P. (1996) Natural products research: perspectives from a major pharmaceutical company. *Journal of Ethnopharmacology* 51, 29–38.

Brown, A.E. and Hamilton, J.T.C. (1992) Indole-3-ethanol produced by *Zygorrhynchus moelleri*, an indole-3-acetic acid analogue with antifungal activity. *Mycological Research* 96, 71–74.

Carvalho, A.O. and Gomes, V.M. (2009) Plant defensins – prospects for the biological functions and biotechnological properties. *Peptides* 30, 1007–1020.

Chakor, N., Dallavalle, S., Musso, L. and Moretti, M. (2008) First total synthesis of the antifungal antibiotic thiobutacin. *Tetrahedron Letters* 49, 5056–5058.

Chang, H.-T., Cheng, Y.-H., Wu, C.-L., Chang, S.-T., Chang, T.-T. and Su, Y.-C. (2008) Antifungal activity of essential oil and its constituents from *Calocedrus macrolepis* var. *formosana* Florin leaf against plant pathogenic fungi. *Bioresource Technology* 99, 6266–6270.

Chu, K.T., Liu, K.H. and Ng, T.B. (2003) Cicerarin, a novel antifungal peptide from the green chickpea. *Peptides* 24, 659–663.

Chu, K.T. and Ng, T.B. (2003) Isolation of a large thaumatin-like antifungal protein from seeds of the Kweilin chestnut *Castanopsis chinensis*. *Biochemical and Biophysical Research Communications* 301, 364–370.

Clark, A.M. (1996) Natural products as a resource for new drugs. *Pharmaceutical Research* 13, 1133–1141.

Dahot, M.U. (1999) Antibacterial and antifungal activity of small protein of *Indigofera oblongifolia* leaves. *Journal of Ethnopharmacology* 64, 277–282.

Dalié, D.K.D., Deschamps, A.M. and Richard-Forget, M. (2010) Lactic acid bacteria – potential for control of mould growth and mycotoxins: a review. *Food Control* 21, 370–380.

De Lucca, A.J. and Walsh, T.J. (2000) Antifungal peptides: origin, activity, and therapeutic potential. *Revista Iberoamericana de Micología* 17, 116–120.

De Lucca, A.J., Jacks, T.J., Takemoto, J., Vinyard, B., Peter, J., Navarro, E. and Walsh, T.J. (1999) Fungal lethality, binding, and cytotoxicity of syringomycin-E. *Antimicrobial Agents and Chemotherapy* 43, 371–373.

de Silva, R.V., Navickiene, H.M.D., Kato, M.J., Bolzani, V.S., Méda, C.I., Young, M.C.M. and Furlan, M. (2002) Antifungal amides from *Piper arboreum* and *Piper tuberculatum*. *Phytochemistry* 59, 521–527.

Derita, M.G., Leiva, M.L. and Zacchino, S.A. (2009) Influence of plant part, season of collection and content of the main active constituent, on the antifungal properties of *Polygonum acuminatum* Kunth. *Journal of Ethnopharmacology* 124, 377–383.

Duraipandiyan, V. and Ignacimuthu, S. (2009) Antibacterial and antifungal activity of flindersine isolated from the traditional medicinal plant, *Toddalia asiatica* (L.) Lam. *Journal of Ethnopharmacology* 123, 494–498.

El-Tarabily, K.A. and Sivasithamparam, K. (2006) Non-streptomycete actinomycetes as biocontrol agents of soil-borne fungal plant pathogens and as plant growth promoters. *Soil Biology and Biochemistry* 38, 1505–1520.

Escalante, A.M., Santecchia, C.B., López, S.N., Gattuso, M.A., Ravelo, A.G., Delle Monache, F., Sierra, M.G. and Zacchino, S.A. (2002) Isolation of antifungal saponins from *Phytolacca tetramera*, an Argentinean species in critic risk. *Journal of Ethnopharmacology* 82, 29–34.

Fernando, W.G.D., Ramarathnam, R., Krishnamoorthy, A.S. and Savchuk, S.C. (2005) Identification and use of potential bacterial organic antifungal volatiles in biocontrol. *Soil Biology & Biochemistry* 37, 955–964.

Freitas, G.C., Kitamura, R.O.S., Lago, J.H.G., Young, M.C.M., Guimarães, E.F. and Kato, M.J. (2009) Caldensinic acid, a prenylated benzoic acid from *Piper caldense*. *Phytochemistry Letters* 2, 119–122.

Galeotti, F., Barile, E., Curir, P., Dolci, M. and Lanzotti, V. (2008) Flavonoids from carnation (*Dianthus caryophyllus*) and their antifungal activity. *Phytochemistry Letters* 1, 44–48.

Ghisalberti, E.L. (2000) Bioactive metabolites from soilborne fungi: natural fungicides and biocontrol agents. In: Atta-ur-Rahman (ed.) *Studies in Natural Products Chemistry, Volume 21, Part 2: Bioactive Natural Products (Part B)*. Elsevier Science B.V.: Amsterdam, pp. 181–250.

Giudici, A.M., Regenthe, M.C. and de la Canal, L. (2000) A potent antifungal protein from *Helianthus annuus* flowers is a trypsin inhibitor. *Plant Physiology and Biochemistry* 38, 881–888.

Grayer, R.J. and Kokubun, T. (2001) Plant-fungal interactions: the search for phytoalexins and other antifungal compounds from higher plants. *Phytochemistry* 56, 253–263.

Guan, Y., Ramalingam, S., Nagegowda, D., Taylor, P.W.J. and Chye, M.-L. (2008) *Brassica juncea* chitinase BjCHI1 inhibits growth of fungal phytopathogens and agglutinates Gram-negative bacteria. *Journal of Experimental Botany* 59, 3475–3484.

Harborne, J.B. and Baxter, H. (1999) *The Handbook of Natural Flavonoids*. Wiley & Sons, Chichester, UK.

Harborne, J.B. and Williams, C.A. (2000) Advances in flavonoid research since 1992. *Phytochemistry* 55, 481–504.

Harper, J.K., Arif, A.M., Ford, E.J., Strobel, G.A., Porco, J.A., Tomer, D.P., Oneill, K.L., Heidere, E.M. and Granta, D.M. (2003) Pestacin: a 1,3-dihydroisobenzofuran from *Pestalotiopsis microspora* possessing antioxidant and antimycotic activities. *Tetrahedron* 59, 2471–2476.

Inamori, Y., Kato, Y., Kubo, M., Yasuda, M., Baba, K. and Kozawa, M. (1984) Physiological activities of 3,3′,4,5′-tetrahydroxystilbene isolated from the heartwood of *Cassia garrettiana* CRAIB. *Chemical & Pharmaceutical Bulletin* (Tokyo) 32, 213–218.

Ioset, J.-R., Marston, A., Gupta, M.P. and Hostettmann, K. (1998) Antifungal and larvicidal meroterpenoid naphthoquinones and a naphtoxirene from the roots of *Cordia linnaei*. *Phytochemistry* 47, 729–734.

Ismet, A., Vikineswary, S.. Paramaswari, S., Wong, W.H., Ward, A., Sek, T., Fiedler, H.P. and Goodfellow, M. (2004) Production and chemical characterization of antifungal metabolites from *Micromonospora* sp. M39 isolated from mangrove rhizosphere soil. *World Journal of Microbiology & Biotechnology* 20, 523–528.

Kajiyama, S.-I., Kanzakt, H., Kawazu, K. and Kobayashi, A. (1998) Nostofungicidine, an antifungal lipopeptide from the field-grown terrestrial blue-green alga *Nostoc commune*. *Tetrahedron Letters* 39, 3737–3740.

Kaur, R., Macleod, J., Foley, W. and Nayudu, M. (2006) Gluconic acid: An antifungal agent produced by *Pseudomonas* species in biological control of take-all, *Phytochemistry* 67, 595–604.

Kavitha, A., Prabhakar, P., Narasimhulu, M., Vijayalakshmi, M., Venkateswarlu, Y., Rao, K.V. and Raju, V.B.S. (2010) Isolation, characterization and biological evaluation of bioactive metabolites from *Nocardia levis* MK-VL_113. *Microbiological Research* 165, 199–210.

Kelemu, S., Cardona, C. and Segura, G. (2004) Antimicrobial and insecticidal protein isolated from seeds of *Clitoria ternatea*, a tropical forage legume. *Plant Physiology and Biochemistry* 42, 867–873.

Ko, W.-H., Tsou, Y.-J., Ju, Y.-M., Hsieh, H.-M. and Ann, P.-J. (2010) Production of a fungistatic substance by *Pseudallescheria boydii* isolated from soil amended with vegetable tissues and its significance. *Mycopathologia* 169, 125–131.

Koketsu, M., Kim, M. and Yamamoto, T. (1996) Antifungal activity against food-borne fungi of *Aspidistra elatior*. *Journal of Agricultural and Food Chemistry* 44, 301–303.

Lee, S.E., Mahoney, N.E. and Campbell, B.C. (2002) Inhibition of aflatoxin B_1 biosynthesis by piperlongumine isolated from *Piper longum* L. *Journal of Microbiology and Biotechnology* 12, 679–682.

Li, E., Jiang, L., Guo, L., Zhang, H. and Che, Y. (2008) Pestalachlorides A-C, antifungal metabolites from the plant endophytic fungus *Pestalotiopsis adusta*. *Bioorganic & Medicinal Chemistry* 16, 7894–7899.

Lin, P. and Ng, T.B. (2008) A novel and exploitable antifungal peptide from kale (*Brassica alboglabra*) seeds. *Peptides* 29, 1664–1671.

Lopes, N.P., Kato, M.J. and Yoshida, M. (1999) Antifungal constituents from roots of *Virola surinamensis*. *Phytochemistry* 51, 29–33.

Lopes, J.L.S., Valadares, N.F., Moraes, D.I., Rosa, J.C., Araújo, H.S.S. and Beltramini, L.M. (2009) Physico-chemical and antifungal properties of protease inhibitors from *Acacia plumose*. *Phytochemistry* 70, 871–879.

Macías-Rubalcava, M. L., Hernández-Bautista, B.E., Jiménez-Estrada, M., González, M.C., Glenn, A.E., Hanlin, R.T., Hernández-Ortega, S., Saucedo-García, A., Muria-González, J.M. and Anaya, A.L. (2008) Naphthoquinone spiroketal with allelochemical activity from the newly discovered endophytic fungus *Edenia gomezpompae*. *Phytochemistry* 69, 1185–1196.

Mishra, B.B., Singh, D.D., Kishore, N., Tiwari, V.K. and Tripathi, V. (2010) Antifungal constituents isolated from the seeds of *Aegle marmelos*. *Phytochemistry* 71, 230–234.

Moretti, M., Gilardi, G., Gullino, M.L. and Garibaldi, A. (2008) Biological control potential of *Achromobacter xylosoxydans* for suppressing *Fusarium* wilt of tomato, *International Journal of Botany* 4, 369–375.

Moyne, A.-L., Shelby, R., Cleveland, T.E. and Tuzun, S. (2001) Bacillomycin D: an iturin with antifungal activity against *Aspergillus flavus*. *Journal of Applied Microbiology* 90, 622–629.

Mukherjee, G. and Sen, S.K. (2006) Purification, characterization, and antifungal activity of chitinase from *Streptomyces venezuelae* P_{10}. *Current Microbiology* 53, 265–269.

Ojika, M., Watanabe, T., Qi, J., Tanino, T. and Sakagami, Y. (2004) Syntheses of cystothiazole A and its stereoisomers: importance of stereochemistry for antifungal activity. *Tetrahedron* 60, 187–194.

Oksman-Caldentey, K.-M. and Inzé, D. (2004) Plant cell factories in the post-genomic era: new ways to produce designer secondary metabolites. *Trends in Plant Science* 9, 433–440.

Ongena, M. and Jacques, P. (2007) *Bacillus* lipopeptides: versatile weapons for plant disease biocontrol, *Trends in Microbiology* 16, 115–125.

Onishi, J.C., Milligan, J.A., Basilio, A., Bergstrom, J., Curotto, J., Huang, L., Meinz, M., Nallin-Omstead, M., Peláez, F., Rew, D., Salvatore, M., Thompson, J., Vicente, F. and Kurtz, M.B. (1997) Antimicrobial activity of viridiofungins. *Journal of Antibiotics* 50, 334–338.

Osbourn, A. (1996a) Saponins and plant defense – a soap story. *Trends in Plant Science* 1, 4–9.

Osbourn, A.E. (1996b) Preformed antimicrobial compounds and plant defense against fungal attack. *Plant Cell* 8, 1821–1831.

Pedras, M.S.C., Ismail, N., Quail, J.W. and Boyetchko, S.M. (2003) Structure, chemistry, and biological activity of pseudophomins A and B, new cyclic lipodepsipeptides isolated from the biocontrol bacterium *Pseudomonas fluorescens*. *Phytochemistry* 62, 1105–1114.

Pedras, M.S.C., Sarwar, M.G., Suchy, M. and Adio, A.M. (2006) The phytoalexins from cauliflower, caulilexins A, B and C: isolation, structure determination, syntheses and antifungal activity. *Phytochemistry* 67, 1503–1509.

Pitt, J.I. and Hocking, A. (2009) *Fungi and Food Spoilage*, 3rd edn. Springer Science, New York, NY 10013, USA.

Prabavathy, V.R., Mathivanan, N. and Murugesan, K. (2006) Control of blast and sheath blight diseases of rice using antifungal metabolites produced by *Streptomyces* PM5. *Biological Control* 39, 313–319.

Quiroga, E.N., Sampietro, D.A., Sgariglia, M.A., Soberón, J.R. and Vattuone, M.A. (2009) Antimycotic activity of 5′-pernylisoflavanones of the plant *Geoffroea decorticans* against *Aspergillus* species. *International Journal of Food Microbiology* 132, 42–46.

Razzaghi-Abyaneh, M., Shams-Ghahfarokhi, M., Yoshinari, T., Rezaee, M.-B., Jaimand, K., Nagasawa, H. and Sakuda, S. (2008) Inhibitory effects of *Satureja hortensis* L. essential oil on growth and aflatoxin production by *Aspergillus parasiticus. International Journal of Food Microbiology* 123, 228–233.

Razzaghi-Abyaneh, M., Shams-Ghahfarokhi, M., Rezaee, M.-B., Jaimand, K., Alinezhad, S., Saberi, R. and Yoshinari, T. (2009) Chemical composition and antiaflatoxigenic activity of *Carum carvi* L., *Thymus vulgaris* and *Citrus aurantifolia* essential oils. *Food Control* 20, 1018–1024.

Razzaghi-Abyaneh, M., Shams-Ghahfarokhi, M., Rezaee, M.-B. and Sakuda, S. (2010) Natural aflatoxin inhibitors from medicinal plants. In: Rai, M. and Varma, A. (eds) *Mycotoxins in Food, Feed and Bioweapons.* Springer-Verlag, Berlin, Heidelberg, pp. 329–352.

Ryazanova, L.P., Stepnaya, O.A., Suzina, N.E. and Kulaev, I.S. (2005) Antifungal action of the lytic enzyme complex from *Lysobacter* sp. XL1. *Process Biochemistry* 40, 557–564.

Salazar, R., Rivas, V., González, G. and Waksman, N. (2006) Antimicrobial activity of coupled hydroxyanthracenones isolated from plant of the genus *Karwinskia. Fitoterapia* 77, 398–400.

Samson, A. R., Hoekstra, E.S., Frisvad, J.C. and Filtenborg, O. (2000) *Introduction to Food and Airborne Fungi*, 6th edn. CBS-Utrecht, The Netherlands.

Schlingmann, G., Milne, L., Borders, D.B. and Carter, G.T. (1999) Strevertenes, antifungal pentaene macrolides produced by *Streptoverticillcum* LL-30F848. *Tetrahedron* 55, 5977–5990.

Sharma, R.R., Singh, D. and Singh, R. (2009) Biological control of postharvest diseases of fruits and vegetables by microbial antagonists: a review. *Biological Control* 50, 205–221.

Singh, S.K., Sarma, B.K., Srivastava, J.S., Singh, U.P. and Ray, A.B. (1999) Antifungal activity of Δ^3-alstovenine, a plant alkaloid isolated from *Alstonia venenata. Folia Microbiologica* 44, 510–512.

Sitohy, M., Doheim, M. and Badr, H. (2007) Isolation and characterization of a lectin with antifungal activity from Egyptian *Pisum sativum* seeds. *Food Chemistry* 104, 971–979.

Skouri-Gargouri, H. and Gargouri, A. (2008) First isolation of a novel thermostable antifungal peptide secreted by *Aspergillus clavatus. Peptides* 29, 1871–1877.

Srivastava, S.K., Srivastava, S.D. and Chouksey, B.K. (2001) New antifungal constituents from *Terminalia alata. Fitoterapia* 72, 106–112.

Stein, A.C., Álvarez, S., Avancini, C., Zacchino, S. and von Poser, G. (2006) Antifungal activity of some coumarins obtained from species of *Pterocaulon* (*Asteraceae*). *Journal of Ethnopharmacology* 107, 95–98.

Stockwell, C. (1985) *Nature's Pharmacy.* Century Hutchinson Ltd, London, United Kingdom.

Tabopda, T.K., Ngoupayo, J., Liu, J., Mitaine-Offer, A.-C., Tanoli, S.A.K., Khan, S.N., Ali, M.S., Ngadjui, B.T., Tsamo, E., Lacaille-Dubois, M.-A. and Luu, B. (2008) Bioactive aristolactams from *Piper umbellatum. Phytochemistry* 69, 1726–1731.

Terras, F.R.G., Schoofs, H.M.E., De Bolle, M.F.C., van Leuven, F., Rees, S.B., Vanderleyden, J., Cammue, B.P.A. and Broekaert, W.F. (1992) Analysis of two novel classes of plant antifungal proteins from radish (*Raphanus sativus* L.) seeds. *Journal of Biological Chemistry* 267, 15301–15309.

Tolouee, M., Alinezhad, S., Saberi, R., Eslamifar, A., Zad, S.J., Jaimand, K., Taeb, J., Rezaee, M.-B., Kawachi, M., Shams-Ghahfarokhi, M. and Razzaghi-Abyaneh, M. (2010) Effect of *Matricaria chamomilla* L. flower essential oil on the growth and ultrastructure of *Aspergillus niger* van Tieghem. *International Journal of Food Microbiology* 139, 127–133.

Tripathi, P. and Dubey, N.K. (2004) Exploitation of natural products as an alternative strategy to control postharvest fungal rotting of fruit and vegetables. *Postharvest Biology and Technology* 32, 235–245.

Välimaa, A.-L., Honkalampi-Hämäläinen, U., Pietarinen, S., Willför, S., Holmbom, B. and von Wright, A. (2007) Antimicrobial and cytotoxic knotwood extracts and related pure compounds and their effects on food-associated microorganisms. *International Journal of Food Microbiology* 115, 235–243.

Van den Berg, K.B.P., Rougé, P., Proost, P., Coosemans, J., Krouglova, T., Engelborghs, Y., Peumans, W.J. and Van Damme, E.J.M. (2004) Synergistic antifungal activity of two chitin-binding proteins from spindle tree (*Euonymus europaeus* L.). *Planta* 219, 221–232.

Verpoorte, R. (1998) Exploration of nature's chemodiversity: the role of secondary metabolites as leads in drug development. *Drug Discovery Today* 3, 232–238.

Walsh, U.F., Morrissey, J.P. and O´Gara, F. (2001) *Pseudomonas* for biocontrol of phytopathogens: from functional genomics to commercial exploitation. *Current Opinion in Biotechnology* 12, 289–295.

Wan, M., Li, G., Zhang, J., Jiang, D. and Huang, H.-C. (2008) Effect of volatile substances of *Streptomyces platensis* F-1 on control of plant fungal diseases. *Biological Control* 46, 552–559.

Wang, H.X. and Ng, T.B. (2007) An antifungal peptide from red lentil seeds. *Peptides* 28, 547–552.

Wang, J., Liu, J., Chen, H. and Yao, J. (2007) Characterization of *Fusarium graminearum* inhibitory lipopeptide from *Bacillus subtilis* IB. *Applied Microbiology and Biotechnology* 76, 889–894.

Weber, W., Anke, T., Bross, A.T., Steglich, W., Strobilurin, D. and Strobilurin, F. (1990) Two new cytostatic and antifungal (E)-13-methoxyacrylate antibiotics from *Cyphellopsis anomala*. *Planta Medica* 56, 446–450.

Wei-Wei, L., Wei, M., Bing-Yu, Z., You-Chen, D. and Feng, L. (2008) Antagonistic activities of volatiles from four strains of *Bacillus* spp. and *Paenibacillus* spp. against soil-borne plant pathogens. *Agricultural Sciences in China* 7, 1104–1114.

Whipps, J.M. (1997) Developments in the biological control of soil-borne plant pathogens. In: Callow, J.A. (ed.) *Advances in Botanical Research*. Academic Press: London, pp. 1–134.

Wicklow, D.T., Roth, S., Deyrup, S.T. and Gloer, J.B. (2005) A protective endophyte of maize: *Acremonium zeae* antibiotics inhibitory to *Aspergillus flavus* and *Fusarium verticillioides*. *Mycological Research* 109, 610–618.

Wicklow, D.T., Jordan, A.N. and Loer, J.B. (2009) Antifungal metabolites (monorden, monocillins I, II, III) from *Colletotrichum graminicola*, a systemic vascular pathogen of maize. *Mycological Research* 113, 1433–1442.

Widmer, T.L. and Laurent, N. (2006) Plant extracts containing caffeic acid and rosmarinic acid inhibit zoospore germination of *Phytophthora* spp. pathogenic to *Theobroma cacao*. *European Journal of Plant Pathology* 115, 377–388.

Yang, L., Xie, J., Jiang, D., Fu, Y., Li, G. and Lin, F. (2008) Antifungal substances produced by *Penicillium oxalicum* strain PY-1-potential antibiotics against plant pathogenic fungi. *World Journal of Microbiology and Biotechnology* 24, 909–915.

Yang, R., Han, Y.C., Li, G.Q., Jiang, D.H. and Huang, H.C. (2007) Suppression of *Sclerotinia sclerotiorum* by antifungal substances produced by the mycoparasite *Coniothyrium minitans*. *European Journal of Plant Pathology* 119, 411–420.

Yen, T.-B., Chang, H.-T., Hsieh, C.-C. and Chang, S.-T. (2008) Antifungal properties of ethanolic extract and its active compounds from *Calocedrus macrolepis* var. *formosana* (Florin) heartwood. *Bioresource Technology* 99, 4871–4877.

You, F., Han, T., Wu, J.-Z., Huang, B.-K. and Qin, L.-P. (2009) Antifungal secondary metabolites from endophytic *Verticillium* sp. *Biochemical Systematics and Ecology* 37, 162–165.

Zhou, C.-X., Liu, J.Y., Ye, W.-C., Liu, C.-H. and Tan, R.X. (2003) Neoverataline A and B, two antifungal alkaloids with a novel carbon skeleton from *Veratrum taliense*. *Tetrahedron* 59, 5743–5747.

14 Application of Plant-based Antimicrobials in Food Preservation

Brijesh Kumar Tiwari, Vasilis P. Valdramidis, Paula Bourke and Patrick Cullen*

Introduction

A number of non-traditional preservation techniques are being developed to satisfy public health policies and consumer demand with regard to nutritional and sensory aspects of foods. Generally, foods are thermally processed by subjecting them to temperatures varying from 60 to 100°C for durations of a few seconds up to some minutes in order to destroy vegetative microorganisms. During this period of treatment a large amount of energy is transferred to the food. However, this energy can trigger unwanted reactions, leading to undesirable organoleptic and nutritional effects. Ensuring food safety and at the same time meeting the demands for retention of nutrition and quality attributes has resulted in increased interest in alternative preservation techniques for inactivating microorganisms and enzymes in foods (Barbosa-Canovas *et al.*, 1997). Quality attributes of importance include; flavour, odour, colour, texture and nutritional value. This increasing demand has resulted in new dimensions for the use of preservatives derived from natural sources such as plants. Extensive research has investigated the potential of natural antimicrobial agents in food preservation. Antimicrobial compounds present in foods can extend shelf life of unprocessed or processed foods by reducing microbial growth rate or viability (Beuchat and Golden, 1989). Originally, spices and herbs were added to change and/or improve taste; they can also enhance shelf life because of their antimicrobial nature. Some of the active components present in plant-based foods that are known to confer antimicrobial activity are also known to contribute to the self-defence of plants against infectious organisms (Deans *et al.*, 1995; Kim *et al.*, 2001). Edible, medicinal and herbal plants and their derived essential oils (EOs) and isolated compounds contain a large number of secondary metabolites that are known to retard or inhibit the growth of bacteria, yeast and moulds (Burt and Reinders, 2003). Many of these compounds are under investigation and are not yet exploited commercially. This chapter outlines the antimicrobials from plant sources and their applications in food preservation. Natural antimicrobials in food preservation can be used alone or in synergistic combinations as additives but also show potential for combination with other non-thermal technologies. Finally, suggested approaches for the quantification of the minimum (and non-inhibitory) concentration of antimicrobials and/or their components are presented.

* Corresponding author.

Sources of Antimicrobial Agents from Plants

The antimicrobial compounds in plant materials are commonly found in the essential-oil fraction of leaves (rosemary, sage, basil, oregano, thyme, marjoram), flower or bud (clove), bulb (garlic, onion), seeds (caraway, fennel, nutmeg, parsley), rhizomes (asafoetida), fruit (pepper, cardamom) or other parts of plants (Nychas and Skandamis, 2003; Gutierrez *et al.*, 2008a). Plant EOs and their constituents have been widely used as flavouring agents in foods since the earliest recorded history and it is well established that many have a wide spectrum of antimicrobial action (Kim *et al.*, 1995; Smith-Palmer *et al.*, 1998; Packiyasothy and Kyle, 2002; Alzoreky and Nakahara, 2002). These compounds may be lethal to microbial cells or they might inhibit the production of secondary metabolites (e.g. mycotoxins) (Davidson, 2001). Plant EOs are generally more inhibitory against Gram-positive than Gram-negative bacteria (Marino *et al.*, 2002; Chorianopoulos *et al.*, 2004; Gutierrez *et al.*, 2008a). While this is true for many EOs, there are some agents which are effective against both groups, such as oregano, clove, cinnamon, citral and thyme (Sivropoulou *et al.*, 1996; Skandamis *et al.*, 2002; Kim and Fung, 2004). The major EO components with antimicrobial effects are found in plants, herbs and spices and are phenolic compounds, terpenes, aliphatic alcohols, aldehydes, ketones, acids and isoflavonoids (Farag *et al.*, 1989; Nychas, 1995; Dorman and Deans, 2000; Lambert *et al.*, 2001; Burt and Reinders, 2003; Lopez-Malo *et al.*, 2005). Some of the selected components of EOs responsible for antimicrobial activity are shown in Fig. 14.1. Chemical analysis of a range of EOs reveals that the principal constituents of many include: carvacrol, thymol, citral, eugenol and their precursors (Juliano *et al.*, 2000; Demetzos and Perdetzoglou, 2001). Simple and complex derivatives of phenol are reported to be the main antimicrobial compounds in EOs from spices (Shelef, 1983). It has been reported that some non-phenolic constituents of EOs are more effective against Gram-negative bacteria, e.g. allyl isothiocyanate (AIT) (Ward *et al.*, 1998) and garlic oil (Yin and Cheng, 2003), respectively. In addition, AIT is also effective against many Gram-positive fungi (Nielsen and Rios *et al.*, 2000). Generally, the antimicrobial efficacy of EOs is dependent on many factors, such as the chemical structure of their components as well as concentration, but other factors important for retaining efficacy in foods include matching the antimicrobial spectrum of activity with the target microorganism, interactions with the food matrix and also the method of application. Many of the antimicrobial compounds present in plants can be part of the plants pre- or post-infectional defence mechanisms for combating infectious or parasitic agents (Rauha *et al.*, 2000). Consequently, plants that manifest relatively high levels of antimicrobial action may be sources of compounds that inhibit the growth of food-borne pathogens (Ibrahim *et al.*, 2006). Compounds are also generated in response to stress from inactive precursors (Sofos *et al.*, 1998), which may be activated by enzymes: hydrolases or oxidases, usually present in the plant tissues (Holley and Patel, 2005). In mustard and horseradish, precursor glucosinolates are converted by the enzyme myrosinase to yield a variety of isothiocynates including the allyl form, which is a strong antimicrobial agent (Delaquis and Mazza, 1995). Several EOs obtained from oils of garlic, cinnamon, thyme, oregano, clove, basil, coriander, citrus peel, laurel, ginger, rosemary and peppermint, among others, have been studied as antimicrobial natural products against both bacteria and moulds (Ayala-Zavala *et al.*, 2009). Table 14.1 lists some of the antimicrobial and aroma characteristics of EOs.

Mechanisms of Antimicrobial Action

The possible modes of action for phenolic compounds (EO fractions) as antimicrobial agents have been previously reviewed (Wilkins and Board, 1989; Beuchat, 1992; Nychas, 1995; Sofos *et al.*, 1998; Lopez-Malo *et al.*, 2000; Davidson, 2001; Lopez-Malo

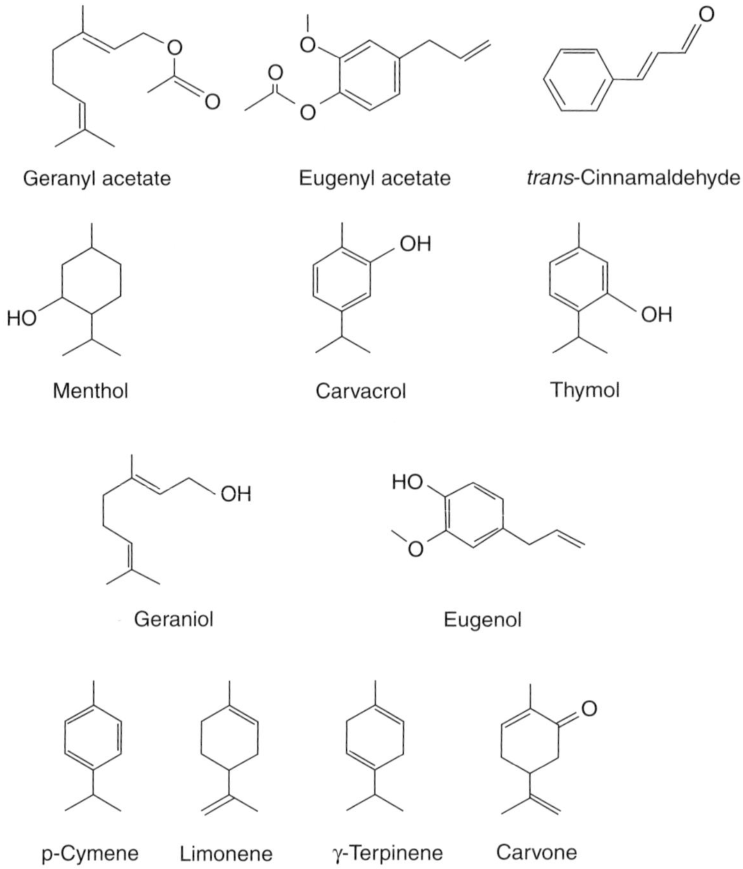

Fig. 14.1. Structural formulae of selected components of EOs (adapted from Burt (2004), with permission).

et al., 2005). However, the exact mechanism of action is not clear. The effect of phenolic compounds can be concentration dependent (Juven *et al.*, 1994). At low concentration, phenols affect enzyme activity, particularly those associated with energy production, while at high concentrations, they cause protein denaturation. The antimicrobial effect of phenolic compounds may be due to their ability to alter microbial cell permeability, thereby permitting the loss of macromolecules from the interior (e.g. ribose, Na glutamate). They could also interfere with membrane function (electron transport, nutrient uptake, protein, nucleic acid synthesis, enzyme activity) (Bajpai *et al.*, 2008) and interact with membrane proteins, causing deformation in structure and functionality (Fung *et al.*, 1977; Rico-Munoz *et al.*, 1987; Kabara and Eklund, 1991).

Delaquis and Mazza (1995) reported that the antimicrobial activity of isothiocyanates derived from onion and garlic is related to inactivation of extracellular enzymes through oxidative cleavage of disulfide bonds and that the formation of the reactive thiocyanate radical was proposed to mediate the antimicrobial effect. Carvacrol, (+)-carvone, thymol and trans-cinnamaldehyde are reported to decrease the intracellular ATP (adenosine triphosphate) content of *Escherichia coli* O157:H7 cells while simultaneously increasing extracellular ATP, indicating the disruptive action of these compounds on the plasma membrane (Helander *et al.*, 1998).

Rojas-Graü *et al.* (2006) and Rojas-Graü *et al.* (2007) have studied the effects of oregano, cinnamon and lemongrass oils and their active components (carvacrol, cinnamaldehyde and citral) which were incorporated into apple puree and alginate apple puree edible films to investigate efficacy against *E. coli* O157:H7. The effectiveness of these antimicrobial agents was evaluated using an agar diffusion method, which is commonly used to evaluate the antimicrobial activity for films. Oregano oil or one of its most active components, carvacrol, showed the greatest efficacy against *E. coli* O157:H7, as reflected in a greater zone of inhibition (Rojas-Graü *et al.*, 2009) (Fig. 14.2). Inactivation of yeasts can be attributed to the disturbance of several enzymatic systems, such as energy production and structural component synthesis (Connor and Beuchat, 1984). The mode of action of EOs is multiple and they have several targets in the microbial cell (Fig. 14.3) which may cause deterioration of cell wall, damage to cytoplasmic membrane, damage to membrane proteins, leakage of cell contents, coagulation of cytoplasm, depletion of protonmotive active sites, inactivation of essential enzymes, and disturbance of genetic material functionality (Burt, 2004; Ayala-Zavala *et al.*, 2008, 2009; Gutierrez *et al.*, 2008a). Similarly Fisher and Phillips (2008) also reported morphological changes in the microbial cell due to the presence of EOs (Fig. 14.4).

Factors Affecting Antimicrobial Activity

Various factors can impact on antimicrobial efficacy. These include the emergence of resistant bacteria, conditions that destabilize the biological activity of antimicrobial agents, binding to food components such as fat particles or protein surfaces, inactivation by other additives, poor solubility and uneven distribution in the food matrix and/ or pH effects on stability and activity of antimicrobial agents (Tiwari *et al.*, 2009). Antimicrobial activity of EOs is influenced by a number of factors including botanical source, time of harvesting, stage of development and method of extraction (Janssen *et al.*, 1986). For example, Chorianopoulos *et al.* (2006) reported that *Satureja* EOs obtained during the flowering period were the most potent with remarkable bactericidal properties. The composition, structure as well as functional groups of the oils play an important role in determining their antimicrobial activity. Usually compounds with phenolic groups are the most effective (Deans *et al.*, 1995; Dorman and Deans, 2000). The high antibacterial activity of phenolic components can be further explained in terms of alkyl substitution into the phenol nucleus (Dorman and Deans, 2000). The formation of phenoxyl radicals which interact with alkyl substituents does not occur with more stable molecules such as the ethers myristicin or anethole, which was related to the relative lack of antimicrobial activity of fennel, nutmeg or parsley EOs (Gutierrez *et al.*, 2008a). Most studies related to antimicrobial efficacy of EOs have been conducted *in vitro* using microbiological media (Ting and Diebel, 1992; Remmal *et al.*, 1993; Pandit and Shelef, 1994; Firouzi *et al.*, 1998; Hammer *et al.*, 1999; Campo *et al.*, 2000; Griffin *et al.*, 2000; Elgayyar *et al..*, 2001; Delaquis *et al.*, 2002; Gómez-Estaca *et al.*, 2010; Tyagi and Malik, 2010). Consequently, there is less understanding related to their efficacy when applied to complex food systems. Key areas requiring further knowledge for optimized application of natural antimicrobials in food include: targeting the microorganism of concern, the intelligent use of combinations to provide a synergy of activity, matching the activity of the compounds to the composition, processing and storage conditions of the food, as well as the effects on organoleptic properties (Roller and Covill, 1999; Nychas and Skandamis, 2003).

Plant EOs of thyme, clove and pimento were tested against *Listeria monocytogenes* and were found to be highly effective in peptone water. However, when the EOs were applied in a food system, Singh *et al.* (2003) concluded that the efficacy of EOs was reduced due to interaction with food components. In general, higher concentrations of EOs are required in foods than in

Table 14.1. Antimicrobial and aroma characteristics of EOs (adapted from Ayala-Zavala *et al.* (2009), with permission).

Essential oil	Major volatile constituents	Antimicrobial effect against	Aroma notes	References
Garlic root (*A. sativum*)	Methyl disulfide, allyl sulfide, allyl disulfide, allyl trisulfide, trimethylene trisulfide, allyl tetrasulfide	*Bacillus cereus, Escherichia coli, Shigella* spp., *Vibrio parahaemolyticus, Yercinia enterolitica, Salmonella enterica,* serovars *Enteritidis, Infantis, Typhimurium, Bacillus subtilis, Enterococcus faecalis, Streptococcus faecalis, Alternaria alternata*	Pungent, spice	(Ross *et al.*, 2001; Ayala-Zavala *et al.*, 2008c)
Cinnamon leaf (*C. zeylanicum*)	Cinnamaldehyde, eugenol, copaene, β-caryophyllene	*Escherichia coli, Pseudomonas aeruginosa, Escherichia faecalis, Staphylococcus aureus, Staphylococcus epidermidis,* methicillin-resistant *Staphylococcus aureus, Klebsiella pneumoniae, Salmonella* spp., *Vibrio parahemolyticus, Alternaria alternata, Aspergillus flavus, Aspergillus niger, Penicillium corylophilum*	Sweet, wood, spice	(Chang *et al.*, 2001; Guynot *et al.*, 2003; Ayala-Zavala *et al.*, 2008c)
Thyme (*T. vulgaris*)	Thymol, p-cymene, γ-terpinene, linalool	*Bacillus cereus, Clostridium botulinum, Escherichia faecalis, Escherichia coli, Staphylococcus aureus, Listeria monocytogenes, Aspergillus flavus, Aspergillus niger, Penicillium corylophilum, Klebsiella pneumoniae, Pseudomonas aeruginosa, Salmonella* spp.	Spice, citrus, wood	(Hammer *et al.*, 1999; Guynot *et al.*, 2003; Lee *et al.*, 2005)
Oregano (*O. vulgare)*	Sabinyl monoterpenes, terpinen-4-ol, γ-terpinene, carvacrol, thymol	*Bacillus cereus, Bacillus subtilis, Clostridium botulinum, Enterococcus faecalis, Escherichia coli, Staphylococcus aureus, Aspergillus niger, Listeria monocytogenes, Klebsiella pneumoniae, Pseudomonas aeruginosa, Salmonella* spp.	Spice, herb	(Charai *et al.*, 1996; Hammer *et al.*, 1999; Elgayyar *et al.*, 2001; Burt, 2004)
Clove (*E. aromaticum*)	Eugenol, eugenyl acetate, caryophyllene	*Bacillus brevis, Bacillus subtilis, Clostridium botulinum, Enteroccus faecalis, Candida* spp., *Aspergillus flavus, Aspergiullus niger, Penicillium corylophilum, Escherichia coli, Klebsiella pneumoniae, Pseudomonas aeruginosa, Staphylococcus aureus, Salmonella* spp., *Listeria monocytogenes*	Sweet, spice, wood	(Akgül and Kivanç 1988; Hammer *et al.*, 1999; Guynot *et al.*, 2003; Burt, 2004)

Basil (*O. basilicum)*	Linalool, methylchalvicol, eugenol, methyl eugenol, methyl cinnamate, 1,8-cineole, caryophyllene	*Bacillus brevis, Escherichia coli, Aspergillus flavus, Aspergillus niger, Penicillium corylophilum, Enterococcus faecalis, Escherichia coli, Klebsiella pneumoniae, Pseudomonas aeruginosa, Staphylococcus aureus, Listeria monocytogenes, Lactobacillus plantarum*	Fresh, sweet, herb, spice	(Hammer *et al.*, 1999; Elgayyar *et al.*, 2001; Guynot *et al.*, 2003; Opalchenova and Obreshkova, 2003)
Coriander (*C. sativum*)	2(E)-decanal, 2(E)dodecenal, linalool	*Escherichia coli, Listeria monocytogenes, Lactobacillus plantarum, Staphylococcus aureus*	Sweet, flower, spice, citrus	(Elgayyar *et al.*, 2001)
Citrus peel (*Citrus sp.*)	Limonene, linalool, citral	*Aspergillus niger, Aspergillus flavus, Penicillium verrucosum, Penicillium chrysogenum*	Sweet, citrus	(Viuda-Martos *et al.*, 2007, 2008)
Laurel (*L. nobilis)*	1,8-cineole, α-terpinyl acetate, linalool, methyl eugenol	*Staphylococcus aureus, Bacillus cereus, Micrococcus luteus, Enterococcus faecalis*	Fresh, herb, spice	(Demo and de las Mercedes Oliva, 2009)
Ginger (*Z. officinale*)	β-sesquiphellandrene, zingiberene	*Aspergillus flavus, Aspergillus niger, Penicillium corylophilum, Enterococcus faecalis, Escherichia coli, Klebsiella pneumoniae, Pseudomonas aeruginosa, Staphylococcus aureus*	Pungent, spice	(Hammer *et al.*, 1999; Guynot *et al.*, 2003)
Rosemary (*R. officinalis*)	Borneol, verbenone, camphor, a-pinene, 1,8-cineole	*Aspergillus flavus, Aspergillus niger, Penicillium corylophilum, Enterococcus faecalis, Escherichia coli, Klebsiella pneumoniae, Pseudomonas aeruginosa, Staphylococcus aureus, Listeria monocytogenes, Lactobacillus plantarum, Salmonella sp., Bacillus cereus*	Fresh, herb, resinous	(Hammer *et al.*, 1999; Elgayyar *et al.*, 2001; Guynot *et al.*, 2003; Burt, 2004)
Peppermint (*M. piperita*)	Menthol, menthone, menthyl acetate, menthofurane	*Bacillus brevis, Stahpylococcus aureus, Vibrio choleraei, Enterococcus faecalis, Escherichia coli, Klebsiella pneumoniae, Pseudomonas aeruginosa, Aspergillus flavus, Aspergillus niger, Penicillium corylophilum*	Fresh, herb	(Hammer *et al.*, 1999; Guynot *et al.*, 2003)

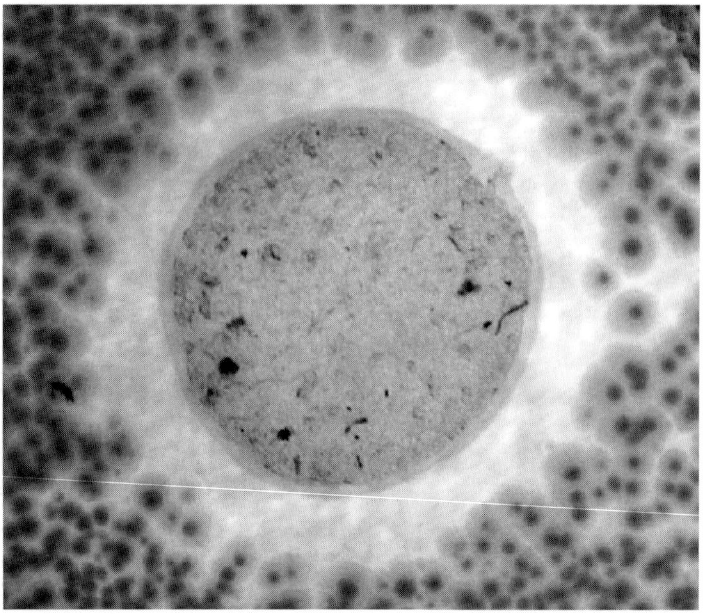

Fig. 14.2. Inhibitory zone (*E. coli* O157:H7 colony-free perimeter) of alginate apple puree edible film containing 0.1% v/v carvacrol oil (Rojas-Graü *et al.* (2009), with permission).

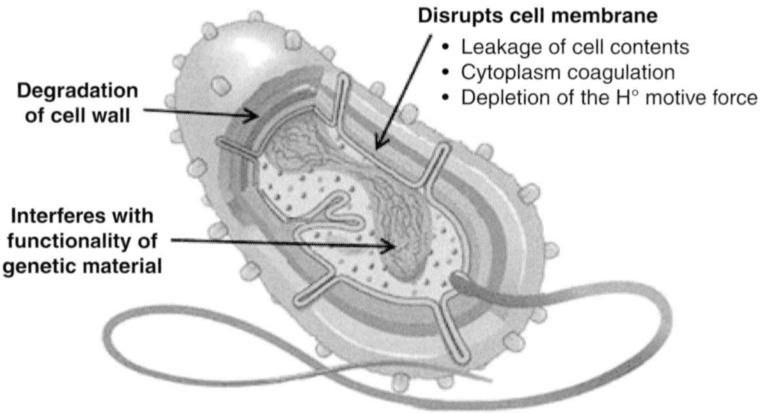

Fig. 14.3. Antimicrobial mode of action of EOs against different cell targets of microorganisms (adapted from Ayala-Zavala *et al.* (2009), with permission).

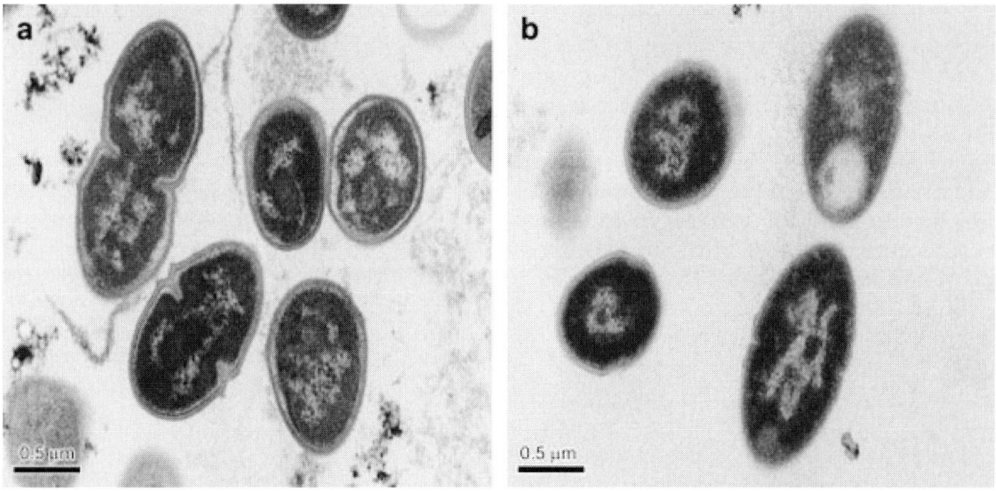

Fig. 14.4. Transmission electron microscopy image of *Enterococcus faecalis* (a) in the absence of EO and (b) after exposure to citrus EO vapour for 1 h (adapted from Fisher and Phillips (2008), with permission).

laboratory media. Combinations of EOs could minimize application concentrations required, thereby reducing any adverse organoleptical impact; however, their application for microbial control may also be affected by food composition (Gutierrez *et al.*, 2008a). The antimicrobial efficacy of EOs was reported to be a function of ingredient manipulation (Gutierrez *et al.*, 2009a). The antimicrobial activity of thyme is increased in high protein concentrations while higher potato starch concentrations decreased the EO antimicrobial activity of oregano and thyme against *L. monocytogenes* in food model systems (Gutierrez *et al.*, 2008a). Concentrations above 5% sugars did not reduce EO efficacy (Gutierrez *et al.*, 2009a). Finally, pH values of 5 seem to have the highest impact on the increase of the antimicrobial effect of EOs on *L. monocytogenes* (Gutierrez *et al.*, 2008a). Accordingly, the challenge for practical application of EOs is to develop optimized low dose combinations which can be delivered in a variety of ways to match product profiles as well as maintaining product safety and shelf life, thereby minimizing the undesirable flavour and sensory changes associated with the addition of high concentrations of EOs.

Quantification of the Minimum and Non-inhibitory Concentration

The use of antimicrobials as preservatives in food systems can be constrained when effective antimicrobial doses exceed organoleptic acceptability levels. This is particularly the case for essential oils due to their potentially high sensory impact. Two specific concentrations appear to be of interest, i.e. the non-inhibitory concentration, NIC, which refers to the concentration above which the inhibitor begins to have a negative effect on growth, and the minimum inhibitory concentration, MIC, which marks the concentration above which no growth is observed by comparison with the control (Carson *et al.*, 1995). Therefore, these concentrations are quantified with the aim of defining the boundaries of sensory acceptability and antimicrobial efficacy of antimicrobials (Lambert *et al.*, 2001). Most of the studies on the calculation of MIC and NIC are semi-quantitative, whilst quantitative approaches have been mainly applied to studies primarily concerned with the antimicrobial activity of plant-origin antimicrobial agents, i.e. essential oils and their components.

The MIC and NIC are dependent on the experimental conditions. The influencing conditions include the incubation temperature, organism and inoculum size and therefore they should be reported in studies where MIC and NIC are evaluated (Lambert and Pearson, 2000). *In vitro* studies for identifying the MIC can be divided into groups such as diffusion, dilutions, impedance and optical density (or absorbance) methods (for examples, see Koutsoumanis *et al.*, 1999; Tassou *et al.*, 2000; Walsh *et al.*, 2003). Most of these evaluations are based on an end-point approach for evaluating the MIC, i.e. end result in which no growth is obtained for a test level of preservative, into which an inoculum of microbes is added. This kind of approach is considered semi-quantitative (Lambert and Pearson, 2000).

Lambert and Pearson (2000) examined the inhibitory activity of single compounds of EOs and developed a fully quantitative approach. This is given by the Lambert–Pearson model (LPM) inspired by a modified Gompertz equation (Equation 1) to evaluate the dose responses of microorganisms against several inhibitors. This modelling approach has already been examined for optical density (Lambert and Pearson, 2000) and impedance microbial measurements (Chorianopoulos *et al.*., 2006).

$$fa = \exp\left[-\left(\frac{x}{P_1}\right)^{P_2}\right] \quad (1)$$

fa is the fractional area, which is defined the ratio of inhibited growth to uninhibited growth as measured by the applied method (impedance, optical density, etc.), x is the inhibitor concentration (mg/l), P_1 is the concentration at maximum slope (of a log x versus *fa* plot) and P_2 is a slope parameter.

The MIC (Equation 2) and the NIC (Equation 3) can then be calculated as the intercept of the concentration axis to the tangent at the maximum gradient of the *fa*/log concentration curve and the intercept of the tangent at the maximum gradient of the *fa*/log concentration curve to the $fa = 1$ contour.

$$MIC = P_1 \cdot \exp\left(\frac{1}{P_2}\right) \quad (2)$$

$$NIC = P_1 \cdot \exp\left(\frac{1-e}{P_2}\right) \quad (3)$$

Guillier *et al.* (2007) developed another approach for evaluating the MIC based on the use of growth rate models. After estimation of the maximum specific growth rates (μ_{max}) from optical density growth kinetics by a modified Gompertz model, they assessed the antimicrobial concentration dependence on μ_{max} (Equation 4).

$$\sqrt{\mu_{\max}} = \sqrt{\mu_{\max}(c=0)\cdot f(c)} \quad (4)$$

With *f(c)* described either as Equation 5, i.e. SR_μ model or Equation 6, i.e. LP_μ model:

$$f(c) = \left(1 - \frac{c}{MIC}\right)^{\beta}, c < MIC \quad or \quad 0, c \geq MIC \quad (5)$$

$$f(c) = \exp - \left(\frac{c}{MIC/\exp\left(\frac{\ln(NIC/MIC)}{-e}\right)}\right)^{-e/(\ln(NIC/MIC))} \quad (6)$$

Where $\mu_{max}(c=0)$ is the growth rate in the absence of the antimicrobial ($c=0$) and β a shape parameter representing the sensitivity of the microorganism to an antimicrobial agent in Equation 5. These two approaches appeared to give equivalent results.

Lambert *et al.* (2001) argued that the majority of antimicrobial activity could be attributed to two components acting independently. Therefore, they also suggested another expression for a mixture of two inhibitors that could be extended in case there are more inhibitors, as presented in Equation 7:

$$fa_{x_i,\ldots,x_k} = \exp\left\{-\left[\left(\frac{x_i}{C_{i,1}}\right)^{C_{i,2}} + \ldots + \left(\frac{x_k}{C_{k,1}}\right)^{C_{k,2}}\right]^{C_Q}\right\} \quad (7)$$

Where parameters $C_{i,1}$ are the concentrations of the x_i inhibitors at the maximum slope. The main difference is that the current expression takes into account interactions between the antimicrobials which means that it could be considered for any additive, antagonistic and synergistic activity between the studied inhibitors. In that case, the MIC is then given by Equation 8:

$$MIC = C_{i,1} \cdot \exp\left(\frac{1}{C_{i,2} + C_Q}\right) \quad (8)$$

Accurate quantitative evaluations of MIC and NIC are important for designing effective preservation methods that are based on the use of the discussed antimicrobials. These quantitative methods can be exploited to give insight to optimal concentrations or combinations for real food systems by direct comparison of the antimicrobial efficacy of different antimicrobials, their individual or combined components or their mixtures and for efficient design of preservation for food products based on the principles of hurdle technology. These approaches have not received much attention for evaluating the MIC or the minimum bactericidal concentration of the antimicrobials of animal and microbial origin but their potential is evident.

Applications of Natural Antimicrobials in Food

The application of plant EOs for controlling the growth of food-borne pathogens and food-spoilage bacteria requires evaluation of the range of activity against the organisms of concern to a particular product, as well as any effects on a food's organoleptic properties. Plant EOs are usually mixtures of several components. Oils with high levels of eugenol (allspice, clove bud and leaf, bay, and cinnamon leaf), cinnamamic aldehyde (cinnamon bark, cassia oil) and citral are usually strong antimicrobials (Lis-Balchin *et al.*, 1998; Davidson and Naidu, 2000). The EOs from *Thymus* spp. possess significant quantities of phenolic monoterpenes, and have reported antiviral (Wild, 1994), antibacterial (Essawi and Srour, 2000; Cosentino *et al.*, 1999) and antifungal (Karaman, *et al.*, 2001; Pina-Vaz *et al.*, 2004) properties. The volatile terpenes carvacrol, p-cymene, γ-terpinene and thymol contribute to the antimicrobial activity of oregano, thyme and savory (Chorianopoulos *et al.*, 2004). The antimicrobial activity of sage and rosemary can be attributed to borneol and other phenolic compounds in the terpene fraction. Davidson and Naidu (2000) reported that the terpene thejone was responsible for the antimicrobial activity of sage, whereas in rosemary a group of terpenes (borneol, camphor, 1,8 cineole, a-pinene, camphone, verbenonone and bornyl acetate) was responsible. Plant EOs such as cumin, caraway and coriander have inhibitory effects against organisms such as *Aeromonas hydrophila, Pseudomonas fluorescens* and *Staphylococcus aureus* (Wan *et al.*, 1998; Fricke *et al.*, 1998), marjoram and basil have high activity against *Bacillus cereus, Enterobacter aerogenes, E. coli* and *Salmonella,* and lemon balm and sage EOs appear to have adequate activity against *L. monocytogenes and S. aureus* (Gutierrez *et al.*, 2008b). Gutierrez *et al.* (2008b) showed that oregano and thyme EOs had comparatively high activity against enterobacteria, lactic acid bacteria, *B. cereus* and *Pseudomonas* spp., although in general *Pseudomonas* species are consistently highly resistant to plant antimicrobials (Matasyoh *et al.*, 2007; Gutierrez *et al.*, 2008b). This may be attributed to the production of exopolysaccharide (EPS) layers forming biofilms that can delay penetration of the antimicrobial agent (Mah and O'Toole, 2001). Thus, the effect of plant-derived antimicrobials against *Pseudomonas* spp. may be improved by combined treatments to permeabilize the EPS and/or the outer membrane (Gutierrez *et al.*, 2008b). Lee *et al.* (2003) investigated the antibacterial activity of vegetables and juices and concluded that green tea and garlic extracts have broad applications as antibacterial agents against a wide range of pathogens. Arrowroot tea extract has reported antimicrobial activity against *E. coli* O157:H7 (Kim and Fung, 2004). Ibrahim *et al.* (2006) reported the potential of caffeine at a concentration of

0.5% or higher as an effective antimicrobial agent for the treatment of *E. coli* O157:H7 infection.

Hao *et al.* (1998) studied the efficacy of a range of plant extracts for inhibition of *Aeromonas hydrophila* and *L. monocytogenes* in refrigerated cooked poultry and found that eugenol reduced pathogen counts by 4 Log_{10} cfu/g over a 14-day-storage trial. Similarly, 1–2% w/w clove oil inhibited the growth of a range of *Listeria* spp. in chicken frankfurters over 2 weeks at 5°C (Mytle *et al.*, 2006). Conversely, Shekarforoush *et al.* (2007) found that EOs of oregano and nutmeg were effective against *E. coli* O157:H7 in a broth system, but had no effect in ready-to-cook chicken. Careaga *et al.* (2003) recorded that 1.5 ml/100 g of capsicum extract was sufficient to prevent the growth of *Salmonella typhimurium* in raw beef, but that 3 ml/100 g was required for a bactericidal effect against *Pseudomonas aeriginosa*. Ahn *et al.* (2007) also found a range of plant extracts to be useful for reduction of pathogens associated with cooked beef and quality maintenance; however, Uhart *et al.* (2006) concluded that, when in direct contact, spices inactivated *S. typhimurium* DT104, but that the activity decreased considerably when added to a complex food system such as ground beef. Gutierrez *et al.* (2008a, 2009) concluded that plant essential oils might be more effective against food-borne pathogens and spoilage bacteria when applied to ready-to-use foods containing a high protein level at acidic pH, as well as lower levels of fats or carbohydrates and moderate levels of simple sugars. The success of plant-derived antimicrobials when applied to fruit and vegetable products is also mixed. Karapinar and Sengun (2007) recommended unripe grape juice as an alternative antimicrobial agent for enhancing the safety of salad vegetables and Martinez-Romero *et al.* (2007) suggested that carvacrol could be applied as a novel approach to control the fungal growth of grapes. Although Valero and Frances (2006) found that low concentrations of carvacrol, cinnamaldehyde or thymol had a clear antibacterial effect against *B. cereus* in a carrot broth, they found that the storage temperature was significant, where only cinnamaldehyde retained activity at 12°C. Gutierrez *et al.* (2009) found that the efficacy of oregano EO was comparable with chlorine as a decontamination treatment for ready-to-eat carrots, while retaining acceptability in terms of sensory quality and appreciation. A novel application of plant extracts is for the production of chocolate; Kotzekidou *et al.* (2007) reported enhanced inhibitory effects of plant extracts against an *E. coli* cocktail at 20°C, indicating potential applications at higher storage temperatures.

Challenges for Food Applications

The extrapolation of results obtained from *in vitro* experiments with laboratory media to food products is not straightforward as foods are complex, multicomponent systems consisting of different interconnecting micro-environments. Though there is vast potential for natural-based antimicrobial agents in food preservation, most of the literature presents inactivation data from model foods or laboratory media. Table 14.2 reports inactivation studies in real food systems. The level of natural preservatives required for sufficient efficacy in food products in comparison with laboratory media may be considerably higher, which may impact negatively on the organoleptic properties of food. To promote optimal practical application, a sequential product or process-specific approach may be necessary. The matching of a range of potential antimicrobials with the spectrum of activity appropriate to the microbiological issues of concern to the product, followed by studies investigating the specific food matrices effects on antimicrobial efficacy coupled with concurrent organoleptic acceptability analysis could provide a more systematic approach to ensuring more successful practical application in the food product or process.

Combination with Antimicrobial Agents

Investigations based on combinations of natural antimicrobials with other non-thermal processing technologies are warranted to counteract any potential organoleptic or textural effects on food products as well as optimizing microbial inactivation. Examples of the synergistic effects that can be obtained using mild traditional preservation techniques in conjunction with novel food-processing technologies are to date better studied *in vitro*, but require further investigation in food products to ensure successful practical application. The combination of plant EOs with modified atmosphere packaging for control of spoilage species was reported by Skandamis and Nychas (2001) and Matan *et al.* (2006). Seydim and Sarikus (2006) also investigated the use of EOs in an active packaging system based on an edible whey protein film and concluded that oregano was the most effective EO against a range of food pathogens. Allyl isothiocyanate was successfully applied to chopped, refrigerated, nitrogen-packed beef for control of *E. coli* at levels in excess of 1000 ppm. Scollard *et al.* (2009) studied the effects of EO treatment, gas atmosphere and storage temperature on *L. monocytogenes* in a model vegetable system and found that increasing CO_2 levels and lowering storage temperatures further enhance these anti-listerial effects of thyme and oregano EOs.

Conclusions and Future Trends

Interest in natural antimicrobials has expanded in recent years in response to consumer demand for 'greener' additives. During the past two decades natural preservatives have been investigated for practical applications. These technologies have been shown to inactivate microorganisms and enzymes without significant adverse effects on organoleptic or nutritional properties. Reported studies have demonstrated that natural antimicrobial agents described in this chapter may offer unique advantages for food processing. In addition to improving the shelf life and safety of foods, natural antimicrobial agents may allow novel food products with enhanced quality and nutritional properties to be introduced to the market. The applications of plant-based antimicrobial agents are likely to grow steadily in the future because of greater consumer demands for minimally processed foods and those containing naturally derived preservation ingredients. More complex considerations arise for combinations of technologies, particularly with respect to optimization of practical applications. Intelligent selection of appropriate systems based on detailed, sequential studies and quantitative approaches to evaluate the efficiency of antimicrobials is necessary. The impact of product formulation, extrinsic storage parameters and intrinsic product parameters on the efficacy of novel applications of combined non-thermal systems requires further study.

Table 14.2. Effect of natural antimicrobial agents on food preservation and quality.

Food product	Antimicrobial agent	Microbial inactivation	Quality attributes	Reference
Fruit yogurt	Vanillin (2000ppm)	Yeast, bacterial (delays growth)	Shelf life (↑)	Penney *et al.*, 2004
Tomato juice	Clove oil (0.1%) Mint extract (1.0%) Nisin (0.004%)	Total plate count (3.9 LR) Total plate count (8.34 LR) Total plate count (↓)	Shelf life (↑), Vitamin C (~)	Nguyen and Mittal, 2007
Ready-to-eat fruit salad	Citral (25–125ppm) Citron (300–900ppm)	Yeasts and lactic acid bacteria (LAB) (delays growth)	Shelf life (↑)	Belletti *et al.*, 2008
	Citron (600ppm)	*S. Enteritidis* E4 (2 LR), *E. coli* 555 (<4.5 LR) *L. monocytogenes* Scott A (4 LR)	Sensory characteristics (~)	
Raspberries	Methyl jasmonate (MJ), Allyl isothiocyanate (AITC) EO of *Melaleuca alternifolia* (tea tree oil)		AC (↑) AC (↓) AC (↑)	Chanjirakul *et al.*, 2006
Lettuce	Thyme oil (1 ml/l)	*E. coli* (6.32LR)		Singh *et al.*, 2002
Baby carrot	Thyme oil (1 ml/l)	*E. coli* (5.57LR)		
Minimally processed carrots	Oregano oil (250 ppm)	Background spoilage microflora TVC (>1 LR) LAB (>1 LR) *Pseudomonas* (<1 LR)	Sensory characteristics (~)	Gutierrez *et al.*, 2009
Minimally processed vegetables	Thyme oil (1%)	*Aeromonas spp* (2 LR) Psychrotrophic aerobic plate count (4.19 LR) Plate count agar (5.44 LR)	Sensory properties (↓), Shelf life (↑)	Uyttendaele *et al.*, 2004
Chicken meat	EOs of mustard oil	*Lactobacillus alimentarius* (~) *Brochothrix thermosphacta* (~) *Lactobacillus alimentarius* (delays growth)	Proximate composition (~), Shelf life (↑)	Lemay *et al.*, 2002

Fish	EOs (0.5% carvacrol + 0.5% thymol)	TVC (2.5 LR)	Shelf life (↑), lipid oxidation (↓) Sensory characteristics (~)	Mahmoud *et al.*, 2006
Red meat	Tea catechins (300 mg/kg)		Shelf life (↑), lipid oxidation (↓)	Tang *et al.*, 2001
Beef hot dog	Clove oil (5 ml/l)	*L monocytogenes* (1.15–1.71 LR)		Singh *et al.*, 2003
	Thyme oil (1 ml/l)	*L. monocytogenes* (0.67–1.05 LR)		
Minced beef	*Capsicum annum* extract	*Salmonella typhimurium* (MLC 15 g/kg) *Pseudomonas aeruginosa* (MLC 30 g/kg)		Careaga *et al.*, 2003
Chicken frankfurter	Clove oil (1% v/w)	*L. monocytogenes* (4.5 LR)		Mytle *et al.*, 2006
Cooked beef	Grape seed extract (1%)	*E. coli* (1.7 LR) *S. Typhimurium* (2.0 LR) *L. monocytogenes* (0.8 LR) *Aeromonas hydrophila* (0.4 LR)	TBA (↓), Colour (~), Lipid oxidation (↓)	Ahn *et al.*, 2007

AU, arbitrary units were defined as the reciprocal of the highest twofold dilution that did not allow the growth of the indicator strain. AC, Anthocyanin content; '↑' and '↓' indicates increase and decrease, respectively, while ~ shows no significant difference. LR, Log reduction; MLC, minimum lethal concentration; TBA, thiobarbituric acid.

References

Ahn, J., Grun, I.U. and Mustapha, A. (2007) Effect of plant extracts on microbial growth, color change, and lipid oxidation in cooked beef. *Food Microbiology* 24, 7–14.

Akgül, A. and Kivanç M. (1988) Inhibitory effects of selected Turkish spices and oregano components on some foodborne fungi. *International Journal of Food Microbiology* 6, 263–268.

Alzoreky, N.S. and Nakahara, K. (2002) Antimicrobial activity of extracts from some edible plants commonly consumed in Asia. *International Journal of Food Microbiology* 80, 223–230.

Ayala-Zavala, J.F., Soto-Valdez, H., Gonzalaz-Leon, A., Alvarez-Parrilla, E., Martin-Belloso, O. and Gonzalez-Aguilar, G.A. (2008) Microencapsulation of cinnamon leaf (*Cinnamomum zeylanicum*) and garlic (*Allium sativum*) oils in β-cyclodextrin. *Journal of Inclusion Phenomena and Macrocyclic Chemistry* 60, 359–368.

Ayala-Zavala, J.F., Gonzalez-Aguilar, G.A. and Del-Toro-Sanchez, L. (2009) Enhancing safety and aroma appealing of fresh-cut fruits and vegetables using the antimicrobial and aromatic power of essential oils. *Journal of Food Science* 74, 84–91.

Bajpai, V.K., Rahman, A., Dung, N.T., Huh, M.K. and Kang, S.C. (2008) *In vitro* inhibition of food spoilage and foodborne pathogenic bacteria by essential oil and leaf extracts of *Magnolia liliflora* Desr., *Journal of Food Science*, 73, 314–320.

Barbosa-Canovas, G.V., Pothakamury, U.H., Palou, E. and Swanson, B.G. (1997). *Nonthermal Preservation of Foods.* Marcel Dekker, New York.

Belletti, N., Lanciotti, R., Patrignani, F. and Gardini, F. (2008) Antimicrobial efficacy of citron essential oil on spoilage and pathogenic microorganisms in fruit-based salads. *Journal of Food Science* 73, M331–M338.

Beuchat, L.R. (1992) Surface disinfection of raw produce. *Dairy Food Environmental Sanitation* 12, 6–9.

Beuchat, L.R. and Golden, D.A. (1989). Antimicrobials occurring naturally in foods. *Food Technology*, 43 (1), 134–142.

Burt S. (2004) Essential oils: their antibacterial properties and potential applications in foods—a review. *International Journal of Food Microbiology* 94, 223–253.

Burt, S.A. and Reinders, R.D. (2003) Antibacterial activity of selected plant essential oils against *Escherichia coli* O157:H7. *Letters in Applied Microbiology*, 36, 162–167.

Campo, J. D., Amiot, M.J. and Nguyen-The, C. (2000) Antimicrobial effect of rosemary extracts. *Journal of Food Protection* 63, 1359–1368.

Careagaa, M., Fernandez, E., Dorantesa, L., Mota, L., Jaramillo, M.E. and Hernandez-Sancheza, H. (2003) Antibacterial activity of Capsicum extract against *Salmonella typhimurium* and *Pseudomonas aeruginosa* inoculated in raw beef meat. *International Journal of Food Microbiology* 83, 331–335.

Carson, C.F., Hammer, K.A. and Riley, T.V. (1995) Broth microdilution method for determining the susceptibility of *Escherichia coli* and *Staphylococcus aureus* to the essential oil of *Melaleuca alternifolia* (tea tree oil). *Microbios* 82, 181–185.

Chang S.T., Chen, P.F. and Chang, S.C. (2001) Antibacterial activity of leaf essential oils and their constituents from *Cinnamomum osmophloeum. Journal of Ethnopharmacology* 77, 123–127.

Chanjirakul, K., Wang, C.Y., Wang, S.Y. and Siriphanich, J. (2006) Effect of natural volatile compounds on antioxidant capacity and antioxidant enzymes in raspberries. *Postharvest Biology Technology* 40,106–115.

Charai, M., Mosaddak, M. and Faid, M. (1996) Chemical composition and antimicrobial activities of two aromatic plants: *Origanum majorana* L. and *O. compactum* Benth. *Journal of Essential Oil Research* 8, 657–664.

Chorianopoulos, N., Kalpoutzakis, E., Aligiannis, N., Mitaku, S., Nychas, G.J. and Haroutounian, S.A. (2004) Essential oils of *Satureja*, *Origanum*, and *Thymus* species: Chemical composition and antibacterial activities against foodborne pathogens. *Journal of Agriculture and Food Chemistry* 52, 8261–8267.

Chorianopoulos, N., Evergetis, E., Mallouchos, A., Kalpoutzakis, E., Nychas, G.J. and Haroutounian, S.A. (2006) Characterization of the essential oil volatiles of *Satureja thymbra* and *Satureja parnassica*: Influence of harvesting time and antimicrobial activity. *Journal of Agriculture and Food Chemistry* 54, 3139–3145.

Connor, D.E. and Beuchat, L.R. (1984) Sensitivity of heat stressed yeasts to essential oils of plants. *Applied Environmental Microbiology* 47, 229–233.

Cosentino, S., Tuberoso, C.I.G., Pisano, B., Satta, M., Mascia, V. and Arzedi, E. (1999) In-vitro antimicrobial activity and chemical composition of Sardinian Thymus essential oils. *Letters in Applied Microbiology* 29, 130–135.

Davidson, P.M. (2001) Chemical preservatives and naturally antimicrobial compounds. In: Doyle, M.P., Beuchat, L.R. and Montville, T.J. (eds) *Food Microbiology. Fundamentals and Frontiers* 2nd edn. ASM Press, Washington DC, pp. 593–628.

Davidson, P.M. and Naidu, A.S. (2000). Phyto-phenols. In: Naidu, A.S. (ed.) *Natural Food Antimicrobial Systems*. CRC Press, Boca Raton, Florida, pp. 265–294.

Deans, S.G., Noble, R.C., Hiltunen, R., Wuryani, W. and Penzes, L.G. (1995) Antimicrobial and antioxidant properties of *Syzygium aromaticum* (L.) Merr. & Perry: impact upon bacteria, fungi and fatty acid levels in ageing mice. *Flavour and Fragrance Journal*, 10, 323–328.

Delaquis, P.J. and Mazza, G. (1995) Antimicrobial properties of isothiocyanates in food preservation. *Food Technology* 49, 73–84.

Delaquis P.J., Stanich K., Girard, B. and Mazza, G. (2002) Antimicrobial activity of individual and mixed fractions of dill, cilantro, coriander and eucalyptus essential oils. *International Journal of Food Microbiology* 74, 101–109.

Demetzos, C. and Perdetzoglou, D.K. (2001) Composition and antimicrobial studies of the oils of *Origanum calcaratum* Juss and *O. scabrum* Boiss from Greece. *Journal of Essential Oil Research* 13, 460–462.

Demo, M.S. and de las Mercedes Oliva, M. (2009) Antimicrobial activity of medicinal plants from South America. In: *Botanical Medicine in Clinical Practice*. CABI, Wallingford, Oxfordshire, UK, pp. 152–163.

Dorman, H.J.D. and Deans, S.G. (2000) Antimicrobial agents from plants: antibacterial activity of plant volatile oils. *Journal of Applied Microbiology* 88, 308–316.

Elgayyar, M., Draughon, F.A., Golden, D.A. and Mount, J.R. (2001) Antimicrobial activity of essential oils from plants against selected pathogenic and saprophytic microorganisms. *Journal of Food Protection* 64, 1019–1024.

Essawi, T. and Srour, M. (2000) Screening of some Palestinian medicinal plants for antibacterial activity. *Journal of Ethnopharmacology* 70, 343–349.

Farag, R.S., Daw, Z.Y., Hewed, F.M. and El-Baroty, G.S.A. (1989) Antimicrobial activity of some Egyptian spice oils. *Journal of Food Protection* 52, 665–667.

Firouzi, R., Azadbakht, M. and Nabinedjad, A. (1998) Anti-listerial activity of essential oils of some plants. *Journal of Applied Animal. Research* 14, 75–80.

Fisher, K and Phillips, C (2008) Potential antimicrobial uses of essential oils in food: is citrus the answer? *Trends in Food Science & Technology* 19, 156–164.

Fricke, G., Hoyer, H., Wermter, R. and Paulus, H. (1998) *Staphylococcus aureus* as an example of the influence of lipophilic components on the microbiological activity of aromatic extracts. *Arch. Lebensmitteltechn.* 49, 107–111.

Fung, D.Y.C., Taylor, S. and Kahan, J. (1977) Effects of butylated hydroxyanisole (BHA) and butylated hydroxitoluene (BHT) on growth and aflatoxin production of *Aspergillus flavus*. *Journal of Food Safety* 1, 39–51.

Gómez-Estaca, J., López de Lacey, A., López-Caballero, M.E., Gómez-Guillén, M.C. and Montero, P. (2010) Biodegradable gelatin-chitosan films incorporated with essential oils as antimicrobial agents for fish preservation. *Food Microbiology* 27, 889–896.

Griffin, S.G., Markham, J.L, and Leach, D.N. (2000) An agar dilution method for the determination of the minimum inhibitory concentration of essential oils. *Journal of Essential Oil Research* 12, 249–255.

Guillier, L., Naser, A.I. and Dubois-Brissonnet, F. (2007) Growth response of *Salmonella Typhimurium* in the presence of natural and synthetic antimicrobials: Estimation of MICs from three different models. *Journal of Food Protection* 70, 2243–2250.

Gutierrez, J., Rodriguez, G., Barry-Ryan, C. and Bourke, P. (2008a) Efficacy of plant essential oils against foodborne pathogens and spoilage bacteria associated with ready-to-eat vegetables: antimicrobial and sensory screening. *Journal of Food Protection* 71, 1846–1854.

Gutierrez, J., Rodriguez, G., Barry-Ryan, C. and Bourke, P. (2008b) The antimicrobial efficacy of plant essential oil combinations and interactions with food ingredients. *International Journal of Food Microbiology* 124, 91–97.

Gutierrez, J., Barry-Ryan, C. and Bourke, P. (2009a) Antimicrobial activity of plant essential oils using food model media: efficacy, synergistic potential and interactions with food components. *Food Microbiology* 26, 142–152.

Gutierrez, J., Bourke P., Lonchamp, J. and Barry-Ryan, C. (2009b) Impact of plant essential oils on microbialogical, organoleptic and quality markers of minimally processed vegetables. *Innov. Food Science Innovative and Emerging Technology* 10, 195–202.

Guynot, M.E., Ramos, A.J., Seto, L., Purroy, P., Sanchis, V. and Marin, S. (2003) Antifungal activity of volatile compounds generated by essential oils against fungi commonly causing deterioration of bakery products. *Journal of Applied Microbiology* 94, 893–899.

Hammer, K.A., Carson, C.F. and Riley, T.V. (1999) Antimicrobial activity of essential oils and other plant extracts. *Journal of Applied Microbiology* 86, 985–990.

Hammer, K.A., Carson, C.F. and Riley, T.V. (1999) Antimicrobial activity of essential oils and other plant extracts. *Journal of Applied Microbiology* 86, 985–990.

Hao, Y.Y., Brackett, R.E. and Doyle, M.P. (1998) Efficacy of plant extracts in inhibiting *Aeromonas hydrophila* and *Listeria monocytogenes* in refrigerated cooked poultry. *Food Microbiology* 15, 367–378.

Helander, I.M., Alakomi, H.L., Latva-Kala, K., Mattila-Sandholm, T., Pol, I., Smid, E.J., Gorris, L.G.M. and von Wright, A. (1998) Characterisation of the action of selected essential oil components on gram-negative bacteria. *Journal of Agricultural and Food Chemistry* 46, 3590–3595.

Holley, R.A. and Patel, D. (2005) Improvement in shelf-life and safety of perishable foods by plant essential oils and smoke antimicrobials. *Food Microbiology* 22, 273–292.

Ibrahim, S.A., Salameh, M.M., Phetsomphou, S., Yang, H. and Seo, C.W. (2006) Application of caffeine, 1,3,7-trimethylxanthine, to control *Escherichia coli* O157:H7. *Food Chemistry* 99, 645–650.

Janssen, A., Scheffer, J. and Baerheim-Svendsen, A. (1986) Antimicrobial activity of essential oils: A 1976–1986 literature review on aspects of test methods. *Planta Medica*, 53, 395–398.

Juliano, C., Mattana, A. and Usai, M. (2000) Composition and *in vitro* antimicrobial activity of the essential oil of *Thymus herba-barona* Loisel growing wild in Sardinia. *Journal of Essential Oil Research* 12, 516–522.

Juven, B.J., Kanner, J., Schved, F. and Weisslowicz, H. (1994) Factors that interact with the antibacterial action of thyme essential oil and its active constituents. *Journal of Applied Bacteriology* 76, 626–631.

Kabara, J.J. and Eklund, T. (1991) Organic acids and esters. In: Russel, N.J. and Gould, G.W. (eds) *Food Preservatives.* Blackie & Son Ltd, Glasgow, pp. 44–71.

Karaman, S., Digrak, M., Ravid, U. and Ilcim, A. (2001) Antibacterial and antifungal activity of the essential oils of *Thymus revolutus* Celak from Turkey. *Journal of Ethnopharmacology* 76, 183–186.

Karapinar, M. and Sengun, I.Y. (2007) Antimicrobial effect of koruk juice against *Salmonella typhimurium* on salad vegetables. *Food Control* 18, 702–706.

Kim, H.Y., Lee, Y.J., Hong, K.-H., Kwon, Y.-K., Sim, K.-C., Lee, J.-Y., Cho, H.-Y., Kim, I.-S., Han, S.-B., Lee, C.-W., Shin, I.-S. and Cho, J.S. (2001) Isolation of antimicrobial substances from natural products and their preservative effects. *Food Science Biotechnology*, 10, 59–71.

Kim, J., Marshall, M.R. and Wei, C. (1995) Antimicrobial activity of some essential oil components against five food borne pathogens. *Journal of Agricultural and Food Chemistry* 43, 2839–2845.

Kim, S. and Fung, D.Y. (2004) Antibacterial effect of water soluble arrowroot (*Puerariae radix*) tea extract on foodborne pathogens in ground beef and mushroom soup. *Journal of Food Protection* 67, 1953–1956.

Kotzekidou, P., Giannakidis, P. and Boulamatsis, A. (2007) Antimicrobial activity of some plant extracts and essential oils against foodborne pathogens *in vitro* and on the fate of inoculated pathogens in chocolate. *LWT – Food Science and Technology* 41, 119–127.

Koutsoumanis, K., Lambropoulou, K. and Nychas, G.J.E. (1999) A predictive model for the non-thermal inactivation of *Salmonella enteritidis* in a food model system supplemented with a natural antimicrobial. *International Journal of Food Microbiology* 49, 67–74.

Lambert, R.J.W. and Pearson, J. (2000) Susceptibility testing: accurate and reproducible minimum inhibitory concentration (MIC) and non-inhibitory concentration (NIC) values. *Journal of Applied Microbiology* 91, 453–462.

Lambert, R.J.W., Skandamis, P.N., Coote, P.J. and Nychas, G.J.E. (2001) A study of the minimum inhibitory concentration and mode of action of oregano essential oil, thymol and carvacrol. *Journal of Applied Microbiology* 91, 453–462.

Lee, D.U., Heinz, V. and Knorr, D. (2003) Effects of combination treatments of nisin and high-intensity ultrasound with high pressure on the microbial inactivation in liquid whole egg. *Innovative Food Science & Emerging Technologies* 4, 387–393.

Lee, S.J., Umano, K., Shibamoto, T. and Lee, K.G. (2005) Identification of volatile components in basil (*Ocimum basilicum* L.) and thyme leaves (*Thymus vulgaris* L.) and their antioxidant properties. *Food Chemistry* 91, 131–137.

Lemay, M.-J., Choquette, J., Delaquis, P.J., Gariepy, C., Rodrigue, N. and Saucier, L. (2002) Antimicrobial effect of natural preservatives in a cooked and acidified chicken meat model. *International Journal of Food Microbiology* 78, 217–226.

Lis-Balchin, M., Deans, S.G. and Eaglesham, E. (1998) Relationship between bioactivity and chemical composition of commercial essential oils. *Flavour and Fragrance Journal* 13, 98–104.

Lopez-Malo, A., Alzamora, S.M. and Guerrero, S. (2000) Natural antimicrobials from plants. In: Alzamora, S.M., Tapia, M.S. and Lopez-Malo, A. (eds) *Minimally Processed Fruits and Vegetables. Fundamentals Aspects and Applications.* Aspen Publishers, Gaithersburg, Maryland pp. 237–264.

Lopez-Malo, A., Alzamora, S.M. and Palou, E. (2005) Naturally occurring compounds – plant sources. In: Davidson, P.M., Sofos, J.N. and Branen, A.L. (eds) *Antimicrobials in Food,* 3rd edn. CRC Press, New York, pp. 429–251.

Mah, T.-F.C. and O'Toole, G.A. (2001) Mechanisms of biofilm resistance to antimicrobial agents. *Trends in Microbiology* 9, 34–39.

Mahmoud, B.S.M., Yamazaki, K., Miyashita, K., Shin, I.I. and Suzuki T. (2006) A new technology for fish preservation by combined treatment with electrolyzed NaCl solutions and essential oil compounds. *Food Chemistry* 99, 656–662.

Marino, M., Bersani, C. and Comi, G. (2002) Impedance measurement to study antimicrobial activity of essential oils from Lamiaceae and Compositae. *International Journal of Food Microbiology* 67, 187–195.

Martinez-Romero, D., Guillen, F., Valverde, J.M., Bailen, G., Zapata, P., Serrano, M., Castillo, S. and Valero, D. (2007) Influence of carvacrol on survival of *Botrytis cinerea* inoculated in table grapes. *International Journal of Food Microbiology* 115, 144–148.

Matan, N., Rimkeeree, H., Mawson, A.J., Chompreeda, P., Haruthaithanasan, V. and Parker, M. (2006) Antimicrobial activity of cinnamon and clove oils under modified atmosphere conditions. *International Journal of Food Microbiology,* 107, 180–185.

Matasyoh, J.C., Kiplimo, J.J., Karubiu, N.M. and Hailstorks, T.P. (2007) Chemical composition and antimicrobial activity of essential oil of *Tarchonanthus camphorates. Food Chemistry* 101, 1183–1187.

Mytle, N., Anderson, G.L., Doyle, M.P. and Smith, M.A. (2006) Antimicrobial activity of clove (*Syzgium aromaticum*) oil in inhibiting *Listeria monocytogenes* on chicken frankfurters. *Food Control* 17, 102–107.

Nguyen, P. and Mittal, G.S. (2007) Inactivation of naturally occurring microorganisms in tomato juice using pulsed electric field (PEF) with and without antimicrobials. *Chemical Engineering and Processing* 46, 360–365.

Nielsen, P.V. and Rios, R. (2000) Inhibition of fungal growth on bread by volatile components from spices and herbs, and their possible application in active packaging, with special emphasis on mustard essential oil. *International Journal of Food Microbiology* 60, 219–229.

Nychas, G.J.E. (1995) Natural antimicrobials from plants. In: Gould, G.W. (ed.) *New Methods of Food Preservation.* Blackie Academic & Professional, Glasgow, pp. 58–89.

Nychas, G.J.E. and Skandamis, P.N. (2003) Antimicrobials from herbs and spices. In: Roller, S. (ed.) *Natural Anti-microbials for the Minimal Processing of Foods.* Woodhead Publishing, Cambridge, UK, pp. 176–200.

Opalchenova, G. and Obreshkova, D. (2003) Comparative studies on the activity of basil – an essential oil from *Ocimum basilicum* L. – against multidrug resistant clinical isolates of the genera *Staphylococcus, Enterococcus* and *Pseudomonas* by using different test methods. *Journal of Microbiological Methods* 54, 105–110.

Packiyasothy, E.V. and Kyle, S. (2002) Antimicrobial properties of some herb essential oils. *Food Australia* 54, 384–387.

Pandit, V.A. and Shelef, L.A. (1994) Sensitivity of *Listeria monocytogenes* to rosemary (*Rosemarinus officinalis* L). *Food Microbiology* 11, 57–63.

Penney, V., Henderson, G., Blum, C. and Johnson-Green, P. (2004) The potential of phytopreservatives and nisin to control microbial spoilage of minimally processed fruit yogurts. *Innovative Food Science and Emerging Technology* 5, 369– 375

Pina-Vaz, C., Gonc, A., Rodrigues, A., Pinto, E., Costa-de-Oliveira, S., Tavares, C. and Salgueiro, L. (2004) Antifungal activity of Thymus oils and their major compounds. *Journal of European Academy of Dermatology and Venereology* 18, 73–78.

Rauha, J.P., Remes, S., Heinonen, M., Hopia, A., Kahkonen, M, Kujala, T., Pihlaja, K., Vuorela, H. and

Vuorela, P. (2000) Antimicrobial effects of Finnish plant extracts containing flavonoids and other phenolic compounds. *International Journal of Food Microbiology* 56, 3–12.

Remmal, A., Bouchikhi, T., Rhayour, K. and Ettayebi, M. (1993) Improved method for the determination of antimicrobial activity of essential oils in agar medium. *Journal of Essential Oil Research* 5, 179–184.

Rico-Munoz, E., Bargiota, E. and Davidson, P.M. (1987) Effect of selected phenolic compounds on the membrane-bound adenosine triphosphate of *Staphylococcus aureus*. *Food Microbiology* 4, 239–249.

Rojas-Graü, M.A., Avena-Bustillos, R. J., Friedman, M., Henika, P. R., Martín-Belloso, O. and McHugh, T.H. (2006) Effect of antimicrobial plant essential oils and oil compounds on physical properties of apple puree-pectin based edible films and coatings. IFT Annual Meeting, Abstract 389.

Rojas-Graü, M.A., Avena-Bustillos, R.J., Olsen, C., Friedman, M., Henika, P.R. and Martin-Belloso, O. (2007) Effects of plant essential oils and oil compounds on mechanical, barrier and antimicrobial properties of alginate – apple puree edible films. *Journal of Food Engineering* 81, 634–641.

Rojas-Graü, M.A., Soliva-Fortuny, R. and Belloso, O. (2009) Edible coatings to incorporate active ingredients to freshcut fruits: a review, *Trends in Food Science and Technology* 20, 438–447.

Roller, S. and Covill, N. (1999) The antifungal properties of chitosan in laboratory media and apple juice. *International Journal of Food Microbiology* 47, 67–77.

Ross, Z.M., O'Gara, E.A., Hill, D.J., Sleightholme, H.V. and Maslin, D.J. (2001) Antimicrobial properties of garlic oil against human enteric bacteria: evaluation of methodologies and comparisons with garlic oil sulfides and garlic powder. *Applied Environmental Microbiology* 67, 475–480.

Scollard, J., Francis, G.A. and O'Beirne, D. (2009) Effects of essential oil treatment, gas atmosphere, and storage temperature on *Listeria monocytogenes* in a model vegetable system. *Journal of Food Protection* 72, 1209–1215.

Seydim, A.C. and Sarikus, G. (2006) Antimicrobial activity of whey protein based edible films incorporated with oregano, rosemary and garlic essential oils. *Food Research International* 39, 639–644.

Shekarforoush, S.S., Nazer, A.H.K., Firouzi, R. and Rostami, M. (2007) Effects of storage temperatures and essential oils of oregano and nutmeg on the growth and survival of *Escherichia coli* O157:H7 in barbecued chicken used in Iran. *Food Control* 18, 1428–1433.

Shelef, L.A. (1983) Antimicrobial effects of spices. *Journal of Food Safety* 6, 29–44.

Singh, N., Singh, R.K., Bhunia, A.K. and Stroshine, R.L. (2002) Efficacy of chlorine dioxide, ozone, and thyme essential oil or a sequential washing in killing *Escherichia coli* O157:H7 on lettuce and baby carrots. *LWT-Food Science Technology* 35, 720–729.

Singh, N., Singh, R.K., Singh, A. and Bhuniab, A.K. (2003) Efficacy of plant essential oils as antimicrobial agents against *Listeria monocytogenes* in hotdogs. *LWT-Food Science Technology* 36, 787–794.

Sivropoulou, A., Papanikolaou, E., Nikolaou, C., Kokkini, S., Lanaras, T. and Arsenakis, M. (1996) Antimicrobial and cytotoxic activities of Origanum essential oils. *Journal of Agricultural and Food Chemistry* 44, 1202–1205.

Skandamis, P. and Nychas, G.-J. (2001) Effect of oregano essential oil on microbiological and physico-chemical attributes of minced meat stored in air and modified atmospheres. *Journal of Applied Microbiology* 91, 1011–1022.

Skandamis, P., Tsigarida, E. and Nychas, G.-J.E. (2002) The effect of oregano essential oil on survival/death of *Salmonella typhimurium* in meat stored at 5 1C under aerobic, VP/MAP conditions. *Food Microbiology*, 19, 97–103.

Smith-Palmer, A., Stewart, J. and Fyfe, L. (1998) Antimicrobial properties of plant essential oils and essences against five important foodborne pathogens. *Letters in Applied Microbiology* 26, 118–122.

Sofos, J., Beuchat, L.R., Davidson, P.M and Johnson, E.A. (1998) Naturally occurring antimicrobials in food. Task force report. *Council of Agricultural Science and Technology*, Ames, IA., 103.

Tang, S., Kerry, J.P., Sheehan, D., Buckley, J. and Morrissey, P.A. (2001) Antioxidative effect of added tea catechins on susceptibility of cooked red meat, poultry and fish patties to lipid oxidation. *Food Research International* 34, 651–657.

Tassou, C.C., Koutsoumanis, K. and Nychas, G.-J.E. (2000) Inhibition of *Salmonella enteritidis* and *Staphylococcus aureus* in nutrient broth by mint essential oil. *Food Research International* 33, 273–280.

Ting, W.T.E. and Diebel, K.E. (1992) Sensitivity of *L. monocytogenes* to spices at two temperatures. *Journal of Food Safety* 12, 129–137.

Tiwari, B.K., Valdramidis, V.P., O'Donnell, C.P., Muthukumarappan, K., Bourke, P. and Cullen, P.J. (2009) Application of natural antimicrobials for food preservation. *Journal of Agricultural & Food Chemistry,* 57, 5987–6000.

Tyagi, A.K. and Malik, A. (2010) Antimicrobial action of essential oil vapours and negative air ions against *Pseudomonas fluorescens. International Journal of Food Microbiology* 143, 205–210.

Uhart, M., Maks, N. and Ravishankar, S. (2006) Effect of spices on growth and survival of *Salmonella typhimurium DT 104* in ground beef stored at 4 and 8°C. *Journal of Food Safety* 26, 115–125.

Uyttendaele, M., Neyts, K., Vanderswalmen, H., Notebaert, E. and Debevere, J. (2004) Control of *Aeromonas* on minimally processed vegetables by decontamination with lactic acid, chlorinated water, or thyme essential oil solution. *International Journal of Food Microbiology* 90, 263–271.

Valero, M. and Francés, E. (2006) Synergistic bactericidal effect of carvacrol, cinnamaldehyde or thymol and refrigeration to inhibit *Bacillus cereus* in carrot broth *Food Microbiology* 23, 68–73.

Viuda-Martos, M., Ruiz-Navajas, Y., Fernandez-Lopez, J. and Perez-Alvarez, J.A. (2007) Chemical composition of the essential oils obtained from some spices widely used in Mediterranean region. *Acta Chimica Slovenica* 54, 921–926.

Viuda-Martos, M., Ruiz-Navajas, Y., Fernandez-Lopez, J. and Perez-Alvarez, J.A. (2008) Antibacterial activity of different essential oils obtained from spices widely used in Mediterranean diet. *International Journal of Food Science & Technology* 43, 526–531.

Walsh, S.E., Maillard, J.-Y., Russell, A.D., Catrenich, C.E., Charbonneau, D.L. and Bartolo, R.G. (2003) Activity and mechanism of action of selective biocidal agents on Gram-positive and -negative bacteria. *Journal of Applied Microbiology* 94, 240–247.

Wan, J., Wilcock, A. and Coventry, M.J. (1998) The effect of essential oils of basil on the growth of *Aeromonas hydrophila* and *Pseudomonas fluorescence. Journal of Applied Microbiology* 84, 152–158.

Ward, S.M., Delaquis, P.J., Holley, R.A. and Mazza, G. (1998) Inhibition of spoilage and pathogenic bacteria on agar and pre-cooked roasted beef by volatile horseradish distillates. *Food Research International* 31, 19–26.

Wild, R. (1994) *The Complete Book of Natural and Medical Cures.* Rodale Press, Inc., Emmaus, USA.

Wilkins, K.M. and Board, R.G. (1989) Natural antimicrobial systems. In: Gould, G.W. (ed.) *Mechanisms of Action of Food Preservation Procedures.* Elsevier, New York, pp. 285–362.

Yin, M.C. and Cheng, W.S. (2003) Antioxidant and antimicrobial effects of four garlic-derived organosulfur compounds in ground beef. *Meat Science* 63, 23–28.

15 Essential Oils and their Components for the Control of Phytopathogenic Fungi that Affect Plant Health and Agri-food Quality and Safety

Caterina Morcia, Martina Spini, Mauro Malnati, A. Michele Stanca and Valeria Terzi*

Introduction

Crop protection plays a key role in ensuring food security for a human population that is projected to be 10 billion people after 2020 (United Nations, 1996). Weeds, animal pests, pathogens and viruses are responsible for competition and destruction of cultivated plants, with the overall result that crop yield can be strongly threatened (Oerke and Dehne, 2004). Fungal pathogens associated with plant diseases account for great losses in crop production. Moreover, some phytopathogenic fungi are able to produce a wide range of mycotoxins, characterized by strongly negative effects on human and animal health, that compromise food and feed safety. For their control, it has been estimated that over 23 million kg of synthetic fungicides are used annually worldwide (Martinez-Romero *et al.*, 2008). The benefits of pesticides to mankind and the environment has been reviewed by Cooper and Dobson (2007), evaluating the complex matrix of benefit interactions in the social, economic and environmental domains. However, the use of synthetic fungicides in crops can result in problems like environmental pollution, phytotoxicity and the selection of resistant pathogen populations (Diaenz *et al.*, 2002). Consequently, in recent years, the use of chemicals has increased consumer concern and their use is becoming more restrictive due to carcinogenic effects and residual toxicity problems (Marìn *et al.*, 2003; Rial-Otero *et al.*, 2005). Therefore, alternative measures have been proposed for crop protection, including mineral salts (Campanella *et al.*, 2002), biological agents (Xue *et al.*, 2008; Khan and Doohan, 2009a,b) and plant extracts, that are expected to have a narrow target range and a highly specific mode of action, to show a brief field persistence, to have a shorter shelf life (Pretorius, 2009).

Essential Oils for Crop Protection

Recent studies have shown the importance of natural plant products as fungicidal agents for crop protection (Table 15.1). Among natural products, both essential oils and the complex mixtures of secondary metabolites, mainly monoterpenes and sesquiterpenes, which characterize their chemical composition, have been investigated both *in vitro* and *in vivo* for their wide spectrum of antimicrobial activity against plant pathogens (Narwal *et al.*, 2004). It is noteworthy that, despite the extensive research on the antimicrobial properties of

* Corresponding author.

natural extracts, standard methods to authenticate the claims of antimicrobial action are not already available (for a review, see Das *et al.*, 2010). The most commonly used *in vitro* antimicrobial assays are based on disc diffusion, well diffusion, agar dilution and broth dilution methods. *In vitro* screening of essential oils for their antifungal properties has been done by the disc-diffusion methods in the works of Amadioha (2000); Soylu *et al.* (2007); and Pawar and Thaker, (2007), in which a paper disc impregnated with essential oil is laid on top of a fungal inoculated agar plate, to evaluate the volatile effect of the essential oils. Another method, frequently used, is the agar-well test, in which the essential oil was mixed with PDA growth media at different concentrations with an emulsifying agent such as Tween (Letessier *et al.*, 2001; Velluti *et al.*, 2004; Oxenham *et al.*, 2005; Terzi *et al.*, 2007; Rongai *et al.*, 2009) or Irol (Terzi *et al.*, 2007), or was dissolved in dimethyl sulfoxide (DMSO) (Kordali *et al.*, 2009; Dambolena *et al.*, 2010). PDA medium was poured into Petri dishes, and in the centre of each test plate a plug of 6–10 day old fungus cultures, for each of the pathogens separately, was inoculated. A plate containing only PDA with emulsifying agent served as negative control, while a plate containing a standard fungicidal compound was used as a positive control to determine the effectiveness of the plant extracts by comparison. Radial mycelial growth was determined after few days (4–7 days), and the result was expressed as percentage of mycelial growth inhibition by the formula (Dc − Dt)/Dc × 100 where Dc = average diameter of fungal growth in negative control plate and Dt = average diameter of fungal growth treated with essential oil. Sometimes the semisolid agar antifungal susceptibility test (SAAS) is used: in this case mycelial growth inhibition was evaluated by visual inspection and using a category scale (Kuzucu *et al.*, 2004; Dambolena *et al.*, 2010). All these methods have been evaluated for reliability by Hood *et al.* (2003), who suggested an optimized broth dilution method with Tween 80 emulsifier as the most accurate one for testing the activity of hydrophobic and viscous essential oils.

In vivo tests are usually carried out under greenhouse controlled conditions and the antimicrobial effects of natural compounds are evaluated on the basis of plant symptoms assessed using an observation-category scale or determining the percentage disease incidences by the formula (plants infected/ plants total) × 100.

Starting from *in vitro* and *in vivo* assays, several experimental works have shown the antifungal activities of some essential oils and of their pure components and their potential applications for crop protection. *In vivo* and *in vitro* inhibition effect against *Pyricularia oryzae* (*Magnaporthe grisea*), the causal agent of rice blast, of different *Azadirachta indica* extracts, a plant traditionally used as a local medicine in Nigeria, has been evaluated (Amadioha, 2000). The water and ethanol extracts obtained from fresh leaves of *A. indica*, together with the essential oil extracted from seeds of mature dehisced fruits, can inhibit the *in vitro* growth of fungus responsible for rice blast disease. Moreover, the treatment of inoculated rice plants in the greenhouse significantly reduced the severity of blast disease 2 days after artificial inoculation. The different *A. indica* extracts exhibited different efficacies in infection reduction in the following ascending order: hot water leaf extract, cold water leaf extract, alcohol leaf extract and seed oil extract treatments. This result is in agreement with the general consideration that the presence of the active components in plant extracts is influenced by several factors, such as method of extraction, age of the plant, time of harvesting and postharvest treatments of plant material, and different extracting solvents (Carrubba and Catalano, 2009). These same extracts showed antifungal activity on *Cochliobolos miyabeanus*, the causal agent of brown spot disease in rice, and confirmed that the oil extracts from seeds and leaf (by means of water and ethanol) were both effective in reducing the radial growth of the pathogen (Amadioha, 2002).

Among nine essential oils, the strong efficacy of *Cymbopogon citratus*, *Ocimum basilicum* and *Ocimum gratissimum* essential oils has been shown, which inhibited the growth of *Fusarium verticillioides*, one of the

Table 15.1. The concentrations of some essential oils effective after *in vitro* assay for the control of relevant crop pathogens are reported. For each oil further indications are shown, as are its major components and the geographical area of growth of the plants from which the oil was extracted.

Essential oils	Growth area	Major components	Pathogens	Concentrations	References
Origanum compactum	Morocco	carvacrol (58,1%), para-cymene (11,4%), thymol (9%)	*Botrytis cinerea*	IC_{50} 35,1 ppm on solid medium Completely inhibited by 100 ppm	Bouchra *et al.*, 2003
Thymus globosus	Morocco	para-cymene (35,7,4%), thymol (43,2%)	*B. cinerea*	IC_{50} 79,1 ppm on solid medium Completely inhibited by 100 ppm	Bouchra *et al.*, 2003
Hyssop officinalis	Greenhouse	Iso-pinocamphone (47,2-57,7%), pinocamphone (11,1%-22.1%)	*Pyrenophora avenae, Pycularia oryzae*	Completely inhibited by 0,4%	Letessier *et al.*, 2001
Agapanthus africanus crude extracts	South Africa	–	*B. cinerea, Fusarium oxysporum, Sclerotium rolfsii, Rhizoctonia solani, Botryosphaeria dothidea, Pythium ultimum, Alternaria alternata, Mycosphaerella pinodes*	*In vitro* MIC 0.8–1.2gl^{-1}	Tegegne *et al.*, 2008
Agapanthus africanus crude extracts	South Africa	–	*M. pinodes*	*In vivo* MIC 0.5gl^{-1} (combined aerial part extract), MIC 1gl^{-1} (flower extract), MIC 2gl^{-1} (root and leaf extract)	Tegegne *et al.*, 2008
Ocimum basilicum	Sagana (Kenya)	Linalool (95.7-98.9%)	*F. verticillioides*	MIC 1–2 µl/ml	Dambolena *et al.*, 2010
Ocimum basilicum	Yatta (Kenya)	Camphor (31-32.6%), Linalool (28.2-29.3%), Terpinen-4-ol (9-12%)	*F. verticillioides*	MIC 1-2 µl/ml	Dambolena *et al.*, 2010
Ocimum basilicum	Yatta (Kenya)	Neral (30.9%), Geranial (49.6%)	*F. verticillioides*	MIC 1–2 µl/ml	Dambolena *et al.*, 2010
Ocimum basilicum methyl chavicol chemotype	Greenhouse	methyl chavicol (76.1%), linalool (18.6%)	*Botrytis fabae*	≥ 250 ppm on solid media	Oxenham *et al.*, 2005
Ocimum basilicum linalol chemotype	Greenhouse	linalool (53%), eugenol (12.4%), eucalyptol (7.7%)	*B. fabae*	≥ 50 ppm on solid media	Oxenham *et al.*, 2005
Ocimum gratissimum	Sagana (Kenya)	Eugenol (64-95.5%), beta-bisabolene (10.2% in flowering tops)	*F. verticillioides*	MIC 0.3 µl/ml	Dambolena *et al.*, 2010

Ocimum gratissimum	Yatta (Kenya)	Eugenol (33.2% in flowering tops -70.1% in leaves), Z-beta-ocimene (34.1% in flowering tops), alfa-humulene (6-11.9%)	*F. verticillioides*	MIC 0.3 µl/ml (leaves)	Dambolena *et al.*, 2010
Cinnamomum zeylanicum	–	Cinnamaldehyde (58.49-64.13%), beta-linalool (10,25% in bark), benzyl benzoate (9.26% in bark), cinnamic acid (16.57% in leaf), benzaldehide (11.8% in leaf)	*F. oxysporum* *Alternaria porri*	5 µl on filter paper discs	Pawar and Thaker, 2007
Cinnamomum cassia	–	Cinnamaldehyde (66.36%), benzyl benzoate (10.24%), beta-linalool (9.16%)	*F. oxysporum, A. porri*	5 µl on filter paper discs	Pawar and Thaker, 2007
Syzygium aromaticum	–	Eugenol (47.64%), Benzyl alcohol (34.10%)	*A. porri*	5 µl on filter paper discs	Pawar and Thaker, 2007
Cymbopogon citratus	–	Citral (29.4%), beta-citral (21.39%), nerol acetate (10.81%)	*F. oxysporum*	5 µl on filter paper discs	Pawar and Thaker, 2007
Cymbopogon citratus	Southern Benin	Myrcene (28%), geranial (27%), neral (20%)	*F. verticillioides*	MIC 1.3µl/ml MIC *in vivo* test >8.0µl/ml	Fandohan *et al.*, 2004
Ocimum basilicum	Southern Benin	Linalol (33%), eugenol (22%), estragol (20%)	*F. verticillioides*	MIC 1.3µl/ml MIC *in vivo* test 6.4µl/ml	Fandohan *et al.*, 2004
Ocimum gratissimum	Southern Benin	Para-cimene (22%), thymol (17%), gamma-terpinene (15%)	*F. verticillioides*	MIC 2.0µl/ml MIC *in vivo* test 4.8µl/ml	Fandohan *et al.*, 2004
Lantana camara	Southern Benin	Beta-caryophyllene (325%), alfa-humulene (11%)	*F. verticillioides*	MIC 4.0µl/ml	Fandohan *et al.*, 2004
Eucalyptus citriodora	Southern Benin	Citronnellal (66%), citronellol (12%)	*F. verticillioides*	MIC 4.7µl/ml	Fandohan *et al.*, 2004
Clausena anisata	Southern Benin	Estragol (93%)	*F. verticillioides*	MIC 6.7µl/ml	Fandohan *et al.*, 2004
Melaleuca quinquenervia	Southern Benin	1,8-cineole (51%), alfa-terpineol (11%), viridiflorol (10%)	*F. verticillioides*	MIC 6.7µl/ml	Fandohan *et al.*, 2004
Xylopia aethiopica	Southern Benin	Sabinene (30%)	*F. verticillioides*	MIC >13.3µl/ml	Fandohan *et al.*, 2004
Origanum syriacum	Eastern Mediterranean Region of Turkey	–	*Sclerotinia sclerotiorum*	Completely inhibited by 0,3 µg ml^{-1} in filter paper disc Completely inhibited by 3.2 µg ml^{-1} on solid media	Soylu *et al.*, 2007
Foeniculum vulgare	Turkey	–	*S. sclerotiorum*	Completely inhibited by 0,2 µg ml^{-1} in filter paper disc Completely inhibited by 1.6 µg ml^{-1} on PDA	Soylu *et al.*, 2007

Continued

Table 15.1. *Continued*

Essential oils	Growth area	Major components	Pathogens	Concentrations	References
Cinnamomum zeylanicum	–	Cinnamaldehyde (70%)	*Fusarium* spp., *Penicillium* spp., *Pythium* spp.	MIC 100–800 μl l^{-1}	Christian and Goggi, 2008
Eugenia caryophyllata	–	Eugenol (60.6%), caryophyllene (26.9%)	*Fusarium* spp., *Penicillium* spp., *Pythium* spp.	MIC 100–800 μl l^{-1}	Christian and Goggi, 2008
Origanum minutiflorum	–	Carvacrol (57.3%), para-cymene (13.6%), gamma-terpinene (11.4%)	*Fusarium* spp., *Penicillium* spp., *Pythium* spp.	MIC 100–800 μl l^{-1}	Christian and Goggi, 2008
Satureja montana	–	gamma-terpinene (37.6%), carvacrol (33%), para-cymene (12.4%)	*Fusarium* spp., *Penicillium* spp., *Pythium* spp.	MIC 100–800 μl l^{-1}	Christian and Goggi, 2008
Thymus vulgaris	–	Linalool (41.4%), lavandulol (10.2%), para-cymene (8.89%)	*Fusarium* spp., *Penicillium* spp., *Pythium* spp.	MIC 100-800 μl l^{-1}	Christian and Goggi, 2008
Cymbopogon citratus	Cameroon	citral	*A. padwickii, B. oryzae, F.moniliforme*	1.5:100 (v/v), 100μl/g of rice seed	Nguefack *et al.*, 2008
Ocimum gratissimum	Cameroon	Thymol (46.2%), terpinene (20%), para-cimene (7%)	*A. padwickii, B. oryzae, F. moniliforme*	4:100 (v/v), 100μl/g of rice seed	Nguefack *et al.*, 2008
Thymus vulgaris	Cameroon	Thymol (27.2%), terpinene (22.7%), para-cimene (23.6%)	*A. padwickii, B. oryzae, F. moniliforme*	4:100 (v/v), 100μl/g of rice seed	Nguefack *et al.*, 2008
Azadirachta indica	Nigeria	–	*Pyricularia oryzae*	1ml on PDA in Petri dish (19 cm diam.); 0.1% on rice plants	Amadioha, 2000

–, means that this information is not available.

predominant pathogen associated with corn ear/kernel rot and tissue destruction on important crops such as maize, wheat and other relevant cereals (Reid *et al.*, 2002; Munkvold, 2003; Fandohan *et al.*, 2004). The *in vivo* test showed that these three essential oils significantly reduced the incidence of the considered pathogen in artificially inoculated corn. This fungus is also able to produce fumonisins, an important group of mycotoxins that causes dangerous diseases such as leukoencephalomalacia in horses, pulmonary edema syndrome in pigs and different types of cancer in humans (Rheeder *et al.*, 1992; Ross *et al.*, 1992; Shim and Woloshuk, 2001; Ghiasian *et al.*, 2006). The effect of essential oils in reducing fumonisin production is effective at concentration > 4.8μl/g.

The infection level of *Fusarium verticillioides* was also reduced by oregano (*Origanum minutiflorum*), clove (*Eugenia caryophyllata*), cinnamon (*Cinnamomum zeylanicum*), lemongrass (*Cymbopogon citratus*) and palmarosa (*Cymbopogon martinii*) oils, with a consequent reduction of mycotoxins (Velluti *et al.*, 2004).

Dambolena *et al.* (2010), by the semisolid agar antifungal susceptibility method (SAAS) (Provine and Hadley, 2000), evaluated the inhibitory effects of *Ocimum basilicum* and *Ocimum gratissimum* collected from Kenya on *F. verticillioides* growth and mycotoxins production. They confirmed the inhibitory effect of these essential oils and the strong dependence with the composition and the concentration of essential oils. Particularly they individuated eugenol as the most important constituent for the fungus inhibition.

O. basilicum essential oil is also effective for the control of *Botrytis fabae* and *Uromyces fabae* that cause diseases on a wide range of legume crops, including *Vicia faba*, *Pisum sativum* and *Lens culinaris*, and of *Trifolium dasyurum* (Oxenham *et al.*, 2005; You *et al.*, 2009). Moreover, *in vivo* and *in vitro* experiments showed that the inhibition of fungal growth is dependent on the two chemotypes of basil oil: methyl chavicol type and linalool type, as well as on the major pure components of these oils. The two oils and their major components (methyl chavicol, linalool, eugenol and eucalyptol) were able to significantly reduce *in vitro* growth of *B. fabae* and to control *in vivo* infection of broad bean (*V. faba*) by *B. fabae* and *U. fabae*. The strong inhibitory effect of cyclic terpenes (especially limonene and thymol) has been demonstrated even on *F. verticillioides*, with the consequent inhibition of toxin biosynthesis (fumonisin B1) at low concentrations (Dambolena *et al.*, 2008).

Pawar and Thaker (2007) screened 75 different essential oils and components *in vitro* for their antifungal effects on pathogenic *Fusarium oxysporum*, that induce wilt or root rots on plants (Gordon and Martyn, 1997) and on *Alternaria porri*, that causes an important disease in *Allium* spp. (Suheri and Price, 2000). Different essential oils presented diverse antifungal activities against the fungal species. The highest activity in *F. oxysporum* inhibition was shown by essential oils of lemongrass (*Cymbopogon citratus*), cumin (*Cuminum cyminum*) and fennel (*Foeniculum vulgare*). Cinnamon (*Cinnamomum zeylanicum*), cassia (*Cinnamomum cassia*) and clove (*Syzygium aromaticum*) oils exhibited the major activity in the control of *A. porri* diseases.

Crude extracts from *Agapanthus africanus*, an evergreen plant indigenous to South Africa, have *in vitro* antifungal activity against eight economically important plant pathogens: *Botrytis cinerea*, *F. oxysporum*, *Sclerotium rolfsii*, *Rhizoctonia solani*, *Botryosphaeria dothidea*, *Pythium ultimum*, *Alternaria alternate* and *Mycosphaerella pinodes* (Tegegne *et al.*, 2008). The results obtained by *in vitro* assays have been confirmed by *in vivo* experiments on the latter pathogen, which causes *Mycosphaerella* blight, the most destructive and serious disease in pea production (Xue *et al.*, 1996; Xue, 2000).

Different concentrations of *Hyssopus officinalis* essential oil and its components have a different impact on *Pyrenophora avenae* and *Pyricularia oryzae* mycelium growth (Letessier *et al.*, 2001). The *in vitro* experiment showed that 0.4% oil (added to PDA) completely inhibited the growth of both fungi. In a further experiment, young barley plants (three leaves) inoculated with

Blumeria graminis, young broad bean plants (two leaves) inoculated with *Uromyces viciae-fabae* and young apple plants (four leaves) inoculated with *Podosphaera leucotricha* were treated with suspensions containing different percentages of hyssop oil. The infection reduction was variable, depending on the time of the treatments. In fact, there were differences in the effects of hyssop oil when used *in vitro* and *in vivo*; the authors explained these results suggesting that the volatile components (responsible for antifungal activity) were able to diffuse away from the plants in the *in vivo* experiments.

Five essential oils (cinnamon, clove, oregano, savory and thyme) have been identified, by *in vitro* experiments, as good inhibitory treatments against three corn pathogens, *Fusarium* spp., *Penicillium* spp. and *Pythium* spp (Christian and Goggi, 2008). However, after *in vivo* experiments, they concluded that essential oil concentrations defined by an *in vitro* approach are not effective to protect the seeds in the open field because the volatile compounds probably diffuse from the surface of the seeds before planting. So they suggested increasing the natural compound's concentration for seed treatments.

Soylu *et al.* (2007) tested, *in vitro* and *in vivo*, the antifungal activities of oregano (*Origanum syriacum*) and fennel (*Foeniculum vulgare*) essential oils against *Sclerotinia sclerotiorum*, the causative agent of *Sclerotinia* stem and root rot of tomato, an important soil-borne disease. The results obtained from an *in vitro* study conducted using the standard disc diffusion assay showed that the volatile phases of these essential oils were more effective than the contact phase against the pathogen. The volatile phases were in fact able to inhibit the mycelial growth at low concentration (0.2–0.3μg ml^{-1} versus 3.2μg ml^{-1}), in agreement with the previous observations reported by Edris and Ferrag (2003) and Soylu *et al.* (2006). In *in vivo* experiments these authors observed that the treatment with oregano (3.2μg ml^{-1}) and fennel (3.2μg ml^{-1}) essential oils significantly improved plant survival in infested soil; moreover, by scanning electron microscopic (SEM) analysis, they observed that both essential oils cause degenerative changes in the hyphal morphology such as shrivelling, blistering and lysis.

Nguefack *et al.* (2008) studied the effects of three essentials oils extracted by hydrodistillation from *Cymbopogon citratus*, *Ocimum gratissimum* and *Thymus vulgaris* for their ability to control four important rice seed-borne infections caused by *P. oryzae* (blast disease), *Bipolaris oryzae* (brown spot disease), *Fusarium moniliforme* (bakanae disease), and *Alternaria padwickii* (stakburn disease). *In vitro* experiments showed that treatment of the seeds with these essential oils significantly reduced the infection and increased the level of rice germination. In particular, the results indicate that the antifungal activities of *O. gratissimum* and *T. vulgaris* essential oils are more effective than *C. citratus*: their superiority could be attributed to their content of thymol (46.2% and 27.2%), terpinene (20% and 22.7%) and para-cymene (7% and 23.6%).

Plant Pathogen Control: Evaluation of Antifungal Activity of Tea Tree Oil and of Single Essential Oil Components

Tea tree oil (TTO) is an essential oil consistently present in leaves and terminal branches of *Melaleuca alternifolia*, a tree belonging to *Myrtaceae* family that occurs in Australia on the north coast and adjacent ranges of New South Wales. This volatile oil is extracted by steam distillation and its composition is defined by international standard ISO 4730 (2004) ('Oil of *Melaleuca*, Terpinen-4-ol type'). Although the terpinen-4-ol is the most frequent chemotype in the *Melaleuca* genus, there is a natural variation in the composition of TTO (Homer *et al.*, 2000). Five out of the eleven populations of *M. alternifolia*, studied by Butcher *et al.* (1994), have more than one chemotype co-occurring. Most of these chemical forms can be considered to represent the appearance of new traits following speciation. However, both common ancestry and gene flow between species can be responsible for similar chemotypes appearing in different species, and in both cases it can be assumed that this would be the result of similar enzymes being present in the biosynthetic pathways (Keszei *et al.*, 2008). The genetic basis of this chemical diversity is not yet completely understood (Shelton *et al.*, 2002) and among significant essential-oil-producing families, *Myrtaceae* is the only one for which there is no molecular information on terpene biosynthesis. However, considerable differences in oil yield and composition are of particular interest to commercial producers. Investigations of genetic diversity in *M. alternifolia* has been conducted utilizing DNA markers and the results have shown that two main genetic provenances exist and that the majority of the genetic variability required by the industry can be obtained from phenotypically selected trees within a limited number of populations from the Clarence catchment (Rossetto *et al.*, 1999).

TTO has a long history of use as topical microbicide in human pharmacology (Carson *et al.*, 2006) and it has been re-evaluated in recent years as an alternative agent to antibiotics (Allen, 2001), to anti-inflammatory and anti-oxidant molecules (Kim *et al.*, 2004). Besides pharmaceutical purposes, this essential oil is now used in wide range of applications that include cosmetics, toiletries, household products and veterinary/pet care. In agriculture, several studies have demonstrated that, among the TTO components, terpinen-4-ol is the most active agent, with contact and fumigant insecticidial action against economically important pests (Isman, 2000). The fungicidal activity of TTO has been demonstrated even on food-borne and phytopathogenic fungi, such as *Aspergillus fumigatus, Fusarium solani, Penicillium expansum, B. cinerea* and *Rhizopus oryzae* (Bishop and Reagan, 1998; Inouye *et al.*, 1998; Bowers and Locke, 2000; Inouye, 2000; Angelini *et al.*, 2006).

The mechanism of action of TTO, studied in a wide range of bacteria and fungi, involves both the loss of membrane integrity accompanied by the release of intracellular material and the inhibition of cellular respiration, with the consequent inability to maintain homeostasis associated with changes in cell morphology (Carson *et al.*, 2006). Straede and Heinisch (2007) evaluated the effect of TTO on cell integrity in *Saccharomyces cerevisiae.* This yeast has a rigid cell wall, the synthesis of which is controlled by a highly conserved MAP kinase signal transduction cascade. In stress conditions a set of sensors activate, through this cascade, the transcription factor Rlm1, which governs expression of many genes encoding enzymes of cell-wall biosynthesis. The effects of TTO and its components have been evaluated on a set of yeast strains transformed with reporter constructs that link activation of a hybrid, Rlm1–lexA, by the MAP kinase Mpk1/Slt2 to the expression of the bacterial *lacZ* gene. The authors found a clear dose-dependent increase in β-galactosidase activities, demonstrating an activation of cell integrity signalling by TTO. Recently, Kapros and McDaniel (2009) have demonstrated that TTO has a major effect on membrane permeability, not only in bacteria and fungi but also in tobacco cells. These authors conclude that *Melaleuca* oils' mode of action at the cellular level can explain the inhibition of germination and seedling growth observed in nature due to the action of the oil.

In our study, the activity of TTO and of its major purified components has been evaluated for the control of three classes of plant pathogenic fungi (*Fusaria, Pyrenophora graminea* and *Blumeria graminis*) that cause relevant yield losses in small grain cereals in Mediterranean environments. These fungi are characterized by different life cycle, pathogenicity behaviour, transmission mode and effects on plants and derived agro-food products.

Fusaria are responsible for fusarium head blight (FHB) symptoms in wheat, barley and oats. These mycotoxigenic fungi represent a major economic concern not only for yield losses, but even more for the lowering of the grain quality and safety. Trichothecenes produced by these fungi, such as deoxynivalenol (DON) and T2-HT2, cause both acute and chronic negative effects on human and animal health (Parent-Massin, 2004). *P. graminea* is a seed-borne pathogen responsible for barley leaf stripe, a common disease in barley districts with a cold sowing season. In susceptible cultivars, the fungus causes brown stripes on the leaves, stunted growth and severe yield reductions (Bulgarelli *et al.*, 2004). *B. graminis* is one of the grass powdery mildew pathogens and widespread biotropic fungus that colonizes plant epidermal tissues, causing large yield losses in several environments characterized by a temperate climate (Zhou *et al.*, 2001).

Effects of Tea Tree Oil and its Major Purified Components on *Fusaria* and *Pyrenophora graminea* Fungal Growth

Inhibition of the mycelium growth of *Fusarium graminearum* and *Fusarium culmorum* has been evaluated on solid PDA medium amended with TTO and its major components in percentages ranging from 0 to 2%. Strains of *F. graminearum* and *F. culmorum* were isolated from bread wheat fields in Italy. Moreover, *Pyrenophora graminea* strain I2, isolated in Northern Italy on barley and characterized by a high level of aggressiveness (Gatti *et al.*, 1992), was used. The experiments were conducted twice in triplicate in 60 mm Petri dishes, inoculated with 8 mm PDA plugs from actively growing cultures. Mycelium growth was evaluated as mean diameter measured each day from fungal inoculation up to 6 days post-inoculation and the results were expressed as growth inhibition, calculated as $I = [(C - T)/C] \times 100$, where C is control and T is treated.

In Fig. 15.1 the *Fusaria* growth inhibition effects of TTO, gamma-terpinene, terpinen-4-ol and 1,8-cineol are reported. The single oil components are more active agents in reducing *in vitro* growth of *F. graminearum* and *culmorum*. Among the tested compounds, terpinen-4-ol is particularly effective in reducing fungal growth, causing complete growth inhibition at 0.1% concentration. The complete inhibition of fungal growth is obtained even in the presence of 0.5% 1,8-cineol, whereas gamma-terpinene inhibition is in the range of 1–2%.

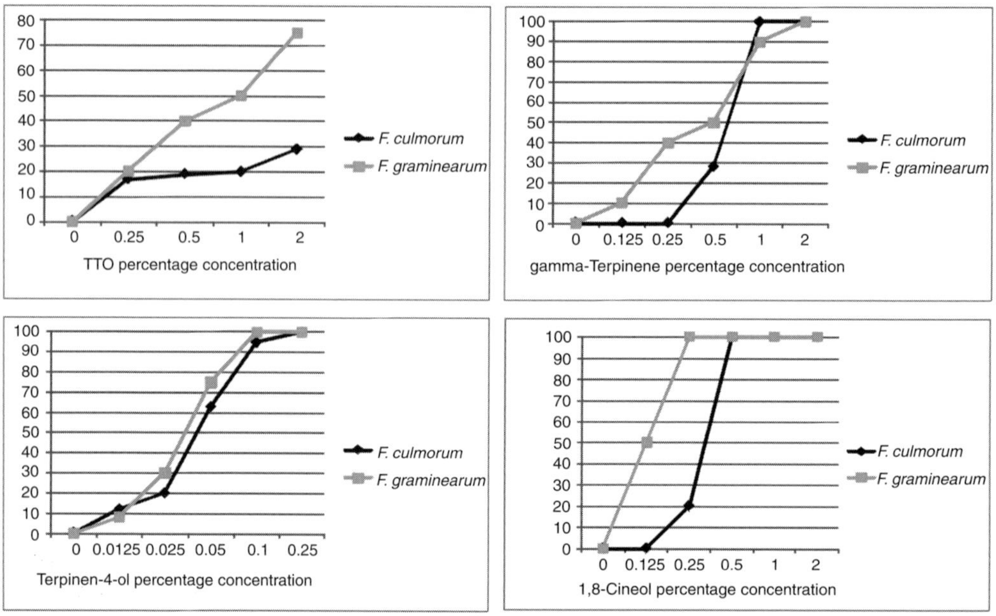

Fig. 15.1. Growth inhibition percentages (y axis) of *Fusarium culmorum* and *Fusarium graminearum* cultured on PDA medium amended with TTO, gamma-terpinene, terpinen-4-ol and 1,8-cineol at concentrations ranging from 0 (control, 0% inhibition of fungal growth) to 2%. The data are the mean of six replicates.

These data are consistent with several studies which demonstrate that terpinen-4-ol is the most active component of TTO on pathogenic microorganisms (Carson and Riley, 1995; Cox *et al.*, 2001; Oliva *et al.*, 2003, Loughlin *et al.*, 2008; Yangui *et al.*, 2009). Antagonistic effects among different TTO constituents can be one of the reasons for the lower level of fungicidal activity shown by the TTO complete formulation (probably due to the reduction of terpinen-4-ol aqueous solubility by non-oxygenated terpenes (Cox *et al.*, 2001).

The effects of TTO, its major components, but even of other compounds present in essential oils have been evaluated *in vitro* even on *P. graminea*, a seed-borne pathogen responsible for barley leaf stripe symptoms, resulting in severe yield losses but not in the production of mycotoxins. The mycelium growth has been evaluated on solid PDA medium amended with TTO and essential oil components in percentages ranging from 0 to 5%. The experiments were conducted twice in triplicate in 60 mm Petri dishes, inoculated with 8 mm PDA plugs from actively growing cultures. Mycelium growth was evaluated as mean diameter measured each days from fungal inoculation up to 6 days post-inoculation and the results were expressed as percentage growth with respect to the control. The results obtained, reported in Fig.15.2, clearly suggest that TTO and all the tested molecules can act as fungicides, even at different concentrations. In more detail, this scale of potency has been individuated: thymol>eugenol>carvon> terpinen-4-ol > cineole > gamma terpinene > TTO.

Effects of Tea Tree Oil and its Major Purified Components on *Blumeria graminis*

Blumeria graminis is an obligate barley leaf pathogen, characterized by a peculiar way of transmission (wind-borne) and site of infection (epidermal tissue). The effect of TTO and oil components was tested with two different experimental approaches: *in vitro* (on barley leaf segments conserved in agar medium) and *in vivo* (on barley plantlets in environmentally controlled conditions).

To evaluate the *in vitro* effect of TTO on *B. graminis* f. sp. *hordei*, five 10-day-old leaf segments (40 mm length) of barley cultivar

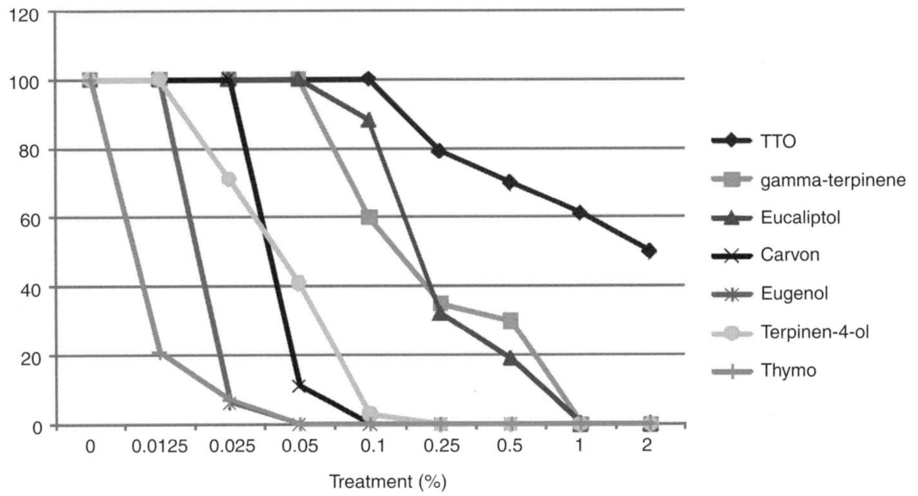

Fig. 15.2. Growth inhibition percentages (y axis) of *Pyrenophora graminea* (strain I2) grown on PDA medium amended with TTO, gamma-terpinene, eucalyptol, carvon, eugenol, terpinen-4-ol and thymol at concentrations ranging from 0 to 5%. The data are the mean of six replicates.

Golden Promise (susceptible to the mildew) were laid out in a Petri dish over a medium containing 0.5% agar. For each experimental condition, a set of four independent Petri dishes was prepared. All the plates were placed on a rotating platform under an inoculation tower (diameter 50 cm, height 100 cm) and inoculated with K1 isolate (kindly provided by Paul Schulze-Lefert, Max-Plank-Institut für Züchtungsforschung, Köln). The leaf segments were then incubated for a total time of 72 h at 20°C in the dark and then sprayed with a control solution containing 0.05% IROL or with a solution containing 0.05% IROL and 0.1% TTO or 0.5% TTO or 1% TTO. After 24 h from the treatment a subset of control and treated leaf samples were observed with Reflected Fluorescence System Olympus BX51. After 7 days of incubation at 20°C the mildew colonies on the leaf segments were counted with the aid of UTHSCSA Image Tool software version 3. The experiment was conducted in triplicate.

The results obtained showed that even a single treatment with a spray solution containing TTO as low as 0.1% completely prevented the colonization of barley leaves. Moreover, 24 h after the single treatment, the mycelium is necrotic, with drastic reduction of sporulating colonies and conidia and with the appearance of about 60% of irregular conidia with fissured cell wall (Fig. 15.3).

To evaluate the *in vivo* effect of TTO on *B. graminis* f. sp. *hordei* plantlets of barley cultivar Golden Promise were grown in pots (five plants per pot) in controlled greenhouse conditions at 20°C until the third-leaf stage.

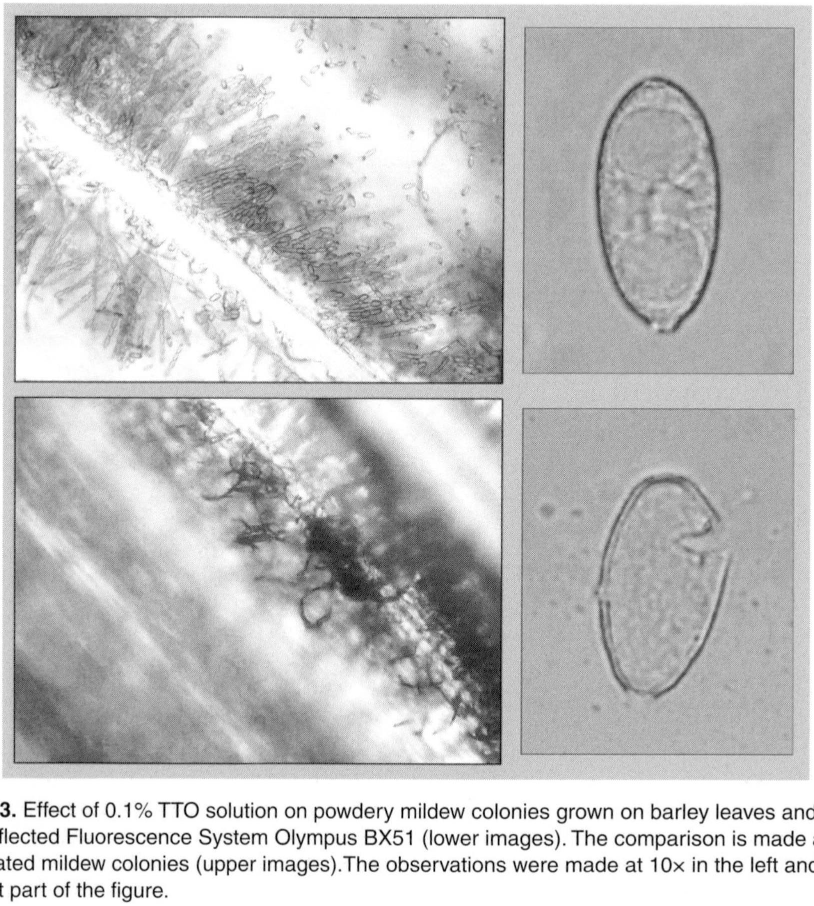

Fig. 15.3. Effect of 0.1% TTO solution on powdery mildew colonies grown on barley leaves and observed with Reflected Fluorescence System Olympus BX51 (lower images). The comparison is made against non-treated mildew colonies (upper images).The observations were made at 10× in the left and 100× in the right part of the figure.

All the pots, with the exception of the controls, were placed on a rotating platform under an inoculation tower (diameter 50 cm, height 100 cm) and sprayed with K1 isolate. The plantlets were then incubated for a total time of 72 h at 20°C and then sprayed with a control solution containing 0.05% IROL or with a solution containing 0.05% Tween 20 and 0.5% TTO or 1% TTO. After 3 days the level of powdery mildew infection was evaluated by a visual score with a scale ranging from 0 (no symptoms) to 9 (maximum of infections). The experiment was conducted in triplicate and the data obtained are reported below:

Treatment	Visual score
0.05% IROL (No treated control)	9
Horizon (control)	0
0.1% TTO	3
0.5% TTO	0
1% TTO	0

The *in vivo* results obtained in the greenhouse trials confirmed the *in vitro* ones: a single treatment with TTO 0.5% is effective in the powdery mildew control.

Moreover, no phytotoxic effects were observed on barley leaves treated with TTO solution and the growth of the sprayed leaves, monitored until the fourth-leaf stage, did not differ from that of the control leaves.

TTO and Chemically Induced Defence Responses

Plants that survive a local infection are often better equipped to fight a second pathogen attack, even in a distal part of the plant, as a result of a phenomenon called induced resistance (IR) (Kogel and Langen, 2005). This genetically programmed plant defence pathway can be mediated by pathogen attack, but even by chemical inducers, such as salicylic acid (SA) and analogues, such as the acibenzolar-S-methyl [benzo(1,2,3) thiadiazole-7- carbothioic acid S-methyl ester (BTH)], commercially available as Bion®. In several dicotyledonous species, SA is necessary for the induction of cascade signalling that activates systemic acquired resistance (SAR) genes, such as the pathogenesis related (PR) gene families. Expression of the gene *pathogenesis-related-1* (*Pr1*) is in fact commonly used as a marker for SAR in tobacco and *Arabidopsis*. However, several studies have suggested that in cereals, with their high SA content, SA is not an effective signal for the activation of defence genes and IR. Chemically, IR in barley is associated with expression of *BCI* genes [for Barley Chemical Induced (Beßer *et al.*, 2000)], a very heterogeneous group of genes, few of which are homologues to any described SAR genes (Maleck *et al.*, 2000). The role of *BCI* genes in barley IR has in fact been evaluated using gene-silencing strategies and one interpretation of the results is that the resistance-inducing chemicals, including the commercial product BTH, are able to activate many genes which are not only directly related to plant defence. A complex interaction of signalling pathways seems, in fact, to be involved in mediating the observed systemic induced resistance in barley. This resistance, although displaying certain similarities to SAR in dicotyledons, is more elaborate and differs in many aspects. Therefore, a variety of defence responses combine together to give rise to IR that confer to the plant the ability to respond more strongly and quickly to secondary pathogen attack.

Starting from this evidence, a set of genes known to be involved in IR or, more generally, in plant pathogen interaction, has been selected and their expression levels have been evaluated both locally and systemically after TTO treatment in comparison with BION treatment. The following genes have been selected for expression studies:

- *BCI-4* considered a marker of chemically induced resistance in barley;
- *CHS* that encodes for chalcone synthase, an enzyme involved in the flavonoid biosynthetic pathway;
- PAL, encoding for phenylalanine ammonia-lyase, a key enzyme in phenylpropanoid biosynthesis;
- *APX*, encoding for ascorbate peroxidase, involved in ROS scavenging;
- *CHN* that encodes for chitinase, a classical

PR involved in the disruption of the fungal cell wall, with the consequent release of elicitors;

- *GLC*, encoding for the pathogenesis-related protein β-1,3-glucanase;
- *PR1*, encoding for a pathogenesis-related protein that can be considered a marker for SAR;
- *SAMS*, encoding for S-adenosyl methionine synthetase, an enzyme involved in methionine biosynthesis; and
- *LOX* that encodes for lipoxygenase, a key enzyme of the jasmonic acid pathway.

Barley plantlets of the cultivar Golden Promise were grown in pots in controlled environment conditions with a 16 h photoperiod at 150–200 μmol m^{-2} s^{-1}, at 20°C day/15°C night, and at a constant 70% relative humidity. Plantlets (7 days) at first-leaf stage were treated with water (control samples) or Bion 100 μM or 0.5% TTO. After 24 h the first leaf was cut for mRNA extraction (local treatment). After 3 days the second leaf of the plantlets appeared and was cut for mRNA extraction (systemic treatment). For each treatment two biological replicates were done. Total RNA was isolated from samples using Trizol (Invitrogen, CA, USA), according to manufacturer instructions. The resulting RNAs were purified using Qiagen RNeasy columns (Qiagen GmbH, Hilden, Germany) following the RNA clean-up protocol. The quality and quantity of RNAs were evaluated with the 2100 Bioanalyzer instrument (Agilent Technologies, Santa Clara, CA, USA). Total RNA samples were reverse transcribed in triplicate following the TaqMan One step RT–PCR protocol (Applied Biosystems, Foster City, CA, USA). Briefly, 100 ng purified RNA with 0.25 U/ml MultiScribe Reverse Transcriptase (Applied Biosystems), 0.4 U/ml Rnase Inhibitor (Applied Biosystems), 150 nM forward and 150 nM reverse primers in a 50 μl final volume of SYBR Green PCR Master Mix 1X (Applied Biosystems) were subjected to the following thermal profile: one step at 48°C for 30 min, one step at 95°C for 10 min, 40 cycles with a denaturation step at 95°C for 15 s and an annealing/extension step at 60°C for 1 min. PCRs were performed in the PE Biosystems GeneAmp 7300 Sequence Detection System (Applied Biosystems) using MicroAmp optical tubes and caps. Primers design was done with Primer Express v.3.0 software (Applied Biosystems) and then the assays were optimized and evaluated for specific product amplification. Primer sequences used in RT–PCR are listed in Table 15.2. A negative control without template was run with every assay to assess the overall specificity. As a reference gene for RTqPCR data normalization TC131363 was used, according to previous work (Faccioli *et al.*, 2007).

The results obtained showed a strong induction of *BCI* and *LOX* genes, known to be involved in IR, and of *PR1* and *GLC* after treatment with Bion mainly in systemic samples, as expected. On the contrary, TTO treatments did not induce any change in the expression of this set of genes. Therefore, it can be hypothesized that TTO has an impact on gene expression different from that of a classical IR inducer like BION. However, further investigations are required to better define the TTO effect on the plant whole transcriptome.

Future Perspectives and Conclusion

Several studies have demonstrated the efficacy of some secondary metabolites for plant protection. In plants, these compounds are not necessarily produced under all conditions and many of them are undoubtedly the result of an adaptive evolutionary process, subjected to natural selection, started about 140 million years ago (Wink, 2003), when the co-existence of plants, herbivores and microorganisms occurred. Convergent, independent and repeated evolution steps brought the development, from ancestor genes, of structurally definite secondary metabolites in different plant groups, reflecting the plasticity of environmental responses and the different adaptive strategy of species (Waterman and Mole, 1989; Harborne and Baxter, 1993). Referred to essential oils, the great variability is also related to the geographical origin of the

Table 15.2. List of the genes studied, reporting for each of them the GenBank accession number, the gene function, the forward and reverse primer sequences, the annealing temperature and the amplicon length.

Gene code	Gene ID	Gene function	Primer sequences	Annealing temperature (°C)	Amplicon length (bp)
LOX	HVU56406	Methyljasmonate-inducible-lipoxygenase	TGGAGGCCCCGGAGAA, F CCAAGATGGCAAGGACTATGG, R	60	75
PR1	X74940	Hv-PR-1b mRNA for a basic PR-1-type pathogenesis-related protein	CGTGAGCTGGAGCACGAA, F TGGAGCTTGCAGTCGTTGAT,R	60	75
PAL	HVPAL2MR	*Hordeum vulgare* partial PAL mRNA for phenylalanine ammonia-lyase	CGCAGACGACCCTTGCA,F AAGGCGTGCTCCACAAGAA,R	60	72
BCI4	HV250283	*H. vulgare* mRNA for putative calcium binding EF-hand protein (bci-4 gene)	CAATCGGTTGCGAGTATGCA,F GAGCAAGGGCACCGTGAA,R	60	63
CHS	X58339	*H.vulgare* CHS gene for chalcone synthase	GCCCTAGAGGAGGCCTTCAA,F TTGGTGCGCTATCCAGAAGA,R	60	69
SAMS	D63835	*H. vulgare* mRNA for S-adenosylmethionine synthetase	ACCTCGATGAGAACACCATCTTC,F GAGGGCCACCGATGACAA,R	60	63
CHN	X8672	*H.vulgare* mRNA for chitinase 2b	CCACGTCTCCACCCTACTATGG,F CCGGGTTGCTCACAAGGT,R	60	62
APX	AJ006358	*H. vulgare* mRNA for ascorbate peroxidase	CTGCGGATGAGAAGGCTTTC, F GCTTCAGCGTACCCCAGTTC,R	60	76
GLCD	AJ271367	*H. vulgare* mRNA for beta-1,3-glucanase	GGCTCTTCAACCCGGACAA,F CTTGGATGCCGCATTACGTA,R	60	117

plant: for example *Ocimum basilicum* genotypes from Italy show three different chemotypes (Marotti *et al.*, 1996), while *O. basilicum* from Turkey shows seven different chemotypes (Telci *et al.*, 2006). Although there are several examples of the use of natural products as commercial products, another interesting opportunity is for them to serve as structural motifs for the development of synthetic fungicides after simplification steps to reduce the complexity of natural compounds (Hillebrand, 2009).

All these research activities focused on the development of sustainable strategies for crop protection are needed because emerging, re-emerging and endemic plant pathogens continue to challenge our ability to safeguard plant health worldwide (Miller *et al.*, 2009). Moreover, it is clear that the benefits deriving from the use of pesticides are many and diverse; however to conjugate the benefits with the minimum human, environmental and economic costs, the use of pesticides must be regulated and optimized (Cooper and Dobson, 2007). Finally, the future strategy of crop protection should lie in a combination of several different tools and methodologies, ranging from the development of durable-resistant and transgenic plants (Collinge *et al.*, 2008; Gust *et al.*, 2010) to the use of soil microbiota as biocontrol agents (Vannacci and Gullino, 2000; Backman and Sikora, 2008) or of synthetic and natural pesticides in an integrated pest management (IPM) approach (Oerke and Dehne, 2004).

References

Allen, P. (2001) Tea tree oil: the science behind the antimicrobial hype. *The Lancet* 358, 1245. DOI:10.1016/S0140-6736(01)06386-3.

Amadioha, A.C. (2000) Controlling rice blast *in vitro* and *in vivo* with extracts of *Azadirachta indica*. *Crop Protection* 19, 287–290.

Amadioha, A.C. (2002) Fungitoxic effects of extracts of *Azadirachta indica* against *Cochliobolus miyabeanusi* causing brown spot disease in rice. *Archives of Phytopathology and Plant Protection* 35, 37–42.

Angelini, P., Pagiotti, R., Menghini, A. and Vianello, B. (2006) Antimicrobial activities of various essential oils against foodborne pathogenic or spoilage moulds. *Annals of Microbiology* 56, 65–69.

Backman, P.A. and Sikora, R.A. (2008) Endophytes: An emerging tool for biological control. *Biological Control* 46, 1–3.

Beßer, K., Jarosch, B., Langen, G. and Kogel, K.H. (2000) Expression analysis of genes induced in barley after chemical activation reveals distinct disease resistance pathways. *Molecular Plant Pathology* 1, 277–286.

Bishop, C.D. and Reagan, J. (1998) Control of the storage pathogen *Botrytis cinerea* on Dutch cabbage (*Brassica oleracea* var. capitata) by the essential oil of *Melaleuca alternifolia*. *Journal of Essential Oil Research* 10, 57–60.

Bouchra, C., Achouri, M., Hassani, L.M.I. and Hmamouchi, M. (2003) Chemical composition and antifungal activity of essential oils of seven Moroccan Labiatae against *Botrytis cinerea* Pers: Fr. *Journal of Ethnopharmacology* 89, 165–169.

Bowers, J.H. and Locke, J.C. (2000) Effect of botanical extracts on the population density of *Fusarium oxysporum* in soil and control of *Fusarium* wilt in the greenhouse. *Plant Disease* 84, 300–305.

Bulgarelli, D., Collins, N.C., Tacconi, G., Dellaglio, E., Brueggeman, R., Kleinhofs, A., Stanca, A.M. and Valè, G. (2004) High-resolution genetic mapping of the leaf stripe resistance gene RDG2a in barley. *Theoretical and Applied Genetics* 108, 1401–1408.

Butcher, P.A., Doran, J.C. and Slee, M.U. (1994) Intraspecific variation in leaf oils of *Melaleuca alternifolia* (*Myrtaceae*). *Biochemical Systematics and Ecology* 22, 419–430.

Campanella, V., Ippolito, A. and Nigro, F. (2002) Activity of calcium salts in controlling Phytophthora root rot of citrus. *Crop Protection* 21, 751–756.

Carrubba, A. and Catalano, C. (2009) Essential oil crops for sustainable agriculture – a review. In: Lichtfouse, E. (ed.) *Climate Change, Intercropping, Pest Control and Beneficial Microorganisms*. Springer Science + Business Media, Dijon, France, 2, pp. 137–187.

Carson, C.F. and Riley, T.V. (1995) Antimicrobial activity of the major components of the essential oil of *Melaleuca alternifolia*. *Journal of Applied Bacteriology* 78, 264–269.

Carson, C.F., Hammer, K.A. and Riley, T.V. (2006) *Melaleuca alternifolia* (tea tree) oil: a review of antimicrobial and other medicinal properties. *Clinical Microbiology Reviews* 19, 50–62.

Christian, E.J. and Goggi, A.S. (2008) Aromatic plant oils as fungicide for organic corn production. *Crop Science* 48, 1941–1951.

Collinge, D.B., Lund, O.S. and Thordal-Christensen, H. (2008) What are the prospects for genetically engineered, disease resistant plants? *European Journal of Plant Pathology* 121, 217–231.

Cooper, J. and Dobson, H. (2007) The benefits of pesticides to mankind and the environment. *Crop Protection* 26, 1337–1348.

Cox, S.D., Mann, C.M. and Markham, J.L. (2001) Interactions between components of the essential oil of *Melaleuca alternifolia*. *Journal of Applied Microbiology* 91, 492–497.

Dambolena, J.S., López, A.G., Cánepa, M.C., Theumer, M.G., Zygadlo, J.A. and Rubinstein, H.R. (2008) Inhibitory effect of cyclic terpenes (limonene, menthol, menthone and thymol) on *Fusarium verticillioides* MRC 826 growth and fumonisin B1 biosynthesis. *Toxicon* 51, 37–44.

Dambolena, J.S., Zunino, M.P., López, A.G., Rubinstein, H.R., Zygadlo, J.A., Mwangi, J.W., Thoithi, G.N., Kibwage, I.O., Mwalukumbi, J.M. and Kariuki, S.T. (2010) Essential oils composition of *Ocimum basilicum* L. and *Ocimum gratissimum* L. from Kenya and their inhibitory effects on growth and fumonisin production by *Fusarium verticillioides*. *Innovative Food Science and Emerging Technologies* 11, 410–414.

Das, K., Tiwari, R.K.S. and Shrivastava, D.K. (2010) Techniques for evaluation of medicinal plant products as antimicrobial agent: Current methods and future trends. *Journal of Medicinal Plants Research* 4, 104–111.

Diaenz, F., Santos, M., Blanco, R. and Tello, J.C. (2002) Fungicide resistance in *Botrytis cinerea* isolate from strawberry crops in Huelva (southwestern Spain). *Phytoparasitica* 30, 529–534.

Edris, A.E. and Ferrag, E.S. (2003) Antifungal activity of peppermint and sweet basil essential oils and their major aroma constituents on some plant pathogenic fungi from the vapour phase. *Nahrung/Food* 47, 117–121.

Faccioli, P., Ciceri, G.P., Provero, P., Stanca, A.M., Morcia, C. and Terzi, V. (2007) A combined strategy of 'in silico' transcriptome analysis and web search engine optimization allows an agile identification of reference genes suitable for normalization in gene expression studies. *Plant Molecular Biology* 63, 679–688.

Fandohan, P., Gbenou, J.D., Gnonlonfin, B., Hell, K., Marasas, W.F.O. and Wingfield, M.J. (2004) Effect of essential oils on the growth of *Fusarium verticillioides* and Fumonisin contamination in corn. *Journal of Agriculture and Food Chemistry* 52, 6824–6829.

Gatti, A., Rizza, F., Delogu, G., Terzi, V., Porta-Puglia, A. and Vannacci, G. (1992) Physiological and biochemical variabilità in a population of *Drechslera graminea. Journal of Genetics and Breeding* 46, 179–186.

Ghiasian, S.A., Maghsood, A.H., Yazdanpanah, H., Shephard, G.S., Van Der Westhuizen, L., Vismer, H.F., Rheeder, J.P. and Marasas, W.F.O. (2006) Incidence of *Fusarium verticillioides* and levels of fumonisins in corn from main production areas in Iran. *Journal of Agriculture and Food Chemistry* 54, 6118–6122.

Gordon, T.R. and Martyn, R.D. (1997) The evolutionary biology of *Fusarium oxysporum. Annual Review of Phytopathology* 35, 111–128.

Gust, A.A., Brunner, F. and Nürnberger, T. (2010) Biotechnological concepts for improving plant innate immunity. *Current Opinion in Biotechnology* DOI 10.1016/j.copbio.2010.02.004

Harborne, J.B. and Baxter, H. (1993) *Phytochemical Dictionary. A Handbook of Bioactive Compounds from Plants.* Taylor & Francis, London.

Hillebrand S. (2009) Natural products as structural motifs for fungicides. *Cornucopia*, spring 2009. AGFD 33.

Homer, L.E., Leach, D.N., Lea, D., Slade Lee, L., Henry, R.J. and Baverstock, P.R. (2000) Natural variation in the essential oil content of *Melaleuca alternifolia* Cheel (*Myrtaceae*). *Biochemical Systematics and Ecology* 28, 367–382.

Hood, J.R., Wilkinson, J.M. and Cavanagh, H.M.A. (2003) Evaluation of common antibacterial screening methods utilized in essential oil research. *Journal of Essential Oil Research* Nov/Dec.

Inouye, S., Watanabe, M., Nishiyama, Y., Takeo, K., Akao, M. and Yamaguchi, H. (1998) Antisporulating and respiration-inhibitory effects of essential oils on filamentous fungi. *Mycoses* 41, 403–409.

Inouye, S., Tsuruoka, T., Watanabe, M., Takeo, K., Akao, M., Nishiyama, Y. and Yamaguchi, H. (2000) Inhibitory effect of essential oils on apical growth of *Aspergillus fumigatus* by vapour contact. *Mycoses* 43, 17–23.

Isman, M.B. (2000) Plant essential oil for pest and disease management. *Crop Protection* 19, 603–608.

Kapros, T. and McDaniel, S.C. (2009) Cytotoxicity of tea tree oil in tobacco cells. *Allelopathy Journal* 23 Print ISSN: 0971-4693. Online ISSN: 0973-5046.

Keszei A., Brubaker, C.L. and Foley, W.J. (2008) A molecular perspective on terpene variation in Australian Myrtaceae. *Australian Journal of Botany* 56, 197–213.

Khan, M.R. and Doohan, F.M. (2009a) Bacterium-mediated control of Fusarium head blight disease of wheat and barley and associated mycotoxin contamination of grain. *Biological Control* 48, 42–47.

Khan, M.R. and Doohan, F.M. (2009b) Comparison of the efficacy of chitosan with that of a fluorescent pseudomonad for the control of Fusarium head blight disease of cereals and associated mycotoxin contamination of grain. *Biological Control* 48, 48–54.

Kim, H.J., Chen, F., Wu, C., Wang, X., Chung, H.Y. and Jin, Z. (2004) Evaluation of antioxidant activity of Australian tea tree (*Melaleuca alternifolia*) oil and its components. *Journal of Agricultural and Food Chemistry* 52, 2849–2854.

Kogel, K.-H. and Langen, G. (2005) Induced disease resistance and gene expression in cereals. *Cellular Microbiology* 7, 1555–1564.

Kordali, S., Cakir, A., Akcin, T.A., Mete, E., Akcin, A., Aydin, T. and Kilic, H. (2009) Antifungal and herbicidal properties of essential oils and *n*-hexane extracts of *Achillea gypsicola* Hub-Mor. and *Achillea biebersteinii* Afan. (*Asteraceae*). *Industrial Crops and Products* 29, 562–570.

Kuzucu, C., Rapino, B., McDermott, L. and Hadley S. (2004) Comparison of the semisolid agar antifungal

susceptibility test with the NCCLS M38-P broth microdilution test for screening of filamentous fungi. *Journal of Clinical Microbiology* 42, 1224–1227.
Letessier, M.P., Svoboda, K.P. and Walters, D.R. (2001) Antifungal activity of the essential oil of Hyssop (*Hyssopus officinalis*). *Journal of Phytopathology* 149, 673–678.
Loughlin, R., Gilmore, B.F., McCarron, P.A. and Tunney, M.M. (2008) Comparison of the cidal activity of tea tree oil and terpinen-4-ol against clinical bacterial skin isolates and human fibroblast cells. *Letters in Applied Microbiology* 46, 428–433.
Maleck, K., Levine, A., Eulgem, T., Morgan, A., Schmid, J., Lawton, K.A., Dangl, J.L. and Dietrich, R.A. (2000) The transcriptome of *Arabidopsis thaliana* during systemic acquired resistance. *Nature Genetics* 26, 403–410.
Marìn, A., Oliva, J., Garcìa, C., Navarro, S. and Barba, A. (2003) Dissipation rates of cyprodinil and fludioxinil in lettuce and table grape in the field and under cold storage conditions. *Journal of Agricultural and Food Chemistry* 51, 4708–4711.
Marotti, M., Piccaglia, R. and Giovanelli, E. (1996) Differences in essential oil composition of basil (*Ocimum basilicum* L.) Italian cultivars related to morphological characteristics. *Journal of Agricultural and Food Chemistry*, 44, 3926–3929.
Martìnez-Romero, D., Serrano, M., Bailén, G., Guillén, F., Zapata, P.J., Valverde, J.M., Castillo, S., Fuentes, M. and Valero, D. (2008) The use of a natural fungicide as an alternative to preharvest synthetic fungicide treatments to control lettuce deterioration during postharvest storage. *Postharvest Biology and Technology* 47, 54–60.
Miller, S.A., Beed, F.D. and Harmon, C.L. (2009) Plant disease diagnostic capabilities and networks. *Annual Review of Phytopathology* 47, 15–38.
Munkvold, G.P. (2003) Epidemiology of *Fusarium* diseases and their mycotoxins in maize ears. *European Journal of Plant Pathology* 109, 705–713.
Narwal S.S., Singh, R. and Walia, R.K. (2004) *Research Methods in Plant Sciences, Allelopathy: Volume 2. Plant Protection.* CABI, Wallingford, Oxfordshire, UK.
Nguefack, J., Leth, V., Lekagne Dongmo, J.B., Torp, J., Amvam Zollo, P.H. and Nyasse, S. (2008) Use of three essential oils as seed treatments against seed-borne fungi of rice (*Oryza sativa* L.). *American-Eurasian Journal of Agricultural & Environmental Sciences* 4, 554–560.
Oerke, E.-C. and Dehne, H.-W. (2004) Safeguarding production – losses in major crops and the role of crop protection. *Crop Protection* 23, 275–285.
Oliva, B., Piccirilli, E., Ceddia, T., Pontieri, E., Aureli, P. and Ferrini, A.M. (2003) Antimycotic activity of *Melaleuca alternifolia* essential oil and its major components. *Letters in Applied Microbiology* 37, 185–187.
Oxenham, S.K., Svoboda, K.P. and Walters, D.R. (2005) Antifungal activity of the essential oil of basil (*Ocimum basilicum*). *Journal of Phytopathology* 153, 174–180.
Parent-Massin, D. (2004) Haematotoxicity of trichothecenes. *Toxicology Letters* 153, 75–81.
Pawar, V.C. and Thaker, V.S. (2007) Evaluation of the anti-*Fusarium oxysporum* f. sp *cicer* and anti-*Alternaria porri* effects of some essential oils. *World Journal of Microbiology and Biotechnology* 23, 1099–1106.
Pretorius, J.C. (2009) Extracts and compounds from '*Agapanthus africanus*' and their use as biological plant protecting agents. *Patent Application Publication*, Pub. No. US 2009/0258097 A1.
Provine, H. and Hadley, S. (2000) Preliminary evaluation of a semisolid agar antifungal susceptibility test for yeasts and molds. *Journal of Clinical Microbiology* 38, 537–541.
Reid, L.M., Woldemariam, T., Zhu, X., Stewart, D.W. and Schaafsma, A.W. (2002) Effect of inoculation time and point of entry on disease severity in *Fusarium graminearum*, *Fusarium verticillioides*, or *Fusarium subglutinans* inoculated maize ears. *Canadian Journal Plant Pathology* 24, 162–167.
Rheeder, J.P., Marasas, W.F.O., Thiel, P.G., Sydenham, E.W., Shephard, G.S. and Van Schalkwyk, D.J. (1992) *Fusarium moniliforme* and fumonisins in corn in relation to human esophageal cancer in Transkei. *Phytopathology* 82, 353–357.
Rial-Otero, R., Arias-Estevez, M., Lopez-Periago, E., Cancho-Grande, B. and Simal-Gandar, J. (2005) Variation in concentrations of the fungicides tebuconazole and dichlofluanid following successive applications to greenhouse-grown lettuces. *Journal of Agricultural and Food Chemistry* 53, 4471–4475.
Rongai, D., Cerato, C. and Lazzeri, L. (2009) A natural fungicide for the control of *Erysiphe betae* and *Erysiphe cichoracearum*. *European Journal of Plant Pathology* 124, 613–619.
Ross, P.F., Rice, L.G., Osweiler, G.D., Nelson, P.E., Richard, J.L. and Wilson, T.M. (1992) A review and

update of animal toxicoses associated with fumonisin-contaminated feeds and production of fumonisins by *Fusarium* isolates. *Mycopathologia* 117, 109–114.
Rossetto, M., Slade, R., Baverstock, P., Henry, R. and Lee, L.S. (1999) Microsatellite variation and assessment of genetic structure in tea tree (*Melaleuca alternifolia — Myrtaceae*). *Molecular Ecology* 8, 633–643.
Shelton, D., Aitken, K., Doimo, L., Leach, D., Baverstock, P. and Henry, R. (2002) Genetic control of monoterpene composition in the essential oil of *Melaleuca alternifolia* (Cheel). *Theoretical and Applied Genetics* 105, 377–383.
Shim, W.-B. and Woloshuk, C.P. (2001) Regulation of fumonisin B1 biosynthesis and conidiation in *Fusarium verticillioides* by a cyclin-like (C-Type) gene, *FCC1*. *Applied and Environmental Microbiology* 67, 1607–1612.
Soylu, E.M., Soylu, S. and Kurt, Ş. (2006) Antimicrobial activities of the essential oils of various plants against tomato late blight disease agent *Phytophthora infestans*. *Mycopathologia* 161, 119–128.
Soylu, S., Yigitbas, H., Soylu, E.M. and Kurt, Ş. (2007) Antifungal effects of essential oils from oregano and fennel on *Sclerotinia sclerotiorum*. *Journal of Applied Microbiology* 103, 1021–1030.
Straede, A. and Heinisch, J.J. (2007) Functional analyses of the extra- and intracellular domains of the yeast cell wall integrity sensors Mid2 and Wsc1. *FEBS Letters* 581, 4495–4500.
Suheri, H. and Price, T.V. (2000) Infection of onion leaves by *Alternaria porri* and *Stemphylium vesicarium* and disease development in controlled environments. *Plant Pathology* 49, 375–382.
Tegegne, G., Pretorius, J.C. and Swart, W.J. (2008) Antifungal properties of *Agapanthus africanus* L. extracts against plant pathogens. *Crop Protection* 27, 1052–1060.
Telci, I., Bayram, E., Yılmaz, G. and Avcı, B. (2006) Variability in essential oil composition of Turkish basils (*Ocimum basilicum* L.). *Biochemical Systematics and Ecology* 34, 489–497.
Terzi, V., Morcia, C., Faccioli, P., Valè, G., Tacconi, G. and Malnati, M. (2007) *In vitro* antifungal activity of the tea tree (*Melaleuca alternifolia*) essential oil and its major components against plant pathogens. *Letters in Applied Microbiology* 44, 613–618.
United Nations (1996) World population prospects: the 1996 revisions. United Nations, New York.
Vannacci, G. and Gullino, M.L. (2000) Use of biocontrol agents against soil-borne pathogens: results and limitations. *ACTA Horticulture* 532, 79–87.
Velluti, A., Sanchis, V., Ramos, A.J. and Marín, S. (2004) Effect of essential oils of cinnamon, clove, lemon grass, oregano and palmarosa on growth of and fumonisin B1 production by *Fusarium verticillioides* in maize. *Journal of the Science of Food and Agriculture* 84, 1141–1146.
Waterman, P.G. and Mole, S. (1989) Extrinsic factors influencing production of secondary metabolites in plants. In: Bernays, E.A. (ed.), *Insect–Plant Interactions.* CRC Press, Boca Raton, Florida, pp.107–134.
Wink, M. (2003) Evolution of secondary metabolites from an ecological and molecular phylogenetic perspective. *Phytochemistry* 64, 3–19.
Xue, A., Chen, Y., Voldeng, H., Savard, M. and Tian, X. (2008) Biological control of *Fusarium* head blight of wheat with *Clonostachys rosea* strain ACM941. 3rd International FHB Symposium, Szeged, Hungary.
Xue, A.G. (2000) Effect of seed-borne *Mycosphaerella pinodes* and seed treatments on emergence, foot rot severity and yield of field pea. *Canadian Journal Plant Pathology* 22, 248–253.
Xue, A.G., Warkentin, T.D., Greeniaus, M.T. and Zimmer, R.C. (1996) Genotypic variability in seedborne infection of field pea by *Mycosphaerella pinodes* and its relation to foliar disease severity. *Canadian Journal of Plant Pathology* 18, 370–374.
Yangui, T., Bouaziz, M., Dhouib, A. and Sayadi, S. (2009) Potential use of Tunisian *Pituranthos chloranthus* essential oils as a natural disinfectant. *Letters in Applied Microbiology* 48, 112–117.
You, M.P., Yang, H.A., Sivasithamparam, K. and Barbetti, M.J. (2009) A new leaf blight disease of *Trifolium dasyurum* caused by *Botrytis fabae*. *European Journal of Plant Pathology* 123, 99–103.
Zhou, F., Kuth, J., Wei, F., Elliott, C., Valè, G., Yahiaoui, N., Keller, B., Somerville, S., Wise, R. and Schulze-Lefert, P. (2001) Cell-autonomous expression of barley Mla1 confers race-specific resistance to the powdery mildew fungus via a Rar1-independent signaling pathway. *Plant Cell* 13, 337–350.

16 Fruit Postharvest Disease Control by Plant Bioactive Compounds

Marta Mari,* Fiorella Neri and Paolo Bertolini

Introduction

A diet rich in fruit and vegetables plays a significant role in improving the health of people around the world. In fact, fresh products are rich in water, fibre, minerals, vitamins and other phytochemicals; in particular, they are a good source of vitamin C, provitamins A and potassium, fundamental in balancing many essential body functions and for correct human growth, while flavonoids provide most of the antioxidant activity involved in prevention of cancer and cardiovascular diseases (Liu, 2003; Prior, 2003). The daily consumption of at least five portions (400 g) of fruits and vegetables is recommended for the prevention of chronic disease and alleviation of micronutrient deficiencies (WHO, 1990); however, consumption of fresh fruits and vegetables is below this level in the USA and many European countries (Agudo *et al.*, 2002; USDA, 2008). An increase of fruit consumption and the promotion of fruit as an alternative to 'junk food', especially among children, are encouraged by the European Community.

In the past few decades, the fruit industry has changed considerably in order to adapt itself to the fast evolution of the market and to the requirements of consumers increasingly concerned about food healthiness and environmental issues. Fruits and vegetables contribute significantly to the economy of the European countries in the Mediterranean area, among the leading exporters of fruit in the world. Transport by road, rail, sea or air allows fruit and vegetables to reach large areas of the world; however, to be distributed over time and space, these products need to be subjected to a short or long retention period, in relation to product characteristics and market demands. For example, strawberries are stored for just a few days (2–3), while pears, apples and kiwifruits sustain longer periods of storage (up to 7 months), after which they reach suitable quality for consumption (Sanzani *et al.*, 2009a). Between harvest and consumption both quantitative and qualitative fruit losses can occur, caused by diseases, disorders and progressive deterioration of fruit quality. Postharvest losses vary in relation to commodities and country; although few up-to-date data are available (Amorim *et al.*, 2008), they can be estimated in the range of 4–8% in countries where refrigeration facilities are well developed, to 50% where these facilities are minimal (Eckert and Ogawa, 1985). Microbial decay is one of the main factors that determine losses, also compromising the

* Corresponding author.

quality of the fresh produce. The use of appropriate postharvest technology procedures, such as low-temperature storage, controlled atmosphere, fungicide treatments, etc., have contributed to extend fruit storage, reducing product losses from production to retail sites in more developed countries (Kader, 2005). However, during the past 20 years, the use of fungicides in postharvest treatments has been considerably reduced, increasing the interest in alternative approaches to control postharvest diseases (Mari *et al.*, 2007). Among these new approaches, plant bioactive compounds could be a good alternative, fitting well with the concept of sustainable agriculture.

Plants produce a myriad of secondary metabolites important for their interaction with the environment; many of them are associated with the defence system and can function as fungal inhibitors (Osbourn, 1996; Treutter, 2005). These compounds are usually concentrated in the outer cell layers of plant organs from which they can spread into the atmosphere or rhizosphere, and are often sequestrated or excreted into extracellular compartments of the plant to avoid self-toxicity (Sirikantaramas *et al.*, 2008). Some antimicrobials, generally stored in secretory glands, ducts or glandular trichomes, are constitutive and occur in healthy plants in their active forms. Others, such as glucosinolates, occur as inactive precursors and are activated in their biologically active form in response to pathogen or tissue damage. Among secondary metabolites involved in defence, many flavour compounds of horticultural products commonly used in the human diet have exhibited a fungicidal activity against fruit postharvest pathogens (Table 16.1) and were found particularly interesting as novel means for decay control because of their safety at low concentrations. Most flavour compounds are also widely used as food additives, and the Joint FAO/WHO Expert Committee on Food Additives expressed no safety concern at current levels of intake for allyl-isothiocyanate (AITC), *p*-anisaldehyde, carvacrol, (-) carvone, *trans*-cinnamaldehyde, hexanal, *trans*-2-hexenal, 2-nonanone, terpineol and thymol, when used as flavouring agents. Although different forms of application (liquid or vapour phase) and measurements of pathogen inhibition (mycelial growth and/or conidial germination) applied in the studies often make it difficult to compare the minimal inhibitory concentrations (MICs) obtained, the most consistent fungicidal activity by plant bioactive compounds was found with some isothiocyanates (ITCs), followed by *trans*-2-hexenal, *trans*-2-nonenal, carvacrol, thymol, citral and *trans*-cinnamaldehyde (Mari *et al.*, 1993; Caccioni and Guizzardi 1994; Andersen *et al.*, 1994; Caccioni *et al.*, 1995; Tsao and Zhou, 2000a; Plotto *et al.*, 2003; Neri *et al.*, 2006a, 2007, 2009a,b; Arroyo *et al.*, 2007). The main factors involved in the antimicrobial activity of the compounds proved to be activity of functional groups, hydrophobicity and vapour pressure (Andersen *et al.*, 1994; Caccioni *et al.*, 1997; Ultee *et al.*, 2002; Arfa *et al.*, 2006). In addition, sub-lethal concentrations can stimulate the spore germination or mycelial growth of pathogens (Eckert and Ratnayake, 1994; Fallik *et al.*, 1998; Palhano *et al.*, 2004; Neri *et al.* 2006 a, 2009a). Some essential oils of edible products have demonstrated an inhibitory activity on postharvest pathogens (Table 16.2), although at concentrations usually higher than single bioactive compounds. In addition to a direct inhibitory effect on pathogens, some plant secondary metabolites, such as flavonoids and jasmonates, seem to be involved in enhancing the defence responses of the host (Ark and Thompson, 1959; Moline *et al.*, 1997). The inhibitory activity of some phenolic compounds (quercetin and umbelliferone) evaluated against *Penicillium expansum* growth was, for example, higher *in vivo* than *in vitro* trials, suggesting that these compounds act mainly by enhancing the fruit defence system (Sanzani *et al.*, 2009b). Similarly, jasmonate treatment proved to control decay in several fruit species; however, it does not have a direct effect on pathogen development.

The *in vitro* inhibition by plant compounds has not always been confirmed in *in vivo* assays. Besides chemical characteristics, other factors proved to influence the effectiveness of antifungal compounds in

Table 16.1. Flavour compounds with antifungal activity against fruit postharvest pathogens.

Compounds	Natural occurrence	Target pathogens	References
Alcohols (unsaturated)			
trans-2-Hexenol	Olive, strawberry, blackberry, kiwi fruit, tea	*Alternaria alternata, Colletotrichum acutatum*	Andersen *et al.*, 1994; Arroyo *et al.*, 2007
Linalool	Peach, apricot, cherry, plum, strawberry, raspberry, coriander, basil, lavender	*Botrytis cinerea, Penicillium expansum*	Tsao and Zhou, 2000a; Plotto *et al.*, 2003; Neri *et al.*, 2009b
α-Terpineol	Orange, mandarin, apricot, mango, marjoram	*B. cinerea, Geotrichum candidum*	Tsao and Zhou, 2000a; Plotto *et al.*, 2003
Alcohols (phenols)			
Carvacrol	Oregano, thyme, summer savory, marjoram	*B. cinerea, G. candidum, Monilinia fructicola, Monilinia laxa, Mucor piriformis, Neofabraea alba, P. expansum, Penicillium italicum, Penicillium digitatum, Rhyzopus stolonifer*	Caccioni and Guizzardi, 1994; Tsao and Zhou, 2000a; Plotto *et al.*, 2003; Neri *et al.*, 2006 a, 2007, 2009a
Eugenol	Allspice, clove, basil	*B. cinerea, M. fructicola, Monilinia fructigena, M. laxa, Neofabraea alba, P. expansum*	Tsao and Zhou, 2000a; Plotto *et al.*, 2003; Neri *et al.*, 2006a, 2007, 2009a
Thymol	Thyme, oregano	*B. cinerea, M. fructicola, G. candidum, Rhyzopus stolonifer*	Tsao and Zhou, 2000a; Plotto *et al.*, 2003
Aldehydes (aromatic)			
p-Anisaldehyde	Anise	*B. cinerea, M. laxa, M. piriformis, N. alba, P. expansum, Penicillium italicum, P. digitatum, R. stolonifer*	Caccioni and Guizzardi, 1994; Neri *et al.*, 2006a, 2007, 2009a
trans-Cinnamaldehyde	Cinnamon, cassia	*Colletotrichum gloeosporioides, Botryodiplodia theobromae, Glicephalotrichum microchlamydosporum, M. laxa, N. alba, P. expansum*	Sivakumar *et al.*, 2002; Neri *et al.*, 2006a, 2007, 2009a
Aldehydes (non-aromatic)			
Citral	Lemon, lime, lemongrass, lemon myrtle	*B. cinerea, C. gloeosporioides, G. candidum, M. fructicola, M. laxa, N. alba, P. expansum, P. italicum, P. digitatum*	Caccioni *et al.*, 1995; Tsao and Zhou, 2000a; Plotto *et al.*, 2003; Wuryatmo *et al.*, 2003; Palhano *et al.*, 2004; Neri *et al.*, 2006a, 2007, 2009a

trans-2-Decenal	Watermelon	*B. cinerea, G. candidum*	Plotto *et al.*, 2003
Hexanal	Olive, banana, endive, tomato, cucumber, kiwi fruit, apple, avocado, blackberry	*A. alternata, B. cinerea, C. acutatum, M. laxa, N. alba, P. expansum, R. stolonifer*	Andersen *et al.*, 1994; Caccioni *et al.*, 1995; Neri *et al.*, 2006a, 2007, 2009a; Arroyo *et al.*, 2007
trans-2-Hexenal	Olive, banana, radish, tea, kiwifruit, strawberry, tomato, cucumber, apple, endive, apricot, peach, pear	*A. alternata, B. cinerea, C. acutatum, C. gloeosporioides, M. laxa, N. alba, P. expansum*	Vaughn *et al.*, 1993; Andersen *et al.*, 1994; Hamilton-Kemp *et al.*, 1992; Fallik *et al.*, 1998; Neri *et al.*, 2006a, 2007, 2009a; Arroyo *et al.*, 2007.
Nonanal	watermelon, citrus	*A. alternata, B. cinerea, P. italicum, P. digitatum*	Hamilton-Kempt *et al.*, 1992; Andersen *et al.*, 1994; Caccioni *et al.*, 1995
trans-2-Nonenal	Cucumber, watermelon, melon, olive	*A. alternata, B. cinerea*	Hamilton-Kempt *et al.*, 1992; Andersen *et al.*, 1994
Esters (lactones)			
γ-Decalactone	Peach, apricot, plum	*P. expansum*	Neri *et al.*, 2009b
Ketones			
(-)-Carvone	Spearmint	*B. cinerea, G. candidum, M. laxa, M. piriformis, N. alba, P. expansum, P. italicum, P. digitatum*	Caccioni and Guizzardi, 1994; Plotto *et al.*, 2003; Neri *et al.*, 2006a, 2007, 2009a
2-Nonanone	Raspberry, strawberry, blackberry, rue	*A. alternata, C. musae, M. laxa, N. alba, P. expansum, P. digitatum*	Andersen *et al.*, 1994; Utama *et al.*, 2002; Neri *et al.*, 2006a, 2007, 2009a
Isothiocyanates (ITCs)			
Ally-ITC (2-Propenyl-ITC)	Black and white mustard, horseradish, Brussels sprouts, cabbage, cauliflower	*A. alternata, B. cinerea, M. laxa, M. piriformis, P. expansum, R. stolonifer*	Mari *et al.*, 1993, 2008; Troncoso-Rojas *et al.*, 2005b
Benzyl-ITC	Cabbage, horseradish	*A. alternata, M. laxa, M. piriformis*	Mari *et al.*, 1993, 2008; Troncoso-Rojas *et al.*, 2005a
3-Butenyl-ITC	Black mustard, cabbage	*M. laxa, M. piriformis*	Mari *et al.*, 1993, 2008
p-Hydroxybenzyl-ITC		*B. cinerea, M. laxa, M. piriformis, P. expansum, R. stolonifer*	Mari *et al.*, 1993
4-(methyltio)-butenyl-ITC	Radish, cabbage, rocket, broccoli	*B. cinerea, M. laxa, M. piriformis, P. expansum, R. stolonifer*	Mari *et al.*, 1993, 2008

Table 16.2. Plant essential oils with antifungal activity against fruit postharvest pathogens.

Plant from which essential oils are extracted	Major component	Target pathogens	References
Cymbopogon citratus (lemongrass)	Citral (81%)	*Colletotrichum gloeosporioides*	Palhano *et al.*, 2004
Cymbopogon flexuosus (East Indian lemongrass)	Citral (68–80%)	*Alternaria alternata, Aspergillus niger, Botrytis cinerea, Gleosporium fructigenum, Penicillium expansum, Penicillium digitatum, Penicillium italicum, Rhyzopus nigricans*	Shahi *et al.*, 2003
Cymbopogon sp. (lemongrass)	Citral (69%)	*B. cinerea, Geotrichum candidum*	Plotto *et al.*, 2003
Cymbopogon martini (palmarosa)		*Alternaria citri, B. cinerea, P. digitatum*	Arras and Usai, 2001
Cinnamomum zeylanicum (cinnamon)		*A. citri, B. cinerea, P. digitatum, P. italicum*	Arras and Usai, 2001
C. zeylanicum		*C. gloeosporioides*	Barrera-Necha *et al.*, 2008
Citrus sinensis (orange)	d-limonene (80%)	*A. alternata, A. niger, B. cinerea, P. expansum*	Sharma and Tripathi, 2006
Laurus nobilis (bay laurel)	1,8-cineole (25%)	*B. cinerea, Monilinia laxa*	De Corato *et al.*, 2010
Lippia scaberrima	(–)-carvone (34%)	*P. digitatum*	Du Plooy *et al.*, 2009
Mentha cardiaca (Scotch spearmint)	(–)-carvone (65%)	*B. cinerea*	Plotto *et al.*, 2003
Mentha spicata (spearmint)	(–)-carvone (>80%)	*P. digitatum*	Du Plooy *et al.*, 2009
Origanum compactum	Carvacrol (58%)	*B. cinerea*	Bouchra *et al.*, 2003
Origanum dictamus (dictamus)	Thymol (78%)	*P. digitatum*	Daferera *et al.*, 2000
O. dictamus	Carvacrol (64%)	*B. cinerea*	Daferera *et al.*, 2003
Origanum vulgare (oregano)		*A. citri, B. cinerea, P. digitatum, P. italicum*	Arras and Usai, 2001
O. vulgare	Thymol (63.3%)	*P. digitatum*	Daferera *et al.*, 2000
O. vulgare	Thymol (63.7%)	*B. cinerea*	Daferera *et al.*, 2003
Syzygium aromaticum, syn. *Eugenia caryophyllata* (clove)		*A. citri, B. cinerea, P. digitatum, P. italicum*	Arras and Usai, 2001
S. aromaticum, syn. *Eugenia caryophyllata*		*C. gloeosporioides*	Barrera-Necha *et al.*, 2008
Thymus capitatus (Spanish oregano)	Carvacrol (81–83%)	*A. citri, B. cinerea, P. digitatum, P. italicum*	Arras and Usai, 2001
T. capitatus	Carvacrol (74%)	*B. cinerea, G. candidum, R. stolonifer*	Plotto *et al.*, 2003
Thymus glandolosus	Thymol (43%)	*B. cinerea*	Bouchra *et al.*, 2003
Thymus vulgaris (white thyme)	Thymol (63.6%)	*P. digitatum*	Daferera *et al.*, 2000
T. vulgaris	Thymol (45%)	*B. cinerea, G. candidum, R. stolonifer*	Plotto *et al.*, 2003

disease control, including treatment conditions (form of application, concentration, temperature, exposure time, time of treatment and formulation) and characteristics of the pathogen (age and form of infection structures, location of pathogen in host tissue) (Neri *et al.*, 2006b,c, 2009a). Moreover, detrimental effects on sensory traits (odour, texture and flavour) or phytotoxic symptoms on fruit have been observed in some studies with treatments effective in disease control (Vaughn *et al.*, 1993; Archbold *et al.*, 1997; Neri *et al.*, 2006c, 2007; Shoelberg and Randall, 2007).

Flavour Compounds

These are volatile components that can be perceived by the olfactory receptors and confer an odour. Some of them, defined 'character impact compounds', possess the typical aroma of the products that contain them, while other compounds only contribute to the overall odour of a product.

***trans*-2-Hexenal** is C_6 an α,β-unsaturated aldehyde naturally occurring in olive oil, tea and most fruits and vegetables (Table 16.1). The compound is also used as a flavouring agent and its estimated daily intake in Europe and the USA (791 and 409 μg/person per day, respectively) is below the threshold of concern (1800 μg/person per day) showing no safety concerns at current levels of intake. The compound is known for its antimicrobial properties and is thought to be involved in the defence mechanism of plants. Its production, together with other C_6 aldehydes and alcohols named 'green leafy volatiles', increases rapidly in damaged plant tissues, as a result of the activation of the lipoxygenase hydroperoxide lyase enzymatic pathway in response to wounding and herbivore or pathogen attack (Matsui, 2006). *trans*-2-Hexenal has shown consistent fungicidal activity against many postharvest pathogens (Table 16.1) and its inhibition has been found particularly marked against the conidial form. The presence of a double bond in the molecule has been considered particularly important for the antimicrobial activity of aldehydes and unsaturated compounds such as *trans*-2-hexenal and *trans*-2-nonenal have been found to exhibit a more consistent spectrum of activity than the corresponding saturated compounds hexanal and nonanal (Hamilton-Kemp *et al.*, 1992). The high electrophilic properties of the carbonyl group adjacent to the double bond make *trans*-2-hexenal particularly reactive with nucleophiles, such as protein sulfydryl and amino groups of the pathogens. The exposure to *trans*-2-hexenal vapour caused severe membrane and cell wall damage of hyphae (Fallik *et al.*, 1998), disruption of plasma membrane and cell wall, and high disorganization of cell components in conidia (Arroyo *et al.*, 2007). In a study where *Botrytis cinerea* was exposed to a radiolabelled mixture of *cis*-2-hexenal and *trans*-2-hexenal, it was demonstrated that fungal proteins, and particularly proteins of the surface, were targets of the C_6 aldehydes (Myung *et al.*, 2007). C_6 aldehydes were preferentially incorporated into conidia rather than mycelia, a result correlated to the greater sensitivity of spore germination than mycelial growth to *trans*-2-hexenal. Postharvest exposure to *trans*-2-hexenal vapour was found to significantly reduce the infections of *B. cinerea, Alternaria alternata* and *Colletotrichum acutatum* in soft fruits (Archbold *et al.*, 1997; Arroyo *et al.*, 2007; Neri *et al.*, 2008), *P. expansum* in pome fruits (Neri *et al.*, 2006a,b) and *Monilinia laxa* in stone fruits (Neri *et al.*, 2007), while it failed to control *Neofabraea alba* in apples (Neri *et al.*, 2009a). Treatment with *trans*-2-hexenal showed a curative activity up to 72 h in blue mould control and was also effective in reducing patulin content in pome fruits (Neri *et al.*, 2006b,c); cold storage temperature after exposure of fruit to *trans*-2-hexenal tested in 'Conference' pears did not affect the activity of the compound in blue mould control (Neri *et al.*, 2006b). Short exposure to *trans*-2-hexenal vapours (4 h) by some cultivars of apples, pears and 'Alba' strawberries was also enough to reduce the development of decay. The timing of treatment was particularly important. Fumigation with *trans*-2-hexenal applied immediately after inoculation (2 h) was effective against *M. laxa* and *B. cinerea,* but was generally not useful to control

P. expansum. The control of *P. expansum* in 'Golden Delicious' apples was particularly interesting, where treatment (12.5 μl l^{-1}) applied 24 h after inoculation greatly reduced blue mould infections (over 90% efficacy) and patulin content in fruit (under the limit of quantification), without causing negative effects on quality traits or *trans*-2-hexenal residue in treated fruit (Neri *et al.*, 2006c). In contrast to results on 'Golden Delicious' apples, concentrations effective in decay control caused phytotoxic effects in apricots, peaches, nectarines, 'Abate Fetel' pears and strawberries, while it affected fruit flavour in plums, strawberries, 'Conference' and 'Bartlett' pears and 'Royal Gala' apples (Neri *et al.*, 2006b,c, 2007, 2008). A suberization of the epidermal tissues surrounding the wound was observed in fruit of different species treated with *trans*-2-hexenal compared with the control (Neri *et al.*, 2007, 2008). Besides a direct effect on fungal pathogens, a stimulation of fruit defence mechanisms was hypothesized in decay inhibition by *trans*-2-hexenal (Neri *et al.*, 2007). Fumigation with the corresponding saturated aldehyde **hexanal** (2 mg l^{-1} for 24 h at 15–20°C) proved to inactivate conidia of *B. cinerea* contaminating 'd'Anjou' pear surface, but treatment carried out just after fruit wounding stimulated grey mould (Shoelberg and Randall, 2007). Treatment applied before fruit wounding was also effective in reducing *P. expansum* infection on 'Gala' apples, while it failed to control blue mould when applied after wounding.

Carvacrol, a monoterpenoid phenol with a warm and pungent odour, is the character-impact constituent of oregano essential oils (*Origanum vulgare, O. onites, Thymus capitatus*), in which it occurs at concentrations of 60–80%. It exhibited fungicidal activity against a wide range of postharvest pathogens and in particular a consistent inhibition of mycelial growth. Comparable fungicidal activity was exhibited by the carvacrol isomer **thymol** (Tsao and Zhou, 2000a; Plotto *et al.*, 2003). This compound has a strongly aromatic, burnt, medicinal odour and occurs mainly in thyme. After fumigation of plums with thymol, the compound was found to be deposited in large crystals on the surface of conidia and hyphae of *Monilinia fructicola*. In addition, disorganization of cellular organelles and inclusion bodies were observed on hyphae adjacent to the germinating spores and in the hyphae located on the surface of plums, while no alterations were observed in hyphae located inside plum tissues (Svircev *et al.*, 2007). The antimicrobial activity of carvacrol and thymol has been ascribed to the hydrophobicity of the compounds, the presence of a phenolic hydroxyl group in these molecules and an adequate system of delocalized electrons (double bonds) that allow the OH group to release its proton (Ultee *et al.*, 2002; Arfa *et al.*, 2006). The chemical structure of these molecules would allow these compounds to act as proton exchangers, reducing the gradient across the cytoplasmic membrane. The resulting collapse of the proton motive force and depletion of the ATP pool eventually lead to cell death. Postharvest fumigations with carvacrol or thymol were effective in controlling *B. cinerea* and *M. fructicola* in cherries (Tsao and Zhou, 2000b) and *M. fructicola* on apricots and plums (Liu, *et al.*, 2002). However, these treatments caused phytotoxic symptoms and off-flavours in cherries and a firmer texture and phytotoxicity in apricots. Phytotoxicity after carvacrol or thymol exposure was also observed in oranges (Arras and Usai, 2001) and tomatoes (Plotto *et al.*, 2003). Addition of a mixture of carvacrol, thymol and eugenol inside active packaging was effective in reducing decay in grapes (Guillén *et al.*, 2007), but, as also confirmed in our unpublished trials on cultivar Italia, fumigation with these compounds caused off-flavours in grapes. Exposure to carvacrol vapours failed to control blue mould in pears and only slightly controlled brown rot in peaches and lenticel rot in apples (Neri *et al.*, 2006a, 2007, 2009a).

Citral, naturally occurring with the isomers neral and geranial, is an acyclic α,β-unsaturated aldehyde mainly contained inside oil glands of lemon and lime peel and in essential oils from lemongrass (*Cymbopogon citratus*) and lemon myrtle (*Backhousia citriodora*). The acceptable daily intake established for citral by the Joint FAO/

WHO Expert Committee on Food Additives (2003) is ≤ 0.5 mg/kg body weight; the compound is generally regarded as safe (GRAS) in the USA. The antifungal activity of citral against several postharvest pathogens has been well documented (Table 16.1). As for *trans*-2-hexenal, the fungicidal activity of citral has been ascribed to the high electrophilic properties of the carbonyl group adjacent to the double bond. In contrast to *in vitro* results, postharvest fumigation with citral showed a low degree of efficacy in blue mould and brown rot control (Neri *et al.*, 2006a and 2007) and failed to control lenticel rot (Neri, 2009a). Treatment with citral caused severe phytotoxicity on tomato fruits, either when evaluated as a pure compound or as the main constituent of lemongrass essential oil (Plotto *et al.*, 2003).

***trans*-Cinnamaldehyde**, an aromatic aldehyde with the typical odour of cinnamon, is the main constituent of essential oils of cinnamon bark and cassia bark and leaves. Its fungicidal activity against postharvest pathogens has been demonstrated in *in vitro* studies (Table 16.1). Exposure of rambutan to *trans*-cinnamaldehyde vapours significantly reduced fungal infections without any negative effect on fruit, while treatment in an aqueous solution of *trans*-cinnamaldehyde caused phytotoxic symptoms (Sivakumar *et al.*, 2002). Vapour treatment with *trans*-cinnamaldehyde failed to control blue mould in pears, brown rot in peaches and lenticel rot in apples (Neri *et al.*, 2006a, 2007, 2009a).

Essential Oils

Essential oils are concentrated mixtures of volatile compounds that can be extracted from fruits, spices or herbs (peel, flowers, buds, leaves, seeds, bark and root). Agro-industrial by-products commercially available in large quantities, such as citrus peel, could also be a good source of essential oils. Most essential oils are characterized by 1–3 main components which impart the characteristic odour or flavour to the oil and are generally the bioactive ingredients; other oils have a large number of major constituents. In general, essential oils and their active ingredients act to make the cell membrane of fungi permeable, causing the contents to leak out. Among the main advantages of the use of essential oils, the possible synergism in the antimicrobial activity among different components and the low risk of resistance development by pathogens because of the variety of functional groups have been indicated (Daferera *et al.*, 2003). Vice versa, the variability in chemical composition of essential oils could be a critical aspect. The composition of essential oils can, in fact, be influenced by many factors, such as climatic and seasonal conditions or harvest period, and the existence of chemical polymorphism in populations could affect the antimicrobial properties of essential oils, as has been observed in thyme (Gonçalves *et al.*, 2010).

In studies comparing the antifungal activity of several essential oils, those containing mainly carvacrol (*T. capitatus* and *O. compactum*) or thymol (*T. glandolosus*, *T. vulgaris* and *O. vulgare*) have been shown to exhibit the highest mycelial growth inhibition of many postharvest pathogens (Table 16.2). Oils of thyme or oregano were found to cause alterations of the morphology of fungal hyphae; severe damage was observed to cell walls, cell membranes and cellular organelles (Arras *et al.*, 1995; Rasooli and Owlia, 2005, Svircev *et al.*, 2007). *Cymbopogon flexuosus* (lemongrass) essential oil, known to be rich in citral, also showed a broad spectrum of antifungal activity at concentrations ranging from 0.2 to 0.5 μl ml^{-1} (Shahi *et al.*, 2003). Vapours of *Citrus sinensis* essential oil, containing 84% of limonene, completely inhibited the growth of several postharvest pathogens (Sharma and Tripathi, 2006). Among nine essential oils tested, cinnamon (*Cinnamomum zeylanicum*) and clove (*Syzygium aromaticum*) oils, known for their high content of *trans*-cinnamaldehyde and eugenol respectively, exhibited the greatest inhibition of *C. gloeosporioides* conidial germination (Barrera-Necha *et al.*, 2008).

Different results with the application of essential oils have been found in *in vivo* assays. Exposure to vapours of thyme or lemongrass oil showed a low reduction of *B. cinerea*, *P. expansum* and *R. stolonifer* infections in peaches; moreover, thyme oil caused severe phytotoxic symptoms on fruit (Arrebola *et al.*, 2010). Lemongrass essential oil sprayed on apples (30 μl ml^{-1}) completely controlled *P. expansum*, *B. cinerea* and *Phoma violacea* infections, without causing phytotoxic effects (Shahi *et al.*, 2003). Spray emulsion of cinnamon essential oil showed better control of crown rot than clove essential oil (0.2%) in bananas, while it failed to control anthracnose disease (Ranasinghe *et al.*, 2005). Vice versa, dip treatment with clove essential oil (50 μg l^{-1}) showed a higher efficacy in reducing natural infections than cinnamon oil in papayas (Barrera-Necha *et al.*, 2008). The incorporation of *Lippia scaberrima* oil (2500 μl l^{-1}), containing 34% of (-) carvone, in Carnauba Tropical coating proved an excellent disease control in citrus, without detrimental effects on fruit (Du Plooy *et al.*, 2009). The advantage of using coatings amended with essential oils could be the close contact between the essential oils and the fruit surface over a long period. Spray application of laurel essential oil (3 mg ml^{-1}), containing several components (mainly 1,8-cineole, linalool, terpineol acetate and methyl eugenol), showed good antifungal activity against *M. laxa* in peaches, while it exhibited less control of *B. cinerea* in kiwifruits and *Penicillium digitatum* in oranges and lemons (De Corato *et al.*, 2010).

Glucosinolates

Glucosinolates (GLs) are a large group of β-thioglucoside N-hydroysulfates that can be found in plants from at least 11 different families of Brassicaceae, Capparidaceae and Resedaceae where a single GL is often dominant (Fenwick *et al.*, 1983). These compounds contained in the vacuoles are biologically active when they are hydrolysed by the enzyme myrosinase (α–thioglucosidase glucohydroalse; EC 3.2.3.1), stored in cell walls, endoplasmatic reticulum, Golgi vesicles and mitochondria but away from vacuoles. On mechanical damage, and infection or pest attack, cellular breakdown exposes the GLs to a degradative enzyme, determining the production of isothiocyanates (ITCs), nitriles, thiocyanates and oxazolidinethiones according to reaction conditions such as pH, iron concentration and the chemical nature of the GL. The ITCs showed a strong activity against a wide range of food pathogens in specific biological tests reviewed by Delaquis and Mazza (1995); however, their fungicidal effect against a given fungus varies depending on the specific ITC. Carter *et al.* (1963) found aromatic ITCs, like benzyl and 2-phenylethyl-ITC, more fungitoxic than those with R groups such as allyl-isothiocyanate (AITC), butenyl and 4-methylthiobutyl-ITC. In contrast, other authors (Stahmann *et al.*, 1943; Mari *et al.*, 2008) found no difference between aromatic phenyl or allyl aliphatic-ITCs. Some ITCs are volatile substances and could potentially be successfully employed as gaseous treatments in a new process defined 'biofumigation'. The term introduced originally to describe soil-borne pest and disease control by GL-containing plants (Kirkegaard *et al.*, 1998) is now used in other examples of exploitation of these compounds such as in pre-cooked roast beef (Delaquis and Mazza, 1995) and grain storage (Worfel *et al.*, 1997), to create a sterile atmosphere in wine during storage and for bread packaging (Nielsen and Rios, 2000). The postharvest phase, characterized by restricted environmental parameters such as temperature, relative humidity and atmosphere gas composition, represents an advantage for biofumigation of fresh or slightly processed fruits and vegetables before storage; it could in fact be a valid alternative to traditional fungicide treatments in controlling fungal pathogens. In *in vitro* tests, some volatile ITCs such as AITC and butenyl-ITC have shown significant inhibition of conidia germination and mycelial growth of *M. laxa* and *P. expansum* in stone fruits and pome fruits, respectively (Mari *et al.*, 1993). In addition, benzyl-ITC inhibited *A. alternata* mycelial growth, a postharvest tomato pathogen (Troncoso-Rojas *et al.*, 2005a), while AITC showed an antifungal effect against *P. notatum* (responsible for food spoilage) alone or combined with sulfur dioxide or cinnamaldehyde (Tunc *et al.*, 2007). The advantage of these combinations was to reduce the concentration of each agent and their relative impact in organoleptic properties. In *in vitro* trials, the ITCs were evaluated one at a time, but in most cases *Brassica* species contain more than one glucosinolate. Few data are available in the literature regarding the antifungal effect of a combination of two or more ITCs. For example, four ITCs (AITC, phenyl, benzyl and 2-phenylethyl -ITC) were detected in the macerated leaves of cabbages and the derived ITC mixture exhibited a synergic effect against *A. alternata*, being more toxic that AITC alone, which has been reported to be highly toxic for fungi (Troncoso-Rojas *et al.*, 2005b). ITCs were found to be active against numerous postharvest pathogens and on different hosts; however *in vivo*, their activity did not always confirm the results obtained in *in vitro* trials, showing that the treatment conditions should often be established not only in relation to the active substance and fungal pathogen, but also to fruit and vegetable response to treatment. A better fungicidal activity was observed with butenyl-ITC applied *in vitro* than on tomato fruit; this behaviour could be related to the more complex composition of the ripe fruit as compared with the agar medium used in the *in vitro* test, and the possible escape of volatile compound during the *in vivo* experiment (Troncoso-Rojas *et al.*, 2005a). Few data reported any activity of ITCs produced *in situ*, although their effectiveness was similar. In fact, *P. expansum* on 'Conference' and 'Kaiser' pears was controlled by synthetic AITC vapours (5 mg l^{-1}) or produced by enzymatic hydrolysis of sinigrin, with equally effective results (Mari *et al.*, 2002). In another study, a treatment based on AITC vapour locally produced by wetted defatted meal of *Brassica carinata* was effective against *M. laxa* on peaches and nectarines, showing results similar to those observed with the synthetic compound (Mari *et al.*, 2008). This is an important aspect, showing that biofumigation could be used for industrial applications; the use of bio-based chemicals obtained from renewable natural resources also fits well with the goals of the European technology Platform 'Plant for the Future'.

The mechanism by which ITCs inhibit fungal growth is not yet completely known. A series of hypotheses have, however, been formulated. One of the most realistic is that a non-specific and irreversible interaction of the ITC with the sulfidryl groups, disulfide bonds and amino groups of proteins and amino acid residues takes place (Kojima and Oawa, 1971; Banks *et al.*, 1986). Lin *et al.* (2000) showed that AITC fumigation caused metabolite leakage in bacteria cells. A change in the physicochemical and enzymatic properties of bromelain, papain, chymotrypsin and trypsin after butenyl-ITC treatment was also observed (Rawel *et al.*, 1998). The cell membrane in eukaryotic cells is the main target of toxic lipophilic compounds such as butenyl-ITC; therefore it is possible that butenyl-ITC could react with some enzymes present at the plasma membrane level, causing fungal growth inhibition or cell death (Sikkema *et al.*, 1995). Recently Wang *et al.* (2010) found that AITC treatment enhanced H_2O_2 production and induced the formation of a hydroxyl radical in blueberries. Since hydroxyl radicals are highly reactive molecules that react with various bio-molecules including lipids, proteins and DNA, it is possible that blueberry fruit treated with AITC produced high amounts of reactive oxygen species that resulted in an intolerable level of oxidative stress in fungal cells, with irreparable DNA damage. Another possible mechanism for the antimicrobial activity of AITC might therefore be via an indirect effect of its pro-oxidant action.

Despite numerous data on antifungal, antibacterial, anti-nematode and anti-insect activities of ITCs, only a few investigations reported their effects on treated fruit quality and residue content. ITCs are natural components present in plant tissues at low concentrations, while their activity against plant pathogens requires higher doses that could influence the flavour of fruits. The postharvest quality of bell peppers represented by general appearance (absence of phytotoxic symptoms), weight losses, firmness, titrable acidity and total soluble solids was not affected by the mixture of ITC treatment (Troncoso-Rojas *et al.*, 2005b). In the same way, the postharvest physiology of tomato fruits was not influenced by the butenyl-ITC treatment (Troncoso-Rojas *et al.*, 2005a). Another positive aspect is related to the residues on fruit treated with ITCs; encouraging results shown by Mari *et al.* (2002) revealed that pears treated with AITC (5 mg l^{-1}) and stored for 7 days at 20°C contained less than 12 $\mu g\ g^{-1}$ in the skin and no detectable residue in the pulp.

The beneficial activity of ITCs on human health is well documented, in particular AITC in wasabi (*Wasabia japonica*) has been shown to have a strong antimicrobial effect against *Escherichia coli, Salmonella typhimurium, Pseudomonas aeruginosa* (Inoue *et al.*, 1983) and other food-borne pathogenic bacteria such as *Helicobacter pylori* (Shin *et al.*, 2004). ITCs are also widely investigated because of their low toxicity and significant chemopreventive efficacy (Jeong and Kong, 2005). In animal and cell culture models, micromole doses or micromolar concentrations of phenethyl-ITC and sulforaphane have been shown to prevent cancer through different mechanisms both *in vivo* and *in vitro* (Hayes *et al.*, 2008).

Jasmonates

Jasmonic acid (JA) and its methyl ester, methyl jasmonate (MJ), collectively referred to as jasmonates, are naturally occurring plant regulators known to regulate plant development and response to environmental stress. JA was first obtained from fungal culture filtrate and MJ was first identified as a component of the *Jasminum* essential oil and other plant species. Both of them are linolenic acid-derived cyclopentanone-based compounds and their level in plants varies as a function of tissue and cell type, developmental stage, and in response to several different environmental stimuli (Creelman and Mullet, 1997). The influence of jasmonates on biochemical and physiological reactions of horticultural products during shelf life was reviewed by Gonzalez-Aguilar *et al.* (2006). The direct inhibitory effect of jasmonates on spore germination and germ tube elongation of fungal pathogens was observed *in vitro* studies on *C. acutatum* (Cao *et al.*, 2008) and *P. expansum* (Yao and Tian, 2005); however, other researchers reported that MJ could not directly influence the growth of *P. digitatum* (Droby *et al.*, 1999) and *M. fructicola* (Tsao and Zhou, 2000b). These different results might be related to different sensitivity levels of fungal species to MJ, and jasmonates could therefore be considered the signal molecules in fruits, inducing defence responses and increasing resistance (Ding *et al.*, 2001). An MJ treatment can reduce postharvest diseases of grapefruits (Droby *et al.*, 1999) and peaches (Yao and Tian, 2005), inhibit decay and enhance antioxidant capacity of raspberries (Chanjirakul *et al.*, 2006), and suppress grey mould rot in strawberries (Moline *et al.*, 1997). MJ can be applied as vapour or in solution. The length of exposure to MJ vapours varies, depending on produce and concentrations. A 24 h treatment with MJ at a concentration ranging from 1 to 100 $\mu mol\ l^{-1}$ is generally effective to significantly reduce anthracnose rot on loquat fruit (Cao *et al.*, 2008) and peach decay (Jin *et al.*, 2006). MJ applied by immersion for a shorter time (0.5–3 min) than that used for vapour treatment reduced decay caused by *P. digitatum* in grapefruits (Droby *et al.*, 1999) and *P. expansum* on pears (Zhang *et al.*, 2009). Vapour treatment seems to be more efficient than dipping in chilling-injury prevention in guavas and papayas (Gonzalez-Aguilar *et al.*, 2003, 2004). The efficacy of jasmonates could also be enhanced by an integrated control strategy, using, for example, a yeast antagonist in order to obtain a synergic effect (Cao *et al.*, 2009a).

It is believed that systemic-acquired resistance is dependent on MJ-mediated signalling and is associated with some signal transduction systems, which induce particular enzymes involved in the biosynthesis of defence compounds such as polyphenols, reactive oxygen species (ROS) or pathogenesis-related (PR) proteins. Exogenous MJ application has been shown to induce the m-RNA accumulation of heat shock proteins in tomato fruit (Ding *et al.*,

2001), to enhance ROS generation on loquat fruit (Cao *et al.*, 2008), to induce anthocyanins, phenols and ethylene antagonists (Gonzalez-Aguilar *et al.*, 2004).

The effect of MJ on quality of treated fruit is produce dependent and could be controversial. A decrease in weight loss associated with a reduction of transpiration was observed in MJ-treated mangoes during low-temperature storage and shelf life (Gonzales-Aguilar *et al.*, 2000). MJ treatments decrease firmness of strawberries (Perez *et al.*, 1997) and 'Kent' mangoes but not of 'Tommy Atkins' mangoes during cold storage (Gonzales-Aguilar *et al.*, 2000). Fruit colour, an important factor in the decision to purchase fruits, was enhanced during the shelf life of different crops treated with MJ, such as 'Golden Delicious' apple (Fan *et al.*, 1998), 'Kent' mangoes (Gonzalez-Aguilar *et al.*, 2001) and 'Sunrise' papayas (Gonzalez Aguilar *et al.*, 2003). This could be attributed to the positive effect of MJ in ethylene synthesis that promotes chlorophyll degradation and biosynthesis of carotenoides (Fan *et al.*, 1998). MJ is an endogenous phytohormone with an important role in regulating physiological and biochemical processes in plants as well as the biosynthesis of secondary metabolites (Creelman and Mullet, 1997). Free sugar and organic acid are natural components of many fruits and they are involved in maintaining fruit quality, determining nutritional value and influencing the organoleptic properties. In a recent study, it was reported that MJ vapour applied to strawberries influenced the composition of its volatile fraction, enhancing in general the production of the most relevant aroma compounds (De la Pena Moreno *et al.*, 2010). In addition, it was demonstrated that MJ treatment in loquat fruit can maintain higher levels of sugars and organic acids during storage by reduction of respiration rate and ethylene production (Cao *et al.*, 2009b). The antioxidant constituents in plants have health-promoting effects in the prevention of chronic problems and are raising interest among scientists and consumers (Surh, 2003). Loquat fruit treated with MJ exhibited a significantly higher radical-scavenging activity than untreated fruit throughout storage. Similar results were also obtained in berry fruits such as raspberries (Chanjirakul *et al.*, 2006), strawberries (Ayala-Zavala *et al.*, 2005) and blackberries (Wang *et al.*, 2008), suggesting that a pre- or postharvest application of MJ will improve the health benefit of these fruits by enhancing their antioxidant activity (Cao *et al.*, 2009b).

Plant Extracts

Plant extracts have recently received attention as one of several non-chemical control options of postharvest fruit diseases. Their potential has long been recognized (Ark and Thompson, 1959), although the use of these compounds to control plant disease is still limited. Extracts of *Allium* or *Capsicum* species showed the greatest antifungal activity against *B. cinerea* among 345 tested (Wilson *et al.*, 1997). Extracts of garlic cloves were active against *P. digitatum* and *P. italicum in vitro* and *in vivo* (Obagwu and Korsten, 2003). The antifungal activity of garlic is attributed to allicin, the main component of garlic extracts, produced, during the crushing of garlic cloves, by the interaction between the non-protein amino acid alliin and the enzyme alliinase (Stoll and Seebeck, 1951). Several aqueous or ethanol extracts of various plants, cultivated and spontaneous, were evaluated against a range of postharvest pathogens as reviewed by Tripathi and Dubey (2004). The major compounds found in *Acacia seyal* and *Withania somnifera* extracts were, respectively, gallic and caffeic acid and they controlled citrus green mould, retaining fruit quality (Mekbib *et al.*, 2009). The mode of action of these two extracts was due to a significant increase in total soluble phenolic concentration at the wound site; in addition a direct interaction of extracts with the pathogen by sticking its spores together at the infection site was also reported (Mekbib *et al.*, 2007). Since the majority of subsistence farmers in African regions are dependent on locally available disease control measures, *A. seyal* and *W. somnifera* extracts can represent important tools to replace synthetic fungicides, being easily obtained and commonly used in traditional medicinal practices in African countries. The presence of the active compounds in the extracts is not stable over time and usually highest in extreme environmental conditions. The plant organ from which extraction of the compounds takes place can affect their composition: extracts from stems are often more effective than those from leaves and flowers. In contrast, leaf extracts of *Annona cherimola*, *Bromelia hemisphaerica* and *Carica papaya* inhibited *Rhizopus* sporulation and rot development on the yellow variety of 'cirulea' fruit better than stem extracts from the same plants (Bautista-Banos *et al.*, 2000). The high variability in composition and concentration of active substances makes it difficult to obtain a standard product for possible formulation; it therefore seems desirable to purify plant extracts, even partially, so as to obtain standardized compounds.

Future Perspectives and Conclusions

Plants are surrounded by a large number of potential enemies (fungi, bacteria, viruses, invertebrates and other plants) in the environment. Because their immobility precludes escape, plants have evolved different strategies to increase their chances of survival. One of these strategies involves the production of secondary metabolites. Tens of thousands of secondary metabolites have been identified in plants (more than 40,000 different terpenes, 20,000 phenolics, 5000 alkaloids, etc.) and the biological role of most of these still remains obscure. A large number of studies have demonstrated the importance of many of these metabolites as plant defence compounds against herbivores, fungal and bacterial diseases or against adverse climatic conditions. Additionally, many natural products serve as signal molecules to mediate pathogenic interactions. As is evident from the amount of research conducted on the antifungal activity of plant bioactive compounds to control fruit postharvest diseases, significant progress in the reduction of fungicides could be achieved.

However, further research is needed to investigate the activity of the compounds in large-scale experiments, their mode of action and their degradation in organisms, which are still not fully understood. For example, AITC is degraded by several different pathways producing many different by-products with a high antimicrobial activity on a broad spectrum of microorganisms. Therefore, before applying AITC, it is necessary to consider how AITC might influence the production of extracellular metabolites which could be induced by AITC to an unwanted level (Nielsen and Rios, 2000). In addition, the different responses found in many studies indicate cultivar specificity in the product–pathogen–compound interaction and a barrier to the use of some plant bioactive compounds may not be efficacy, but rather the off-odours caused in fruits and/or phytotoxicity. The jasmonates, endogenous phytohormones which induce a defence response and increase resistance, could exceed these limits, but their efficacy must be increased by an integrated control strategy, using, for example, biocontrol agents. Finally, some crucial points (practical and economic feasibility) still have to be considered before plant bioactive compounds can be routinely applied in the postharvest phase. There are still very few studies on the effect of treatment on sensory quality, and more investigations are required to avoid the detrimental effect of plant bioactive compounds on the texture and flavour of fruits and vegetables. The absence of any harmful effects on fruit quality is required in order to achieve a practical application of novel control means.

The impressive chemical diversity occurring in plants provides us with an enormous source of active compounds for pharmaceutical and agricultural use. In particular, plant secondary metabolites could have an effective role in sustainable agriculture. In the future, the use of bioactive compounds could play a much greater role in postharvest protection of food in developing countries, where postharvest losses are often severe due to the lack of adequate control measures and fungicides are too expensive and not always available.

In this context, plant extracts are easily obtained and already commonly used in traditional medicinal practices and healing of human aliments in many African countries, although in an empirical way. The development of simple technology, for example, the incorporation of the leaf extract in the coating wax, could be a novel post-harvest treatment for citrus fruit destined for the domestic market. In addition, by-products of plant food processing could be exploited as sources of bioactive compounds. Hulls of rice and pistachio, peels of almonds, citrus fruits and apples, waste water of olive mill and artichoke are, for example, good sources of phenolic compounds.

The need for new and safe fungicides makes plant bioactive compounds interesting for an eco-chemical approach in postharvest disease control and it is expected that natural products could substitute traditional fungicides in the near future.

References

Agudo, A., Slimani, N., Ocké, M.C., Naska, A., Miller, A.B., Kroke, A., Bamia, C., Karalis, D., Vineis, P., Palli, D., Bueno-de-Mesquita, H.B., Peeters, P.H.M., Engeset, D., Hjartaker, A, Navarro, C., Martınez Garcia, C., Wallstrom, P., Zhang, J.X., Welch, A.A., Spencer, E., Stripp, C., Overvad, K., Clavel-Chapelon, F., Casagrande, C. and Riboli, E. (2002) Consumption of vegetables, fruit and other plant foods in the European Prospective Investigation into Cancer and Nutrition (EPIC) cohorts from 10 European countries. *Public Health Nutrition* 5(6B), 1179–1196.

Amorim, L., Martins, M.C., Lourenco, S.A., Gutierrez, A.S.D., Abreu, F.M. and Goncalves, F.P. (2008) Stone fruit injuries and damage at the wholesale market of Sao Paulo, Brazil. *Postharvest Biology and Technology* 47, 353–357.

Andersen, R.A., Hamilton-Kemp, T.R., Hildebrand, D.F, McCracken, C.T., Collins, R.W. and Fleming, P.D. (1994) Structure–antifungal activity relationships among volatile C_6 and C_9 aliphatic aldehydes, ketones, and alcohols. *Journal of Agricultural and Food Chemistry* 42, 1563–1568.

Archbold, D.D., Hamilton-Kemp, T.R., Barth, M.M. and Langlois, B.E. (1997) Identifying natural volatile compounds that control gray mold (*Botrytis cinerea*) during postharvest storage of strawberry, blackberry and grape. *Journal of Agricultural and Food and Chemistry* 45, 4032–4037.

Arfa, A.B., Combes, S., Preziosi-Belloy, L., Gontard, N. and Chalier, P. (2006) Antimicrobial activity of carvacrol related to its chemical structure. *Letters in Applied Microbiology* 43, 149–154.

Ark, P.A. and Thompson, J.P. (1959) Control of certain diseases of plants with antibiotics from garlic (*Alium sativum* L.). *Plant Disease Reporter* 43, 276–282.

Arras, G and Usai, M (2001) Fungitoxic activity of 12 essential oils against four postharvest citrus pathogens: chemical analysis of *Thymus capitatus* oil and its effect in subatmospheric pressure conditions. *Journal of Food Protection* 64, 1025–1029.

Arras, G., Agabbio, M., Piga A., D'hallewin, G. (1995) Fungicide effect of volatile compounds of *Thymus capitatus* essential oil. *Acta Horticulturae* 379, 593–600.

Arrebola, E., Sivakumar, D., Bacigalupo, R. and Kosten, L. (2010) Combined application of antagonist *Bacillus amyloliquefaciens* and essential oils for the control of peach postharvest diseases. *Crop Protection* 29, 369–377.

Arroyo, F.T., Moreno, J., Daza, P., Boianova, L. and Romero, F. (2007) Antifungal activity of strawberry fruit volatile compounds against *Colletotrichum acutatum*. *Journal of Agricultural and Food and Chemistry* 55, 5701–5707.

Ayala-Zavala, J.F., Wang, S.Y., Wang, C.Y. and Gonzalez-Aguilar, G.A. (2005) Methyl jasmonate in conjunction with ethanol treatment increases antioxidant capacity, volatile compounds and postharvest life of strawberry fruit. *European Food Research and Technology* 221, 731–738.

Banks, J.G., Board, R.G. and Sparks, N.H.C. (1986) Natural antimicrobial systems and their potential in food preservation of the future. *Applied Biochemistry and Biotechnology* 8, 103–107.

Barrera-Necha, L., Bautista Banos, S., Flores-Moctezuma, H.E. and Estudillo, A.R. (2008) Efficacy of essential oils on the conidial germination, growth of *Colletotrichum gloeosporioides* (Penz.) Penz. and Sacc and control of postharvest diseases in papaya (*Carica papaya* L.). *Plant Pathology Journal* 7, 174–178.

Bautista-Banos, S., Hernandez-Lopez, M., Diaz-Perez, J.C. and Cano-Ocha, C.F. (2000) Evaluation of the fungicidal properties of plant extracts to reduce *Rhizopus stolonifer* of 'cirulea' fruit (*Spondias purpurea* L.) during storage. *Postharvest Biology and Technology* 20, 99–106.

Bouchra, C., Achouri, M., Hassani, L.M.I. and Hmamouchi, M. (2003) Chemical composition and antifungal activity of essential oils of seven Moroccan Labiatae against *Botrytis cinerea* Per: Fr. *Journal of Ethopharmacology* 89, 165–169.

Caccioni, D.R.L and Guizzardi, M. (1994) Inhibition of germination and growth of fruit and vegetable postharvest pathogenic fungi by essential oil components. *Journal of Essential Oil Research* 6, 174–179.

Caccioni, D.R.L., Deans, S.G. and Ruberto, G. (1995) Inhibitory effect of citrus fruit essential oil components on *Penicillium italicum* and *Penicillium digitatum*. *Petria* 5, 177–182.

Caccioni, D.R.L, Gardini, F., Lanciotti, R. and Guerzoni, E. (1997) Antifungal activity of natural compounds in relation to their vapour pressure. *Sciences des Aliments* 17, 21–34.

Cao, S., Zheng, Y., Yang, Z., Tang, S., Jin, P., Wang, K. and Wang, X. (2008) Effect of methyl jasmonate on the inhibition of *Colletotrichum acutatum* infection in loquat fruit and the possible mechanisms. *Postharvest Biology and Technology* 49, 301–307.

Cao, S., Zheng, Y., Wang, K, Tang, S. and Rui, H. (2009a) Effect of yeast antagonist in combination with methyl jasmonate treatment on postharvest anthracnose rot of loquat fruit. *Biological Control* 50, 73–77.

Cao, S., Zheng, Y., Yang, Z., Wang, K. and Rui, H. (2009b) Effect of methyl jasmonate on quality and antioxidant activity of postharvest loquat fruit. *Journal Science of Food Agriculture* 89, 2064–2070.

Carter, G.A., Garraway, J.L., Spencer, D.M. and Wain, R.L. (1963) Investigation of fungicides. VI. The antifungal activity of certain dithicarbamic and hydroxydithioformic acid derivates. *Annals of Applied Biology* 51, 135–161.

Chanjirakul, K., Wang, S.Y., Wang, C.Y. and Siriphanich, J. (2006) Effect of natural volatile compounds on antioxidant capacity and antioxidant enzymes in raspberries. *Postharvest Biology and Technology* 40, 106–115.

Creelman, R.A. and Mullet, J.E. (1997) Biosynthesis and action of jasmonates in plants. *Annual Review of Plant Physiology* 48, 355–381.

Daferera, D., Ziogas, B.N. and Polissiou, M.G. (2000) GC-MS analysis of essential oils from Greek aromatic plants and their fungitoxicity on *Penicillium digitatum*. *Journal of Agricultural and Food and Chemistry* 48, 2576–2581.

Daferera, D., Ziogas, B.N. and Polissiou, M.G. (2003) The effectiveness of plant essential oils on the growth of *Botrytis cinerea*, *Fusarium* sp. and *Clavibacter michiganensis* subsp. *michiganensis*. *Crop Protection* 22, 39–44.

De Corato, U., Maccioni, O., Trupo, M. and Di Sanzo, G. (2010) Use of essential oil of *Laurus nobilis* obtained by means of a supercritical carbon dioxide technique against post harvest spoilage fungi. *Crop Protection* 29, 142–147.

De la Pena Moreno, F., Blanch, G.P., Flores, G. and Ruiz del Castillo, M.L. (2010) Impact of postharvest methyl jasmonate treatment on the volatile composition and flavonol content of strawberries. *Journal Science of Food Agriculture* 90, 989–994.

Delaquis, P.J. and Mazza, G. (1995) Antimicrobial properties of isothiocyanates in food preservation. *Food Technology* 49, 73–84.

Ding C.K., Wang, C.Y., Gross, K.C. and Smith, D.L. (2001) Reduction of chilling injury and transcript accumulation of heat shock proteins in tomato fruit by methyl jasmonate and methyl salicylate. *Plant Science* 161, 1153–1159.

Droby, S., Porat, R., Cohen, L., Weiss, B., Shapio, B., Philosoph-Hadas, S. and Meir, S. (1999) Suppressing green mould decay in grapefruit with postharvest jasmonate application. *Journal of the American Society of Horticultural Science* 124, 184–188.

Du Plooy, W., Regnier, T. and Combrinck, S. (2009) Essential oil amended coatings as alternatives to synthetic fungicides in citrus postharvest management. *Postharvest Biology and Technology* 53, 117–122.

Eckert, J.W. and Ogawa, J.M. (1985) The chemical control of postharvest diseases: subtropical and tropical fruits. *Annual Review of Phytopathology* 23, 421–454.

Eckert, J.W. and Ratnayake, M. (1994) Role of volatile compounds from wounded oranges in induction of germination of *Penicillium digitatum* conidia. *Phytopathology* 84, 746–750.

Fallik, E., Archbold, D.D., Hamilton-Kemp, T.R., Clements, A. M. and Collins, R.W. (1998) (E)-2-hexenal can stimulate *Botrytis cinerea* growth *in vitro* and on strawberries *in vivo* during storage. *Journal of the American Society for Horticultural Science* 123, 875–881.

Fan, X., Mattheis, J.P. and Fellman, J.K. (1998) Responses of apples to postharvest jasmonate treatments. *Journal of the American Society for Horticultural Science* 123, 421–425.

Fenwick, G.R., Heaney, R.R. and Mullin, W.J. (1983) Glucosinolates and their breakdown products in food and food plants. *Critical Reviews in Food Science and Nutrition* 18, 123–201.

Gonçalves, M.J., Cruz, M.T., Cavaleiro, C., Lopes, M.C. and Salgueiro, L. (2010) Chemical, antifungal and cytotoxic evaluation of the essential oil of *Thymus zygis* subsp. *sylvestris*. *Industrial Crops and Products* 32, 70–75.

Gonzalez-Aguilar, G.A., Fortiz, J., Cruz, R., Baez, R. and Wang, C.Y. (2000) Methyl jasmonate reduces chilling injury and maintains postharvest quality of mango fruit. *Journal of Agricultural and Food Chemistry* 48, 511–519.

Gonzalez-Aguilar, G.A., Buta, J.G. and Wang, C.Y. (2001) Methyl jasmonate reduces chilling injury symptoms and enhances colour development of 'Kent' mangoes. *Journal of the Science of Food and Agriculture* 81, 1244–1249.

Gonzalez-Aguilar, G.A., Buta, J.G. and Wang, C.Y. (2003) Methyl jasmonate and modified atmosphere packaging (MAP) reduce decay and maintain postharvest quality of papaya 'Sunrise'. *Postharvest Biology and Technology* 28, 361–370.

Gonzalez-Aguilar, G.A., Tiznado-Hernandez, M., Zavaleta-Gaica, R. and Martinez-Tellez, M.A. (2004) Methyl jasmonate treatment reduce chilling injury and activate the defense response of guava fruits. *Biochemical and Biophysical Research Communications* 313, 704–711.

Gonzalez-Aguilar, G.A., Tiznado-Hernandez, M. and Wang, C.Y. (2006) Physiological and biochemical responses of horticultural products to methyl jasmonate. *Stewart Postharvest Review* 1, 1.

Guillén, F., Zapata, P.J., Martinez-Romero, D., Castillo, S., Serrano, M. and Valero, D. (2007) Improvement of the overall quality of table grapes stored under modified atmosphere packaging in combination with natural antimicrobial compounds. *Journal of Food Science* 72, S185–S190.

Hamilton-Kemp, T.R., McCracken, C.T., Loughrin, Jr J.H., Andersen, R.A. and Hildebrand, D.F. (1992) Effects of some natural volatile compounds on the pathogenic fungi *Alternaria alternata* and *Botrytis cinerea*. *Journal of Chemical Ecology* 18, 1083–1091.

Hayes, J.D., Kelleher, M.O. and Eggleston, I.M. (2008) The cancer chemopreventive actions of phytochemicals derived from glucosinoltes. *European Journal of Nutrition* 47 (Suppl 2), 73–88.

Inoue, S., Goi, H., Miyauchi, K., Muraki, S., Ogihara, M. and Iwatani, Y. (1983) Inhibitory effect of volatile constituents of plants on the proliferation of bacteria. Antibacterial activity of plant volatiles. *Journal of Antibacterial Antifungal Agents* 11, 600–615.

Jeong, W.S. and Kong, A.N. (2005) Chemopreventive functions of isothiocyanates. *Drug News & Perspectives* 18, 445-451.

Jin, P., Zheng, Y.H., Cheng, C.M., Gao. H.Y., Chen, W.X. and Chen, H.J. (2006) Effect of methyl jasmonate treatment on fruit decay and quality in peaches during storage at ambient temperature. *Acta Horticulturae* 712 II, 711–716.

Kader, A.A. (2005) Increasing food availability by reducing postharvest losses of fresh produce. *Acta Horticulturae* 282, 2169– 2176.

Kirkegaard, J.A., Sarwar, M. and Matthiessen, J.N. (1998) Assessing the biofumigation potential of crucifers. *Acta Horticulturae* 459, 105–111.

Koijma, M. and Oawa, K. (1971) Studies on the effect of isothiocyanates and their analogues on microorganisms. (I). Effects of isothiocyanates on the oxygen uptake of yeast. *Journal of Fermentation and Technology* 49, 740–746.

Lin, C.M., Kim, J.M., Du, W.X. and Wei, C.I. (2000) Bacterial activity of isothiocyanate against pathogens on fresh produce. *Journal of Food Protection* 63, 25–30.

Liu, R.H. (2003) Health benefits of fruit and vegetable are from additive and synergistic combinations of phytochemicals. *American Journal of Clinical Nutrition* 78, 517S–520S.

Liu, W.T., Chu, C.L. and Zhou, T. (2002) Thymol and acetic acid vapors reduce postharvest brown rot of apricots and plums. *HortScience* 37, 151–156.

Mari, M., Iori, R., Leoni, O. and Marchi, A. (1993) *In vitro* activity of glucosinolaes-derived isothyocianates against postharvest fruit pathogens. *Annals of Applied Biology* 123, 155–164.

Mari, M., Leoni, O., Iori, R. and Cembali, T. (2002) Antifungal vapour-phase of allyl-isothiocyanate against *Penicillium expansum* on pears. *Plant Pathology* 51, 231–236.

Mari, M., Neri, F. and Bertolini, P. (2007) Novel approaches to prevent and control postharvest disease of fruits. *Stewart Postharvest Review* 6, 4.

Mari, M., Leoni, O., Bernardi, R., Neri, F. and Palmieri, S. (2008) Control of brown rot on stone fruit by synthetic and glucosinolate-derived isothiocyanates. *Postharvest Biology and Technology* 47, 61–67.

Matsui, K. (2006) Green leaf volatiles: hydroperoxide lyase pathway of oxylipin metabolism. *Current Opinion in Plant Biology* 9, 274–280.

Mekbib, S.B., Regnier, T.J.C. and Korsten, L. (2007) Control of *Penicillium digitatum* on citrus fruit using two plant extracts and study of their mode of action. *Phytoparasitica* 35, 264–276.

Mekbib, S.B., Regnier, T.J.C., Sivakumar, D. and Korsten, L. (2009) Evaluation of Ethiopian plant extracts, *Acacia seyal* and *Withania somnifera*, to control green mould and ensure quality maintance of citrus (*Citrus sinensis* L.). *Fruits* 64, 285–294.

Moline, H.E., Buta, J.G., Saftner, R.A. and Maas, J.L. (1997) Comparison of three volatile natural products for the reduction of postharvest decay in strawberries. *Advances in Strawberry Research* 16, 43–48.

Myung, K, Hamilton-Kemp, T. and Archbold, D. (2007) Interaction with and effects on the profile of proteins of *Botrytis cinerea* by C_6 aldehydes. *Journal of Agricultural and Food Chemistry* 55, 2182–2188.

Neri, F., Mari, M. and Brigati, S. (2006a) Control of *Penicillium expansum* by plant volatile compounds. *Plant Pathology* 55, 100–105.

Neri, F., Mari, M., Menniti, A.M. and Brigati, S. (2006b) Activity of *trans*-2-hexenal against *Penicillium expansum* in "Conference" pears. *Journal of Applied Microbiology* 100, 1186–1193.

Neri, F., Mari, M., Menniti, A.M., Brigati, S. and Bertolini, P. (2006c) Control of *Penicillium expansum* disease in pears and apples by *trans*-2-hexenal vapours. *Postharvest Biology and Technology* 41, 101–108.

Neri, F., Mari, M., Menniti, A.M., Brigati, S. and Bertolini, P. (2007) Fungicidal activity of plant volatile compounds for controlling *Monilinia laxa*. *Plant Disease* 91, 30–35.

Neri, F., Mari, M., Menniti, A.M., Brigati, S. and Bertolini, P. (2008) Control of fruit postharvest decay with *trans*-2-hexenal: perspectives and problems. Commission of the European Communities COST action 924 International Congress, Bologna 4 May 2007, 335–342.

Neri, F., Mari, M., Brigati, S. and Bertolini, P (2009a) Control of *Neofabraea alba* by plant volatile compounds and hot water. *Postharvest Biology and Technology* 51, 425–430.

Neri, F., Veronesi, F. and Mari, M. (2009b) Influence of fruit aroma compounds on *Penicillium expansum* development. *Journal of Plant Pathology* 91, 4, Supplement S 4.76.

Nielsen, P.V. and Rios, R. (2000) Inhibition of fungal growth on bread by volatile components from spices and herbs, and the possible application in active packaging, with special emphasis on mustard oil. *International Journal of Food Microbiology* 60, 219–229.

Obagwu, J. and Korsten, L. (2003) Control of citrus green and blue molds with garlic extracts. *European Journal of Plant Pathology* 109, 221–225.

Osbourn, A.E. (1996) Preformed antimicrobial compounds and plant defense against fungal attack. *Plant Cell* 8, 1821–1831.

Palhano, F.L., Vilches, T.T.B., Santos, R.B., Orlando, M.T.D., Ventura, J.A. and Fernandes, P.M.B. (2004) Inactivation of *Colletotrichum gloeosporoides* spores by high hydrostatic pressure combined with citral or lemongrass essential oil. *International Journal of Food Microbiology* 95, 61–66.

Perez, A.G., Sanz, C. and Olias, J.M. (1997) Effect of methyl jasmonate on *in vitro* strawberries ripening. *Journal of Agricultural and Food Chemistry* 45, 3733–3737.

Plotto, A., Roberts, R.G. and Roberts, D.D. (2003) Evaluation of plant essential oils as natural postharvest control of tomato (*Lycopersicon esculentum*). *Acta Horticulturae* 628, 737–745.

Prior, R.L. (2003) Fruits and vegetables in the prevention of cellular oxidative damage. *American Journal of Clinical Nutrition* 78, 570S–578S.

Ranasinghe, L., Jayawardena, B. and Abeywickrama, K. (2005) An integrated strategy to control postharvest decay of Emul banana by combing essential oils with modified atmosphere packaging. *International Journal of Food Science and Technology* 40, 97–103.

Rasooli, I. and Owlia, P. (2005) Chemoprevention by thyme oils of *Aspergillus parasiticus* growth and aflatoxin production. *Phytochemistry* 66, 2851–2856.

Rawel, H.M., Kroll, J., Riese-Schneider, B. and Haebel, S. (1998) Physicochemical and enzymatic properties of benzyl isothiocyanate derivatized proteinases. *Journal of Agriculture and Food Chemistry* 46, 5043–5051.

Sanzani, S.M., Nigro, F., Mari, M. and Ippolito, A. (2009a) Innovations in the control of postharvest diseases of fresh fruit and vegetables. *Arab Journal of Plant Protection* 27, 241–244.

Sanzani, S.M., De Girolamo, A., Schena, L., Solfrizzo, M., Ippolito, A. and Visconti, A. (2009b) Control of *Penicillium expansum* and patulin accumulation on apples by quercetin and umbelliferone. *European Food Research and Technology* 228, 381–389.

Shahi, S.K., Patra, M., Shukla, A.C. and Sikshit, A. (2003) Use of essential oil as botanical-pesticide against postharvest spoilage in *Malus pumilo* fruits. *BioControl* 48, 223–232.

Sharma, N. and Tripathi, A. (2006) Fungitoxicity of the essential oil of *Citrus sinensis* on the post-harvest pathogens. *World Journal of Microbiology & Biotechnology* 22, 587–593.

Shin, I.S., Masudaa, H. and Naohide, K. (2004) Bacterial activity of wasabi (*Wasabia japonica*) against *Helicobacter pylori*. *International Journal of Food Microbiology* 94, 255–261.

Shoelberg, P. and Randall, P. (2007) Fumigation of stored pome fruit with hexanal reduces blue and gray mold decay. *HortScience* 42, 611–616.

Sikkema, J., de Bont, J.A. and Poolman, B. (1995) Mechanisms of membrane toxicity of hydrocarbons. *Microbiological Review* 59, 201–222.

Sirikantaramas, S., Yamazaki, M. and Saito, K. (2008) Mechanisms of resistance to self-produced toxic secondary metabolites in plants. *Phytochemistry Reviews* 7, 467–477.

Sivakumar, D., Wijeratnam, R.S.W., Wijesundera, R.L.L. and Abeyesekere, M. (2002) Control of postharvest diseases of rambutan using cinnamaldehyde. *Crop Protection* 21, 847–852.

Stahmann, M.A., Link, K.P. and Walker, J.C. (1943) Mustard oils in cruciferous and their relation to resistance to clubroot. *Journal of Agricultural Research* 67, 49–63.

Stoll, A. and Seebeck, E. (1951) Chemical investigation on alliin, the specific principle of garlic. *Advances in Enzymology* 11, 377–400.

Surh, Y.J. (2003) Cancer chemioprevention with dietary phytochemicals. *Nature* 3, 768–780.

Svircev, A.M., Smith, R.J., Zhou, T., Hernandez, M., Liu, W. and Chu, C.L. (2007) Effects of thymol fumigation on survival and ultrastructure of *Monilinia fructicola*. *Postharvest Biology and Technology* 45, 228–233.

Treutter, D. (2005) Significance of flavonoids in plant resistance and enhancement of their biosynthesis. *Plant Biology* 7, 581–591.

Tripathi, P. and Dubey, N.K. (2004) Exploitation of natural products as an alternative strategy to control postharvest fungal rotting of fruit and vegetables. *Postharvest Biology and Technology* 32, 235–245.

Troncoso-Rojas, R., Sanchez-Estrada, A., Ruelas, C., Garcia, H.S. and Tiznado-Hernandez, M.E. (2005a) Effect of benzyl isothiocyanate on tomato fruit infection development by *Alternaria alternata*. *Journal of the Science of Food and Agriculture* 85, 1427–1434.

Troncoso-Rojas, R., Espinoza, C., Sanchez-Estrada, A., Tiznado-Hernandez, M. and Garcia, H.S. (2005b) Analysis of the isothiocyanates present in cabbage leaves extract and their potential application to control *Alternaria* rot in bell peppers. *Food Research International* 38, 701–708.

Tsao, R. and Zhou, T (2000a) Antifungal activity of monoterpenoids against postharvest pathogens *Botrytis cinerea* and *Monilia fructicola*. *Journal of Essential Oil Research* 12, 113–121.

Tsao, R. and Zhou, T. (2000b) Interaction of monoterpenoids, methyl jasmonate, and Ca^{2+} in controlling postharvest brown rot of sweet cherries. *HortScience* 35, 1304–1307.

Tunc, S., Chollet, E., Chalier, P., Preziosi-Belloy, L. and Gontard, N. (2007) Combined effect of volatile antimicrobial agents on the growth of *Penicillium notatum*. *International Journal of Food Microbiology* 113, 263–270.

Ultee, A., Bennik, M.H.J. and Moezelaar, R. (2002) The phenolic hydroxyl group of carvacrol is essential for action against the food-borne pathogen *Bacillus cereus*. *Applied and Environmental Microbiology* 68, 1561–1568.

USDAC (United States Department of Agriculture) (2008) *Fruit and Tree Nuts Yearbook*. Available at: www.ers.usda.gov/Publications/fts/Yearbook08/FTS2008.pdf.

Vaughn, S.F. and Boydston, R.A. (1997) Volatile allelochemicals released by crucifer green manures. *Journal of Chemical Ecology* 23, 2107–2116.

Vaughn, S.F., Spencer, G.F. and Shasha, B.S. (1993) Volatile compounds from raspberry and strawberry fruit inhibit postharvest decay fungi. *Journal of Food Science* 58, 793–796.

Wang, S.Y., Bowman, L. and Ding, M. (2008) Methyl jasmonate enhances antioxidant capacity and flavonoid content in blackberries (*Rubus* sp.) and promotes antiproliferation of human cancer cells. *Food Chemistry* 107, 1261–1269.

Wang, S.Y., Chen, C.T. and Yin, J.J. (2010) Effect of allyl isothiocyanate on antioxidants and fruit decay of blueberries. *Food Chemistry* 120, 199–204.

bark extract showed better antimicrobial effect. The major phytochemicals present in the extracts are flavonoids, saponins, tannins, alkaloids, anthraquinones, terpenes and cardiac glycosides.

Lasienthera africana Beauv

This plant belongs to the family Icacinacea, order Celestrales. *L. africana* is consumed as a vegetable in soup in South-eastern Nigeria. Unorthodox medical practitioners use extracts from *L. africana* for prophylactic and therapeutic purpose. Aqueous and ethanolic extracts from *L. african* show significant antimicrobial effect against clinical isolates of *E. coli, S. typhi, Klebsiella pneumoniae* and *S. aureus* (Adegoke and Adebayo-Tayo, 2009). The phytochemicals present in *L. africana* are alkaloids, saponins, tannins, cardiac glycosides, anthraquinones and cyanogenic glycosides (Adegoke and Adebayo-Tayo, 2009).

Morinda lucida Benth

Morinda lucida belongs to the family Rubiaceae. It is a medium-sized tree which grows up to 15 m high. Locally, the leaves are used in the preparation of fever tea which is used in the treatment of malaria fever and as an analgesic (Olajide *et al.*, 1998). A schizonticidal effect of the leaf juice of the plant had been reported against *Plasmodium berghei* in mice (Obih *et al.*, 1985). Asuzu and Chineme (1990) also reported trypanocidal activity of the methanolic extract of the leaf. The presence of various types of anthraquinones and anthraquinols is a characteristic feature of *M. lucida*. The antimicrobial activity of some anthraquinones isolated from *M. lucida* has been tested for treating fascioliasis and schistosomiasis (Adewunmi and Adesogan, 1983).

Tapinanthus dodoneifolius (DC) Danser

Tapinanthus dodoneifolius belongs to the family Loranthaceae. They are well-known hemiparasites of a variety of different gymnosperms and angiosperms (Deeni and Sadiq, 2002). Ethnomedicinal uses of this plant include treatment of many human and animal ailments such as cancer, gastroenteritis and wound infections (Hussain and Karatela, 1989; Deeni, 1989). A well-structured study by Deeni and Sadiq (2002) revealed that the methanolic extract from the plant has a wide spectrum of antimicrobial activities against certain multiple-drug-resistant bacterial and fungal isolates of farm animals. The organisms inhibited include *Agrobacterium tumefaciens, Bacillus* species, *E. coli, Salmonella* species, *Proteus* species and *Pseudomonas* species. Phytochemicals present in *T. dodoneifolius* include anthraquinones, saponins and tannins. Traces of alkaloids had also been reported (Deeni and Sadiq, 2002).

Gongronema latifolium Benth

Gongronema latifolium belongs to the family Asclepiadaceae. It is commonly called 'utazi' and 'arokeke' in the South-western and South-eastern parts of Nigeria, respectively. This tropical rainforest plant is primarily used as spice and vegetable in traditional folk medicine (Ugochukwu and Babady, 2002). Methanolic extract of this plant was found to inhibit the growth of *S. enteritidis* and *Pseudomonas aeruginosa*, while the aqueous extract was able to inhibit *P. aeruginosa* and *Listeria monocytogenes* (Eleyinmi, 2007). Phytochemicals present in this plant include saponin, tannins and flavonoids (Morebise and Fafunso, 1998).

Diospyros mespiliformis Hochst.

Diospyros mespiliformis is also known as jakkalsbessie (also jackalberry and African ebony). It is a large deciduous tree found mostly in the savannahs of Africa. It is a member of the family Ebenaceae. Jackals are fond of the fruits, hence the common name jackalberry. The fruit is edible for humans; its flavour has been described as lemon-like, with a chalky consistency. They are sometimes preserved by drying and ground into flour. The flour can be used in brewing local beer and brandy. The plant can grow to an average of 4 to 6 m in height, though occasionally trees may be up to 25 m in height. Ethnomedicinal uses of the leaves, bark and roots of the plant include the treatment of leprosy, febrifuge and as a styptic (Lajubutu *et al*., 2006). Diosquinone and plumbagin isolated from the root of have been reported to have wide-ranging antibacterial activity. Some of the organisms inhibited include *S. aureus* NCTC 6571, *S. aureus* E3T, *E. coli* KL16 and *P. aeruginosa* NCTC 6750 (Lajubutu *et al*., 2006).

Heinsia crinata **(Afzel.) G. Taylor**

Heinsia crinata belongs to the family Rubiaceae. It is a shrub with woody stems and branches (Hutchinson and Dalziel, 1972). It is indigenous to West Africa, especially the eastern part of Nigeria. The leaf juice is used to treat various diseases and wounds as well as to treat other gastrointestinal disorders. Two triterpenoid saponins have been isolated from the leaves of the plants. Andy *et al*. (2008) reported the antimicrobial effect of *H. crinata* against *E. coli, S. typhi* and *C. albicans*. Phytochemicals present in the plant include alkaloids and cardiac glycosides (Hussain and Deeric, 1991; Otung, 1998). Traces of saponins, anthraquinones and tannins had also been reported in the plant (Otung, 1998).

Mallotus oppositifolius **(Geisel) Mull –Arg.**

This plant belongs to the family Euphorbiaceae. It is a shrub which is up to 13.5 cm long and 2.5–10 cm wide. *M. oppositifolius* twig is used as chewing sticks for cleaning the teeth; the stem is used as yam stakes. Phytochemical screening of *M. oppositifolius* revealed the presence of secondary metabolites such as alkaloids, phenols, flavonoids, anthraquinones and cardenolides. Ethnomedicinal uses include the treatment of dysentery and as a vermifuge. The leaves are used in preparing antimalarial and anti-inflammatory remedies (Burkhill, 1994). Farombi *et al*. (2001) reported a higher concentration of these phytochemicals resides in the leaves than in the root. The antifungal property of aqueous and ethanol extracts of *M. oppositifolius* against fungi such as *Aspergillus flavus, C. albicans, Microsporum audouinii, Penicillium* sp., *Trichophyton mentagrophytes, Trichoderma* sp. *and Trichosporon cutaneum* was reported by Adekunle and Ikumapayi (2006).

Funtumia **species**

The *Funtumia* species belong to the family Apocynaceae. The genus consists of two common species: *F. elastica* (female) and *F. africana* (male). The decoction of the leaf is used as a cure for mouth and venereal diseases (Sofowora, 1982). Preliminary phytochemical studies of *F. elastica* extracts by Adekunle and Ikumapayi (2006) revealed that they contain anthocyanins, butacyanin, flavonoids, steroids and tannins. Antifungal property of aqueous and ethanol extracts of *F. elastica* against fungi such as *A. flavus, C. albicans, M. audouinii, Penicillium* sp., *T. mentagrophytes, Trichoderma* sp. *and T. cutaneum* have been reported (Adekunle and Ikumapayi, 2006).

Phyllanthus discoideus **Baill**

This plant belongs to the family Euphorbiaceae. It is a small tree widely used in tropical West Africa. In South-west Nigeria, the bark extract is used locally to cure stomach ache and lumbago. It is also useful in the treatment of helminthic infections. Crude aqueous and ethanolic extracts of the plant were found to be active on strains of pathogens such as *S. enteritidis, E. coli* and *S. aureus* at different concentrations with ethanol extract exhibiting more inhibitory activity (Akinyemi *et al*., 2006). The phytochemicals present in the plant include alkaloids, tannins, saponins, flavonoids and anthraquinones.

Cymbopogon citratus Stapf

Cymbopogon citratus, popularly called lemon grass, belongs to the family Graminae. It is native to India but found growing naturally in tropical grassland (Anon, 2007). *C. citratus* is an aromatic, perennial grass. It is a tropical plant, grown as an ornamental in many temperate areas with a maximum height of about 1.8 m and its leaves are 1.9cm wide covered with a whitish bloom. It is used in Ayurvedic medicine to help bring down fevers and the threat of infectious diseases. In Nigeria *C. citratrus* is boiled with other medicinal plants to treat chronic malaria fever. Okigbo and Mmeka (2008) reported the antimicrobial activities of ethanol, cold water and hot water extracts of *C. citratrus* against *S. aureus, E. coli* and *C. albicans*. The phytochemical analysis of *C. citratus* revealed the presence of saponins, tannins, fats and oils, alkaloids and phenol (Nwachukwu *et al.*, 2008).

Garcinia kola Heckel

Garcinia kola belongs to the family Guttiferae. It is also called bitter kola. It is a plant of West and Central Africa origin (Iwu, 1993). Ethnomedicinal uses involve the use of the seed as a masticatory substance and a good preventive agent of dysentery, while the fruit pulp is used for the treatment of jaundice or high fever, and the stem bark is used in medicinal preparations to heal various ailments (Adebisi, 2007). The antibacterial effect of ethanol, cold- and hot-water extracts of *G. kola* against *S. aureus* and *E. coli* has been reported (Okigbo and Mmeka, 2008). The following phytochemical compounds were reported to be present in the seed extract of *Garcinia kola*: flavonoids, tannins, cardiac glycoside, saponins, steroids and reducing sugars (Adegboye *et al.*, 2008).

Bridella ferruginea Benth

This plant belongs to the family Euphorbiaceae. It is commonly found in savannah regions. *B. ferruginea* is a gnarled shrub which sometimes reaches the size of a tree. Ethnomedicinal uses of *B. ferruginea* include the treatment of diabetes. It is also used as purgative and a vermifuge (Cimanga *et al.*, 1999). Irobi *et al.* (1999) tested the water and ethanol extract of the plant against hospital strains of *S. aureus, C. albicans, Staphylococcus epidermidis, E. coli, Streptococcus lactis, Proteus vulgaris, Proteus mirabilis, Streptococcus pyogenes* and *Klebsiella* sp. The zones of inhibition produced by the extracts in agar diffusion assays against the test microorganisms ranged from 4 to 20 mm. In another recent study, Adebayo and Ishola (2009) recorded a significant inhibition against *S. typhi, E. coli, S. aureus, P. mirabilis* and *C. albicans* by crude methanol extracts of the root, stem bark and leaves of *B. ferruginea*. Preliminary phytochemical analysis of the plant extracts showed the presence of phenols and tannins. Sesquiterpenes, anthroquinones and saponins were not detected in the extracts (Irobi *et al.*, 1999).

Calotropis procera Ait. f

This plant belongs to the family Asclepiadaceae. It is commonly called Sodom apple. It is unbranched with a soft wooden trunk, yellowish brown stem bark and the slash exudes caustic latex that turns yellow on exposure to air (Aliero *et al.*, 2001). The juice from the plant is used in cuddling milk in the production of a local milk product called wara. Ethnomedicinal uses of the plant include the treatment of fungal diseases, convulsion, asthma, cough and inflammation (Hassan *et al.*, 2006). Organic solvents and aqueous extracts of the stem bark, leaves and roots were found to significantly inhibit common dermatophytes such as *Trichophyton rubrum* and *Microsporum gypseum*, and a common spoilage fungus, *Aspergillus niger* (Hassan *et al.*, 2006). Alkaloids, flavonoids, tannins, steroids, triterpenoids, saponins and saponin glycosides have been reported in the leaves and roots extracts while stem bark contains flavonoids, triterpenoids and saponins (Hassan *et al.*, 2006).

***Parkia* species**

Parkia species belong to the family Mimosaceae. Two important species of this genus are *Parkia biglobosa* (Jacq) Benth and *Parkia bicolor* A. Chev. *P. biglobosa* (Jacq) also known as African locust bean tree is widely used in Nigeria as sources of timber, food and medicines. The fermented seed is a very common seasoning in Nigerian traditional cuisine. Ethnomedicinal uses of *Parkia* species include the use of the pulverized bark of *P. bicolor* in wound healing, and *P. biglobosa* provides an ingredient that is used for the treatment of leprosy and hypertension (Ajaiyeoba, 2002). Antimicrobial activity of hexane, ethyl acetate, ethanol and water extracts of *P. biglobosa* and *P. bicolor* against standard strains of *P. aeruginosa, A. niger and Candida utilis* have been reported (Ajaiyeoba, 2002). Extracts of *P. bicolor* were slightly more active than those of *P. biglobosa.* El-Mahmood and Ameh (2007) also reported the antimicrobial effect of *P. biglobosa* on *E. coli, S. aureus, K. pneumoniae* and *P. aeruginosa* obtained from urinary tract infections. It was observed that aqueous solutions are more potent than methanolic solutions on the indicator bacteria, and activity is concentration dependent. The major phytochemicals present in *Parkia* species are cardiac glycosides, steroids, tannins and alkaloids (Ajaiyeoba, 2002).

***Mitragyna stipulosa* (DC.) Kuntze**

The plant belongs to the family Rubiaceae. It is a tree which grows to about 3 to 4 m in height. It is widely distributed in East Asia and subtropical Africa. Decoction of the bark of *M. stipulosa* is taken as a heart tonic. The major bioactive compounds in a *M. stipulosa* decoction are mitraphylline and mitraversine. Aboaba *et al.* (2006) recorded significant inhibition of ten strains of *E coli* O157:H7 (EHEC) by ethanol extract of *M. stipulosa* stem bark using the agar diffusion method. A report on phytochemical screening revealed the presence of saponins, tannins and glycosides (Aboaba *et al.*, 2006). Further phytochemical investigation of *M. stipulosa* bark extract has led to the isolation of a series of triterpenoids mainly consisting of quinovic acid and its glycoside derivatives (Naheed *et al.*, 2002).

***Gynandropsis gynandra* L. (Briq.)**

Gynandropsis gynandra belongs to the family Capparidaceae. It is an herb indigenous to the tropical and pantropical regions. This herb is edible and grows up to about 60 cm high (Dalziel, 1937; Burkhill, 1985). *G. gynandra* leaves with a high percentage of vitamin C are taken as a pot herb in soups, fresh or dried (Watt and Breyer-Brandwijk, 1962). The leaves are used as disinfectants. Inhalation of the leaves also relieves headaches; leaf juice and oil are used for earache and eyewash (Oliver, 1960). Ruptured seeds of the plant are reputed to have anthelmintic properties and the oil is used as fish poison. Ajaiyeoba (2000) reported that hexane and methanolic extracts of *G. gynandra* demonstrated significant antimicrobial properties against 11 clinical strains of human pathogenic microorganisms which include six bacteria (*B. cereus, B. subtilis, S. aureus, E. coli, P. aeruginosa* and *Streptococcus faecalis* and five fungi (*C. albicans, Penicillium* sp., *Fusarium oxysporum, A. flavus* and *A. niger*). The main secondary metabolites present in *G. gynandra* are alkaloids, cyanogenetic glycosides and steroidal ring, while anthraquinones were slightly detected (Ajaiyeoba, 2000).

***Leptadenia hastata* (Pers.) Decne**

This wild edible plant belongs to the family Asclepiadaceae, used as food by many African populations (Hutchinson and Dalziel, 1954). It is commonly used as a vegetable and is considered as a famine food due to its high content of valuable nutrients in Niger (Freiberger *et al.*, 1998). Information on the plant also revealed that the leaves are chewed by shepherds against polydipsia and mouth dryness. Aliero *et al.* (2001) reported that the ethnomedicinal application of the plant involves the use of leaf extract for the treatment of stomach upset in children in some parts of northern Nigeria. In a well-structured study, Aliero and Wara (2009) showed that an aqueous extract of *L. hastate* markedly inhibited the growth of *Salmonella paratyphi* and *E. coli* at 30 mg/ml and *P. aeruginosa* at 60 mg/ml, while ethanol and acetone extracts show weak antimicrobial activities against the test organisms. Antimycotic assay by Aliero and Wara (2009) also revealed that methanol extract suppressed the growth of *F. oxysporum* and *A. niger* at 80 mg/ml with inhibition percentages ranging from 58.89 to 73.30%. Acetone extract has low activity with 40 and 50% inhibition on the growth of *A. niger* and *F. oxysporum*, respectively. The major chemical compounds found in *L. hastata* were: triterpenes, fatty acids, amino acids, poly-oxypregnane, lutein and β-carotene (Aquino *et al*., 1996).

Buchholzia coriacea (A. Chev) Engl

Buchholzia coriacea belongs to the plant family Capparidaceae. It is commonly known as the Musk tree. It is an evergreen 'under storey' of the lowland rainforest, usually attaining a height of about 20 m (Ajaiyeoba, 2000). Ethnomedicinal uses of the plant include the inhalation of the bark to relieve the symptoms of headache, sinusitis and nasal congestion. The bark sap is also applied for chest pain, bronchitis and kidney pains. Fresh bark is used in some regions for earache and a decoction of the bark is used for washing of smallpox patients (Kerharo and Bouquet, 1950; Irvine, 1961; Bouquet and Debray, 1974). The leaves have been used for the treatment of boils; fruits for fever; oil of fruits for fish poison; fruits as an anthelmintic (Dalziel, 1937; Walker, 1953). Ajaiyeoba (2000) reported that hexane and methanolic extracts of *B. coriacea* at a concentration of 200mg/ml demonstrated significant antimicrobial properties against 11 clinical strains of human pathogenic microorganisms. Organisms inhibited include six bacteria (*B. cereus, B. subtilis, S. aureus, E. coli, P. aeruginosa* and *S. faecalis*) and five fungi (*C. albicans, Penicillium* sp., *F. oxysporum, A. flavus* and *A. niger*). The main secondary metabolites present in *B. coriacea* are alkaloids, cyanogenetic glycosides and steroidal nucleus, while anthraquinones were slightly detected.

Massularia acuminata (G. Don) Bullock ex Hoyl

This plant belongs to the family Rubiaceae. In South-west Nigeria, it is popularly called 'pako ijebu'. It is a small tropical plant found as undergrowth of closed moist forest. It is a tree, growing up to 5 m high and is distributed from Sierria Leone through Zaire to Nigeria (Yakubu *et al*., 2008). The stem is used as a chewing stick for oral hygiene in Southern Nigeria (Ndukwe *et al*., 2004). The decoction or infusion of the stem is also claimed to be used as an aphrodisiac and anticancer agent (Gill, 1992). *Massularia acuminata* stem at a concentration of less than 10% is capable of inhibiting the growth of *Bacteriodes gingivalis, Bacteriodes assacharolyticus* and *Bacteriodes melaninogenicus* (Rotimi *et al*., 1988; Aderinokun *et al*., 1999). Phytochemical screening of the aqueous extract of *M. acuminata* stem revealed the presence of alkaloids (0.22%), saponins (1.18%), anthraquinones (0.048%), flavonoids (0.032%), tannins (0.75%) and phenolics (0.066%) (Yakubu *et al*., 2008).

Landolphia owariensis P. Beauv

This plant belongs to the Apocynaceae. It is commonly called vine rubber and known by various names in different parts of Nigeria. In the south east, it is called Eso/Utu in Ibo, in south west it is called Mba by the Yoruba while the Hausas in the north call it Ciwa. Ethnomedicinal uses of the plant involve the use of the decoction in the treatment of many ailments (Owoyele *et al*., 2002). The decoction of its leaves is used as a purgative, and to cure malaria and in the treatment of gonorrhoea infection (Gill, 1992). The latex is drunk or used in French Equatorial Africa as an enema for intestinal worms (Irvine, 1961). The latex is also used as a natural preservative (Anthony, 1995). Nwaogu *et al.*

(2007) reported the inhibition of *Staphylococcus* sp., *Proteus* sp. and *E. coli* at concentrations between 700 and 1800 mg/ml. The major phytochemicals present in *L. owariensis* are alkaloids, flavonoids, tannins and saponins.

Spondias mombin L.

This plant belongs to the family Anacardiaceae. It is popularly called Hog plum. The plant tree is erect, stately, to 65 ft (20 m) tall, with a somewhat buttressed trunk and thick bark; often, in young trees, bearing many blunt-pointed spines or knobs. Ripe fruits are eaten out-of-hand, or stewed with sugar. The extracted juice is used to prepare ice cream, cool beverages and jelly. Young leaves are cooked as greens. The fruit juice is drunk as a diuretic and febrifuge. The leaves and the flower are used to brew tea which is used to relieve stomach ache, biliousness, urethritis, cystitis and eye and throat inflammation. The juice of crushed leaves and the powder of dried leaves are used as poultices on wounds and inflammations. The gum is employed as an expectorant and to expel tapeworms (Olugbuyiro *et al.*, 2009). A semi-pure triterpenoid portion of *Spondias mombin* demonstrated potency of 92.8% inhibition against *Mycobaterium tuberculosis* (Mtb) at a concentration of 64 μg/ml (Olugbuyiro *et al.*, 2009). The major phytochemicals in the stem bark are cardiac glycosides, flavonoids, tannins and anthraquinones.

Paullinia pinnata L.

This plant belongs to the family Sapindaceae. It is an African woody vine whose fruits are widely eaten (FAO, 1968), and its leaves are used in traditional medicine for the treatment of malaria (Chihabra *et al.*, 1991). The root decoction is drunk in the case of nausea and vomiting (Annan and Houghton, 2004). Pharmacists at King's College London have discovered that the roots of the shrub *P. pinnata* possess antibacterial properties that are effective against methicillin-resistant *S. aureus* (MRSA). Annan and Houghton (2004) reported the antimicrobial properties of the leaves, stems and roots of *P. pinnata* prepared following the traditional method of preparation against Gram-positive *B. subtilis* (NCTC 10073), *S. aureus* (NCTC 4163), *M. flavus* (NCTC 7743), *S. faecalis* (NCTC 775), *Clostridium tetani* and Gram-negative *P. aeruginosa* (MCIMB 1042). Phytochemical investigations have shown the presence of triterpene saponins and cardiotonic catechol tannins in this plant. Two new flavonoid glycosides had been isolated from the aerial parts of *P. pinnata*.

Cleome rutidosperma DC

It is a common annual weed that belongs to the Capparaceae family. It attains about 90 cm in height and occurs in West and East Africa. The leaves are edible and have alleged medicinal uses. *C. rutidosperma*, like some other *Cleome* species, has been suspected of having antimicrobial activity (Sudhakar *et al.*, 2006). The seeds are bean shaped and are enclosed in a row of two to five seeds per fruit by a hairy, allergy-causing and skin-irritating fruit coat. Thus, the fruit is used ethnobotanically to punish truants or scare off intruders to farms or economic trees (Burkhill, 1985). Phytochemical screening revealed the presence of alkaloids, steroids, pentose, reducing sugars, tannins, flavonoids and cardiac glycosides in *C. rutidosperma* seed (Ojiako and Igwe, 2007).

Tetracarpidum conophorum (Arg) Mull.

This wild edible plant belongs to the family Euphorbiaceae. It is also known as Conophor. The leaves are used for the treatment of dysentery and to improve fertility in males. Some of the local names include 'ukpa' (Igbo) and 'awusa' or 'asala' (Yoruba). The oil from the nut has found use in the formulation of wood varnish, stand oil, vulcanized oil for rubber and as a leather substitute. The leaf, stem bark, kernel and root methanol extracts as well as the hexane, chloroform, ethyl acetate and methanol fractions of the leaf of *T. conophorum* showed significant inhibition of clinical strains of four human pathogenic bacteria made up of two Gram-positive (*S. aureus* and *B. subtilis*) and two Gram-negative bacteria (*P. aeruginosa* and *E. coli*) and antifungal activity was also observed against one yeast (*C. albicans*) and a mould (*A. flavus*) (Ajaiyeoba and Fadare, 2006).

Conclusions and Future Perspective

Utilization of wild edible plant food had been in the decline in most African homes. This is as a result of the advent of processed foods which are now a common feature in the homes of the rich and medium-income earners. Some researchers at the University of Zimbabwe working on Wild Food of Africa clearly stated that 'the biggest barrier to the widespread use of wild plant foods are social stigmas in which people consuming traditional foods are perceived as **'poor'** as well as lack of promotion and research'. In essence, awareness of the health importance of these wild edible plant foods must therefore be created through research. One obvious disadvantage of processed foods is the attendant disorders such as obesity, carcinogenesis, diabetes, cardiovascular diseases and a host of other diseases which are now rampant. A report by WHO (2003) summarized the links between diet and obesity, diabetes, cardiovascular disease (CVD), cancer and osteoporosis. In essence, diet and nutrition are major determinants of chronic diseases. The argument may be that the older generations live a healthier life because they consume these wild plant foods. This is because most of these disorders were not common in the past. Nature has been so magnanimous not only in providing nutrients in wild edible plants but also in spicing them with phytochemicals which are known to have positive effects, such as antioxidant, anti-tumour, antibacterial, antiviral, immunomodulating, anticholestrolaemic, etc. Some that had been investigated but not listed above include *Lannae acida, Ageratum conyzoides, Vernonia colorata, Uvaria chamae, Pisidium guajava, Alchornea laxiflora, Hibiscus sabdariffa, Piper guineense, Ocimum gratissimum, Pentaclethra macrophylla, Irvingia gabonensis, Afzelia africana, Prosporis africana* and *Monodora myristica*, etc. Apart from those mentioned above, several others still abound as yet to be investigated and utilized.

Most studies on the antimicrobial effect of edible wild plants had been by *in vitro* investigations. Also, in most cases, whole extracts which contain a lot of bioactive compounds are used. Hence, antimicrobial effect observed may be as a result of synergy between two or more bioactive compounds in the extract. In the future, studies on isolated constituents of these extracts are expedient to know the specific active agent responsible for inhibition. Moreover, a combination of well-structured *in vitro* and *in vivo* studies involving animal and human studies need to be performed to verify whether *in vitro* observation can be replicated *in vivo* in whole-organism systems. The effect of extracts and isolated compounds from wild edible plants on beneficial normal gastrointestinal microbiota also needs to be verified. Microbial balance in the gastrointestinal tract is one major mechanism of maintaining sound gut health. Another issue that needs to be considered in the full utilization of wild edible plants is: in what form will it be delivered so that the concentration of bioactive(s) that will bring about antimicrobial or other positive effect will be manifest. Conclusively, wild edible plants with antimicrobial properties hold promise in tackling the problem of resistance posed by microorganisms to synthetic antibiotics. They may also be useful as biopreservative agents in place of synthetic preservatives that are known to have serious side effects. These edible wild plants which hitherto had been underutilized food sources, can contribute to food security, agricultural diversification and income generation. The health-promoting effect of these edible plants will also help consumers to live a healthier life.

References

Abdullahi, A.L., Agho, M.O., Amos, S., Gamaniel, K.S. and Wambebe, C. (2001) Antidiarrhoeal activity of the aqueous extract of *Terminalia avicennoides* roots *Phytotherapy Research* 15, 431–434.

Abo, K.A., Adeyemi, A.A. and Jegede, I.A. (2000) Spectrophotometric estimation of anthraquinone content and antimicrobial potential of extracts of some *Cassia* species used in herbal medicine in Ibadan. *Science Forum* 3, 57–63.

Aboaba, O.O., Smith, S.I. and Olude, F.O. (2006) Antibacterial effect of edible plant extract on *Escherichia coli* 0157:H7. *Pakistan Journal of Nutrition* 5, 325–327.

Adebayo, E.A. and Ishola, O.R. (2009) Phytochemical and antimicrobial screening of the crude extracts from the root, stem bark and leaves of *Brideliaferruginea. African Journal of Biotechnology.* 8, 650–653.

Adebayo-Tayo, B.C. and Ajibesin, K.K. (2008) Antimicrobial activities of *Coula edulis. Research Journal of Medicinal Plant* 2, 86–91.

Adebisi, A.A. (2007) Bitter cola: the African wonder nut. http://www.Cifor.Cgiar.org/publications/pdf-files/books/Restution-Africa-case/NTFP-Africa-case-part 2.pdf.

Adegboye, M.F., Akinpelu, D.A. and Okoh, A.I. (2008) The bioactive and phytochemical properties of *Garcinia kola* (Heckel) seed extract on some pathogens. *African Journal of Biotechnology* 7, 3934–3938.

Adegoke, A.A. and Adebayo-Tayo, B.C. (2009) Antibacterial activity and phytochemical analysis of leaf extracts of *Lasienthera africana. African Journal of Biotechnology*, 8, 077–080.

Adekunle, A.A. and Ikumapayi, A.M. (2006) Antifungal property and phytochemical screening of the crude extracts of *Funtumia Elastica* and *Mallotus Oppositifolius. West Indian Medical Journal* 55, 219–223.

Aderinokun, G.A., Lawoyin, J.O. and Onyeaso, C.O. (1999) Effect of two common Nigerian chewing sticks on gingival health and oral hygiene, *Odontostomatologie Tropicale* 2, 3–18.

Adesina, S.K., Idowu, O., Ogundaini, A.O., Oladimeji, H., Olugbade, T.A., Onawunmi, G.O. and Pais, M. (2000) Antimicrobial constituents of the leaves of *Acalypha wilkesiana* and *Acalypha hispida. Phytotherapy Research* 14, 371–374.

Adewunmi, C.O. and Adesogan, E.K. (1983) Anthraquinones and oruwacin from *Morinda lucida* as possible agents in fascioliasis and schistosomiasis control. *Fitoterapia* 55, 259–263.

Agoha, B.C. (1960) Malaria. In: Oliver, O.B. (ed). *Medicinal Plants in Nigeria*. University Press, Ibadan, p. 22.

Ajaiyeoba, E.O. (2000) Phytochemical and antimicrobial studies on *Gynandropsis gynandra* and *Buchholzia coriaceae* extracts. *African Journal of Biomedical Research* 3, 161–165.

Ajaiyeoba, E.O. (2002) Phytochemical and antibacterial properties of *Parkia biglobosa* and *Parkia bicolor* leaf extracts. *African Journal of Biomedical Research* 5, 125–129.

Ajaiyeoba, E.O. and Fadare, D.A. (2006) Antimicrobial potential of extracts and fractions of the African walnut–*Tetracarpidium conophorum. African Journal of Biotechnology* 5, 2322–2325.

Ajali, U. (2000) Antibacterial activity of *Alchornea cordifolia* stem bark. *Fitoterapia* 71, 436–438.

Akinde, B.E. (1986) Phytochemical and microbiological evaluation of the oils from the leaves of *Acalypha wilkesiana.* In: Sofowora, A. (ed.) *The State of Medicinal Plant Research in Nigeria.* University of Ibadan Press, Nigeria, pp. 362–363.

Akinyemi, K.O., Oluwa, O.K. and Omonigbehin, E.O. (2006) Atimicrobial activity of crude extracts of three medicinal plants used in South-West Nigerian folk medicine on some foodborne bacterial pathogen. *African Journal of Traditional, Complementary and Alternative Medicines*, 3, 13–22.

Aliero, A.A. and Wara, S.H. (2009) Validating the medicinal potential of *Leptadenia hastate. African Journal of Pharmacy and Pharmacology* 3, 335–338.

Aliero, B.L., Umaru, M.A., Suberu, H.A. and Abubakar, A. (2001) *A Handbook of Common Plants in North Western Nigeria.* Usman Danfodiyo University Press, Kano.

Andy, I.E., Eja, M.E. and Mboto, C.I. (2008) An evaluation of the antimicrobial potency of *Lasianthera africana* (BEAUV) and *Heinsia crinata* (G.Taylor) on *Escherichia coli, Salmonella typhi, Staphylococcus aureus* and *Candida albicans. Malaysian Journal of Microbiology* 4, 25–29.

Annan, K. and Houghton, P.J. (2004) Antibacterial activity of *Paullinia pinnata* extracts. Abstract of Ninth International Congress, International Society of Ethnobiology, Canterbury, UK.

Anonymous (2007) *Cymbopogon citratus.* www.floridata.com/ref/C/Cymb-cit.cfm.

Anthony, C.D. (1995) Natural preservatives from *Landolphia owariensis. African Development Journal* 2, 21–22.

Aquino, R., Peluso, G., de Tommassi. N., de Simone, F. and Pizza, C. (1996) New polyoxypregnane ester derivates from *Leptadenia hastata*. *Journal of Natural Products* 59, 555–564.
Asuzu, I.U. and Chineme, C.N. (1990) Effects of *Morinda lucida* leaf extract on *Trypanosoma brucei brucei* infections in mice. *Journal of Ethnopharmacology* 30, 307–313.
Atawodi, S.E. (2004) Antioxidant potential of African medicinal plants. *African Journal of Biotechnology* 4, 128–133.
Ayo, R.G, Amupitan, J.O. and Zhao, Y. (2007) Cytotoxicity and antimicrobial studies of 1,6,8-trihydroxy-3-methyl-anthraquinone (emodin) isolated from the leaves of *Cassia nigricans* Vahl. *African Journal of Biotechnology* 6, 1276–1279.
Bako, S.P., Bakfur, M.J., John, I. and Bala, E.I. (2005) Ethnomedicinal and phytochemical profile of some savanna plant species in Nigeria. *International Journal of Botany* 1, 147–150.
Bouquet, A. and Debray, M. (1974) Plantes Medicinalis de Côte d´Ivoire. Document Orstrom 32, 63.
Burkhill, H.M. (1985) *The Useful Plants of West Tropical Africa, Vol. 1,* 2nd edn. Royal Botanic Gardens, Kew.
Burkhill, H.M. (1994) *The Useful Plants of West Tropical Africa, Vol 2.* Royal Botanic Gardens, Kew, pp. 144–150.
Chihabra, S.C., Mahunnah, R.L.A. and Mshiu, E.N. (1991) Plants used in traditional medicine in Eastern Tanzania. V. Angiosperms (Passifloraceae to Sapindaceae). *Journal of Ethnopharmacology* 33, 43–157.
Cimanga, K., DeBruyne, T., Apers, S., Dieters, L., Totte, J., Kambu, K., Tona, L., Bakana, P., Van Ufford, L.Q., Beukelman, C., Labadie, R. and Vlietinck, A.J. (1999) Complement-inhibiting constituents of *Bridelia ferruginea* stem bark. *Plant Medicine* 65, 213–217.
Cowan, M.M. (1999) Plant products as antimicrobial agents. *Clinical Microbiology Reviews* 12, 564–582.
Dalziel, J.M. (1937) *The Useful Plants of Tropical West Africa,* Crown agents for the colonies, London, pp. 18–22.
Deeni, Y.Y. (1989) Plants in Kano ethnomedicine: screening for antimicrobial activity and alkaloids, Bayero University, Kano, Nigeria, MSc. thesis.
Deeni, Y.Y. and Sadiq, N.M. (2002) Antimicrobial properties and phytochemical constituents of the leaves of African mistletoe (*Tapinanthus dodoneifolius* (DC) Danser) (Loranthaceae): an ethnomedicinal plant of Hausaland, Northern Nigeria. *Journal of Ethnopharmacology* 83, 235–240.
Ebi, G.C. (2001) Antimicrobial activities of *Alchornea cordifolia. Fitoterapia* 72, 69–72.
Eleyinmi, A.F. (2007) Chemical composition and antibacterial activity of *Gongronema latifolium. Journal of Zhejiang University Science (B)* 8, 352–358.
El-Mahmood, A.M. and Ameh, J.M. (2007) *In vitro* antibacterial activity of *Parkia biglobosa* (Jacq.) root bark extract against some microorganisms associated with urinary tract infections. *African Journal of Biotechnology,* 6, 1272–1275.
FAO (2006) Corporate Document Repository, Agriculture Department, 1968. Food Composition Table for Use in Africa http://www.fao.org/docrep/003/x6877E/x6877E00.htm
Farombi, E.O., Ogundipe, O.O., Moody, J.O. (2001) Antioxidant and anti-inflammatory activities of *Mallotus oppositifolius* in model systems. *African Journal of Medical Science* 30, 213–215.
Freiberger, C.E., Vandergagt, D.J., Pastuszyn, A., Glew, R.S., Mounkarla, G., Millson, M. and Glew, R.H. (1998) Nutrient content of edible leaves of seven wild plants from Niger. *Plant Food for Human Nutrition* 53, 57–69.
Geissman, T.A. (1963) Flavonoid compounds, tannins, lignins and related compounds. In: M. Florkin and E. H. Stotz (eds) *Pyrrole Pigments, Isoprenoid Compounds and Phenolic Plant Constituents, vol. 9.* Elsevier, New York, p. 265.
Gill, L.S. (1992) *Ethnomedical Uses of Plants in Nigeria.* Uniben Press, Benin, Nigeria, p. 157.
Hassan, S.W., Bilbis, F.L., Ladan, M.J., Umar, R.A., Dangoggo, S.M., Saidu, Y., Abubakar, M.K. and Faruk, U.Z. (2006) Evaluation of antifungal activity and phytochemical analysis of leaves, root and stem barks extracts of *Calotropis procera* (Asclepiadaceae). *Pakistan Journal of Biological Sciences* 9, 2624–2629.
Hussain, H.S.N. and Karatela, Y.Y. (1989) Traditional medicinal plants used by Hausa tribe of Kano State of Nigeria. *International Journal of Crude Drug Research* 27, 211–216.
Hussain, H.S.N. and Deeric, Y.Y. (1991) Plants in Kano ethnomedicine: screening for antimicrobial activity and alkaloids. *International Journal of Pharmaceutics* 29, 51–56.
Irobi, O.N. (1997) Antibiotic properties of ethanol extract of *Chromolaena odorata* (Asteriaceae). *Phamaceutical Biology* 35, 111–115.

Irobi, O.N., Moo-Young, M., Anderson, W.A. and Daramola, S.O. (1999) Antimicrobial activity of bark extracts of *Bridella ferruginea* (Euphorbiaceae). *Journal of Ethnopharmacology* 43, 185–190.

Irvine, F.R. (1961) *The Woody Plants of Ghana.* Oxford University Press, London.

Iwalokun, B.A, Gbenle, G.O, Adewole, T.A. and Akinsinde, K.A. (2001) Shigellocidal properties of three Nigerian medicinal plants: *Ocimum gratissimum, Terminalia avicennoides,* and *Momordica balsamina. Journal of Health Population and Nutrition* 19, 331–335.

Iwu, M.M. (1993) *Handbook of African Medical Plants.* CRC Press, Boca Raton, Florida.

Kerharo, J. and Bouquet, J. (1950) *Plantes Medicinalis de la Côte d'Ivoire et Haute Volta*, Paris, pp. 289.

Lajubutu, B.A., Pinney, R.J., Roberts, M.F., Odelola, H.A. and Oso, B.A. (2006) Antibacterial activity of diosquinone and plumbagin from the root of *Diospyros mespiliformis* (Hostch) (Ebenaceae): Review Paper. *Phytotherapy Research* 9, 346–350.

Mason, T.L. and Wasserman, B.P. (1987). Inactivation of red beet betaglucan synthase by native and oxidized phenolic compounds. *Phytochemistry* 26, 2197–2202.

Morebise, O. and Fafunso, M.A. (1998) Antimicrobial and phytotoxic activities of saponin extracts from medicinal plants. *Biokemistri* 8, 69–77.

Morton, J.F. (1961) West Africa lilies and orchids. In: Savory, H.J. (ed.) *The Useful Plants of West Africa.* Longmans, London, p.196.

Naheed, F., Tapondjou, L.A., Lontsi, D., Sondengam, B.L., Atta-Ur-Rahman, A-R. and Choudhary, I.M. (2002) Quinovic acid glycosides from *Mitragyna stipulosa* – first examples of natural inhibitors of snake venom hosphodiesterase I. *Natural Product Research*, 16, 389–393.

Ndukwe, K.C., Lamikanra, A. and Okeke, I.N. (2004) Antibacterial activity in plants used as chewing sticks in Africa. *Drugs of the Future* 29, 1221.

Nwachukwu, I. N., Allison, L.N., Chinakwe, E.C. and Nwadiaro, P. (2008) Studies on the effects *Cymbopogon citratus, Ceiba pentandra* and *Loranthus bengwelensis* extracts on species of dermatophytes. *Journal of American Science*, 4 (1545–1003).

Nwaogu, L.A., Alisi, C.S., Ibegbulem, C.O. and Igwe, C.U. (2007) Phytochemical and antimicrobial activity of ethanolic extract of *Landolphia owariensis* leaf. *African Journal of Biotechnology.* 6, 890–893.

Obih, P.O., Makinde, J.M. and Laoye, J.O. (1985) Screening of various extracts of *Morinda lucida* for antimalarial action on *Plasmodium berghei berghei* in vitro. *African Journal of Medical Science.* 14, 51–54.

Ogundaini, A.O. (2005) From greens into medicine: taking a lead from nature. An inaugural lecture delivered at Oduduwa Hall, Obafemi Awolowo University, Ile-Ife, Nigeria. Inaugural Lecture Series 176. OAU Press Limited, Ile-Ife, Nigeria, pp. 12–15.

Ogundare, A.O., Adetuyi, F.C. and Akinyosoye, F.A. (2006) Antimicrobial activities of *Vernonia tenoreana. African Journal of Biotechnology* 5, 1663–1668.

Ojiako, O.A. and Igwe, C.U. (2007) Nutritional and anti-nutritional compositions of *Cleome rutidosperma, Lagenaria siceraria,* and *Cucurbita maxima* seeds from Nigeria. *Journal of Medicinal Foods* 10, 735–738.

Okeke, I.N., Ogundaini, A.O., Ogungbamila, F.O. and Lamikanra, A. (1999) Antimicrobial spectrum of *Alchornea cordifolia* leaf extracts. *Phytotherapy Research* 13, 67–69.

Okigbo, R.N. and Mmeka, E.C. (2008) Antimicrobial effects of three tropical plant extracts on *Staphylococcus aureus, Escherichia coli* and *Candida albicans. African Journal of Traditional, Complementary and Alternative Medicines* 5, 226–229.

Oladunmoye, M.K. (2006) Comparative evaluation of antimicrobial activities and phytochemical screening of two varieties of *Acalypha wilkesiana. Trends in Applied Science Research* 1, 538–541.

Oladunmoye, M.K. and Akinyosoye, F.A. (2004) Phytochemical screening and antimicrobial activities of *Cassia alata. Journal of Science and Technology Research* 3, 42–44.

Olajide, O.A., Awe, S.O. and Makinde, J.M. (1998) The effects of *Morinda lucida* Benth (Rubiaceae) extract on the gastrointestinal tract of rodents. *Phytotherapy Research* 12, 439–441.

Oliver, B. (1960) *Medicinal Plants in Nigeria.* College of Arts, Science and Tech., Lagos, p. 72.

Olorode, O. (1984) *Taxonomy of West African Flowering Plants.* 1st edn. Longman, UK, pp. 98–100.

Olugbuyiro, J.A.O., Moody, J.O. and Hamann, M.T. (2009) Antiicrobial activity of triterpenoid-rich fractions from *Spondias mombin* L. *African Journal of Biotechnology* 8, 1807–1809.

Otung, I. (1998) The effect of combining the extracts of *Lasianthera Africana* (Beauv) (Editan) and *Heinsia crinata* (G.Taylor) (Atama) with each other and with ampicillin and chloramphenicol on some pathogenic organisms. M.Sc. Thesis, University of Calabar, Nigeria.

Owoyele, B.Y., Olayele, S.B. and Elegbe, R.A. (2002) Anti-inflammatory and analgesic activities of leaf extracts of *Landolphia owariensis. African Journal of Biomedical Research* 4, 131–133.

Oyetayo, V.O. and Oyetayo, F.L. (2006) Phytochemicals screening and antibacterial properties of siam weed, *Chromolaena odorata*, leaf against some aerobic isolates of wound. *Journal of Applied and Environmental Science.* 2, 7–11.

Oyetayo, V.O. (2007) Comparative studies of the phytochemical and antimicrobial properties of the leaf, stem and tuber of *Anchomanes difformis. Journal of Pharmacology and Toxicology* 2, 407–410.

Puupponen-Pimia, R., Aura, A-M., Karppinen, S., Oksman-Caldentey, K-M. and Pontanen, K. (2004) Interactions between plant bioactive food ingredients and intestinal flora – effects on human health: Review. *Bioscience Microflora* 3, 67–80.

Rios, J.L. and Recio, M.C. (2005) Medicinal plants and antimicrobial activity. *Journal of Ethnopharmacology*, 100, 80–84.

Rolfe, R.D. (2000) The role of probiotic cultures in the control of gastrointestinal health. *Journal of Nutrition* 130, 396S-402S.

Rotimi, V., Laughon, B.E., Bartlett, J.G. and Mosadomi, H.A. (1988) Activities of Nigerian chewing stick extracts against *Bacteroides gingivalis* and *Bacteroides melaninogenicus*, *Antimicrobial Agents and Chemotherapy* 32, 598–600.

Schultes, R.E. (1978) The kingdom of plants. In: Thomson, W.A.R. (ed.) *Medicines from the Earth.* McGraw-Hill Book Co., New York, p. 208.

Sofowora, A.E. (1982) *Medicinal Plants and Traditional Medicine in Africa Vol 1.* John Wiley and Sons, New York.

Sudhakar, M., Rao, Ch.V., Rao, P.M. and Raju, D.B. (2006) Evaluation of antimicrobial activity of *Cleome viscose* and *Gmelina asiatica. Fitoterapia* 77, 47–49.

Ugochukwu, N.H. and Babady, N.E. (2002) Antioxidant effects of *Gongronema latifolium* in hepatocytes of rat models of non-insulin dependent diabetes mellitus. *Fitoterapia* 73, 7–8.

Velioglu, Y.S., Mazza, G., Gao, L. and Oomah, B.D. (1998) Antioxidant activity and total phenolics in selected fruits, vegetables, and grain products. *Journal of Agriculture and Food Chemistry* 46, 4113–4117.

Walker, A.R. (1953) Usages Pharmaceutiques des plantes spontanus du Gabon. *Institut d´Etudes Centra Fracaines* 9, 13–26.

Watt, J.M. and Breyer-Brandwijk, M.G. (1962) *The Medicinal and Poisonous Plants of Southern and Eastern Africa.* 2nd edn, Livingstone, London, pp. 160–163.

WHO (World Health Organization) (2003) Diet, Nutrition and the Prevention of Chronic Diseases. Report of a Joint WHO/FAO Expert Consultation. WHO Technical Report Series p. 916, Geneva.

Yakubu, M.T., i Akanji, M.A., Oladiji, A.T. and Adesokan, A.A. (2008) Androgenic potentials of aqueous extract of *Massularia acuminata* (G. Don) Bullock ex Hoyl. stem in male Wistar rats. *Journal of Ethnopharmacology* 118, 508–513.

18 Natural Antimicrobial Compounds to Preserve Quality and Assure Safety of Fresh Horticultural Produce

Gustavo A. González-Aguilar,* J. Fernando Ayala-Zavala, Emilio Alvarez-Parrilla, Laura de la Rosa, G.I. Olivas, Basilio Heredia and María Muy-Rangel

Introduction

Fresh fruits and vegetables are highly perishable products because of their intrinsic characteristics. Microbial growth, sensorial attributes, decay and loss of nutrients are amongst the major causes that compromise quality and safety of fresh produce (Ayala-Zavala *et al.*, 2008a). Chemical synthetic additives can reduce the decay rate, but consumers are concerned about chemical residues in the product, which could affect their health and cause environmental pollution (White and McFadden, 2008), thereby giving rise to the need for developing alternative methods for controlling fresh fruit and vegetable decay. One of the major emerging technologies for the control of postharvest diseases is the application of natural additives.

In recent years, the interest in natural antimicrobial compounds has increased and numerous studies on the antimicrobial activity of a wide range of natural compounds have been reported (Ayala-Zavala *et al.*, 2008b). Many pathogenic microorganisms that can be the cause of food-borne diseases and/or fresh-food decay can be inhibited using natural compounds (Fisher and Phillips, 2008). Among these, several essential oils, alcohols, organic acids and aromatic compounds have proven to be biologically active against microbial growth.

The main reason for promoting the application of natural products in fresh fruits and vegetables is the consumer's demand for natural and/or organic methods to preserve foods. There is an increasing portion of consumers choosing convenient and ready-to-use fruits and vegetables with a fresh-like quality, containing only natural ingredients (Roller and Lusengo, 1997). Different studies have been focused on improving the efficiency of natural compounds as emerging technologies to preserve fresh-fruit safety and quality (Ayala-Zavala *et al.*, 2008a,b,c). However, regulatory actions on the use of natural alternative additives are still being analysed. Demands from increasingly mistrustful consumers have led to numerous legislation reviews, which are expected to result in well-planned laws regarding regulations on natural food additives.

Microbial Growth as a Quality and Safety Concern in Fresh Fruit and Vegetables

On the basis of nutrient content, fruits and vegetables can be a good source of nutrients for bacteria, yeast and mould growth (Ayala-

* Corresponding author.

Zavala *et al.*, 2008a). The micro-ecology of fresh produce is especially important to consider because such produce can change the microenvironment through their metabolic activity. Natural microbial flora may be present at about 10^4–10^5 colony-forming units per gram (CFU/g) of fresh produce (Busta *et al.*, 2003). The commonly encountered microflora of fruits and vegetables are *Pseudomonas* spp., *Erwinia herbicola, Enterobacter agglomerans*, lactic acid bacteria, and moulds and yeasts (Busta *et al.*, 2003). Although this microflora is largely responsible for the spoilage of fresh produce, it can vary greatly for each product, depending on the medium's pH, nutrient availability, water activity, storage conditions, among other factors. Temperature can play an important role in determining the outcome of the final microflora found in refrigerated fruits and vegetables (Gonzalez-Aguilar *et al.*, 2004). The humidity at which fruits and vegetables are stored can also affect microbial development; low humidity will discourage bacteria from growing on the surface of fruits and vegetables (Ayala-Zavala *et al.*, 2008a).

The fact is that fresh fruits and vegetables are complex and active systems, in which microbiological, enzymatic and physico-chemical reactions are simultaneously taking place (Artés *et al.*, 2007). Fresh fruit deterioration is dependent on the understanding of these reactions and their respective mechanisms. Fruit composition involves different compounds that will affect the shelf life of the product. Key factors include moisture (water activity), sugar, acid content and pH (Brecht, 2006). Water activity (a_w) is directly related to the relative humidity (RH) equilibrium of a given food, and describes the degree at which water is 'bound' in the system, controlling its availability to act as a solvent and participating in chemical/biochemical reactions and growth of microorganisms (Ayala-Zavala *et al.*, 2008a). Fungi are the most important microorganisms causing postharvest wastage of fresh fruit, where the relatively acid conditions tend to suppress bacterial growth (Frazier and Westhoff, 1993). However, in vegetables, bacterial infections are more common, due to their high pH. Bacteria are also important as agents of both spoilage and food-borne diseases. These important properties can be used to predict the stability and safety of food with respect to microbial growth, rates of deteriorative reactions and chemical/physical properties.

Natural Antimicrobial Compounds with the Potential to be Applied to Fresh-cut Fruit and Vegetables

Essential oils

An EO is a concentrated, hydrophobic liquid containing a mixture of antimicrobial volatile aroma compounds commonly derived from plant tissues (Ha *et al.*, 2008). EOs are 'essential' in the common sense that they carry a distinctive scent of the source plant and are generally extracted by steam distillation and other processes like maceration, cold pressing or solvent extraction. They are used in perfumes, cosmetics, bath products, scenting incense, household cleaning products, as well as for flavouring food and drink (Berger, 2007). The aromatic constituents of known EOs are mostly compounds with short hydrocarbon chains, complemented with oxygen, nitrogen and sulfur atoms attached at various positions of the chain (Braca *et al.*, 2008). Such mixtures of different scented molecules with highly reactive atoms give EOs different functional properties, which could be considered for many food applications.

Several EOs such as oils of garlic, cinnamon, thyme, oregano, clove, basil, coriander, citrus peel, laurel, ginger, rosemary and peppermint, among others, have been studied as antimicrobial natural products against both bacteria and moulds (Arcila-Lozano *et al.*, 2004; Edris and Farrag, 2003; Jugl-Chizzola *et al.*, 2006; Mukherjee and Datta, 2007; Novak *et al.*, 2003; Ravi *et al.*, 2007; Senhaji *et al.*, 2007; Wang, 2006; Zeller and Rychlik, 2007). The inherent aroma and antimicrobial activity of EOs are commonly related to their chemical composition, the proportions in which their components are present and to interactions between them, affecting their bioactive properties (Fisher and Phillips, 2008). Some studies have concluded that whole EOs have a greater antibacterial activity than their major constituents separately (Chang *et al.*, 2008, Marino *et al.*, 1999, 2001), which suggests that the components at lower concentration are critical to the activity and the entire mixture must have a synergistic effect. Considering the complex mixture of EOs constituents it is difficult to attribute the antimicrobial mode of action to one specific mechanism, and several targets in the microbial cell are reported. EOs may cause deterioration of cell wall, damage to cytoplasmic membrane, damage to membrane proteins, leakage of cell contents, coagulation of cytoplasm, depletion of proton motive active sites, inactivation of essential enzymes and disturbance of genetic material functionality (Ayala-Zavala *et al.*, 2008b; Burt, 2004; Gutierrez *et al.*, 2008).

There are several volatile compounds present in the EOs that possess antimicrobial activity and flavouring properties. Some of the most representative and studied are terpenoids, organic sulfur compounds, aldehydes, alcohols, among others (Berger, 2007).

Among the studied compounds, terpenoids are used extensively for their aromatic qualities and significant contribution to the scent of coriander, cinnamon, oregano, rosemary, cloves, thyme and citrus oils, among others (Berger, 2007). Well-known terpenoids include citral, menthol, camphor, geraniol, eugenol, carvacrol, thymol and cinnamaldehyde that have also been found to inhibit the *in vitro* growth of bacteria and fungi, and form the characteristic odour classes of the different EOs (Chaumont and Léger, 1992). Alpha-terpinene inhibited several bacterial species and its odour perception is lemon like (Baranauskiene *et al.*, 2005). Cinnamaldehyde is known to inhibit the growth of *Escherichia coli* O157:H7 and *Salmonella typhimurium* and it is perceived as a sweet cinnamon–honey odour (Aggarwal *et al.*, 2002; Zhou *et al.*, 2007). Carvone presents a warm herb-like character and suppresses the specific growth rate of *E. coli*, *Streptococcus thermophilus* and *Lactococcus lactis*, suggesting that it acts by disturbing the metabolic energy status of cells (Aggarwal *et al.*, 2002). Different studies have reported the possible mode of action of specific compounds, but considering the wide variety of these compounds more basic research is needed.

Organic compounds containing either nitrogen or sulfur are commonly found in garlic, onion and leek oils, which have demonstrated antimicrobial activity against a wide variety of microorganisms. Organosulfur compounds from garlic such as diallyl mono-, di- and trisulfide have been reported to have antimicrobial activity against fungi and bacteria, presenting a characteristic smell to sulfur spices (Chung *et al.*, 2007). For example, diallyl sulfide and dimethyl trisulfide are volatile constituents of garlic oil that presented antimicrobial activity against *Enterobacter aerogenes*, *E. coli*, *Salmonella* sp., *Listeria monocytogenes* and *Yersinia enterocolytica*, are perceived as sulfur onion, garlic and/or cabbage-like odourants (Benkeblia, 2004). A recent study reported that the main compounds detected in garlic oil were allyl disulfide, allyl trisulfide and allyl tetrasulfide (Ayala-Zavala *et al.*, 2008c).

EO and/or their components have been registered by the legislation of different regions as flavourings in foodstuffs (Ayala-Zavala *et al.*, 2008a, Fisher and Phillips, 2008). The flavourings registered are considered generally to present no risk to human health of the consumer and include carvacrol, carvone, cinnamaldehyde, citral, *p*-cymene, eugenol, limonene, menthol and thymol (CFSAN/FDA, 2006). Estragole and methyl eugenol were deleted from the list in 2001 because they were found to be genotoxic (Zeller and Rychlik, 2007). The European Union registered flavourings that also appear on the 'Everything Added to Food in the US' (EAFUS) list, which means that the United States Food and Drug Administration (FDA) has classified the substances as generally recognized as safe (GRAS) or as approved food additives.

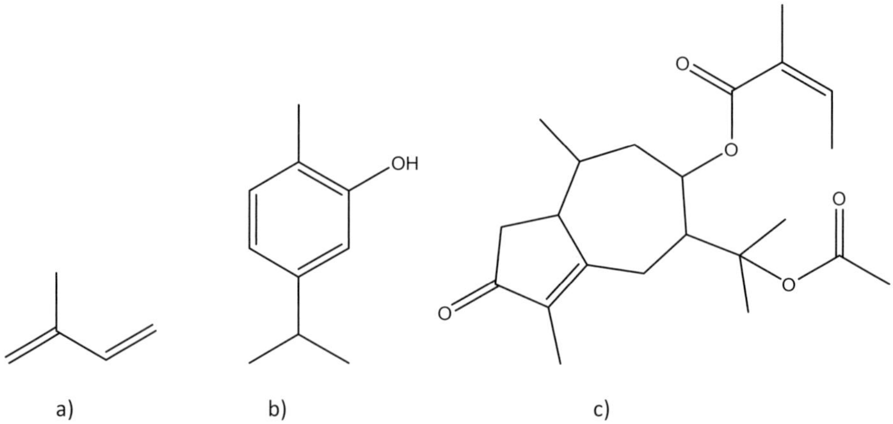

Fig 18.1. Structure of: (a) isoprene, the biosynthetic unit of terpenoids; (b) carvacrol, an antimicrobial monoterpenoid from oregano essential oil; and (c) torilin, a sesquiterpenoid isolated from *Torilis japonica* fruits.

Terpenoids

Terpenoids are secondary metabolites synthesized by plants, marine organisms and fungi by head-to-tail joining of isoprene units (Fig. 18.1a). They are classified according to the number of isoprene units involved in their biosynthesis (Bhat *et al.*, 2009). Mono- and sesquiterpenoids, made up of two and three isoprene units, respectively, are abundant in essential oils and are known for their antimicrobial properties (Saleem *et al.*, 2010).

Carvacrol (Fig. 18.1b) and thymol are monoterpenic phenols, they are the major components of oregano and thyme essential oils, and are responsible for their antimicrobial activities (Burt, 2004). Carvacrol is probably the most studied and one of the most effective antimicrobial terpenoids. Its antimicrobial properties have been documented by *in vitro* tests against cultured strains of food-borne microorganisms (Cosentino *et al.*, 1999; Lambert *et al.*, 2001; Zhou *et al.*, 2007) and by the inhibition of microbial growth on horticultural food products (Roller and Seedhar, 2002; Kisko and Roller, 2005; Obaidat and Frank, 2009b).

Other antimicrobial monoterpenoids include citral, linalool, β-pinene, all of them constituents of essential oils from various plants, herbs and spices (Burt, 2004; Tiwari *et al.*, 2009; Belleti *et al.*, 2010).

Sesquiterpenoids and higher terpenoids (containing more than three isoprene units in their structure) also possess antimicrobial properties. For example, torilin (Fig. 18.1c), a sesquiterpene naturally found in the fruits of *Torilis japonica*, showed a strong antimicrobial activity against *Bacillus subtilis* spores and vegetative cells (Cho *et al.*, 2008).

Generally speaking, terpenoids' bactericidal and fungicidal activities are related to their ability to interact with the plasma membrane and alter its permeability (Burt, 2004; Tiwari *et al.*, 2009). The mechanism of action of carvacrol, and its isomer thymol, has been most extensively studied. Both are able to increase the permeability of bacterial membranes inducing ion leakage and membrane depolarization in several bacterial species, including *E. coli* and *Bacillus cereus* (Ultee, 1999; Xu, 2008). The presence of the hydroxyl group in these two molecules was shown to be essential for their microbial action, probably due to its action as a proton exchanger, leading to transmembrane pH gradient collapse and ATP depletion (Ultee, 2002). Citral and torilin were capable of inactivating fungus and bacterial spores, respectively, by disrupting and disorganizing their structure, as shown by electron microscope analysis (Palhano *et al.*, 2004; Cho, 2008). Citral has also been described as an alkylating agent capable of modifying numerous cellular processes through irreversible covalent binding (Witz, 1989). Other sesquiterpenoids have been shown to enhance bacterial permeability and susceptibility to exogenous antimicrobial compounds (Brehm-Stecher and Johnson, 2003).

Despite their proven antimicrobial activities, effective use of terpenoids, or their source essential oils, as preservatives of fresh horticultural produce has been somewhat hampered by the fact that usually high concentrations are required to assure microbiological safety, which have an adverse organoleptic impact (Burt, 2004; Tiwari *et al.*, 2009). However, careful selection of concentrations, combination of agents with synergic effects or with other novel preservation technologies, different methods of application and use of compounds which confer desirable organoleptic attributes to a specific food product can allow the successful use of these natural products. For example, treatment with 1 mM of carvacrol delayed spoilage of fresh-cut kiwifruit and honeydew melon at chill temperatures without adverse sensory consequences (Roller and Seedhar, 2002). Use of citral and citron essential oil (containing citral and other terpenoids, mainly limonene) was able to prolong the microbial shelf life of the fruit-based salads and, while the citral negatively affected colour, texture and flavour of the fruit salads, citron essential oil gave excellent results, avoiding the undesirable sensory effects attributable to citral (Belleti *et al.*, 2008). Carvacrol has been used in the vapour phase to delay spoilage of leafy vegetables and tomato (Obaidat and Frank, 2009a; 2009b). However, more studies analysing sensory and overall quality changes in produce treated with these compounds are needed in order to fully understand and take advantage of their antimicrobial potential.

Phenol compounds

Plants are excellent sources of phenolic metabolites. In particular, phenolic antioxidants from food-grade plants have potential for long-term chemo-preventive and therapeutic applications against oxidation-linked diseases (Rice-Evans *et al.*, 1995) and increasingly have antimicrobial potential (Shetty and Lin, 2005). Phenolic phytochemicals have been shown to be associated with antioxidative action in biological systems, acting as scavengers of singlet oxygen and free radicals (Jorgensen *et al.*, 1999). They serve as effective antioxidants due to their ability to donate hydrogen from hydroxyl groups positioned along the aromatic ring to terminate free radical oxidation of lipids and other biomolecules (Jorgensen *et al.*, 1999). Phenolic antioxidants therefore short circuit a destructive chain reaction that ultimately degrades cellular membranes and in prokaryotes such phenolic antioxidants have potential for antimicrobial activity.

Plant phenolic extracts that impart flavour and aroma also have potential for inhibiting pathogenic microorganisms (Shetty and Lin, 2005). These phenolic secondary metabolites are defensive antimicrobials produced against invading pathogens and different types of stress. In certain cases the induction is associated with action of diphenolic oxidases and resulting modified compounds can have antimicrobial activity (Walker, 1994). In other situations dihydroxy phenolics are oxidized to quinones, which can interact with the proteins of the invading pathogens, forming melanoid polymers. The quinones are sources of stable free radicals and complex irreversibly with nucleophilic amino acids and proteins leading to inactivation of protein and loss of function (Stern *et al.*, 1996). Therefore potential antimicrobial benefits of quinones are substantial. The potential targets for quinones in the bacterial cells are surface adhesions, cell-wall polypeptides and membrane-bound enzymes (Stern *et al.*, 1996).

Hydroxylated phenols, such as catechol and pyrogallol, are known to be toxic to microorganisms (Cowan, 1999). The site and number of hydroxyl groups is linked to the antimicrobial effect and, in some cases, oxidized forms show higher activity.

The mode of action of phenolics against bacterial pathogens has not been defined or clearly understood. Hydrophobic phenolics, like thymol from certain essential oils, may act by inducing changes in membrane permeability. Membrane-localized hyperacidity from water- and ethanol-soluble phenolics such as benzoic acid and rosmarinic acid may affect proton motive force across the membrane and therefore energy depletion may take place (Tassou *et al.*, 2004). It is suggested that pH gradient and electrical potential of the proton motive force are disrupted as a result of damage to structural and functional properties of membranes (Burt, 2004). The disruption of the proton motive force and reduction of ATP pool lead to cell death. Further, leakage of ions, nucleic acid and amino acids can occur (Burt, 2004). It is also proposed that enzyme inhibition by the oxidized compounds through reaction with sulfhydryl groups, or through nonspecific interactions with membrane proteins, may also be the reason for inhibition of microbial growth (Cowan, 1999). Other proposed mechanisms include formation of Schiff's bases with membrane proteins by the reaction of aldehyde groups from phytochemicals, which prevent cell wall biosynthesis (Friedman, 1999), and interaction of ferrous iron with phenolic compounds, which can damage membranes by enhancing oxidative stress (Patte, 1996). More research is needed in order to elucidate the mode and mechanism of action of these compounds.

Chitosan

Chitosan is primarily produced from chitin, which is widely distributed in nature, mainly as the structural component of the exoskeletons of arthropods (including crustaceans and insects), in marine diatoms and algae, as well as in some fungal cell walls. Figure 18.2 shows the chemical structure of chitin and chitosan.

It has been observed that chitosan and its derivatives have a wide spectrum of antimicrobial activity against filamentous fungi, yeasts and bacteria, being more active against Gram-positive than Gram-negative bacteria (Moller *et al.*, 2004; No *et al.*, 2002).

The mechanism of the antimicrobial activity of chitosan has not yet been fully elucidated, but several hypotheses have been proposed. The most feasible hypothesis is a change in cell permeability due to interactions between the positively charged chitosan molecules and the negatively charged microbial cell membranes. This interaction leads to the leakage of ions and other intracellular constituents (Sudarshan *et al.*, 1992; Fang *et al.*, 1994). Other mechanisms are the interaction of diffused hydrolysis products with microbial DNA, which leads to the inhibition of the mRNA and protein synthesis (Sudarshan *et al.*, 1992), and the chelation of metals, spore elements and essential nutrients (Cuero *et al.*, 1991).

The overall mechanism(s) of action of an antimicrobial may be defined according to the target component of the bacterial cell against which it has its main activity. Thus, three levels of interaction can be described: (i) interaction with outer cellular components, (ii) interaction with the cytoplasmic membrane and (iii) interaction with cytoplasmic constituents. The mechanisms underlying the antimicrobial activity of chitosan have only been studied comparatively recently and the amount of information available is limited, although increasing. Several studies claim to have identified such mechanisms, but only few were supported by experimental evidence.

As related by Cooksey (2001), the positive charges of chitosan interfere with the negatively charged residues of the macromolecules at the bacterial cell surface, compete with calcium for electronegative sites on the membrane, and compromise the membrane's integrity, thus causing leakage of intracellular material, leading to bacterial cell death.

Chitosan generally has a stronger antimicrobial activity against bacteria rather

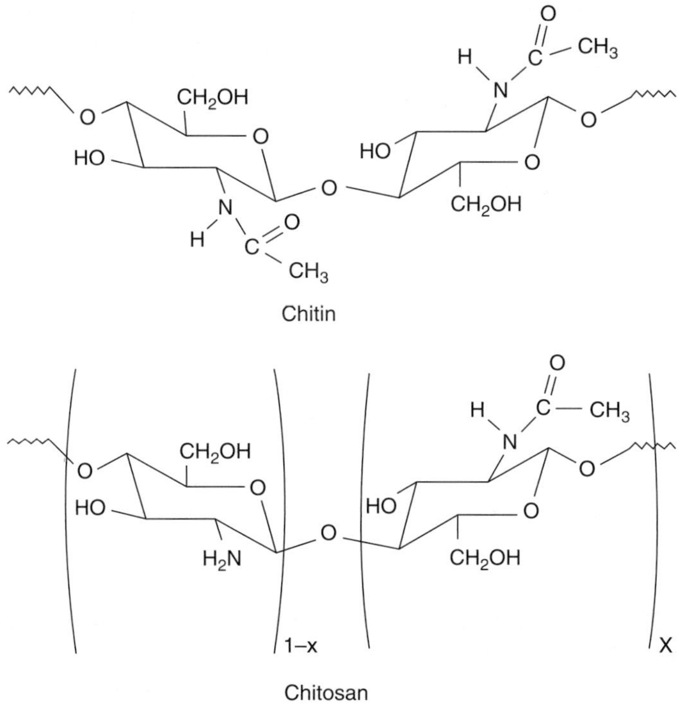

Fig 18.2. A schematic representation of chitin and chitosan.

than against fungi (Tsai *et al.*, 2002). Recent studies on antibacterial activity of chitosan and chitosan oligomers have revealed that chitosan is more effective in inhibiting growth of bacteria than are chitosan oligomers (Zheng, 2003). Furthermore, the antibacterial effects of chitosan and chitosan oligomers are reported to be dependent on its molecular weight (Jeon *et al.*, 2001; No *et al.*, 2002), degree of deacetylation (DD) (Tsai *et al.*, 2002), and the type of bacterium (No *et al.*, 2002). More extensive information on the antibacterial activity of chitosan is available (Sudarshan *et al.*, 1992; Roller, 2003; No *et al.*, 2002).

The antimicrobial properties of chitosan have been reported widely in the literature but mainly based on the *in vitro* trials. Most foods are a mixture of different compounds (for example, carbohydrate, protein, fat, minerals, vitamins and salts, and many of them may interact with chitosan and lead to loss or enhancement of antibacterial activity. Recently, Devlieghere *et al.* (2004) extensively studied the influence of different food components (starch, protein, oil and NaCl) on the antimicrobial effect of chitosan. For this, the media was inoculated with *Candida lambica* (2 log CFU/ml) and incubated at 7°C with varying chitosan concentrations (43 kDa, DD = 94%; 0%, 0.005% and 0.01%), and with the separate addition of the following food components: starch (0%, 1% and 30% water-soluble starch), proteins (0%, 1% and 10% whey protein isolate), oil (0%, 1% and 10% sunflower oil) and NaCl (0%, 0.5% and 2%). Results showed that starch, whey protein and NaCl had a negative effect on the antimicrobial activity of chitosan. Oil, conversely, had no influence.

Sulfur compounds

Several natural compounds containing sulfur in their structure, such as glucosinolates, thiosulfinates (allicin), isothiocyanates and amino acids, have been used as antimicrobial and antifungal agents for the preservation of fresh horticultural produce. The following

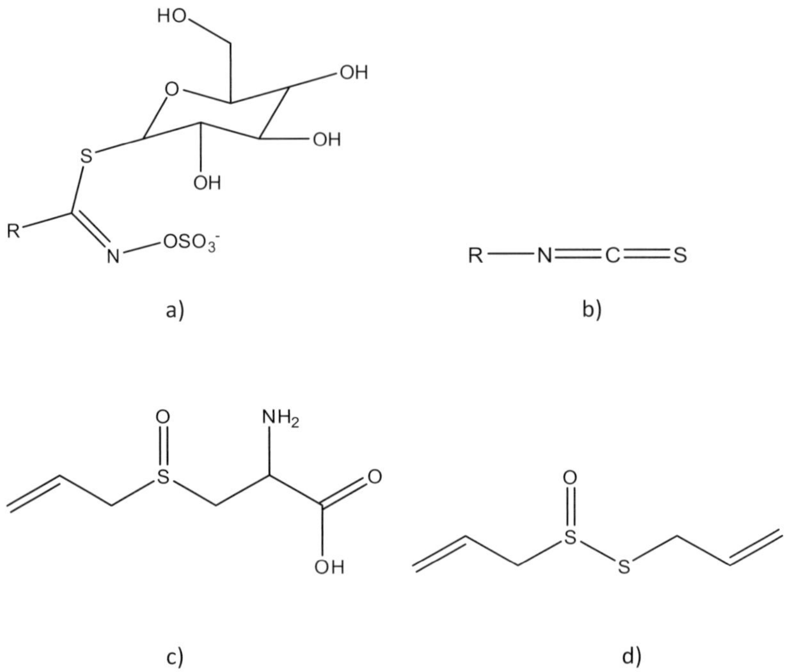

Fig. 18.3. Schematic representations of (a) the glucosinolate skeleton, (b) isothiocyanate, (c) alliin and (d) allicin.

section describes the common natural source, mode of action and effect of these compounds on the overall quality of the treated produce.

Glucosinolates and isothiocyanates

Glucosinolates are β-thioglucoside N-hydroxysulfates synthesized mainly by the Brassicaceae family (cruciferous), even though they have been identified in 16 families of dicotyledonous angiosperms (Fahey *et al.*, 2001). Glucosinolate content in Brassica vegetables is about 1% dry weight; however, it can reach 10% in some seeds. There are more than 120 known glucosinolates, all of them with the same skeleton (Fig. 18.3a) and different side chain (aliphatic, ω-methylthioalkyl, aromatic and heterocyclic) (Fahey *et al.*, 2001). Several studies have shown that native glucosinolates don't possess any anti-microbial activity (Pal Vig *et al.*, 2009); however, after wounding or damage of the plant tissue, they are transformed to isothiocyanates (Fig. 18.3b) by the enzyme mirosinase. This enzyme releases the glucosyl moiety, producing an unstable intermediate, which rearranges to form the isothiocyanates (ITCs), as well as other breakdown products (thiocyanates, nitriles, among others). ITCs have shown several biological activities including antioxidative, antibacterial, antifungal, anti-nematode and anti-insect activities (Fahey *et al.*, 2001).

Several studies have shown that ITCs inhibit pathogenic and deteriorative fungal and bacterial growth in different fruits and vegetables (Shin *et al.*, 2004; Mari *et al.*, 2002, 2008; Troncoso-Rojas *et al.*, 2009; Wang *et al.*, 2010). Aromatic ITCs show higher activity than aliphatic ITCs, and among aliphatic ITCs, their activity decreases as the length of the aliphatic chain increases (Tierens *et al.*, 2001). However, the biochemical mechanism of action is still unclear. Among the biochemical mechanisms involved in the antimicrobial activity of ITCs, inhibition of oxygen uptake in the mitochondria, membrane integrity and cellular structure disruption have been reported (Pal Vig *et al.*, 2009). Luciano and Holley (2009) described that the inhibition of *E. coli* O157:H7 by ITCs showed a pH-dependent pattern, being 20 times higher at pH 4.5 compared to 8.5. In this study, the authors suggested that ITC inhibits *E. coli* by a complex mechanism, which involves both membrane damage as well as thioredoxin reductase and acetate kinase enzymatic inhibition.

Allium vegetables such as garlic and onions contain high amounts of cysteine sulfoxide derivatives, among them, alliin (S-allyl-L-cysteine sulfoxide; Fig. 18.3c) is the main compound found in garlic. As in the case of glucosinolates, cysteine sulfoxide derivatives don't possess any antimicrobial activity; however, after cutting, they are hydrolysed by the enzyme allinase (a cysteine sulfoxide lyase) to form the active thiosulfinate compounds (Kim *et al.*, 2004; Tajkarimi *et al.*, 2010). Allicin (allyl 2-propenylthiosulfinate; Fig. 18.3d) is the main derivative found in chopped garlic, and possesses a strong activity against Gram-negative bacteria. After processing, allicin suffers non-enzymatic degradation to produce further sulfur compounds such as ajoene, diallyl sulfide, diallyl disulfide, diallyl trisulfide and diallyl tetrasulfide (Kyung and Lee, 2001; Kim *et al.*, 2004). Since initial studies wrongly suggested that allicin was the only compound that presented antimicrobial activity, and that it was rapidly degraded in the presence of water or organic solvents, for a period of time, no studies were carried on using garlic derivatives (garlic oil and garlic juice) for antimicrobial purposes (Kim *et al.*, 2004). However, further studies showed that the sulfide derivatives found in garlic oil presented high antimicrobial activity against moulds, fungus and pathogenic and deteriorative bacteria, and that this activity increases with the number of sulfurs in the structure: tetrasulfide > trisulfide > disulfide > sulfide (Kim *et al.*, 2004; Park and Shin, 2005; Ayala-Zavala *et al.*, 2009). Tetrasulfide derivatives showed high antimicrobial activity, against both yeast and bacteria, comparable to those of ITCs. As in the case of ITC, the activity of the sulfides decreases as the number of carbons in the alk(en)yl side chain increases (Kim *et al.*, 2004). Interestingly, garlic derivatives possess strong activity against *Helicobacter pilory*, the cause of gastric cancer, and it has been suggested that a high consumption of garlic reduces the risk for this cancer (Ankri, 1999; Sivam, 2001).

The mechanism of action of thiosulfinates against microorganisms involves structural and functional damages to the bacterial cell membrane, as well as inhibition of thiol-containing enzymes such as cysteine proteinases, alcohol dehydrogenases and thioredoxin reductases of different organisms by the reaction between the thiosulfinate and the SH peptides found in the enzymes, through the formation of a S(O)S linkage (Ankri, 1999; Kyung and Lee, 2001; Kim *et al.*, 2004). For this reason, antimicrobial activity of thiosulfinates is inhibited by the presence of L-cysteine. Microbiological activity of thiosulfinates has also been reported due to inhibition of RNA synthesis (Sivam, 2001).

There are some sulphur-containing peptides that present antimicrobial activity. Nisin is a natural antimicrobial peptide produced from skimmed or soy milk by *L. lactis* that effectively inhibits both Gram-positive and Gram-negative bacteria, alone or in combination with EDTA or other chelating agents, as well as spores of *Bacilli* and *Clostridia*. It has been used for the preservation of canned and fresh-cut fruits and vegetables (de Arauz *et al.*, 2009).

There are only few studies in which the effect on the overall quality of fresh horticultural produce after treatment with sulphur-containing compounds has been considered. One of the main concerns is that the intense sensory attributes of these compounds (strong smell) may be an impediment for the preservation of fresh horticultural products. Troncoso-Rojas *et al.* (2009) observed that the external quality of netted melon was not affected by treatment with ITCs. However, Wang *et al.* (2010) observed that ITCs reduced decay of blueberries due to an increased amount of ROS, which ultimately inhibits microbiological growth, but at the same time this increase in the ROS compounds reduced the total phenolics and antioxidant capacity. Ayala-Zavala *et al.* (2008b) found that garlic oil in the vapour phase conferred desirable sensory attributes to fresh-cut tomatoes, which made them more acceptable than non-treated tomatoes.

Disadvantages of the Use of Natural Antimicrobials as Fresh-Produce Additives

The chemical reactivity of natural antimicrobial agents with fresh fruits and vegetables and package matrix could significantly affect the antimicrobial activity and sensorial properties of the produce. Reactions with lipids, proteins, carbohydrates and other additives can result in an overall decrease in the activity of the antimicrobial compound (Ayala-Zavala *et al.*, 2008c). The ability of most antimicrobials, including natural compounds, to inhibit microorganisms can be overcome on extended storage (Ayala-Zavala *et al.*, 2008a). Depending on the time and temperature of storage, these antimicrobials can be volatilized or become inactive (Gutierrez *et al.*, 2008). Normally direct application of antimicrobials to food must be done at high concentrations to achieve activity against target microorganisms on extended storage of fresh fruit and vegetables. Obviously, antimicrobial compounds that negatively affect flavour and odour would be unacceptable. In addition to adverse effects on flavour, odour or texture, it would also be unacceptable for a natural antimicrobial to mask perceptible spoilage, because this kind of spoilage may protect the consumer from ingesting food-borne pathogens.

Appropriate or compatible use of natural antimicrobial agents would involve using these compounds to add positive sensory characteristics in addition to improve food safety and/or extending shelf life of fresh fruits and vegetables (Ayala-Zavala *et al.*, 2009; Gonzalez-Aguilar *et al.*, 2010). Essential oils are effective antimicrobials; however, their aromatic volatile constituents can be absorbed by the food product. By choosing the right combination of aromas between the antimicrobial essential oil and the fresh-cut produce, quality involving safety and flavour can be improved. However, the problem of the extra addition still remains, so alternatives to solve the addition problems of natural antimicrobial compounds must be contemplated.

Actual and Future Trends for the Optimization of Natural Antimicrobials as Fresh Fruit and Vegetable Additives

Future research to optimize the use of natural antimicrobials to preserve fresh fruit and vegetables' quality and safety should be focused on including: the sensorial appeal of the natural antimicrobial treated products; characterization of the antimicrobial efficacy of new natural compounds; physicochemical interactions between the natural antimicrobial and the packaging material; and physicochemical, physiological and biochemical effects of the natural antimicrobial on the treated produce. Another interesting topic that should be highlighted as a promising research area is the design of active packaging as a delivery system for natural antimicrobials and antioxidants to avoid oxidation, as well microbial spoilage of fresh produce. All this research will be useful to allow the development of new practical methods to preserve quality of fresh produce, accomplishing consumer demands.

Conclusions

Synthetic additives can reduce the decay rate of fresh produce, but they come with major concerns in terms of chemical residues found in foods as well as causing environmental pollution. Therefore, emerging technologies such as natural additives (i.e. essential oils, terpenoids, phenol compounds, chitosan and sulfur compounds) are gaining more consumer and scientific interest. Consumers demand natural and/or organic methods to preserve foods, while researchers are committed to investigate and fulfil those requirements.

Countless possibilities can be found in nature of the outstanding role of antimicrobial compounds that can be used as effective treatments to extend shelf life of fresh produce by delaying postharvest decay. In this context, research scientists are contributing to the practical use of such resources, by innovating through designing active packaging as natural antimicrobial delivery systems in order to preserve quality of fresh produce.

However, some limitations should be considered when researching natural additives. Their chemical reactivity with lipids, proteins, carbohydrates and the package matrix from fruits and vegetables could significantly affect the antimicrobial activity and sensorial properties of the produce. Amongst them, the aromatic volatile constituents from essential oils can be absorbed by the food product. Alternatives to solve the addition problems of natural antimicrobial compounds must be contemplated.

References

Aggarwal, K.K., Khanuja, S.P.S., Ahmad, A., Kumar, T.R.S., Gupta, V.K. and Kumar, S. (2002) Antimicrobial activity profiles of the two enantiomers of limonene and carvone isolated from the oils of *Mentha spicata* and *Anethum sowa*. *Flavour Fragrance Journal* 17, 59–63.

Ankri, S. and Mirelman, D. (1999) Antimicrobial properties of allicin from garlic. *Microbes and Infection* 1, 125–129.

Arcila-Lozano, C.C., Loarca-Pina, G., Lecona-Uribe, S. and De Mejia, E.G. (2004) Oregano: properties, composition and biological activity. *Archivos Latinoamericanos de Nutrición* 54, 100–111.

Artés, F., Gómez, P., Artés-Hernández, F., Aguayo, E. and Escalona, V. (2007) Improved strategies for keeping overall quality of fresh-cut produce. *Acta Horticulturae (ISHS)* 746, 245–258.

Ayala-Zavala, J.F., Del-Toro-Sanchez, L., Alvarez-Parrilla, E. and Gonzalez-Aguilar, G.A. (2008a) High relative humidity in-package of fresh-cut fruits and vegetables: advantage or disadvantage considering microbiological problems and antimicrobial delivering systems? *Journal of Food Science* 73, R41–R47.

Ayala-Zavala, J.F., Oms-Oliu, G., Odriozola-Serrano, I., Gonzalez-Aguilar, G.A., Alvarez-Parrilla, E. and Martin-Belloso, O. (2008b) Bio-preservation of fresh-cut tomatoes using natural antimicrobials. *European Food Research and Technology* 226, 1047–1055.

Ayala-Zavala, J.F., Soto-Valdez, H., Gonzalaz-Leon, A., Alvarez-Parrilla, E., Martin-Belloso, O. and Gonzalez-Aguilar, G.A. (2008c) Microencapsulation of cinnamon leaf (*Cinnamomum zeylanicum*) and garlic (*Allium sativum*) oils in β-cyclodextrin. *Journal of Inclusion Phenomena and Macrocyclic Chemistry* 60, 359–368.

Ayala-Zavala, J.F., Gonzalez-Aguilar, G.A. and Del-Toro-Sanchez, L. (2009) Enhancing safety and aroma appealing of fresh-cut fruits and vegetables using the antimicrobial and aromatic power of essential oils. *Journal of Food Science* 74, R84–R91.

Baranauskiene, R., Venskutonis, P.R. and Demyttenaere, J.C.R. (2005) Sensory and instrumental evaluation of sweet marjoram (*Origanum majorana* L.) aroma. *Flavour and Fragrance Journal* 20, 492–500.

Belleti, N., Lanciotti, R., Patrignani, F. and Gardini, F. (2008) Antimicrobial efficacy of citron essential oil on spoilage and pathogenic microorganisms in fruit-based salads. *Journal of Food Science* 73, M331–M338.

Belleti, N., Kamdem, S.S., Tabanelli, G., Lanciotti, R. and Gardini, F. (2010) Modeling of combined effects of citral, linalool and -pinene used against *Saccharomyces cerevisiae* in citrus-based beverages subjected to a mild heat treatment. *International Journal of Food Microbiology* 136, 283–289.

Benkeblia, N. (2004) Antimicrobial activity of essential oil extracts of various onions (*Allium cepa*) and garlic (*Allium sativum*). *Lebensmittel Wissenschaft und Technologie* 37, 263–268.

Berger, R.G. (2007) *Flavours and Fragrances, Chemistry, Bioprocessing and Sustainability,* Springer, Berlin.

Bhat, S.V., Nagasampagi, B.A. and Meenakshi, S. (2009) *Natural Products Chemistry and Applications.* Alpha Science International Ltd, Oxford, UK. p. 596.

Braca, A., Siciliano, T., D'arrigo, M. and Germano, M.P. (2008) Chemical composition and antimicrobial activity of *Momordica charantia* seed essential oil. *Fitoterapia* 79, 123–125.

Brecht, J.K. (2006) Shelf-life limiting quality factors in fresh-cut (sliced) tomatoes: Anti-ethylene treatment and maturity & variety selection to ensure quality retention. *Oral presentation on the 2006 Tomato Breeders Round Table & Tomato Quality Workshop.*

Brehm-Stecher, B.F. and Johnson, E.A. (2003) Sensitization of *Staphylococcus aureus* and *Escherichia coli* to antibiotics by the sesquiterpenoids nerolidol, farnesol, bisabolol, and apritone. *Antimicrobial Agents Chemotherapy* 47, 3357–3360.

Burt, S. (2004) Essential oils: their antibacterial properties and potential applications in foods—a review. *International Journal of Food Microbiology* 94, 223–253.

Busta, F.F., Suslow, T.V., Parish, M.E., Beuchat, L.R., Farber, J.N., Garrett, E.H. and Harris, L.J. (2003) The use of indicators and surrogate microorganisms for the evaluation of pathogens in fresh and fresh-cut produce. *Comprehensive Reviews in Food Science and Food Safety* 2, 179–185.

CFSAN/FDA (2006) *Food Additives List* [Online]. Food and Drug Agency-Center for Food Safety and Applied Nutrition. Available: http://www.foodsafety.gov/~dms/opa-appa.html [Accessed Jun 7 2008].

Chang, H.T., Cheng, Y.H., Wu, C.L., Chang, S.T., Chang, T.T. and Su, Y.C. (2008) Antifungal activity of essential oil and its constituents from *Calocedrus macrolepis* var. formosana Florin leaf against plant pathogenic fungi. *Bioresource Technology* 99, 6266–6270.

Chaumont, J.P. and Léger, D. (1992) Elimination of allergenic moulds in dwellings. Antifungal properties of vapours of essential oil of Geranium Bourbon, citronellol, geraniol and citral. *Annales Pharmaceutiques Francaises* 50, 156–166.

Cho, W.I., Choi, J.B., Lee, K., Chung, M.S. and Pyun, Y.R. (2008) Antimicrobial activity of torilin isolated from *Torilis japonica* fruit against *Bacillus subtilis. Journal of Food Science* 73, M37–M46.

Chung, I., Kwon, S.H., Shim, S.T. and Kyung, K.H. (2007) Synergistic antiyeast activity of garlic oil and allyl alcohol derived from alliin in garlic. *Journal of Food Science* 72, M437–M440.

Cooksey, K. (2001) Antimicrobial food packaging materials. *Additives for Polymers* 8, 6–10.

Cosentino, S., Tuberoso, C.I., Pisano, B., Satta, M., Mascia, V., Arzedi, E. and Palmas, F. (1999) In-vitro antimicrobial activity and chemical composition of Sardinian *Thymus* essential oils. *Letters of Applied Microbiology* 29, 130–135.

Cowan, M.M. (1999) Plant products as antimicrobial agents. *Clinical Microbiology Reviews* 12, 564–582.

Cuero, R.G., Osuji, G. and Washington, A. (1991) N-carboxymethyl chitosan inhibition of aflatoxin production: role of zinc. *Biotechnology Letters* 13, 441–444.

De Arauz, L.J., Jozala, A.F., Mazzola, P.G. and Vessoni Penna, T.C. (2009) Nisin biotechnological production and application: a review. *Trends in Food Science and Technology* 20, 146–154.

Devlieghere, F., Vermeulen, A. and Debevere, J. (2004) Chitosan: antimicrobial activity, interactions with food components and applicability as a coating on fruit and vegetables. *Food Microbiology* 21, 703–714.

Edris, A.E. and Farrag, E.S. (2003) Antifungal activity of peppermint and sweet basil essential oils and their major aroma constituents on some plant pathogenic fungi from the vapor phase. *Nahrung* 47, 117–121.

Fahey, J.W., Zalcmann, A.T. and Talalay, P. (2001) The chemical diversity distribution of glucosinolates and isothiocyanates among plants. *Phytochemistry* 56, 5–51.

Fang, S.W., Li, C.F. and Shih, D.Y.C. (1994) Antifungal activity of chitosan and its preservative effect on low-sugar candied kumquat. *Journal of Food Protection.* 57, 136–140.

Fisher, K. and Phillips, C. (2008) Potential antimicrobial uses of essential oils in food: is citrus the answer? *Trends in Food Science & Technology* 19, 156–164.

Frazier, W.C. and Westhoff, D.C. (1993) *Microbiología de los alimentos,* Acribia, Zaragoza, Spain.

Friedman, M. (1999) Chemistry nutrition and microbiology of d-amino acids. *Journal of Agricultural of Food Chemistry* 47, 3457–3479.

Gonzalez-Aguilar, G.A., Ayala-Zavala, J.F., Ruiz-Cruz, S., Acedo-Felix, E. and Diaz-Cinco, M.E. (2004) Effect of temperature and modified atmosphere packaging on overall quality of fresh-cut bell peppers. *Lebensmittel-Wissenschaft und-Technologie-Food Science and Technology* 37, 817–826.

Gonzalez-Aguilar, G.A., Ayala-Zavala, J.F., Olivas, G.I., De La Rosa, L.A. and Alvarez-Parrilla, E. (2010) Preserving quality of fresh-cut products using safe technologies. *Journal Fur Verbraucherschutz und Lebensmittelsicherheit-Journal of Consumer Protection and Food Safety* 5, 65–72.

Gutierrez, J., Barry-Ryan, C. and Bourke, R. (2008) The antimicrobial efficacy of plant essential oil combinations and interactions with food ingredients. *International Journal of Food Microbiology* 124, 91–97.

Ha, H.K.P., Maridable, J., Gaspillo, P., Hasika, M., Malaluan, R. and Kawasaki, J. (2008) Essential oil from lemongrass extracted by supercritical carbon dioxide and steam distillation. *Philippine Agricultural Scientist* 91, 36–41.

Jeon, Y.J., Park, P.J. and Kim, S.K. (2001) Antimicrobial effect of chitooligosaccharides produced by bioreactor. *Carbohydrate Polymer* 44, 71–76.

Jorgensen, L.V., Madsen, H.L., Thomsen, M.K., Dragsted, L.O. and Skibsted, L.H. (1999) Regulation of phenolic antioxidants from phenoxyl radicals: an ESR and electrochemical study of antioxidant hierarchy. *Free Radical Research* 30, 207–220.

Jugl-Chizzola, M., Ungerhofer, E., Gabler, C., Hagmuller, W., Chizzola, R., Zitterl-Eglseer, K. and Franz, C. (2006) Testing of the palatability of *Thymus vulgaris* L. and *Origanum vulgare* L. as flavouring feed additive for weaner pigs on the basis of a choice experiment. *Berliner und Munchener Tierarztliche Wochenschrift* 119, 238–243.

Kim, J.W., Huh, J.E., Kyung, S.H. and Kyung, K.H. (2004) Antimicrobial activity of alk(en)yl sulfides found in essential oils of garlic and onion. *Food Science and Biotechnology* 13, 235–239.

Kisko, G. and Roller, S. (2005) Carvacrol and p-cymene inactivate *Escherichia coli* O157:H7 in apple juice. *BMC Microbiology* 5, 36.

Kyung, K.H. and Lee, Y.C. (2001) Antimicrobial activities of sulfur compounds derived from S-alk(en)yl-L-cysteine sulfoxides in Allium and Brassica. *Food Review International* 17, 183–198.

Lambert, R.J., Skandamis, P.N., Coote, P.J. and Nychas, G.J. (2001) A study of the minimum inhibitory concentration and mode of action of oregano essential oil, thymol and carvacrol. *Journal of Applied Microbiology* 91, 453–462.

Luciano, F.B. and Holley, R.A. (2009) Enzymatic inhibition by isothiocyanate and factors affecting its antimicrobial action against *Escherichia coli O157:H7. International Journal of Food Microbiology* 131, 240–245.

Mari, M., Leoni, O., Lori, R. and Cembali, T. (2002) Antifungal vapour-phase activity of allyl-isothiocyanate against *Penicillium expansum* on pears. *Plant Pathology* 51, 231–236.

Mari, M., Leoni, O., Bernardi, R., Neri, F. and Palmieri, S. (2008) Control of brown rot on stonefruit by synthetic and glucosinolates-derived isothiocyanates. *Posharvest Biology and Technology*, 61–67.

Marino, M., Bersani, C. and Comi, G. (1999) Antimicrobial activity of the essential oils of *Thymus vulgaris* L. measured using a bioimpedometric method. *Journal of Food Protection* 62, 1017–1023.

Marino, M., Bersani, C. and Comi, G. (2001) Impedance measurements to study the antimicrobial activity of essential oils from *Lamiacea* and *Compositae. Journal of Food Microbiology* 67, 187–195.

Moller, H., Grelier, S., Pardon, P. and Coma, V. (2004) Antimicrobial and physicochemical properties of chitosan -HPMC-based films. *Journal of Agricultural and Food Chemistry* 52, 6585–6591.

Mukherjee, M. and Datta, A.K. (2007) The basils – A review. *Plant Arch* 7, 473–483.

No, H.K., Park, N.A., Lee, S.H. and Meyers, S.P. (2002) Antibacterial activity of chitosans and chitosan oligomers with different molecular weights. *International Journal of Food Microbiology* 74, 65–72.

Novak, I., Zambori-Nemeth, E., Horvath, H., Seregely, Z. and Kaffka, K. (2003) Study of essential oil components in different *Origanum* species by GC and sensory analysis. *Acta Alimentaria* 32, 141–150.

Obaidat, M.M. and Frank, J.F. (2009a) Inactivation of *Escherichia coli O157:H7* on the intact and damaged portions of lettuce and spinach leaves by using allyl isothiocyanate, carvacrol, and cinnamaldehyde in vapor phase. *Journal of Food Protection* 72, 2046–2055.

Obaidat, M.M. and Frank, J.F. (2009b) Inactivation of *Salmonella* and *Escherichia coli* O157:H7 on sliced and whole tomatoes by allyl isothiocyanate, carvacrol, and cinnamaldehyde in vapor phase. *Journal of Food Protection* 72, 315–324.

Pal Vig A., Rampal, G., Singh Thind, T. and Arora, S. (2009) Bio-protective effects of glucosinolates. *LWT-Food Science and Technology* 42, 1561–1572.

Palhano, F.L., Vilches, T.T.B., Santos, R.B., Orlando, M.T.D., Ventura, J.A. and Fernandez, P.M.B. (2004) Inactivation of *Colletotrichum gloeosporioides* spores by high hydrostatic pressure combined with citral or lemongrass essential oil. *International Journal of Food Microbiology* 95, 61–66.

Park, I.K. and Shin, S.C. (2005) Fumigant activity of plant essential oils and components from garlic (*Allium sativum*) and clove bud (*Eugenia caryophyllata*) oils against the Japanese termite (*Reticulitermes speratus* Kolbe). *Journal of Agricultural and Food Chemistry* 53, 4388–4392.

Patte, J. (1996) Biosynthesis of threonine and lysine. In: Fredeci, M., Neidhardt, F. (eds) *Escherichia coli and Salmonella*, 2nd edn. ASM Press, Washington, DC, pp. 528–541.

Ravi, R., Prakash, M. and Bhat, K. K. (2007) Aroma characterization of coriander (*Coriandrum sativum* L.) oil samples. *European Food Research and Technology* 225, 367–374.

Rice-Evans, C.A., Miller, N.J., Bolwell, P.G., Bramley, P.M. and Pridham, J.B. (1995) The relative antioxidant activities of plant-derived polyphenolic flavonoids. *Free Radical Research* 22, 375–383.

Roller, S. (2003) Chitosan: new food preservative or laboratory curiosity? In: Roller, S. (ed.) *Natural Antimicrobials for the Minimal Processing of Foods.* Woodhead Publishing, Cambridge, UK, pp. 158–175.

Roller, S. and Lusengo, J. (1997) Developments in natural food preservatives. *Agro Food Industry Hi-Tech* 8, 22–25.

Roller, S. and Seedhar, P. (2002) Carvacrol and cinnamic acid inhibit microbial growth in fresh-cut melon and kiwifruit at 4° and 8° C. *Letters of Applied Microbiology* 35, 390–394.

Saleem, M., Nazir, M., Ali, M.S., Hussain, H., Lee, Y.S., Riaz, N. and Jabbar, A. (2010) Antimicrobial natural products: an update on future antibiotic drug candidates. *Natural Products Report* 27, 238–254.

Senhaji, O., Faid, M. and Kalalou, I. (2007) Inactivation of *Escherichia coli* O157:H7 by essential oil from *Cinnamomum zeylanicum. Brazilian Journal of Infectious Diseases* 11, 234–236.

Shetty, K. and Lin, Y.-T. (2005) Phenolic antimicrobials from plants for control of bacterial pathogens. In: Clay, D. and Pierce, F.J. (eds) *Functional Foods and Biotechnology,* CRC Press, Boca Raton, Florida.

Shin, I.S., Masuda, H. and Naohide, K. (2004) Bacterial activity of wasabi (*Wasabia japonica*) against *Helicobacter pylori. International Journal of Food Microbiology* 94, 255–261.

Sivam, G.P. (2001) Recent advances on the nutritional effects associated with the use of garlic as supplement. *Journal of Nutrition* 1106S–1108S.

Stern, J.L., Hagerman, A.E., Steinberg, P.D. and Mason, P.K. (1996) Phlorotannins-protein interactions. *Journal of Chemical Ecology* 22, 1877–1899.

Sudarshan, N.R., Hoover, D.G. and Knorr, D. (1992). Antibacterial action of chitosan. *Food Biotechnology* 6, 257–272.

Tajkarimi, M., Ibrahim, S.A. and Cliver, D.O. (2010) Antimicrobial herb and spice compounds in food. *Food Control* 21, 1199–1218.

Tassou, C.C., Nychas, G.J.E. and Skandamis, P.N. (2004) Herbs and spices and antimicrobials. In: Peter, K.V. (ed.) *Handbook of Herbs and Spices*, CRC Press, Boca Raton, Florida. pp. 23–40.

Tierens, K, F.M.J., Thomma, B.P.H.J., Brouwer, M., Schmidt, J., Kistner, K., Porzel, A., Mauch-Mani, B., Cammue, B.P.A. and Broekaert, W.F. (2001) Study of the role of antimicrobial glucosinolates-derived isothiocyanates in resistance of Arabidopsis to microbial pathogens. *Plant Physiology* 1688–1699.

Tiwari, B.K., Valdramidis, V.P., O'Donell, C.P., Muthukumarappan, K., Bourke, P. and Cullen, P.J. (2009) Application of natural antimicrobials for food preservation. *Journal of Agricultural and Food Chemistry* 57, 5987–6000.

Troncoso-Rojas, R., Corral-Acosta, Y., Sánchez-Estrada, A., García-Estrada, R., Aguilar-Valenzuela, A., Ojeda-Contreras, J. and Tiznado-Hernández, M.E. (2009) Postharvest treatment of isothiocyanates to control alternaría rot in detted melon. *Phytoparasitica* 445–451.

Tsai, G.J., Su, W-H., Chen, H-S. and Pan, C.L. (2002) Antimicrobial activity of shrimp chitin and chitosan from different treatments and applications of fish preservation. *Fisheries Science* 68, 170–177.

Ultee, A., Kets, E.P.W. and Smid, E.J. (1999) Mechanisms of action of carvacrol on the food-borne pathogen *Bacillus cereus. Applied and Environmental Microbiology* 65, 4606–4610.

Ultee, A., Bennink, M.H.J. and Moezelaar, R. (2002) The phenolic hydroxyl group of carvacrol is essential for action against the food-borne pathogen *Bacillus cereus.* Appl Environ Microb 68, 1561–1568.

Walker, J.R.L. (1994) Antimicrobial compounds in food plants. In: Board, R.G. and Dillon, V. (eds), *Natural Antimicrobial Systems and Food Preservation.* CAB International, Wallingford, UK, pp. 181–204.

Wang, C. (2006) The use of essential oils as natural preservatives for berry fruits. *Hortscience* 41, 1042–1043.

Wang, S.Y., Chen, Ch.T. and Yin, J.J. (2010) Effect of allyl isothiocyanate on antioxidants and fruit decay of blueberries. *Food Chemistry* 120, 199–204.

White, J.M.L. and McFadden, J.P. (2008) Contact allergens in food ingredients and additives: atopy and the hapten-atopy hypothesis. *Contact Dermatitis* 58, 245–246.

Witz, G. (1989) Biological interaction of α,β-unsaturated aldehydes. *Free Radical Biology and Medicine* 7, 333–349.

Xu, F., Zhou, B.P., Ji, R.S., Pei and N. Xu (2008) The antibacterial mechanism of carvacrol and thymol against *Escherichia coli. Letters in Applied Microbiology* 47, 174–179.
Zeller, A. and Rychlik, M. (2007) Impact of estragole and other odorants on the flavour of anise and tarragon. *Flavour Fragrance Journal* 22, 105–113.
Zheng, L.Y and Zhu. J.F. (2003) Study on antimicrobial activity of chitosan with different molecular weights. *Carbohydrate Polymers 54,* 527–530.
Zhou, F., Ji, B., Zhang, H., Jiang, H., Zhiwei, Y., Li, J., Li, J. and Yan, W. (2007) The antibacterial effect of cinnamaldehyde, thymol, carvacrol, and their combinations against the foodborne pathogen *Salmonella typhimurium. Journal of Food Safety* 27, 124–133.

19 Biological Approaches for Control of Human Pathogens on Produce

William F. Fett, Ching-Hsing Liao and Bassam A. Annous*

Introduction

Fresh and minimally processed fruits and vegetables are an important part of a healthy diet as they provide needed nutrients, fibre and antioxidants. At a time when two-thirds of all Americans are overweight or obese, increasing the consumption of fruits and vegetables is highly desirable in order to fight obesity and associated diseases such as cardiovascular disease and type 2 diabetes (USDHHS/USDA, 2005). Greater awareness of the health benefits of increased intake of produce has led to a sharp increase in per capita consumption of fresh produce in the USA and around the world (Kaufman *et al.*, 2000; ERS, 2004).

Unfortunately, along with the increase in produce consumption there has been a sharp increase in the number of food-borne outbreaks due to fresh fruits and vegetables contaminated with a variety of human pathogens. The outbreaks associated with fresh produce doubled between the periods 1973 to 1987 and 1988 to 1992 (Buck *et al.*, 2003). Produce-related outbreaks accounted for 6% of all reported food-borne outbreaks in the 1990s compared to only 0.7% in the 1970s (FDA, 2004; Sivapalasingam *et al.*, 2004). A recent analysis of food-borne outbreaks occurring in the USA between 1990 and 2003 indicated that contaminated fresh produce caused the most illnesses and the second most number of outbreaks (CSPI, 2005).

Conventional washing and sanitizing technologies are not very effective in reducing the populations of human pathogens on fresh produce, resulting in reductions of only 1 to 2 log cfu/g produce (Sapers, 2006). Since complete elimination of sources of contamination on the farm is not feasible, more effective intervention strategies are needed. Such strategies may include the use of biological-based interventions (Gombas, 1989) for suppressing the populations of pathogens and/or inhibiting the outgrowth of survivors after other chemical or physical interventions are applied (the multiple hurdle approach).

A number of commercial products containing various types of microbes are commercially available for use in combating plant diseases (OSU, 2005) and for controlling the colonization of newly hatched chicks with *Salmonella* or shedding of *Escherichia coli* O157:H7 by cattle (Schneitz, 2005). These products consist of either a single microorganism or an undefined mixture of microorganisms. Mechanisms responsible for biological control activity could be due to competition for nutrients or

* Corresponding author.

space, production of antimicrobial substances, and induction of host defence responses (Haas and Défago, 2005; Anderson *et al.*, 2006). The development of biological control agents for reducing pathogen contamination on fresh and fresh-cut produce represents a relatively unexplored area of research. In this chapter the potential of using microbial agents for controlling the growth of food-borne pathogens on fresh and fresh-cut produce will be examined.

Native and Spoilage Microflora on Fresh Produce

Produce surfaces harbour a wide variety of microbes, including bacteria, yeasts and fungi. On fresh and minimally processed produce populations of total aerobic bacteria ranging from 10^2 to 10^9 cfu/g of tissue have been reported, with the highest levels found on seed sprouts (Nguyen-the and Carlin, 1994; Fett *et al.*, 2006). The majority of bacteria recovered are Gram-negative rods and most belong to the genera *Pseudomonas, Enterobacter, Erwinia* and *Pantoea*. Other Gram-negative genera isolated include *Alcaligenes, Chromobacterium, Chryseomonas, Citrobacter, Flavobacterium, Klebsiella, Rhanella, Serratia* and *Xanthomonas*. Gram-positive bacteria isolated include *Bacillus, Micrococcus, Paenibacillus, Sarcina* and lactic acid bacteria (LAB) such as *Leuconostoc mesenteroides*. LAB usually constitute a minor component of the microbial community on the surfaces of non-spoiled vegetables (Nguyen-the and Carlin, 1994). Certain members of native bacteria can exist internally in plants and are referred to as endophytes (Lund, 1992). Numerous genera of yeasts including *Candida, Cryptococcus, Rhodotorula, Trichosporon, Pichia, Sporobolomyces* and *Torulaspora* have also been isolated from produce (Nguyen-the and Carlin, 2000).

Bacteria and fungi capable of causing diseases or spoilage can be readily found on the surfaces of fresh produce (ICMSF, 2005a,b). Pectolytic bacteria including *Pseudomonas, Erwinia, Cytophaga, Xanthomonas, Clostridium* and *Bacillus* account for a major proportion of soft rot or spoilage of fresh vegetables at markets (Liao *et al.*, 2003). Plant-pathogenic fungi including *Alternaria, Aspergillus, Botrytis, Cladosporium, Geotrichum, Penicillium* and *Rhizopus* are often found associated with spoilage of acidic fruits such as orange, berry and apple. The diversity and populations of microbes on plant surfaces are often underestimated as indicated by recent studies using cultivation-independent methods (Yang *et al.*, 2001).

Presence of Human Pathogens on Fresh Produce

A wide variety of human pathogens can be isolated from the surfaces of raw produce (Beuchat, 1996), where the pathogens often survive for extended periods of time (Francis *et al.*, 1999). Although most of produce-related outbreaks were caused by *Salmonella*, other bacterial pathogens such as *E. coli* O157:H7, *Shigella, Listeria monocytogenes, Campylobacter* and *Yersinia enterocolitica* were also involved. Contamination of fresh produce with viruses (hepatitis A and norovirus) or protozoa (*Cyclospora cayetanensis, Cryptosporidium parvum* and *Giardia lamblia*) has also been reported (Silvapalasingam *et al.*, 2004). Fresh produce most often involved in food-borne outbreaks from 1973 to 1997 included mixed salads, lettuce, melon, tomato, juices, sprouts and berries. In most cases human pathogens are probably surface borne or within substomatal cavities; however, internalization of human pathogens within the vascular system of growing fruits or vegetables has been reported (Warriner *et al.*, 2003; Solomon *et al.*, 2006).

Interactions between Native Microflora, Spoilage Microorganisms and Human Pathogens

Native microflora associated with fresh produce can enhance, decrease or show no effect on the survival and growth of human pathogens. Some members of native microflora are antagonistic against human pathogens when tested using *in vitro* or *in situ* assays (Schuenzel and Harrison, 2002). Aytec and Gorris (1994) reported that *L. monocytogenes* grew well on chicory endive stored for 7 days under moderate vacuum at 6.5°C, but it grew poorly on mung bean sprouts stored under the same conditions. The difference observed was attributed to a much higher population of native microflora on mung bean sprouts compared to chicory. Carlin *et al.* (1996) also reported that *L. monocytogenes* grew more rapidly on the surface of endive leaves at 10°C if the indigenous microflora on the leaves was first reduced by treatment with 10% hydrogen peroxide. In addition, the inhibitory effect of native microflora on the survival or growth of *L. monocytogenes, Vibrio cholerae, Salmonella typhi* and *E. coli* O157:H7 on shredded lettuce, endive leaves, cantaloupe and alfalfa sprouts have been previously reported (Bennik *et al.*, 1996; Francis and O'Bierne, 1997; Castro-Rosas and Escartin, 2000; Schoeller *et al.*, 2002; Ukuku *et al.*, 2004). In contrast to the studies mentioned above, Beuchat and Brackett (1991) found that elimination of native microflora from tomatoes by chlorine treatment did not affect the survival and growth of *L. monocytogenes*. Native microflora associated with pre-peeled baby carrot also inhibited the growth of four food-borne pathogens and one spoilage pseudomonad (*L. monocytogenes, Y. enterocolitica, Salmonella enterica, E. coli* O157:H7 and *Pseudomonas fluorescens*) on bell pepper slices by 3 to 4 log units (Fig.19.1; Liao, unpublished data).

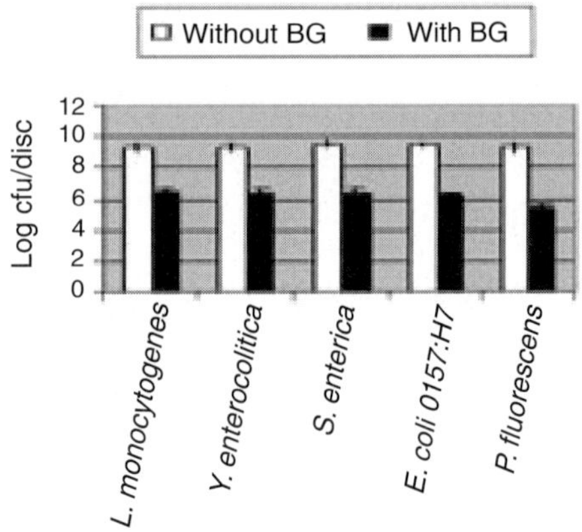

Fig.19.1. Inhibition of the growth of *Listeria monocytogenes*, *Yersinia enterocolitica*, *Salmonella enterica*, *Escherichia coli* O157:H7 and *Pseudomonas fluorescens* in carrot tissue homogenate by native microflora isolated from pre-peeled baby carrot and designated as BG.

The spoilage of fresh fruits and vegetables can be caused by different types of bacteria and fungi (Tournas, 2005). The interaction of spoilage organisms with human pathogens can also affect the growth of human pathogens on produce surfaces. Wells and Butterfield (1997; 1999) reported that soft-rotted portions of fruits and vegetables were more likely to harbour *Salmonella* than healthy, intact tissues. They also found that the population of *Salmonella* was 3 to 10-fold greater on potato, carrot and pepper discs co-inoculated with pectolytic bacteria such as *Pseudomonas viridiflava* or *Erwinia carotovora* than on disks inoculated with *Salmonella* alone. In a similar study, Wells and Butterfield (1999) found that growth of *Salmonella* on tomato, potato and onion was greatly enhanced in the presence of two pectolytic fungi (*Botrytis* and *Rhizopus*) but not in the presence of two non-pectolytic fungi (*Alternaria* and *Geotrichum*). They suggested that maceration of plant tissues by pectolytic fungi favoured the growth of *Salmonella*. However, not all of pectolytic microorganisms associated with plants promote the growth of human pathogens on fresh produce. Liao and Sapers (1999) reported that the growth of *L. monocytogenes* on potato slices was inhibited in the presence of pectolytic fluorescent pseduomonads such as *P. fluorescens* and *P. viridiflava*, possibly due to the production of iron-chelating fluorescent pigments.

Co-inoculation of wounded apple surfaces with the pathogenic fungus *Glomerella cingulata* favoured survival and growth of *E. coli* O157:H7 (Riordan *et al.*, 2000) or *L. monocytogenes* (Conway *et al.*, 2000), whereas co-inoculation with *Penicillium expansum* led to a rapid decline in pathogen populations. The increased growth of *E. coli* O157:H7 or *L. monocytogenes* was attributed to a rise in pH from 4.1 to 6.8 and the decreased growth due to a decline in pH from 4.1 to 3.0 in the infected tissue due to the differential fungal activity of *G. cingulata* or *P. expansum*. Wade and Beuchat (2003) also reported that co-inoculation with proteolytic moulds (*Alternaria alternata, Cladosporium herbarum* and *C. cladosporioides*) enhanced the growth of *Salmonella* in tomatoes due to an increased pH in infected tissue. Inoculation of wounds in cantaloupe rinds with *C. cladosporioides* also aided the survival and migration of *Salmonella* into the internal tissues (Richards and Beuchat, 2005). In contrast, co-inoculation of intact spinach leaves with *E. coli* O157:H7 and plant pathogen *Pseudomonas syringae* DC3000 did not increase the survival of *E. coli* O157:H7 (Hora *et al.*, 2005).

Biological Control of Bacterial Human Pathogens

Use of LAB for preserving foods including fruits and vegetables by fermentation has been employed for thousands of years (Breidt and Fleming, 1997). Use of microbial antagonists as a means for controlling plant diseases has been studied quite intensively since the early 1920s (Janisiewicz and Korsten, 2002; Haas and Défago, 2005). Recently, application of beneficial microorganisms to a variety of food animals (e.g. chicks, swine and calves), either as a competitive exclusion or probiotic treatment, has been used to prevent the establishment of zoonotic pathogens such as *Salmonella* spp., *E. coli* O157:H7 and *Campylobacter jejuni* in the intestinal tract (Anderson *et al.*, 2006). Commercial biocontrol products to be used for controlling human pathogens on fresh produce are currently unavailable. Studies with the microbial antagonists for inhibition of human pathogens on vegetables, seed sprouts and fruits have become a subject of extensive investigations in several laboratories around the world (Table 19.1).

Vegetables

A large number of studies on the use of biological control for human pathogens on fresh produce utilized LAB. Vescovo and co-workers (Vescovo *et al.*, 1995; Vescovo *et al.*, 1996; Torriani *et al.*, 1997) evaluated the ability of LAB to ensure the microbiological safety of refrigerated ready-to-use vegetables. Initially, four strains of LAB (two strains of *Lactobacillus casei* and one strain each of *Pediococcus pentosacceus* and *Pediococcus* spp.) were tested based on their ability to produce antimicrobial substances and to grow at 8°C *in vitro*. They found that these four strains, especially two *L. casei* strains, were very effective in reducing the growth of coliforms and enterococci between days 3 and 8 of storage at 8°C (Vescovo *et al.*, 1995). The authors suggested that inhibition was due to the production of antibacterial substances along with a slight decrease in pH.

Later, five psychrotrophic strains of LAB, including two strains of *L. casei*, two strains of *L. plantarum* and one strain of *Pediococcus* spp. were isolated from mixed salad and tested for their ability to inhibit *Aeromonas hydrophila, L. monocytogenes, Salmonella* and *Staphylococcus aureus in vitro* and in mixed salad (Vescovo *et al.*, 1996). *L. casei* strain IMPC LC34 was most effective against four pathogens mentioned above at 8 and 37°C. In mixed salad, inoculation with *L. casei* strain IMPC LC34 led to the elimination of *A. hydrophila, Salmonella, S. aureus* and *L. monocytogenes*. The inhibitory activity of *L. casei* strain IMPC LC34 against *A. hydrophila* on mixed salad vegetables, containing carrots, endive, garden rocket and green chicory was also confirmed by Torriani *et al.* (1997). The authors suggested that *in situ* inhibitory activity of strain IMPC LC34 was due to production of lactic acid and an uncharacterized active agent.

Carlin *et al.* (1996) tested 10 strains of Gram-negative rods in the families of Pseudomonadaceae and Enterobacteriaceae for their activity against *L. monocytogenes* on endive. Intact leaves were first surface sanitized by treatment with 10% hydrogen peroxide, cut into pieces, and then inoculated with *P. fluorescens* F2 or *P. chlororaphis* C1 to give 10^6 to 10^7 cfu/g and also *L. monocytogenes* to give 10^4 to 10^5 cfu/g. Following incubation at 10°C for 7 days,

Table 19.1. Biological agents reported to inhibit the growth and/or survival of human pathogens on produce.

Antagonists	Targeted pathogens	Fresh produce	References
Yeasts			
Candida sp.	*Listeria monocytogenes*	Apple	Leverentz *et al.*, 2006
Discosphaerina fagi			
Metschnikowia pulcherrima			
Lactic acid bacteria			
Lactobacillus casei	*Aeromonas hydrophila*	Mixed salad	Vescovo *et al.*, 1996
Lactobacillus platarum	*Listeria monocytogenes*		
Pedoococcus spp.	*Salmonella enterica*		
	Staphylococcus aureus		
Lactobacillus lactis	*Listeria monocytogenes*	Alfalfa sprout	Palmai and Buchanan, 2002a
Lactobacillus lactis	*Listeria monocytogenes*	Alfalfa sprout	Wilderdyke *et al.*, 2004
Enterococcus mundtii	*Listeria monocytogenes*	Mung bean sprout	Bennik *et al.*, 1999
Pseudomonadaceae			
Pseudomonas syringae	*Escherichia coli* O157:H7	Apple	Janisiewicz *et al.*, 1999
Pseudomonas fluorescens	*Salmonella enterica*	Alfalfa sprout	Fett, 2006
Pseudomonas chlororaphis			
Pseudomonas aureofaciens			
Pseudomonas fluorescens	*Salmonella enterica*	Bell pepper	Liao and Fett, unpublished
	Listeria monocytogenes		
	Yersinia enterocolitica		
	Escherichia coli O157:H7		
Pseudomonas fluorescens	*Listeria monocytogenes*	Endive	Carlin *et al.*, 1996
Pseudomonas chlororaphis			
Enterobacteriaceae			
Enterobacter cloacae	*Listeria innocua*	Lettuce	Francis and O'Beirne, 1997
Enterobacter agglomerans			
Acetobacteriaceae			
Gluconobacter asaii	*Listeria monocytogenes*	Apple	Leverentz *et al.*, 2006
Bacillaceae			
Bacillus amyloliquefaciens	*Salmonella enterica*	Bell pepper	Liao, unpublished
	Listeria monocytogenes		
	Yersinia enterocolitica		
	Escherichia coli O157:H7		
Bacteriophages	*Salmonella enterica*	Honeydew melon	Leverentz *et al.*, 2001, 2003
	Listeria monocytogenes		
Undefined microbial mixture	*Salmonella enterica*	Alfalfa sprout	Matos and Garland, 2005

growth of *L. monocytogenes* on endive was reduced by approximately 1 log unit at day 7 as compared to the controls, possibly due to the production of iron-chelating siderophores but not due to the lowering of pH.

Francis and O'Beirne (1997) investigated the effect of native microflora on the growth of *Listeria innocua* on shredded lettuce. Shredded lettuce was inoculated with 10^4 to 10^5 cfu/g of *L. innocua* and 10^3 to 10^7 (low) cfu/g of native bacteria. The inoculated lettuce was sealed in polypropylene bags and stored at 8°C for up to 16 days. The survival or growth of *L. innocua* on lettuce was not affected by the densities of native bacteria on the surfaces of lettuce. However, growth of *L. innocua* in a liquid model medium was slightly inhibited by some individual strains of resident bacteria. The antagonistic isolates potentially useful for biological control including *Enterobacter cloacae*, *Enterobacter agglomerans* and two LAB (*Leuconostoc citreum* and *Lactobacillus brevis*) were isolated. The inhibitory activity of native bacteria is not dependent on the total number of bacteria present but is dependent on the presence of specific antagonistic strains (Francis and O'Beirne, 1997).

Use of modified atmosphere packaging can affect the structure and composition of native microflora and the interactions of native microflora with plant pathogens and human pathogens. Francis and O'Beirne (1997) studied the effect of modified atmospheres on survival and growth of *L. monocytogenes* in the presence or absence of native microflora originating from lettuce on the surfaces of the model agar medium. Inoculated plates were stored in air or under one of the five modified atmospheres. Inhibition of *L. monocytogenes* by *E. cloacae* or *E. agglomerans* was greatly reduced when agar plates were incubated in high concentrations of CO_2. In contrast, inhibition of *L. monocytogenes* by *L. citreum* was enhanced at elevated CO_2 concentrations. The authors (Francis and O'Beirne, 1997) suggested that the use of modified atmosphere packaging may lead to an increase in the growth of *L. monocytogenes* either by changing the composition and population of native microflora and specific antagonists. It should also be noted that the use of modified atmosphere to control microbial growth is effective only when the temperature is kept at 5°C or but not at 10°C (Babic and Watada, 1996).

Liao and Fett (2001) tested 128 strains of native bacteria from a variety of produce for their ability to inhibit the growth of *Salmonella* and *L. monocytogenes* on an agar medium. Two antagonistic isolates including a *P. fluorescens* strain and a yeast were further tested for their ability to inhibit the growth of *Salmonella* and *L. monocytogenes* on green pepper discs. The discs were inoculated with a pathogen and an antagonist to give concentrations of 10^5 and 10^7 cfu/g, respectively. After incubation at 20°C for 3 days, the population of *Salmonella* and *L. monocytogenes* on bell pepper discs was reduced by approximately 1 log unit. Whole microbial communities isolated from green pepper were not effective against either pathogen in the *in situ* bioassay.

Sprouts

Due to numerous outbreaks of food-borne illness associated with contaminated sprouts, seed sprouts are currently listed as a hazardous food in the U.S. Food and Drug Administration's 2005 Food Code (FDA, 2005). To avoid food-borne illness, it is recommended that the public should avoid consuming raw sprouts (FDA, 2004). There is an urgent need for a more effective intervention strategy to ensure the safety of sprouts. Both LAB and fluorescent pseudomonads have been studied as biological control agents for food-borne pathogens on sprouts. Palmai and Buchanan (2002a) isolated 40 LAB from alfalfa sprouts, of which 16 produced large inhibition zones against *L. monocytogenes* on agar media. The majority of isolates were identified as *Lactococcus lactis* subsp. *lactis*. In a model 'sprout juice' study, one isolate designated SP26 reduced the growth of *L. monocytogenes* by 2 to 3 log cycles. However, this isolate was not very effective in reducing the growth of *L. monocytogenes* during actual sprouting, causing only one log unit reduction (Palmai and Buchanan, 2002b). Similarly, 58 strains of LAB were isolated from alfalfa sprouts by Wilderdyke *et al.* (2004). One strain identified as *L. lactis* subsp. *lactis* L7 inhibited the growth of *L. monocytogenes*, *Salmonella* and *E. coli* O157:H7 on agar media. A *Pediococcus acidilactici* strain (D3) isolated from ham was also found to be inhibitory against the growth of all three pathogens mentioned above in broth media.

Bennik *et al.* (1999) tested a bacteriocinogenic strain of *Enterococcus mundtii* for controlling the growth of *L. monocytogenes* on mung bean sprouts stored under a modified atmosphere (1.5% O_2/ 20% CO_2/78.5% N_2) at 8°C. They found that *E. mundtii* did not inhibit the growth of *L. monocytogenes* on bean sprouts but inhibited the growth of this pathogen on vegetable agar. A bacteriocin (mundticin) produced by *E. mundtii* was reported to have potential as a biopreservative of mung bean sprouts (Bennik *et al.*, 1999). Five strains of Enterobacteriaceae isolated by Enomoto (2004) were also found to be effective in controlling the growth of spoilage bacteria and presumably human pathogens as well.

Matos and Garland (2005) tested *P. fluorescens* strain 2-79 as well as undefined microbial communities isolated from market or laboratory-grown alfalfa sprouts for controlling outgrowth of *Salmonella* from inoculated alfalfa seed. Small-scale laboratory assays were employed where the microbes were added to the seed soak water prior to germination. Strain 2-79, a rhizosphere bacterium previously studied as a biological control agent of the fungal root disease of wheat (Thomashow *et al.*, 1990), caused a 4 to 4.5 log reduction at days 1 and 3. The undefined community from market sprouts caused a 7 log unit reduction at day 7. Interestingly, the undefined community from laboratory-grown sprouts caused a 2.5 log unit reduction only at day 1 of sprouting. The authors concluded that the microbial community present on commercially grown sprouts was more robust with greater antagonistic or competitive community properties. Even though results using the undefined microbial community from market sprouts were very promising, maintaining such a community in a highly effective state for commercial use presents a challenge. Also, the absence of plant-pathogenic microbes in the community needs to be confirmed (Matos and Garland, 2005).

Fett (2006) confirmed the usefulness of *P. fluorescens* 2-79 as a biological control agent for controlling outgrowth of *Salmonella* from sprouting alfalfa seed using similar small-scale bioassays. In this study, strain 2-79 was demonstrated to be highly antagonistic towards growth of five strains of *Salmonella* on agar and in broth media due to a diffusible, heat-stable factor. Several other fluorescent pseudomonads were tested and were found to be less antagonistic towards *Salmonella* both *in vitro* and *in situ*. The application of some of these additional bacterial strains to seed inhibited subsequent seed germination and growth. By use of mutant strains, antibiosis by strain 2-79 was shown not to be due to the two primary antimicrobial metabolites (iron-binding siderophore and phenazine-1-carboxylic acid) known to be produced by this bacterium. In addition, these studies indicated that production of the antimicrobial substance(s) by strain 2-79 was controlled by the GacS/GacA two-component system. The non-pectolytic, non-phytotoxic nature of strain 2-79 makes it suitable for further evaluation as a biocontrol agent for controlling the outgrowth of *Salmonella* on sprouting seed. The effectiveness of both whole microbial communities and *P. fluorescens* 2-79 as biological control agents needs to be confirmed in larger-scale laboratory and pilot plant studies.

Fruits

Janisiewicz *et al.* (1999) evaluated the ability of *P. syringae* L-59-66 to inhibit the growth of *E. coli* O157:H7 on wounded apple tissue. The activity was demonstrated only when the antagonist was co-inoculated with the pathogen or was introduced 24 to 48 h before the inoculation of the pathogen. Using inocula of *P. syringae* L-59-66 containing a 10- to 1000-fold higher number of cells than the pathogen, growth of *E. coli* O157:H7 was completely inhibited over the 48 h storage period at 24°C. A commercial formulation of *P. syringae* L-59-66 (Bio-save 110, EcoScience Corp., Longwood, Florida) sold for controlling postharvest fungal decay of apple and pear was also effective against *E. coli* O157:H7. The antagonism was possibly due to competition for space or nutrients as secondary metabolites produced by this antagonist showed no inhibitory activity against bacteria.

Seventeen antagonists originally isolated for their ability to inhibit fungal postharvest pathogens on fruits were tested for their ability to inhibit *L. monocytogenes* and *Salmonella* on fresh-cut apple (Leverentz *et al.*, 2006). Four strains including three yeasts (*Candida* sp., *Discosphaerina fagi* and *Metschnikowia pulcherrima*) and one bacterium (*Gluconobacter asaii*) effective against *L. monocytogenes in vitro* were selected for further testing. All four antagonists reduced the populations of *L. monocytogenes* on fresh-cut apple during storage at 10°C or at 25°C. With the exception of *Candida* sp., all four strains were also effective in reducing the growth of *Salmonella* on apple slices at 25°C for 2 days. However, *Salmonella* growth resumed after 5 days on apple slices treated with *G. asaii* and *M. pulcherrima*. *Salmonella* exhibited slight growth during storage and none of the four antagonists was effective in reducing the growth of *Salmonella* at 10°C or below.

In addition to the use of bacterial and yeast antagonists, use of bacteriophages as biocontrol agents in animal food products has recently been reviewed (Hudson *et al.* 2005). Leverentz *et al.* (2001; 2003; 2004) studied the use of bacteriophages for controlling *Salmonella* on fresh-cut fruit. In their initial study (Leverentz *et al.*, 2001) a mixture of four lytic phages specifically against *S. enterica* serovar Enteritidis were evaluated for their ability to control the growth of this pathogen. The phage mixture (5×10^6 pfu) was co-inoculated with *Salmonella* (2.5×10^4 cfu) to fresh-cut honeydew melon and then stored in air at 5, 10 and 20°C for up to 7 days. On fresh-cut melon, the phage mixture reduced populations of *Salmonella* by 3.5 log units at 5 and 10°C, and 2.5 log units at 20 °C. However, the phage mixture was not effective in inhibiting the pathogen on fresh-cut Red Delicious apple, possibly due to the inactivation of the phage particles by the low pH on apple surfaces (pH = 4.2) as compared to the melon surfaces (pH = 5.8).

Leverentz *et al.* (2003) evaluated the inhibitory effect of two phage mixtures in the presence of nisin on the survival and growth of *L. monocytogenes*. Fresh-cut honeydew melon and Red Delicious apples were incubated with a mixture of 1.25×10^4 cfu of the pathogen and 1.25×10^6 pfu of phage mixture containing or lacking 400 IU of nisin. Inoculated fruits were incubated in air at 10°C for 7 days. On honeydew melon both phage mixtures were equally effective and resulted in an approximate 5 log unit reduction in pathogen populations at day 7 of storage. Both phage mixtures alone (in the absence of nisin) were ineffective against *L. monocytogenes* on apple, resulting in less than one log unit reduction. The application of a combination of phage mixture and nisin was more effective, resulting in a 5.7 log unit reduction on melon and a 2.3 log unit reduction on apple. Leverentz *et al.* (2004) recently showed that application of a phage cocktail at the concentration of 8 log pfu/ml 1 h prior to the challenge with *L. monocytogenes* was most effective, resulting in 6.8 log unit reduction in the population of the pathogen on fresh-cut melon after 7 days of storage at 10°C.

Conclusions and Future Research Needs

The candidates for development of commercial biological control agents for use on produce include LABs, fluorescent pseudomonads, bacteriophages and yeast. Endophytic bacteria may also be useful for the control of human pathogens that enter into the vascular system of sprouts during propagation (Cooley *et al.*, 2003). Fluorescent pseudomonads have been proposed as potential probiotics for preventing a variety of fish diseases (Gram *et al.*, 1999; Irianto and Austin, 2002) and may be used to control human pathogens on fresh produce.

The mechanisms of antagonistic actions mentioned above can be attributed to several factors including the decrease in pH, competition for nutrients, competition for colonization sites, production of antimicrobial compounds or stimulation of host defences. However, in most instances, these conclusions were based on the results of experiments conducted *in vitro*. Further experimentation is necessary for confirmation of the mode of action responsible for pathogen inhibition *in situ*. Presence of a sub-population of the pathogen resistant to biological control treatment is an issue of concern. The use of mixtures of two or more defined antagonists or whole microbial communities with more than a single mode of action appears to be desirable. In addition, biological control strategies that are not based on the production of antimicrobial substances may more easily gain regulatory approval.

Microbes antagonistic towards plant or fish pathogens *in vitro* have often been demonstrated to be ineffective *in situ* (Lindow, 1988; Irianto and Austin, 2002). Another example of conflicting results between *in vitro* screening and *in situ* bioassays were the studies by Palmai and Buchanan (2002a,b) concerning the use of LABs for controlling the growth of *L. monocytogenes* on sprouting alfalfa. Initial screening for potential biological control agents might be more effective for conducting tests using *in situ* rather than *in vitro* bioassays. The use of modified atmospheres has the potential to increase or decrease antibiosis of native microflora or applied antagonists towards the survival and growth of human pathogens.

As sanitizer treatment becomes a standard procedure in produce processing, the potential for increased safety risk due to the drastic reduction in the number of competitive microflora and post-contamination of human pathogens is of concern. Application of antagonists to fresh and fresh-cut produce after chemical or physical interventions as an additional hurdle may inhibit the outgrowth of any survivors. Antagonists that are psychotrophic would be most suitable for post-processing applications to allow the control of psychrotrophic pathogens such as *L. monocytogenes*, *Aeromonas* spp. and *Yersinia* spp.

References

Anderson, R.C., Genovese, K.J., Harvey, R.B., Callaway, T.R. and Nisbet, D.J. (2006) Preharvest food safety applications of competitive exclusion cultures and probiotics In: Goktepe, I., Juneja, V.K. and Ahmedna, M. (eds) *Probiotics in Food Safety and Human Health*, CRC Press, Boca Raton, Florida. pp. 273–284.

Aytec, S.A. and Gorris. L.G.M. (1994). Survival of *Aeromonas hydrophila* and *Listeria monocytogenes* on fresh vegetables stored under moderate vacuum. *World Journal of Microbiology and Biotechnology* 10, 610–672.

Babic, I. and Watada, W.E. (1996) Microbial populations of fresh-cut spinach leaves affected by controlled atmospheres. *Postharvest Biology and Technology* 9, 187–193.

Bennik, M.H.J., Peppelenbos, H.W., Nguyen-the, C., Carlin, F., Smid, E.J. and Gorris, L.G.M. (1996) Microbiology of minimally processed, modified atmosphere packaged chicory endive. *Postharvest Biology and Technology* 9, 209–221.

Bennik, M.H.J., Van Overbeek, W., Smid, E.J. and Gorris, L.G.M. (1999) Biopreservation in modified atmosphere stored mungbean sprouts: the use of vegetable-associated bacteriocinogenic lactic acid bacteria to control the growth of *Listeria monocytogenes*. *Letters in Applied Microbiology* 28, 226–232.

Beuchat, L.R. (1996) Pathogenic microorganisms associated with fresh produce. *Journal of Food Protection* 59, 204–216.

Beuchat, L.R. and Brackett, R.E. (1991) Behavior of *Listeria monocytogenes* inoculated into raw tomatoes and processed tomato products. *Applied and Environmental Microbiology* 57, 1367–1371.

Breidt, F. and Fleming, H.P. (1997) Using lactic acid bacteria to improve the safety of minimally processed fruits and vegetables. *Food Technology* 51, 44–49.

Buck, J.W., Walcott, R. and Beuchat, L.R. (2003) Recent trends in microbiological safety of fruits and vegetables. Available at: http://www.apsnet.org/online/feature/safety/. Accessed 20 August 2009.

Carlin, F., Nguyen-the, C. and Morris, C.E. (1996) Influence of background microflora on *Listeria monocytogenes* on minimally processed broad-leaved endive (*Cichorium endivia* var. *latifolia*). *Journal of Food Protection* 59, 698–703.

Castro-Rosas, J. and Escartin, E.F. (2000) Survival and growth of *Vibrio cholerae* O1, *Salmonella typhi*, and *Escherichia coli* O157:H7 in alfalfa sprouts. *Journal of Food Science* 65, 162–165.

Conway, W.S., Leverentz, B., Saftner, R.A., Janisiewicz, W.J., Sams, C.E. and Leblanc, E. (2000) Survival and growth of *Listeria monocytogenes* on fresh-cut apple slices and its interaction with *Glomerella cingulata* and *Penicillium expansum*. *Plant Disease* 84, 177–181.

Cooley, M.B., Miller, W.G. and Mandrell, R.E. (2003) Colonization of *Arabidopsis thaliana* with *Salmonella enterica* and enterohemorrhagic *Escherichia coli* O157:H7 and competition by *Enterobacter asburiae*. *Applied and Environmental Microbiology* 69, 4915–4926.

CSPI (Center for Science in the Public Interest) (2005) Outbreak Alert. Available at: http://www.cspinet.org/foodsafety/outbreak_report.html. Accessed 20 August 2009.

Enomoto, K. (2004) Use of bean sprout Enterobacteriaceae isolates as biological control agents of *Pseudomonas fluorescens*. *Journal of Food Science* 69, FMS17–FMS21.

ERS (Economic Research Service) (2004) Food consumption (per capita) data system. Available at: http://www.ers.usda.gov/Data/FoodConsumption. Accessed 20 August 2009.

FDA (U.S. Food and Drug Administration) (2004) Produce safety from production to consumption: 2004 action plan to minimize foodborne illness associated with fresh produce consumption. Available at: http://www.cfsan.fda.gov/~dms/prodpla2.html. Accessed 14 March 2005.

FDA (U.S. Food and Drug Administration). (2005) FDA Food Code. Available at: http://www.cfsan.fda.gov/~dms/foodcode.html. Accessed 20 August 2009.

Fett, W.F. (2006) Inhibition of *Salmonella enterica* by plant-associated pseudomonads in vitro and on sprouting alfalfa seed. *Journal of Food Protection* 69, 719–728.

Fett, W.F., Fu, T-J. and Tortorello, M.L. (2006) Seed sprouts: the state of microbiological safety. In: Matthews, K.R. (ed.) *Microbiology of Fresh Produce*. ASM Press, Washington, DC. pp. 167–219.

Francis, G.A. and O'Beirne, D. (1997) Effects of gas atmosphere, antimicrobial dip and temperature on the fate of *Listeria innocua* and *Listeria monocytogenes* on minimally processed lettuce. *International Journal of Food Science and Technology* 32, 141–151.

Francis, G.A., Thomas, C. and O'Beirne, D. (1999) The microbiological safety of minimally processed vegetables. *International Journal of Food Science and Technology* 34, 1–22.

Gombas, D.E. (1989) Biological competition as a preservation mechanism. *Journal of Food Safety* 10, 107–117.

Gram, L., Melchiorsen, J., Spanggaard, B., Huber, I. and Nielsen, T.F. (1999) Inhibition of *Vibrio anguillarum* by *Pseudomonas fluorescens* AH2, a possible probiotic treatment of fish. *Applied and Environmental Microbiology* 65, 969–973.

Haas, D. and Défago, G. (2005) Biological control of soil-borne pathogens by fluorescent pseudomonads. *Nature Reviews Microbiology* 3, 307–319.

Hora, R., Warriner, K.B., Shelp, J. and Griffiths, M.W. (2005) Internalization of *Escherichia coli* O157:H7 following biological and mechanical disruption of growing spinach plants. *Journal of Food Protection* 68, 2506–2509.

Hudson, J.A., Billington, C., Carey-Smith, G. and Greening, G. (2005) Bacteriophages as biocontrol agents in food. *Journal of Food Protection* 68, 426–437.

ICMSF (International Commission on Microbiological Specifications for Foods) (2005a) Vegetables and vegetable products. In: *Microrganisms in Foods 6 – Microbial Ecology of Food Commodities,* 2nd edn, Kluwer Academic/Plenum Publishers, New York. pp. 277–325.

ICMSF (International Commission on Microbiological Specifications for Foods) (2005b) Fruits and fruit products. In: *Microrganisms in Foods 6 – Microbial Ecology of Food Commodities,* 2nd edn, Kluwer Academic/Plenum Publishers, New York. pp. 326–359.

Irianto, A. and Austin, B. (2002) Probiotics in aquaculture. *Journal of Fish Diseases* 25, 633–642.

Janisiewicz, W.J. and Korsten, L. (2002) Biological control of postharvest diseases of fruits. *Annual Review of Phytopathology* 40, 411–441.

Janisiewicz, W.J., Conway, W. S. and Leverentz, B. (1999) Biological control of postharvest decays of apple can prevent growth of *Escherichia coli* O157:H7 in apple wounds. *Journal of Food Protection* 62, 1372–1375.

Kaufman, P.R., Handy, C.R., McLaughlin, E.W., Park, K. and Green, G.M. (2000) Understanding the dynamics of produce markets: consumption and consolidation grow. Food and Rural Economics Division, Economic Research Service, U.S. Department of Agriculture. Agriculture Information Bulletin No. 758.

Leverentz, B., Conway, W.S., Alavidze, Z., Janisiewicz, W.J., Fuchs, Y., Camp, M.J., Chighladze, E. and Sulakvelidze, A. (2001) Examination of bacteriophage as a biocontrol method for *Salmonella* on fresh-cut fruit: a model study. *Journal of Food Protection* 64, 1116–1121.

Leverentz, B., Conway, W.S., Camp, M.J., Janisiewicz, W.J., Abuladze, T., Yang, M., Saftner, R. and Sulakvelidze, A. (2003) Biocontrol of *Listeria monocytogenes* on fresh-cut produce by treatment with lytic bacteriophages and a bacteriocin. *Applied and Environmental Microbiology* 69, 4519–4526.

Leverentz, B., Conway, W.S., Janisiewicz, W.J. and Camp, M.J. (2004) Optimizing concentration and timing of a phage spray application to reduce *Listeria monocytogenes* on honeydew melon tissue. *Journal of Food Protection* 67, 1682–1686.

Leverentz, B., Conway, W.S., Janisiewicz, W., Abadias, M., Kurtzman, C.P. and Camp, M.J. (2006) Biocontrol of the food-borne pathogens *Listeria monocytogenes* and *Salmonella enterica* serovar Poona on fresh-cut apples with naturally occurring bacterial and yeast antagonists. *Applied and Environmental Microbiology* 72, 1135–1140.

Liao, C.-H. and Fett, W.F. (2001) Analysis of native microflora and selection of strains antagonistic to human pathogens on fresh produce. *Journal of Food Protection* 64, 1110–1115.

Liao, C.-H. and Sapers, G.M. (1999) Influence of soft rot bacteria on growth of *Listeria monocytogenes* on potato tuber slices. *Journal of Food Protection* 62, 343–348.

Liao, C.-H., McEvoy, J.L. and Smith, J.L. (2003) Control of bacterial soft rot and foodborne human pathogens on fresh fruits and vegetables. In: Huang, H.C. and Aharya, S.N. (eds) *Advances in Plant Disease Management*, Research Signpost, Kerala, India. pp. 165–193.

Lindow, S.E. (1988) Lack of correlation of *in vitro* antibiosis with antagonism of ice nucleation active bacteria on leaf surfaces by non-ice nucleation active bacteria. *Phytopathology* 78, 444–450.

Lund, B.M. (1992) Ecosystems in vegetable foods. *Journal of Applied Bacteriology Symposium Supplement* 73, 115S–126S.

Matos, A. and Garland, J.L. (2005) The effects of community versus single strain inoculants on the bio-control of *Salmonella* and microbial dynamics in alfalfa sprouts. *Journal of Food Protection* 68, 40–48.

Nguyen-the, C. and Carlin, F. (1994) The microbiology of minimally processed fresh fruits and vegetables. *Critical Reviews in Food Science and Nutrition* 34, 371–401.

Nguyen-the, C. and Carlin, F. (2000) Fresh and processed vegetables. In: Lund, B.M., Baird-Parker, T.C. and Gould, G.W. (eds) *The Microbiological Safety and Quality of Food*, Aspen Publishers, New York. pp. 620–684.

OSU (Ohio State University) (2005) Commercial biocontrol products available for use against plant pathogens, Available at: http://www.oardc.ohio-state.edu/apsbcc/productlist.htm. Accessed 28 January 2005.

Palmai, M. and Buchanan, R.L. (2002a) The effect of *Lactococcus lactis* on the growth characteristics of *Listeria monocytogenes* in alfalfa sprout broth. *Acta Alimentaria* 31, 379–392.

Palmai, M. and Buchanan, R.L. (2002b) Growth of *Listeria monocytogenes* during germination of alfalfa sprouts. *Food Microbiology* 19, 195–200.

Richards, G.M. and Beuchat, L.R. (2005) Infection of cantaloupe rind with *Cladosporium cladosporioides* and *Penicillium expansum*, and associated migration of *Salmonella poona* into edible tissues. *International Journal of Food Microbiology* 103, 1–10.

Riordan, D.C.R., Sapers, G.M. and Annous, B.A. (2000) The survival of *Escherichia coli* O157:H7 in the presence of *Penicillium expansum* and *Glomerella cingulata* in wounds on apple surfaces. *Journal of Food Protection* 63, 1637–1642.

Sapers, G.M. (2006) Washing and sanitizing treatments for fruits and vegetables. In: Sapers, G.M., Gorny, J.R. and Yousef, A.E. (eds) *Microbiology of Fruits and Vegetables*, CRC Press, Boca Raton, Florida. pp. 375–400.

Schneitz, C. (2005) Competitive exclusion in poultry – 30 years of research. *Food Control* 16, 657–667.

Schoeller, N.P., Ingham, S.C. and Ingham, B.H. (2002) Assessment of the potential for *Listeria monocytogenes* survival and growth during alfalfa sprout production and use of ionizing radiation as a potential intervention treatment. *Journal of Food Protection* 65, 1259–1266.

Schuenzel, K.M. and Harrison, M.A. (2002) Microbial antagonists of foodborne pathogens on fresh, minimally processed vegetables. *Journal of Food Protection* 65, 1909–1915.

Sivapalasingam, S., Friedman, C.R., Cohen, L. and Tauxe, R.V. (2004) Fresh produce: a growing cause of outbreaks of foodborne illness in the United States, 1973 through 1997. *Journal of Food Protection* 67, 2342–2353.

Solomon, E.B., Brandl, M.T. and Mandrell, R.E. (2006) Biology of foodborne pathogens on produce. In: Matthews, K.R. (ed.) *Microbiology of Fresh Produce.* ASM Press, Washington, D.C. pp. 55–83.

Thomashow, L.S., Weller, D.M., Bonsall, R.F. and Pierson, L.S. III (1990) Production of the antibiotic phenazine-1-carboxylic acid by fluorescent *Pseudomonas* species in the rhizosphere of wheat. *Applied and Environmental Microbiology* 56, 908–912.

Torriani, S., Orsi, C. and Vescovo, M. (1997) Potential of *Lactobacillus casei*, culture permeate, and lactic acid to control microorganisms in ready-to-use vegetables. *Journal of Food Protection* 60, 1564–1567.

Tournas, V.H. (2005) Spoilage of vegetable crops by bacteria and fungi and related health hazards. *Critical Reviews in Microbiology* 31, 33–44.

Ukuku, D.O., Fett, W.F. and Sapers, G.M. (2004) Inhibition of *Listeria monocytogenes* by native microflora of whole cantaloupe. *Journal of Food Safety* 24, 129–146.

USDHHS/USDA (U.S. Department of Health and Human Services and U.S. Department of Agriculture). (2005) Dietary Guidelines for Americans 2005. Available at: http://www.healthierus.gov/dietaryguidelines. Accessed 20 August 2009.

Vescoco, M., Orsi, C., Scolari, G. and Torriani, S. (1995) Inhibitory effect of selected lactic acid bacteria on microflora associated with ready-to-use vegetables. *Letters in Applied Microbiology* 21, 121–125.

Vescovo, M., Torriani, S., Orsi, C., Macchiarolo, F. and Scolari, G. (1996) Application of antimicrobial-producing lactic acid bacteria to control pathogens in ready-to-use vegetables. *Journal of Applied Bacteriology* 81, 113–119.

Wade, W.N. and Beuchat, L.R. (2003) Metabiosis of proteolytic moulds and *Salmonella* in raw, ripe tomatoes. *Journal of Applied Microbiology* 95, 437–450.

Warriner, K., Ibrahim, F., Dickinson, M., Wright, C. and Waites, W.M. (2003) Internalization of human pathogens within growing salad vegetables. *Biotechnology and Genetic Engineering Reviews* 20, 117–134.

Wells, J.M. and Butterfield, J.E. (1997) *Salmonella* contamination associated with bacterial soft rot of fresh fruits and vegetables in the marketplace. *Plant Disease* 81, 867–872.

Wells, J.M. and Butterfield, J.E. (1999) Incidence of *Salmonella* on fresh fruits and vegetables affected by fungal rots or physical injury. *Plant Disease* 83, 722–726.

Wilderdyke, M.R., Smith, D.A. and Brashears, M.B. (2004) Isolation, identification, and selection of lactic acid bacteria from alfalfa sprouts for competitive inhibition of foodborne pathogens. *Journal of Food Protection* 67, 947–951.

Yang, C., Crowley, D.E., Borneman, J. and Keen, N.T. (2001) Microbial phyllosphere populations are more complex than previously realized. *Proceedings of the National Academy of Science USA* 98, 3889–3894.

20 Antimicrobial and Other Biological Effects of *Garcinia* Plants Used in Food and Herbal Medicine

Govind J. Kapadia* and G. Subba Rao

Introduction

The plants belonging to the genus *Garcinia* (family, *Guttiferae*; subfamily, *Clusiaceae*) are native to tropical and subtropical Asia, Africa, Australia, South America and Polynesia. While their edible fruits are considered a delicacy, the fruit rinds and seeds of some species are used as food flavours and colourants in regional cooking, and seed oils are utilized in the production of confections, cosmetics and medicinal products (Kamat, 2006; Bodeker, 2009; Padhye *et al*., 2009). The natural dyes (red, yellow, green and black) from various *Garcinia* plants are used in foods, drugs, cosmetics and textiles (Bhaskaran, 2004; Seth, 2004; Siva, 2007). However, the major utilization of the *Garcinia* plants is in traditional herbal medical practice which is popular throughout Asia and Africa, with products ranging from chewing sticks for oral hygiene to the control of various microbial infections, asthma, arthritis, haemorrhoids, diabetes, cancer, cardiovascular and other common diseases, and obesity (Iwu, 1993; Jantan, 2004; Toromanyan *et al*., 2007; Pedraza-Chaverri *et al*., 2008; Upadhyaya *et al*., 2009; Wiart, 2009; Khan and Yadava, 2010). It is estimated that two-thirds to three-quarters of the world's population currently relies on medicinal plants as their primary source of medicine (Suksamram, *et al*., 2003). Extensive phytochemical studies of the *Garcinia* species used in the herbal medicinal preparations indicate that most of the antimicrobial and other biological activities observed could be attributed to the derivatives of xanthone, flavone and benzophenone, the three major bioactive chemical classes common to this genus (Iwu, 1986; Iinuma *et al*., 1996a–c; Ito *et al*., 2003a,b; Rukachaisirikul *et al*., 2003a–d, 2005a–d; Nguyen *et al*., 2005; Chen *et al*., 2006, 2009; Obolskiy *et al*., 2009). An exception to this is (-)-hydroxycitric acid (Fig. 20.1) (HCA) which has gained much popularity as a potential bioactive *Garcinia* chemical constituent for treating obesity (Jena *et al*., 2002b; Pittler and Ernst, 2004; Preus *et al*. 2004).

Currently, there is renewed interest in the antimicrobial constituents of *Garcinia* plants in the fight against the emerging drug-resistant strains of tuberculosis, malaria, HIV and other infectious microbes (Lin *et al*., 1999; Rukachaisirikul *et al*., 2003a; Suksamram *et al*., 2003; Hay *et al*., 2004b; Rukachaisirikul *et al*., 2005a,b; Chen *et al*., 2006; Hay *et al*., 2008; Elfita *et al*., 2009). Also, the potential utilization of these natural antimicrobials by the global food industry in

* Corresponding author.

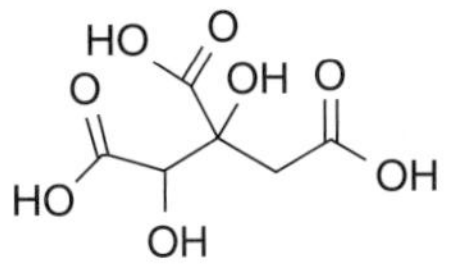

Fig. 20.1. (-)-Hydroxycitric acid.

the preservation and safety of processed foods is gaining attention (Joseph *et al.*, 2005; Negi *et al.*, 2008; Zhong *et al.*, 2009). Further, drug, cosmetic, confection, cereal and soft-drink manufacturers are now exploring the beneficial uses of *Garcinia* plant-derived products, such as fruit pulps, seeds, seed oils, barks and resins in their products (Yamaguchi *et al.*, 1999; Chomnawang *et al.*, 2005; Adeyeye *et al.*, 2007; Hsu *et al.*, 2007; Nanditha and Prabhasankar, 2009). Accordingly, a comprehensive review of the chemical constituents of *Garcinia* plants and their antimicrobial and other biological effects are presented in this chapter.

Habitat and Cultivation of *Garcinia* Plants Around the World

The plant genus *Garcinia* contains about 300 species, the exact number being in dispute. They consist of large rainforest trees as well as medium to smaller size evergreens and flowering shrubs. In tropical and subtropical countries around the world the large trees are utilized as ornamental and shade trees. Table 20.1 lists the native habitat of various *Garcinia* (*G*) species that have been subjected to chemical and biological investigations.

At the present time, only a small number of *Garcinia* plant species are cultivated for food and commerce. *G. cambogia, G. dioica, G. hombroniana, G. indica, G. kola, G. madruno* (lemondrop mangosteen), *G. mangostana* (purple mangosteen), *G. multiflora, G. parvifolia, G. pedunculata, G. prainiana* (button mangosteen) and *G. xanthochymus* are the most widely cultivated species for their fruits followed by, to a lesser extent, *G. forbessii, G. intermidia, G.manni* and *G. purpurea*. Tropical Africa, South-east Asia, India and Sri Lanka are the main regions where these species are cultivated for food and commercial use. *G. mangostana* fruit is considered 'the queen of tropical fruits' and widely consumed in South-east Asia. Recently, it has been introduced as an exotic fruit in the gourmet restaurants of New York City (David Karp, Food Critic, *The New York Times*, Food Section, November 9, 2006). Currently, efforts are being made to grow *G. mangostana* and other mangosteens in Florida for marketable fruits (Campbell and Ledesma, 2002). The fruit juice from *G. mangostana,* commonly known as Mangosteen, widely promoted in Asia for its health benefits attributed to antioxidant constituents (Marcason, 2006), is now available in health-food stores in Western countries including the USA. In Indonesia and Malaysia the fruits of *G. diocia* are eaten while the young leaves are used as a vegetable. *G. indica* and *G. cambogia* along the western coast of India and *G. multiflora* in Vietnam are the commonly grown species of *Garcinia* for their fruit rinds and seeds used in traditional cooking for flavour and colour. In Southern Nigeria, *G. kola* is cultivated for the consumption of its fruit pulps and seeds while its twigs along with those of *G. manni* are utilized as chewing sticks to maintain oral hygiene (Taiwo *et al.*, 1999; Addai *et al.*, 2002). The fruit extracts and raw seeds of *G. pedunculata* and *G. xanthochymus* are popular herbal remedies of asthma in the Manipur district in Nortrh-east India (Khan and Yadava, 2010).

Among the currently cultivated *Garcinia* plants in India, *G. indica* which is regionally known as kokum, is of significant commercial interest with more than 100 US patents covering such diverse uses as food colourants (red and yellow), spices, cosmetics and sun screens to exotic wines, pH indicators and biosensors (Kamat, 2006) to its credit. The oil from *G. indica* seeds (with a yield of ~25%), which remains solid at room temperature, is being utilized by a growing number of pharmaceutical, cosmetic and confection industries in India and other Asian countries. The cosmetic industry has incorporated *G. indica* and *G. mangostana* plant products in the formulation of antibacterial soaps, lotions, lip balms, hair-care products, and antiage and acne treatment creams (Chomnawang *et al.*, 2005; Hsu *et al.*, 2007; Padhye *et al.*, 2009; Upadhyaya *et al.*, 2009). A US patent has been

Table 20.1. Native habitat of *Garcinia* species subjected to chemical and biological investigations.

Garcinia species	Native habitat	References
G. afzelii	West Africa	Kamdemwaffo *et al.*, 2006
G. andamanica	Andaman Isles	Alam *et al.*, 1986
G. aristata	Cuba	Cuesta-Rubino *et al.*, 2001
G. assigu	Papua New Guinea	Ito *et al.*, 2003a
G. atroviridis	Malaysia, Thailand	Permanaa *et al.*, 2001
G. bancana	Thailand	Rukachaisirikul *et al.*, 2005a
G. benthami	Asia, Australia	Elya *et al.*, 2006
G. brasiliensis	Brazil, Paraguay	Martins *et al.*, 2008; Naldoni *et al.*, 2009; Neves *et al.*, 2007
G. brevipedicellata	Cameroon, Nigeria	Ngoupayo *et al.*, 2008
G. bracteata	China	Thoison *et al.*, 2000
G. cambogia	India, South-east Asia	Rao *et al.*, 1980; Hayamizu *et al.*, 2003 Lewis and Neelakantan, 1965
G. cantleyana	Malaysia	Shadid *et al.*, 2007
G. celebica	Indonesia, Sri Lanka	Elfita *et al.*, 2009
G. cornea	Indonesia, East Indies	Elfita *et al.*, 2009
G. cowa	Singapore	Pattalung *et al.*, 1994; Likhitwitayawuid *et al.*, 1998b; Jena *et al.*, 2002a; Mahabusarakam *et al.*, 2005
G. cuneifolia	Malaysia	Hui, 2005
G. cymosa	Indonesia, Australia	Elfita *et al.*, 2009
G. dioica	Indonesia, Malaysia	Iinuma *et al.*, 1996b
G. dulcis	Australia, Indonesia Papua New Guinea	Deachathai *et al.*, 2005, 2006, 2008; Iinuma *et al.*, 1996a; Likhitwitayawuid *et al.*, 1998a
G. epunctata	West Africa	Mbafor *et al.*, 1989
G. eugeniafolia	Philippines	Hartati *et al.*, 2008
G. forbesii	South-east Africa	Harrison *et al.*, 1993
G. fusca	Thailand	Ito *et al.*, 2003b
G. gardneriana	Brazil	Castardo *et al.*, 2008
G. gerrardii	South Africa	Sordat-Diserens *et al.*, 2004
G. griffithii	Indonesia	Elfita *et al.*, 2009; Nguyen *et al.*, 2005
G. hanburyi	Cambodia, Vietnam, Thailand	Panthong *et al.*, 2007; Wang *et al.*, 2008b; Tao *et al.*, 2009
G. hombroniana	Malaysia, Indonesia	Rukachaisirikul *et al.*, 2005c
G. huillensis	Central and Southern Africa	Bakana *et al.*, 1987
G. indica	India	Jayaprakhasha and Sakariah, 2002; Yamaguchi *et al.*, 2000
G. intermedia	Central America, Mexico	Abe *et al.*, 2004
G. kola	West Africa	Hussain *et al.*, 1982; Iwu, 1985; Iwu *et al.*, 1987, 1990; Kapadia *et al.*, 1994; Niwa *et al.*, 1994; Taiwo *et al.*, 1999; Okunji *et al.*, 2002
G. lancilimba	China	Yang *et al.*, 2007; Han *et al.*, 2008a
G. linii	Taiwan	Chen *et al.*, 2006

Garcinia species	Native habitat	References
G. livingstonei	Southeast Africa	Mbwambo *et al.*, 2006; Kaikabo *et al.*, 2009; Sordat-Diserens *et al.*, 1992
G. lucida	Cameroon	Fotie *et al.*, 2007
G. macrophylla	Malaysia, Suriname	Williams *et al.*, 2003
G. madruno	Brazil	Suffredini *et al.*, 2006
G. maingayi	Thailand	Jabit *et al.*, 2009
G. malaccensis	Thailand	Jabit *et al.*, 2009
G. mangostana	India, Southeast Asia	Schmid, 1855; MacLeod and Pieris, 1982; Sundaram *et al.*, 1983; Gopalakrishnan *et al.*, 1997; Nilar, 2002; Jung *et al.*, 2006; Fu *et al.*, 2007; Doi *et al.*, 2009; Han *et al.*, 2009
G. manni	West Africa	Addai *et al.*, 2002
G. merguensis	Vietnam	Nguyen *et al.*, 2003; Kijjoa *et al.*, 2008
G. morella	India	Sani and Rao, 1966
G. multiflora	South China, Taiwan, Vietnam	Lin *et al.*, 1997, 1999; Chen *et al.*, 2009
G. neglecta	New Caledonia	Ito *et al.*, 2001
G. nervosa	Indonesia, Taiwan	Ilyas *et al.*, 1994; Parveen *et al.*, 2004; Ampofo and Waterman, 1986
G. nigrolineata	Malaysia, Myanmar	Rukachaisirikul *et al.*, 2003a,d, 2005d
G. oblongifolia	China	Hamed *et al.*, 2006; Huang *et al.*, 2009
G. opaca	Malaysia	Goh *et al.*, 1991; Jabit *et al.*, 2009
G. ovalifolia	West Africa	Waterman and Crichton, 1980
G. parvifolia	Malaysia, Singapore	Xu *et al.*, 2001; Rukachaisirikul *et al.*, 2006, 2008
G. pedunculata	India, China	Jayaprakasha *et al.*, 2003; Negi *et al.*, 2008
G. penangiana	Malaysia	Jabit *et al.*, 2007, 2009
G. polyantha	Cameroon	Lannang *et al.*, 2005, 2008; Louh *et al.*, 2008; Ampofo and Waterman, 1986
G. prainiana	Malaysia, Indonesia	Jabit *et al.*, 2009
G. pseudoguttifera	Fiji	Ali *et al.*, 2000
G. puat	New Caledonia	Ito *et al.*, 2001
G. purpurea	India	Inuma *et al.*, 1996d
G. pyrifera	Malaysia	Roux *et al.*, 2000; Ampofo and Waterman, 1986
G. rigida	Indonesia	Elya *et al.*, 2008
G. rostrata	Malaysia	Jabit *et al.*, 2009
G. scortechinii	Malaysia	Rukachaisirikul *et al.*, 2003c, 2005b; Sukpondma, 2005
G. semseii	Tanzania	Magadula *et al.*, 2008
G. smeathmanni	Siera Leone	Komguem *et al.*, 2005
G. solomonsis	Papua New Guinea	Carroll *et al.*, 2009

Continued

Table 20.1. *Continued*

Garcinia species	Native habitat	References
G. speciosa	Thailand	Rukachaisirikul *et al.*, 2003b; Vieira *et al.*, 2004
G. spicata	India, Southeast Asia	Konoshima and Ikeshiro, 1970
G. staudtii	Cameroon	Ngoupayo *et al.*, 2009
G. subelliptica	Japan, Taiwan	Fukuyama *et al.*, 1991; Minami *et al.*, 1994; Iinuma *et al.*, 1994, 1996d; Lin *et al.*, 1996; Chung *et al.*, 1998; Abe *et al.*, 2003; Weng *et al.*, 2004; Wu *et al.*, 2005
G. tetralanta	China	Wang *et al.*, 2008a
G. urophylla	Thailand	Jabit *et al.*, 2009
G. vieillardii	New Caledonia	Hay *et al.*, 2004a,b, 2008
G. vilersiana	Vietnam	Nguyen and Harrison, 2000
G. virgata	New Caledonia	Merza *et al.*, 2006
G. xanthochymus	India, Malaysia, Myanmar, Thailand	Baggett *et al.*, 2005 Chanmahasathien *et al.*, 2003
G. xipshuanbannaensis	China	Zhou *et al.*, 2008; Han *et al.*, 2008b
G. yunnanensis	China	Xu *et al.*, 2008

granted for the use of *G. indica, G. cambogia* and *G. purpurea* constituents in health foods (Yamaguchi *et al.*, 1999). The nuts of *G. kola,* known as 'bitter cola', are used to enhance the protein quality of cereals (Adeyeye *et al.*, 2007). The *Garcinia* plant extracts are utilized as an antioxidant in bakery products (Nanditha and Prabhasankar, 2009). The barks of *G. spicata* (Konoshima and Ikeshiro, 1970) and *G. subelliptica* (Fukuyama *et al.*, 1997) trees, and the gamboge resin from *G. dulcis* (Ansari *et al.*, 1976) are used as commercial source of yellow dye, while the bark of *G. dulcis* also yields green dye (Seth, 2004). Similarly, red and black dyes are extracted from the fruits of *G. indica* (Bhaskaran, 2004) and *G. mangostana* (Siva, 2007), respectively, for use in the textile industry. Recently, chemical industries have started extracting HCA, the purported anti-obesity constituent (Heymsfield *et al.*, 1998; Lenz and Hamilton, 2003; Hasani-Ranjbar *et al.*, 2009), from the fruit rinds of *G. indica* and *G.cambogia.*

Most herbal medicinal practitioners in Africa and other tropical countries utilize *Garcinia* trees and shrubs growing wild in the countryside and rainforests. There are very few botanical gardens in the developing countries of the world that have even a modest collection of *Garcinia* species and access to these for herbal medicine is quite limited. The following is a short list of most frequently used *Garcinia* plants in current herbal medicine: *G. atroviridis, G.cambogia, G. cowa, G. dulcis, G. huillensis, G. indica, G. kola, G. mangostana, G. manni, G. pedunculata, G. subelliptica* and *G. xanthochymus.*

Biologically Active Chemical Constituents of *Garcinia* Plants

Interest in the chemical constituents of *Garcinia* plants dates back to the middle of the 19th century. Mangostin (now designated α-mangostin) (Fig. 20.2) from the fruit hulls of *G. mangostana* appears to be the first biologically significant chemical constituent to be isolated from a *Garcinia* plant. This pioneering work was published in 1855 by Dr W. Schmid, a German chemist. Because of the bright yellow colour of mangostin, he coined the word xanthone, based on the Greek word xanthos for yellow, to name the new chemical class he had discovered during its structural investigation and reported mangostin as the first naturally occurring xanthone derivative. Since then active investigations into the isolation, characterization and biological effects of a wide variety of novel chemical constituents from *Garcinia* species have continued to-date unabated. Antimicrobial and other biological effects of the chemical constituents isolated from various *Garcinia* plant species are summarized in Table 20.2 and Table 20.3, respectively.

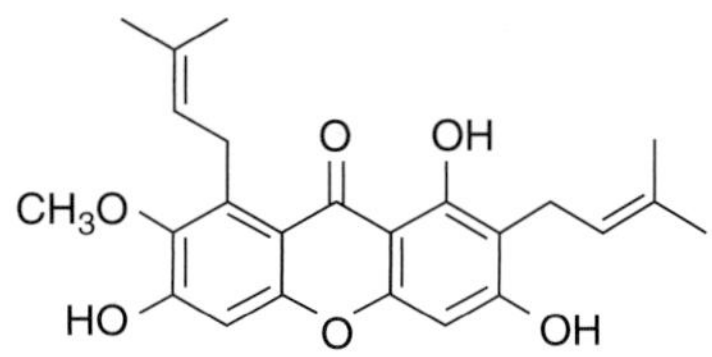

Fig. 20.2. α-Mangostin (Mangostin).

Table 20.2. Antimicrobial effects of chemical constituents isolated from *Garcinia* plants.

Garcinia species (Plant part)	Chemical constituents (Chemical class)[a]	Microbes tested	References
Antibacterial effect			
G. atroviridis (root)	Atrovirinone (BZQ)	*Bacillus cereus, Staphylococcus aureus*	Permanaa *et al.*, 2001
G. bancana (leaf, twig)	BPH	Methicillin-resistant *S. aureus* (MRSA)	Rukachaisirikul *et al.*, 2005a
G. brasiliensis (fruit)	BZPs	Gram-(+) bacteria	Naldoni *et al.*, 2009
G. cowa	Cowanol, cowaxanthone	*S. aureus*, Gram-(+) bacteria	Pattalung *et al.*, 1994 Negi *et al.*, 2008
G. cuneifolia (stem bark)	Extract	MRSA	Hui, 2005
G. dioica (bark)	XTHs	MRSA	Iinuma *et al.*, 1996b,c
G. dulcis (fruit, flower, seed)	Dulcixanthones A–G (XTHs)	MRSA	Deachathai *et al.*, 2005, 2006, 2008
G. hanburyi (gamboge)	Resin	Gram-(+/-) bacteria	Cowan, 1999
G. hombroniana (leaf)	Garcihombronane (TRP)	*S. aureus*	Rukachaisirikul *et al.*, 2005c
G. huillensis (stem bark)	Garcinol and xanthochymol (BZPs)	Gram-(+/–) cocci, micobacteria	Bakana, *et al.*, 1987
G. kola (fruit)	Kolanone (BZP)	Gram-(+/–) bacteria	Hussain *et al.*, 1982
G. livingstonei (leaf)	Amentoflavones (FLVs)	*E. coli, S. aureus, Enterococcus faecalis, Pseudomonas aeruginosa*	Kaikabo *et al.*, 2009
G. madruno (stem)	Extract	Gram-(+/–) bacteria	Suffredini *et al.*, 2006
G. mangostana (fruit)	XTHs	MRSA	Iinuma *et al.*, 1996c; Sundaram *et al.*, 1983
G. morella (seed)	Guttiferins (BZPs)	Various microbes	Santhanam and Rao, 1969
G. nigrolineata (leaf, twig)	Nigrolineaxanthones, nigrolineabenzopyran A, nigrolineabiphenyls A,B	MRSA	Rukachaisirikul *et al.*, 2003a,d; 2005d
G. pedunculata (fruit)	Rind extract	Gram-(–) bacteria	Negi *et al.*, 2008
G. parvifolia (twig, leaf)	Parvifolins A-G (PHLs), parvidepsidones A,B parvixanthones A–C	MRSA	Rukachaisirikul *et al.*, 2006, 2008
G. purpurea	Garcinol, isogarcinol (BZP)	MRSA	Iinuma *et al.*, 1996d
G. scortechinii (latex, fruit, stem bark)	Scortechinone B, XTHs, TRPs	MRSA	Rukachaisirikul *et al.*, 2005b; Sukpondma *et al.*, 2005
G. smeathmanni (stem bark)	Smeathxanthones A,B		Komguem *et al.*, 2005
G. staudtii (twig)	Staudtiixanthones A–D	MRSA	Ngoupayo *et al.*, 2009
G. subelliptica	Xanthochymols (BZPs)	MRSA	Iinuma *et al.*, 1996d

Continued

Table 20.2. *Continued*

Garcinia species (Plant part)	Chemical constituents (Chemical class)[a]	Microbes tested	References
Antifungal effect			
G. atroviridis (fruit)	HCA derivative	*Cladosporium herbarum*	Mackeen *et al.*, 2002
G. cowa	Extracts	*Aspergillus flavus* (antiaflatoxigenic)	Joseph *et al.*, 2005
G. gerrardii (root bark)	Garcigerrins (XTHs)	*Cladosporium cucumerinum*	Sordat-Diserens *et al.*, 2004
G. huillensis (stem bark)	Garcinol	Various fungi	Bakana *et al.*, 1987
G. indica	Garcinol	*A. flavus* (antiaflatoxigenic)	Selvi *et al.*, 2003
G. mangostana (fruit hull)	Mangostins, garcinone D (XTHs)	*Fusarium oxysporum vasinfactum, Alternaria tenuis, Dreschlera oryzae*	Gopalakrishnan *et al.*, 1997
G. pedunculata	Extract	*A. flavus* (antiaflatoxigenic)	Joseph *et al.*, 2005
Antiparasitic effect			
G. brasiliensis (fruit)	BZPs	Leishmanicidal (*Leishmania amazonensis*)	Pereira *et al.*, 2009
G. celebica (leaf)	Garcihombronane D (TRP)	*Plasmodium falciparum*	Elfita *et al.*, 2009
G. cowa (bark)	Cowa xanthones	*P. falciparum*	Likhitwitayawuid *et al.*, 1998b
G. cuneifolia (stem bark)	Extract	*Aedes aegypti*	Hui, 2005
G. cymosa (stem bark)	Glutinenoic acid derivative (TRP)	*P. falciparum*	Elfita *et al.*, 2009
G. dulcis (bark)	Dulcis XTHs	*P. falciparum*	Likhitwitayawuid *et al.*, 1998a
G. griffithii (stem bark)	XTH	*P. falciparum*	Elfita *et al.*, 2009
G. indica (fruit)	Extract	Anthelmintic	Jena, *et al.*, 2002b
G. intermedia (leaf)	Guttiferone A (BZP)	*Trypanozoma cruzi*	Abe *et al.*, 2004
G. kola (seed)	FLVs	Molluscidal	Okunji and Iwu, 1991
Extracts		*P. falciparum*	Tona *et al.*, 1999
G. linii (root)	XTHs	*Mycobacterium tuberculosis*	Chen *et al.*, 2006
G. livingstonei (root bark)	FLVs, XTHs	Various parasites	Mbwambo *et al.*, 2006
G. lucida (stem bark)	Benzo-O-phenanthridine alkaloids	*Trypanosoma brucei brucei, Leishmania donovani*	Fotie *et al.*, 2007
G. mangostana (fruit)	Mangostins, XTHs	*M. tuberculosis*	Suksamram *et al.*, 2003
G. morella (seed coat)	Morellins (FLVs)	Various protozoa	Sani and Rao, 1966
G. polyantha (root bark)	Isoxanthochymol (BZP)	*P. falciparum*	Lannang *et al.*, 2008
G. subelliptica (stem bark)	XTHs	Trypanocidal	Abe *et al.*, 2003
G. vieillardii (stem bark)	Trihydroxyxanthones, polyphenols	Chloroquin-resistant *P. falciparum*, antileishmanial	Hay *et al.*, 2004b, 2008
Antiviral effect			
G. livingstonei	BZPs	HIV-1	Gustafson *et al.*, 1992
G. multiflora (leaf, twig)	Morelloflavone (FLV)	HIV-1, influenza A,B, parainflueza type 3, adenovirus type 5, herpes virus	Lin *et al.*, 1997, 1999
G. mangostana (fruit peel)	Mangostins (XTHs)	HIV-1	Chen *et al.*, 1996
G. ovalifolia	BZPs	HIV-1	Gustafson *et al.*, 1992
G. speciosa (trunk bark)	Digeranyl BZP	HIV-1	Rukachaisirikul *et al.*, 2003b
	Garciosaterpenes A,B		

[a]BPH, biphenyl; BZF, benzofuran; BPY, bezopyran; BZP, benzophenone; BZQ, benzoquinone; DPD, depsidone; FLV, flavonoid; PHL, phloroglucinol; TRP, terpenoid; XTH, xanthone.

Table 20.3. Other biological effects of chemical constituents isolated from *Garcinia* plants.

Garcinia species (Plant part)	Chemical constituents (Chemical class)[a]	Test system	References
Antidiabetic effect			
G. cambogia (fruit)	Extract	In mice	Hayamizu *et al.*, 2003
G. kola (fruit, seed)	Kolaviron	In rats	Adaramoye and Adeyemi,
	(FLV mixture)	In rabbits	2006; Iwu *et al.*, 1990
Anti-inflammatory effect			
G. bancana (stem)	Extract	NO inhibition	Jabit *et al.*, 2009
G. brasiliensis (fruit peel)	Volatile oil	Rat paw oedema	Martins *et al.*, 2008
G. gardneriana (leaf)	FLVs	Mouse paw oedema	Castardo *et al.*, 2008
G. hanburyi (resin)	Resin extract	Rat paw oedema	Panthong *et al.*, 2007
G. cowa (stem)	Extract	NO inhibition	Jabit *et al.*, 2009
G. indica	Garcinol (BZP)	NO inhibition	Padhye *et al.*, 2006
G. kola	Kolaviron (FLV)	Knee osteoarthritis	Adegbehingbe *et al.*, 2008
G. malaccensis (stem)	Extract	NO inhibition	Jabit *et al.*, 2009
G. mangostana	Mangostins (XTHs)	In rats	Shankaranarayanan *et al.*, 1979; Chen *et al.*, 2008
G. multiflora (fruit)	BZPs	Superoxide anion	Chen *et al.*, 2009
G. nervosa (leaf)	Extract	NO inhibition	Jabit *et al.*, 2009
G. prainiana (leaf)	Extract	NO inhibition	Jabit *et al.*, 2009
G. rostrata (stem, leaf)	Extracts	NO inhibition	Jabit *et al.*, 2009
G. subelliptica (seed)	PHLs, TRPs	Peritoneal mast cells	Weng *et al.*, 2004
Antiobesity effect			
G. cambogia (fruit)	HCA	Clinical trials	Toromanyan *et al.*, 2007; Preus *et al.*, 2004; Lenz and Hamilton, 2003; Lewis and Neelakantan, 1965; Heymsfield *et al.*, 1998
G. cowa (leaf, fruit)	HCA		Jena *et al.*, 2002b
G. indica (leaf, fruit)	HCA		Jayaprakasha and Sakariah, 2002
G. pedunculata (leaf, fruit)	HCA		Jayaprakasha *et al.*, 2003
Antioxidant/radical scavenging effects			
G. brasiliensis (fruit peel)	Volatile oil	Free radical	Martins *et al.*, 2008
G. cambogia (fruit) `	FLVs	Hypolipidaemic in rats	Koshy and Vijayalakshmi, 2001
G. cowa (latex, fruit rind)	Cowagarcinones A-E (XTHs), extract	Radical scavenger, PMo complex formation	Mahabusarakam *et al.*, 2005; Joseph *et al.*, 2005
G. indica (fruit rind)	Garcinol (BZP)	Antiglycination	Yamaguchi *et al.*, 2000
G. kola (seed)	FLVs, garcinol	Free radical scavenger	Okoko, 2009; Padhye *et al.*, 2009; Sang *et al.*, 2002;
		Antilipid peroxidation	Adegoke *et al.*, 1998
G. mangostana (pericarp)	Mangostin (XTH)	Human LDL oxidative damage protection	Williams *et al.*, 1995
	XTHs, isoflavones	Peroxynitrite	Jung *et al.*, 2006;
	XTHs	Antioxidant capacity in humans	Kondo *et al.*, 2009
G. pedunculata	Extract	PMo complex formation	Joseph *et al.*, 2005
G. subelliptica (wood)	Garciniaxanthones	Superoxide, free radical, lipid peroxidation	Minami *et al.*, 1995
G. vieillardii (stem bark)	XTHs	DPPH free radical	Hay *et al.*, 2004a
G. xanthochymus (bark)	Garcinone A-E (XTHs)	DPPH free radical	Zhong *et al.*, 2009

Continued

Table 20.3. *Continued*

Garcinia species (Plant part)	Chemical constituents (Chemical class)[a]	Test system	References
Cytotoxic/chemopreventive effects			
G. assigu	Polyprenylated BZPs	Raji cells	Ito et al., 2003a
G. atroviridis (roots)	Atrovirinone (BZQ), atrovirisidones (DPD)	Human breast MCF-7, prostate DU-145, lung H-460 cells	Permanaa *et al.*, 2001, 2005
G. bancana (stem)	Extract	Lung tumour cells	Jabit *et al.*, 2009
G.bracteata (leaf)	Bractins, isobractins (prenylated XTHs)	KB cells	Thoison *et al.*, 2000
G. cantleyana (trunk bark)	Cantleya XTHs	MCF-7, HeLa cells	Shadid *et al.*, 2007
G. fusca (stem bark)	Fuscaxanthones A-H	Raji cells	Ito *et al.*, 2003b
G. hanburyi (resin)	TRPs	Human leukemia cells	Wang *et al.*, 2008b
	Gambogic acids (XTHs)	K562 cells	Han *et al.*, 2006
	Polyprenylated XTHs		Tao *et al.*, 2009; Asano *et al.*, 1996
G. lancilimba (stem bark)	Polyprenylated XTHs	HeLa cells	Han *et al.*, 2008a
G. maingayi (stem)	Extract	MCF-7 cells	Jabit *et al.*, 2009
G. malaccensis (leaf)	Extract	Lung tumour cells	Jabit *et al.*, 2009
G. mangostana (pericarp, fruit, stem bark)	Panaxanthone (~80% α – and ~20% γ-mangostin)	Mammary cancer mouse model	Doi *et al.*, 2009
	α-Mangostin	Breast cancer cells, mouth cancer cells	Suksamram *et al.*, 2006
		HL60 cells	Matsumoto *et al.*, 2004
	XTHs	Human colon cancer	Akao *et al.*, 2008
		DLD-1 cells, etc.	Han et al., 2009; Ha *et al.*, 2009
	Garunone E (XTH)	Liver cancer cells	Ho *et al.*, 2002
G. macrophylla (twigs)	Guttiferones A,G (BZPs)	Ovarian cells	Williams *et al.*, 2003
G. merguensis (wood)	Rubraxanthone	MCF-7 cells	Kijjoa *et al.*, 2008
G. neglecta	DPDs	Raji cells	Ito *et al.*, 2001
G. oblongifolia (bark)	XTHs, PHLs	HeLa-C3 cells	Huang *et al.*, 2009
G. opaca (fruit)	Extract	MCF-7 cells	Jabit *et al.*, 2009
G. penangiana (leaf)	XTHs	MCF-7 cells	Jabit *et al.*, 2007, 2009
G. puat	DPDs	Raji cells	Ito *et al.*, 2001
G. pyrifera (fruit)	BZPs (guttiferone E and xanthochymol)	KB cells	Roux *et al.*, 2000
G. rigida (leaf)	Yahyaxanthone	L1210 cells	Elya *et al.*, 2008
G. subelliptica (pericarp)	Garcinielliptone (TRP)	Cell culture	Wu *et al.*, 2005
G. urophylla (leaf)	Extract	Lung tumour cells	Jabit *et al.*, 2009
G. virgata (stem bark)	Guttiferones I,J (BZPs)	KB cells	Merza *et al.*, 2006
G. xanthochymus (fruit)	BZPs	Colon cancer	Baggett *et al.*, 2005
G. *xipshuanbannaensis*	XTHs	HeLa cells	Han *et al.*, 2008b
G. yunnanensis (pericarp)	Oblongifolin C (BZP) Garciyunnanins A,B (XTHs)	Cervical cancer HeLa-C3 cells	Xu et al., 2008
Miscellaneous effects			
G. brasiliensis	7-Epiclusianone (BZP)	Antihistamine	Neves *et al.*, 2007
G. brevipedicellata (leaf)	Brevipsidones A–D (DPDs)	α-Glucosidase inhibitor	Ngoupayo *et al.*, 2008
G. hanburyi	Gamboge extract	Analgesic, antipyretic	Panthong *et al.*, 2007
G. kola (seed)	Kolaviron, biflavonoids	Antihepatotoxic Tyrosinase inhibitor	Iwu, 1985; Akintonwa and Essien 1990
		Rat lens aldose reductase inhibitor, broncodilator	Okunji *et al.*, 2007; Iwu *et al.*, 1990; Orie and Ekon, 1993

Garcinia species (Plant part)	Chemical constituents (Chemical class)[a]	Test system	References
G. *mangostana* (fruit, pericarp)	Mangosteen (40% ethanol extract)	Histamine release, prostaglandin E2 inhibitor	Nakatani *et al.*, 2002
	Mangostin (XTH) Polysaccharide	Antiulcer in rats Immunological	Shankaranarayanan *et al.*, 1979; Chanarat *et al.*, 1997
G. ovalifolia (bark)	Macluraxanthone	Mosquito larvae (malaria, yellow fever)	Wolfrom *et al.*, 1964
		Anti-termite	Harbone *et al.*, 1999
G. solomonsis	Guttiferones O,P (BZPs)	Phosphorylation inhibitor	Carroll *et al.*, 2009
G. staudtii (twig)	Staudtiixanthones A-D	Immune modulator	Ngoupayo *et al.*, 2009
G. subelliptica	Garciniaxanthones A,B, garsubellin A (PHL)	Enhance choline acetyltransferase activity	Fukuyama *et al.*, 1991, 1997
G. xanthochymus (wood)	Prenylated XTHs	Nerve growth factor potentiator	Chanmahasathien *et al.*, 2003

[a]BZF, benzofuran; BZP, benzophenone; BZQ, benzoquinone; DPD, depsidone; FLV, flavonoid; PHL, phloroglucinol; TRP, terpenoid; XTH, xanthone.

The bioactive chemical constituents of *Garcinia* species generally belong to three major chemical classes: xanthone, benzophenone and flavone. However, this species is also known to contain, to a lesser extent, biologically active compounds belonging to a wide variety of other chemical classes, such as biphenyl, benzofuran, benzopyran, benzoquinone, depsidone, phloroglucinol, terpenoid and organic acid, as listed in Tables 20.2 and 20.3. Interestingly, there is a report of the isolation of dehydrochelerythrine (benzo-*O*-phenanthridine) alkaloids with trypanocidal and leishmanicidal activities from the stem bark of *G. lucida* (Fotie *et al.*, 2007). Also, D-galacturonic-acid-based polysaccharides with immunopharmacological activity (phagocystic cell stimulation with intracellular bacteria destruction) have been isolated from the pericarp of *G. mangostana* (Chanarat *et al.*, 1997). Additionally, the gamboge resin from *G. hanburyi* exhibits antibacterial (Cowan, 1999), analgesic and antipyretic activities (Panthong *et al.*, 2007). For separation, identification and quantification of these polar compounds, high-speed countercurrent (Kapadia *et al.*, 1994; Okunji *et al.*, 2007) and high-performance liquid chromatography (Jayaprakasha *et al.*, 2003; Li *et al.*, 2008; Xu *et al.*, 2008), particularly in conjunction with mass spectrometry (Chattopadhyay and Kumar, 2007; Zhou *et al.*, 2008; Kumar *et al.*, 2009) are remarkably efficient and powerful tools.

It is of interest to note that the antimicrobial constituents are present across the entire *Garcinia* genus and are represented by all three major chemical classes, described above, in this diverse plant species. Further, the antimicrobial activity spectrum is broad and includes antibacterial, antifungal, antiparasitic and antiviral effects. Also, the antimicrobial principles are found in all parts of the plant, such as flower, latex, pericarp, fruit, seed, leaf, twig, wood, bark and root. This information should be of interest in considering the cultivation of *Garcinia* plants for potential commercial use.

Among the benzophenone class, kolanone (Fig. 20.3), isolated from the *G. kola* fruit, has potent antimicrobial activity (Hussain *et al.*, 1982) while garcinol (Fig. 20.4), found in several species of *Garcinia* (*e.g. G. cambogia, G. huillensis, G. indica* and *G. purpurea*) and xanthochymol (Fig. 20.5), from the pericarps of *G. huillensis* and *G. subelliptica*, are active against methicillin-resistant *S. aureus* (MRSA) (Iinuma *et al.*, 1996d). Xanthochymol had the lowest minimum effective concentration at 3.1–12.5 μg/ml. This concentration is nearly equal to that of the vancomycin activity profile. In addition, garcinol (also known as camboginol) is active against several mycobacteria and fungi (Bakana *et al.*, 1987), and exhibits anti-inflammatory, antioxidant, anti-ulcer and anti-HIV effects (Padhye *et al.*, 2006). Also, it is noteworthy that the beneficial effect of tumour prevention by garcinol on the human colorectal cancer cell line HT-29 has been recently reported (Liao *et al.*, 2005). This *Garcinia* constituent, labelled as 'the nature's wonder agent', has been the subject of a recent comprehensive review (Padhye *et al.*, 2009).

The xanthone class of compounds is widely distributed in the *Garcinia* species. α-Mangostin (Fig. 20.2), the first xanthone to be isolated from the fruit hulls of *G. mangostana* (Schmid. 1855), exhibits anti-inflammatory activity (Shankaranarayanan *et al.*, 1979), and has an antimicrobial profile against MRSA (Iinuma *et al.*, 1996c), malaria (Tona *et al.*, 1999) and tuberculosis (Suksamram *et al.* 2003). Macluraxanthone (Fig. 20.6) from *G. ovalifolia* is more toxic than the pesticide rotenone when tested against the larvae of both malaria and yellow fever mosquitoes (Wolfrom *et al.*, 1964).

Macluraxanthone also exhibits anti-termite activity (Harbone *et al.*, 1999). Antimalarial activity comparable to that of pyrimethamine is demonstrated by xanthones isolated from *G. cowa* and *G. dulcis* (Likhitwitayawuid *et al.*, 1998a,b). Garciniaxanthone C (Fig. 20.7) and its analogues from *G. subelliptica* are efficient antioxidants, inhibit lipid peroxidation, and are useful for scavenging free radicals and the superoxide anion (Minami *et al.*, 1994, 1995). Another constituent of this species, 1,4,5-trihydroxyxanthone (Fig. 20.8) has a profound antioxidation profile as it is capable of scavenging ~90% of superoxide anion at concentration as low as 5μg/ml.

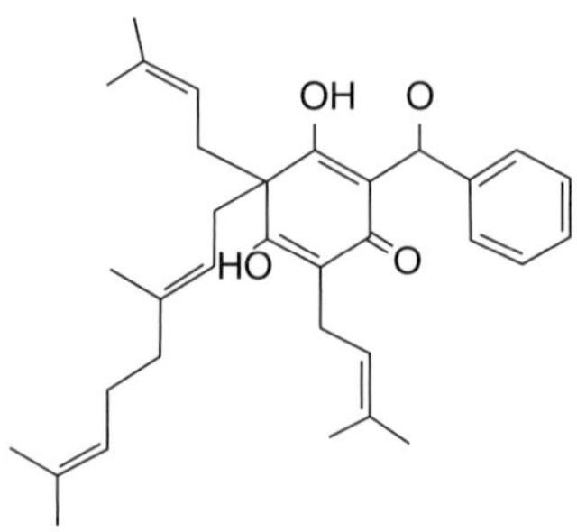

Fig. 20.3. Kolanone.

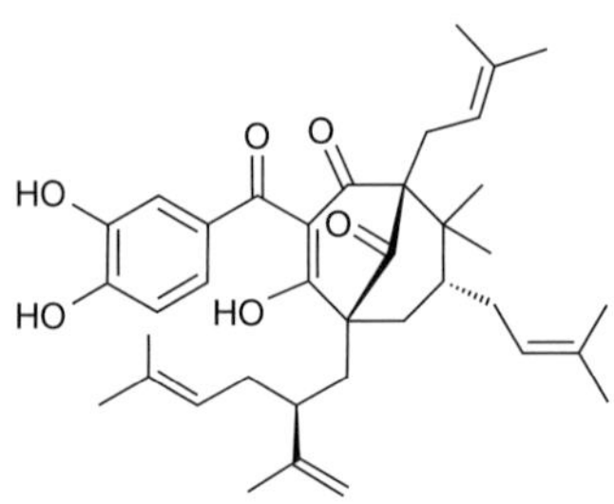

Fig. 20.4. Garcinol.

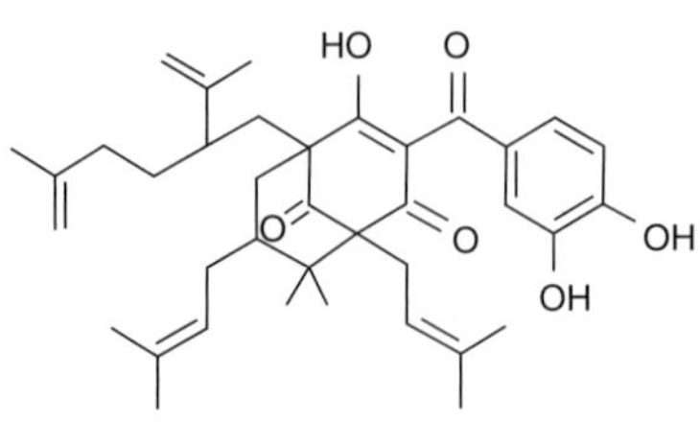

Fig. 20.5. Xanthochymol.

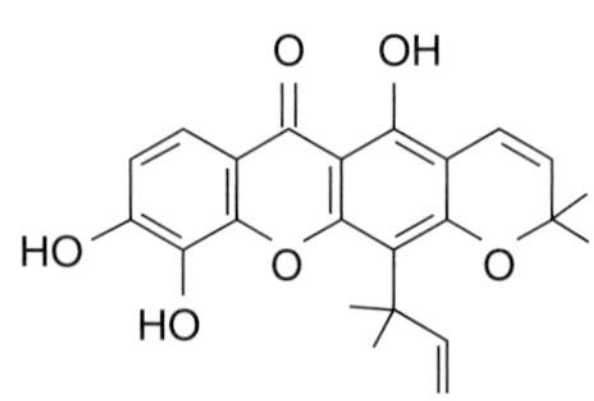

Fig. 20.6. Macluraxanthone.

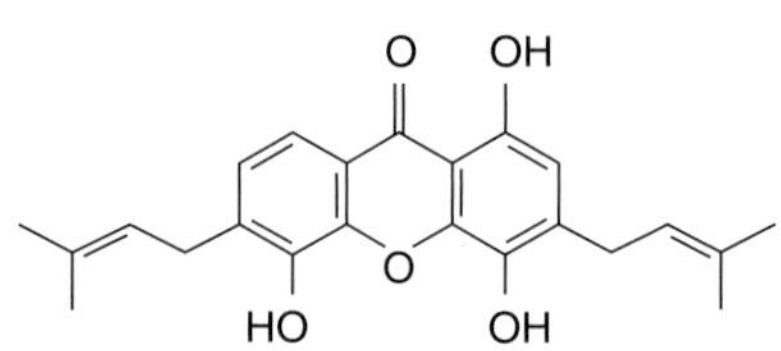

Fig. 20.7. Garciniaxanthone C.

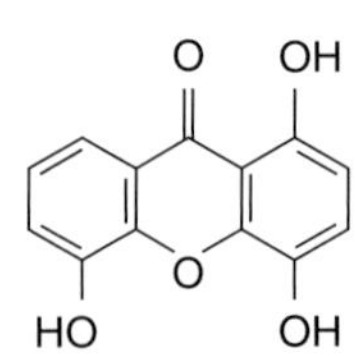

Fig. 20.8. 1,4,5-Trihydroxyxanthone.

Kolaflavanone (Fig. 20.9) and its derivatives, isolated from the fruit of *G. kola*, belong to the flavone chemical class. They exhibit antihepatotoxic and antidiabetic effects, as well as aldose reductase inhibitory activity in the rat lens (Iwu *et al.*, 1990). Among other bioactive flavones, morelloflavone (Fig. 20.10) from *G. morella* has a wide antiviral activity range which includes HIV-1, influenza, herpes and adeno viruses (Lin *et al.*, 1997,1999), while amentoflavone (Fig. 20.11) from *G. livingstonei* is active against a broad spectrum of bacteria (Kaikabo *et al.*, 2009).

Many *Garcinia* chemical constituents belonging to the benzophenone and xanthone chemical classes are found to be potent cytotoxic and chemopreventive agents in cancer cell culture and in animal studies. Representative examples are atrovirinone (Fig. 20.12) from *G. atroviridis*, which is toxic to the HeLa human cervical cancer cells (Permanaa *et al.*, 2001), and α-mangostin from *G. mangostana*, which is active against mouse mammary cancer cells (Jung *et al.*, 2006), both *in vitro*. Further, in a mouse test model, panaxanthone (a natural mixture of α- and γ-mangostins, Figs 20.2 and 20.13, respectively), from the pericarp of *G. mangostana*, is found it to be active against mammary cancer (Doi *et al.*, 2009).

Recently, isolation and biological effects of several noteworthy, new, *Garcinia* compounds belonging to other chemical classes have been reported. For example, garsubellin A (Fig. 20.14), a polyprenylated phloroglucin derivative from *G. subelliptica*, is found to increase choline esterase activity in the rat septal neuron cultures (Fukuyama *et al.*, 1997). Deficiency of this key enzyme in the synthesis of the neurotransmitter acetylcholine is implicated in dementia. Thus, this novel *Garcinia* constituent should be of interest in developing a new agent for the treatment of Alzheimer's disease.

The sustained interest in the efficacy of HCA (Fig. 20.1) in the treatment of obesity is noteworthy (Jena *et al.*, 2002b; Pittler and Earnst, 2004; Toromanyan *et al.*, 2007). Since *in vitro* and animal studies indicated that

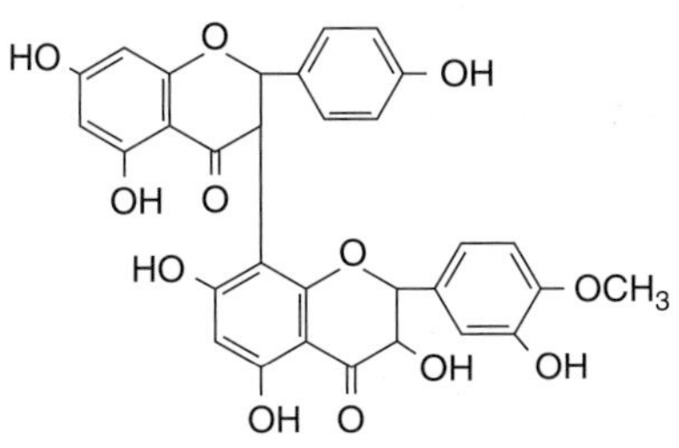

Fig. 20.9. Kolaflavanone.

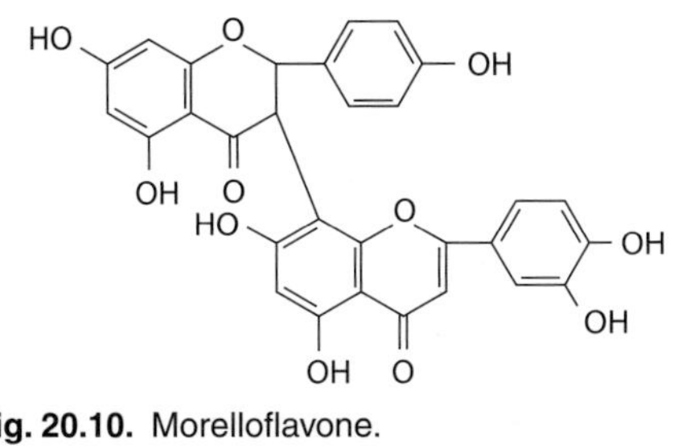

Fig. 20.10. Morelloflavone.

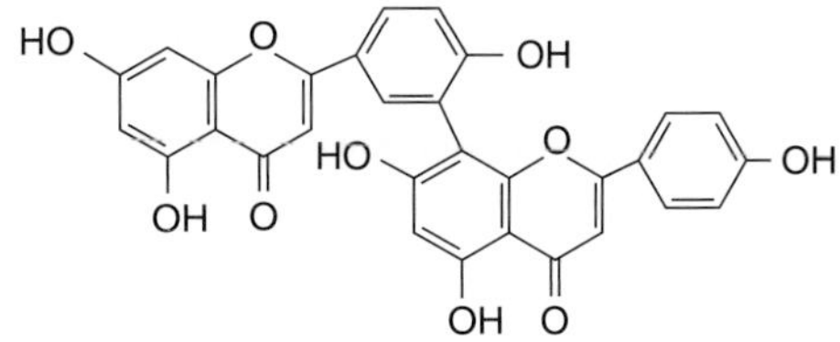

Fig. 20.11. Amentoflavone.

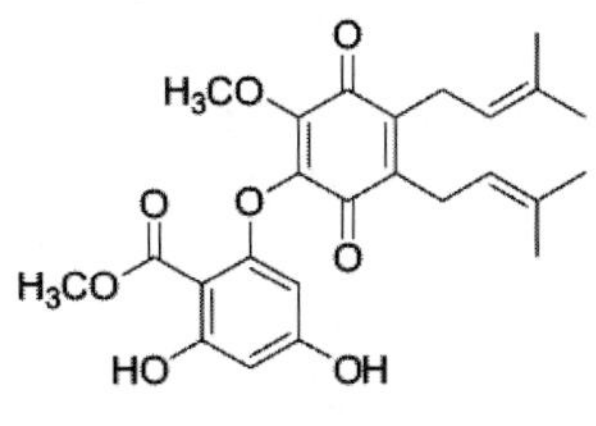

Fig. 20.12. Atrovirinone.

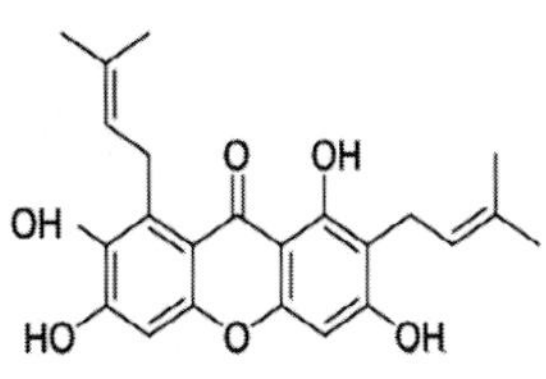

Fig. 20.13. γ-Mangostin.

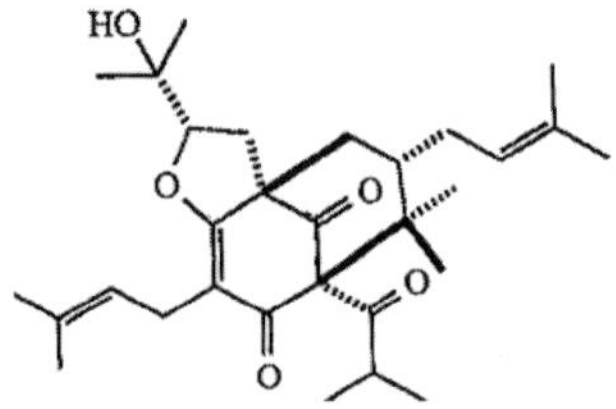

Fig. 20.14. Garsubellin A.

HCA has a potential for modulation of lipid metabolism and suppressing fat accumulation, it is currently used as an ingredient in several weight-loss products. However, to-date there is no clinical evidence for body-weight loss or reduction of fat mass by HCA in human subjects (Heymsfield *et al.*, 1998; Lenz and Hamilton, 2003; Hasani-Ranjbar *et al.*, 2009). Interestingly, an isomer of HCA, identified as (2*S*,3*R*)-HCA, is found to inhibit pancreatic α-amylase and intestinal α-glucosidase leading to a reduction in carbohydrate metabolism *in vitro* (Yamada *et al.*, 2007). But currently there appears to be no clinical study available on this isomer of HCA.

The chemical structure and biological activity studies with xanthones and benzophenones isolated from *Garcinia* plants suggest that ring hydroxyl groups are essential since their alkylation reduced observed biological activities. Thus, alkylating the C-3 and C-6 hydroxyl groups in α-mangostin (Fig. 20.2) reduced its antifungal activity (by about half for methyl substitution) and replacement with alkyl groups of increasing chain length correlated with decreasing inhibittory activity (Gopalakrishnan *et al.*, 1997). Similar structure–activity relationships (SAR) were observed with α-mangostin and its derivatives in their apoptosis inducing potency in the human leukemia HL-60 cells (Matsumoto *et al.*, 2004) as well as with other cytotoxic geranylated xanthones and *O*-alkylated derivatives of α-mangostin (Ha *et al.*, 2009). The prenylated xanthone derivatives from the fruit of *G. mangostana* also showed similar SAR with their inhibitory effect against the tuberculosis bacterium (Suksamram *et al.*, 2003). Also, the inhibitory effect of the benzophenones from *G. pyrifera* on the tubulin/microtubule system followed a similar SAR pattern (Roux *et al.*, 2000). Additionally, the unsaturated, isoprenyl side chain appears to be important for biological activity (Ito *et al.*, 2003a) and, in the case of garcinol (Fig. 20.4), it is shown that the double bond of the isopentenyl group is a principal site for its antioxidant activity (Sang *et al.*, 2002). A study with antimicrobial activity of benzophenones from *G. brasiliensis* showed a direct relationship between the lipophilic character of the chemical structure and its activity against the test bacteria (Naldoni *et al.*, 2009).

Potential toxicity of *Garcinia* plant products has been addressed by several studies in test animals and humans. For example, ulcer patients should avoid the *G. kola* nut due to its high amino acid content (Adeyeye *et al.*, 2007), while habitual chewing of the nut is implicated in oral cancer due to its nitrosatable amine content leading to carcinogenic nitrosamide formation *in situ* (Atawodi *et al.*, 1995). The frequent use of *G. cambogia* fruit-derived tamarind in cooking may increase the risk of gallstone formation (Jayanthi *et al.*, 2005) and high doses of HCA may have the potential for testicular and liver toxicity (Saito *et al.*, 2005; Stevens *et al.*, 2005; Stohs *et al.*, 2009). Further, severe lactic acidosis can occur in susceptible individuals who drink the juice of *G. mangostana* fruit (Wong and Klemmer, 2008).

Future Directions and Conclusions

The *Garcinia* species have been a fertile source for phytochemical studies in the past two centuries and the trend continues even in this century. Over the years, the *Garcinia* plants have yielded scores of new compounds with unique chemical structures and medically important biological activities. Tables 20.2 and 20.3 summarize these achievements. The range of chemical classes and the spectrum of biological activities of compounds isolated from this species are probably unmatched in the annals of medicinal plant investigations. This could explain why the *Garcinia* plants have such an important place in the present day herbal medical practice around the globe and they continue to be popular remedies for a variety of common ailments. The isolation and identification of bioactive compounds from these indigenous medicinal plants give much credence to their continued use over the past several centuries in the tropical countries of the world.

For example, recent clinical studies have shown that chewing twigs of *G. kola* and *G. manni* which have been used for centuries for oral hygiene, do provide clinically proven beneficial antibacterial effect in maintaining oral hygiene (Taiwo *et al.*, 1999; Ndukwe *et al.*, 2005) and may prevent caries formation (Addai *et al.*, 2002). Similarly, mouthwash and gel containing *G. mangostana* pericarp were effective in treating halitosis and as an adjunct to periodontal treatment (Rassameemasmaung *et al.*, 2007, 2008), respectively. In another front, there are potentials for utilizing the antimicrobials isolated from the *Garcinia* plants in combating the current drug-resistant strains of malaria, leischmaniasis, tuberculosis, HIV and other microbial infections. Also, the novel cytotoxic and chemopreventive principles from the *Garcinia* species could serve as templates in the discovery of new anticancer drugs. It should be noted that the US Patent Office has recently issued a patent for the use of gambogic acid (Fig. 20.15), the principal pigment of gamboge resin from several *Garcinia* species and its analogs as activators of caspases and inducers of apoptosis (Cai *et al.*, 2009).

The commercial use of *Garcinia* plant products by global food, cosmetic and drug industries has been sparse to date. But this may change as the current trend of going 'green' and 'organic' should give fresh incentives to these manufacturers in utilizing the natural antimicrobials from *Garcinia* fruit pulps, seeds, seed oils, barks, resins, etc. in processed food preservation and as disinfectants in the manufacturing plants. In this context, a manufacturing technology report from China on the selection of the technology for processing steamed *G. hunburyi* constituents based on the indexes of anti-inflammatory, bactericidal and anti-tumour effects, and gambogic acid (Fig. 20.15) content (Ye and Kong, 1996) is encouraging. Recently, the extracts from the fruit rinds of *G. cowa* and *G. pedunculata* were found to exhibit antibacterial activity against food-borne pathogens and spoilage bacteria (Negi *et al.*, 2008). Also, antiaflatoxigenic and antioxidation activities of *Garcinia* extracts (Joseph *et al.*, 2005) and antimicrobial activity in cultures of endophytic fungi isolated from five common *Garcinia* species (Phongpaichit *et al.*, 2006) have been reported. Further, addition of the *G. cambogia* extracts as spice while processing fish may block histamine formation and thus prevent allergic reactions in susceptible consumers (Thadhani *et al.*, 2002).

Thus, there is potential for increased use of *Gracinia* plant constituents by drug, cosmetic and food industries, such as in 'natural' antibacterial soaps and disinfectant sprays, acne and antiageing creams, sun screens, hair-care products, lip balms, tooth-pastes, oral gels, mouthwashes, 'organic' protein-enhanced cereals, and in the prevention of food allergy, preservation and safety of processed foods, disinfecting manufacturing equipment and production lines, etc. Perhaps in the near future, efforts will be made by global food, cosmetic and drug industries to develop novel products based on the efficient utilization of the natural antimicrobial and other bioactive constituents from *Garcinia* plants which could be economically cultivated in the tropical countries around the world.

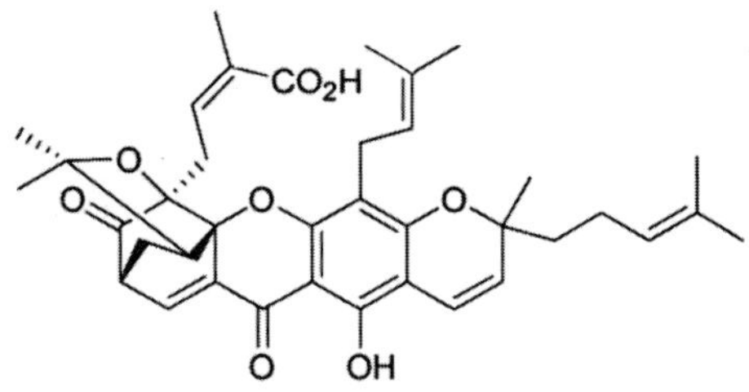

Fig. 20.15. Gambogic acid.

References

Abe, F., Nagafuji, S., Okabe, H., Higo, H. and Akahane, H. (2003) Trypanocidal constituents in plants. 2. Xanthones from the stem bark of *Garcinia subelliptica*. *Biological and Pharmaceutical Bulletin* 26, 1730–1733.

Abe, F., Nagafuji, S., Okabe, H., Akahane, H., Estrada-Muniz, E., Huerta-Reyes, M. and Reyes-Chilpa, R. (2004) Trypanocidal constituents in plants. 3. Leaves of *Garcinia intermedia* and heartwood of *Calophyllum brasiliense*. *Biological and Pharmaceutical Bulletin* 27, 141–143.

Adaramoye, O.A. and Adeyemi, E.O. (2006) Hypoglycemic and hypolipidaemic effects of fractions from kolaviron, a biflavonoid complex from *Garcinia kola* in streptozotocin-induced diabetes mellitus rats. *Journal of Pharmacy and Pharmacology* 58, 121–128.

Addai, F.K., Nuamah, I.K. and Parkins, G.E. (2002) Brief chewing of *Garcinia manni* stick reverses reduced saliva pH after a glucose rinse. *Medical Science Monitor: International Medical Journal of Experimental and Clinical Research* 8, CR746–750.

Adegbehingbe, O.O., Adesanya, S.A., Idowu, T.O., Okimi, O.C., Oyelami, O.A. and Iwalewa, E.O. (2008) Clinical effects of *Garcinia kola* in knee osteoarthritis. *Journal of Orthopaedic Surgery and Research* 3, 34.

Adegoke, G.O., Kumar, M.V., Sambaiah, K. and Lakesh, B.R. (1998) Inhibitory effect of *Garcinia kola* on lipid peroxidation in rat liver homogenate. *Indian Journal of Experimental Biology* 36, 907–910.

Adeyeye, E.I., Asaolu, S.S. and Aluko, A.O. (2007) Amino acid composition of two masticatory nuts (*Cola acuminata* and *Garcinia kola*) and a snack nut (*Anacaradium occidentale*). *International Journal of Food Science and Nutrition* 58, 241–249.

Akao, Y., Nakagawa, Y., Iinuma, M. and Nozawa, Y. (2008) Anticancer effects of xanthone from pericarps of mangosteen. *International Journal of Molecular Sciences* 3, 355–370.

Akintonwa, A. and Essien, A.R. (1990) Protective effects of *Garcinia kola* seed extract against paracetamol-induced hepatotoxicity in rats. *Journal of Ethnopharmacology* 29, 207–211.

Alam, M.S., Qasim, M.A., Kamil, M. and Ilyas, M. (1986) Sorbifolin 6-galactoside from *Garcinia andamanica*. *Phytochemistry* 25, 2900–2901.

Ali, S., Goudar, R., Sotheeswaran, S., Beaulieu, C. and Spino, C. (2000) Benzophenones of *Garcinia pseudoguttifera* (Clusiaceae). *Phytochemistry* 53, 281–284.

Ampofo, S.A. and Waterman, P.G. (1986) Xanthones from three *Garcinia* species. *Phytochemistry* 25, 2351–2355.

Ansari, W.H., Rahman, W., Barraclough, D., Maynard, R. and Scheinmann, F. (1976) Biflavonoids and a flavanone-chromone from the leaves of *Garcinia dulcis* (Roxb.) Kurz. *Journal of the Chemical Society Perkins Transactions 1* 1976, 1458–1463.

Asano, J., Chiba, K., Tada, M. and Yoshi, T. (1996) Cytotoxic xanthones from *Garcinia hanburyi*. *Phytochemistry* 41, 815–820.

Atawodi, S.E., Mende, P., Pfundstein, B., Preussmann, R. and Spiegelhalder, B. (1995) Nitrosatable amines and nitrosamide formation in natural stimulants: *Cola acuminata, C. nitida* and *Garcinia kola*. *Food and Chemical Toxicology* 33, 625–630.

Baggett, S., Protiva, P., Mazzola, E.P., Yang, H., Ressler, E.T., Basile, M.J., Weinstein, I.B. and Kennelly, E.J. (2005) Bioactive benzophenones from *Garcinia xanthochymus* fruits. *Journal of Natural Products* 68, 354–360.

Bakana, P., Claeys, M., Totte, J., Pieters, L.A.C., Hoof, L. van., Tamba-Vemba, Van den Berghe, D.A. and Vlietinck, A.J. (1987) Structure and chemotherapeutical activity of a polyisoprenylated benzophenone from the stem bark of *Garcinia huillensis*. *Journal of Ethnopharmacology* 21, 75–84.

Bhaskaran, S. (2004) Stabilized anthocyanin extract from *Garcinia indica*. US Patent No. 7261769.

Bodeker, G. (2009) *Health and Beauty from the Rainforest: Malysian Traditions of Ramuan,* Didier Millet, Kuala Lumpur, Malaysia.

Cai, S.X., Jiang, S. and Zhang, H-Z. (2009) Gambogic acid and analogs as activators of caspases and inducers of apoptosis. US Patent No. 7622497.

Campbell, C.W. and Ledesma, N. (2002) The acquisition and introduction of Asian tropical fruit with potential for South Florida. *Proceedings of the Florida State Horticultural Society* 115, 87–88.

Carroll, A.R., Suraweera, L., King, G., Rali, T. and Quinn, R.J. (2009) Guttiferones O and P, prenylated benzophenone MAPKAPK-2 inhibitors from *Garcinia solomonsis*. *Journal of Natural Products* 72, 1099–1701.

Castardo, J.C., Prudente, A.S., Ferreira, J., Guimaraes, C.L., Monache, F.D., Filho, V.C., Otuki, M.F. and

Cabrini, D.A. (2008) Anti-inflammatory effects of hydroalcoholic extract and two biflavonoids from *Garcinia gardneriana* leaves in mouse paw aedema. *Journal of Ethnopharmacology* 118, 405–411.
Chanarat, P., Chanarat, N., Fujihara, M. and Nagumo, T. (1997) Immunopharmacological activity of polysaccharide from the pericarp of mangosteen garcinia: Phagocytic intracellular killing activities. *Journal of the Medical Association of Thailand* 80, 5149–5154.
Chanmahasathien, W., Li, Y., Satake, M., Oshima, Y., Rhuangrungsi, N. and Ohizumi, Y. (2003) Prenylated xanthones with NGF-potentiating activity from *Garcinia xanthochymus. Phytochemistry* 64, 981–986.
Chattopadhyay, S.K. and Kumar, S. (2007) A rapid liquid chromatography – tandem mass spectrometry method for quantification of a biologically active molecule camboginol in the extract of *Garcinia cambogia. Biomedical Chromatography* 1, 55–66.
Chen, J.J., Peng, C.F., Huang, H.Y. and Chen, I.S. (2006) Benzopyrans, biphenyls and xanthones from the root of *Garcinia linii* and their activity against *Mycobactirium tuberculosis. Planta Medica* 72, 473–477.
Chen, J.J., Ting, C.W., Hwang, T.L. and Chen, I.S. (2009) Benzophenone derivatives from the fruits of *Garcinia multiflora* and their anti-inflammatory activity. *Journal of Natural Products* 72, 253–258.
Chen, L.G., Yang, L.L. and Wang, C.C. (2008) Anti-inflammatory activity of mangostins from *Garcinia mangostana. Food and Chemical Toxicology* 46, 688–693.
Chen, S.X., Wan, M. and Loh, B.N. (1996) Active constituents against HIV-1 protease from *Garcinia mangostana. Planta Medica* 62, 381–382.
Chomnawang, M.T., Surassmo, S., Nukoolkarn, V.S. and Gritsanapan, W. (2005) Antimicrobial effects of Thai medicinal plants against acne-inducing bacteria. *Journal of Ethnopharmacology* 101, 330–333.
Chung, M-I., Su, H-J. and Lin, C-N. (1998) A novel triterpenoid from *Garcinia subelliptica. Journal of Natural Products* 61, 1015–1016.
Cowan, M.M. (1999) Plant products as antimicrobial agents. *Clinical Microbiology Reviews* 12, 564–582.
Cuesta-Rubino, O., Padron, A., Castro, H.V., Pizza, C. and Rastrelli, L. (2001) Aristophenones A and B. A new tautomeric pair of polyisoprenylated benzophenones from *Garcinia aristata. Journal of Natural Products* 64, 973–975.
Deachathai, S., Mahabusarakam, W., Phongpaichit, S. and Taylor, W.C. (2005) Phenolic compounds from the fruit of *Garcinia dulcis. Phytochemistry* 66, 2368–2375.
Deachathai, S., Mahabusarakam, W., Phongpaichit, S., Taylor, W.C., Zhang, Y.J. and Yang, C.R. (2006) Phenolic compounds from the flowers of *Garcinia dulcis. Phytochemistry* 67, 464–469.
Deachathai, S., Phongpaichit, S. and Mahabusarakam, W. (2008) Phenolic compounds from the seeds of *Garcinia dulcis. Natural Product Research* 22, 1327–1332.
Doi, H., Shibata, M.A., Morimoto, J., Akao, Y., Iinuma, M., Tanigawa, N. and Otsuki, Y. (2009) Panaxanthone isolated from pericarp of *Garcinia mangostana* L. suppresses tumor growth and metstasis of a mouse model of mammary cancer. *Anticancer Research* 29, 2485–2495.
Elfita, E., Muharni, M., Latief, M., Darwati, D., Widiyantoro, A., Supriyanta, S., Bahti, H.H., Dachriyanus, D., Cos, P., Maes, L., Foubert, K., Apers, S. and Pieters, L. (2009) Antiplasmodal and other constituents from four *Garcinia* spp. *Phytochemistry* 70, 907–912.
Elya, B., He, H.P., Kosela, S., Hanafi, M. and Hao, X.J. (2006) A new benzophenone from the stem bark of *Garcinia benthami. Natural Products Research* 20, 1059–1062.
Elya, B., He, H.P., Hanafi, M. and Hao, X.J. (2008) A new cytotoxic xanthone from *Garcinia rigida. Fitoterapia* 79, 182–184.
Fotie, J., Bohle, D.S., Olivier, M., Adelaida-Gomez, M. and Nzimiro, S. (2007) Trypanocidal and antileishmanial dehydrochelerythrine derivatives from *Garcinia lucida. Journal of Natural Products.*70, 1650–1653.
Fu, C., Loo, A.E., Chia, F.P. and Huang, D. (2007) Oligomeric proanthocyanidins from mangosteen pericarps. *Journal of Agricultural and Food Chemistry* 55, 7689–7694.
Fukuyama, Y., Kamiyama, A., Mima, Y. and Kodama, M. (1991) Prenylated xanthones from *Garcinia subelliptica. Phytochemistry* 30, 3433–3436.
Fukuyama, Y., Kuwayama, A. and Minami, H. (1997) Garsubellin A, a novel polyprenylated phloroglucin derivative, increasing choline acetyltransferase (ChAT) activity in postnatal rat septal neuron cultures. *Chemical and Pharmaceutical Bulletin* 45, 947–949.
Goh, H., Jantan, I., Gray, A.I. and Waterman, P.G. (1991) Prenylated xanthones from *Garcinia opaca. Phytochemistry* 31, 1383–1386.
Gopalakrishnan, G., Banumathi, B. and Suresh, G. (1997) Evaluation of the antifungal activity of natural xanthones from *Garcinia mangostana* and their synthetic derivatives. *Journal of Natural Products* 60, 519–524.

Gustafson, K.R., Blunt, J.W., Munro, M.H.G., Fuller, R.W., McKee, T.C., Cardellinia II, J.H., McMohan, J.B., Graff, G.M. and Boyd, M.R. (1992) The guttiferones, HIV-inhibitory benzophenones from *Symphonia globulifera, Garcinia livingstonei, Garcinia ovalifolia* and *Clusia rosea. Tetrahedron* 48, 10093–10102.

Ha, L.D., Hanse, P.E., Vang, O., Duus, F., Pham, H.D. and Nguyen, L.H. (2009) Cytotoxic geranylated xanthones and *O*-alkylated derivatives of alpha-mangostin. *Chemical and Pharmaceutical Bulletin* 57, 830–834.

Hamed, W., Brajeul, S., Mahateau-Betzer, F., Thoison, O., Mons, S., Delpech, B., Nguyen, V.H., Sevenet, T. and Merazano, C. (2006) Oblongifolins A-D, polyprenylated benzophloroglucinol derivatives from *Garcinia oblongifolia. Journal of Natural Products* 69, 774–777.

Han, A.R., Kim, J.A., Lantvitt, D.D., Kardono, L.B., Riswan, S., Chai, H., Carcache de Blanco, E.J., Farnsworth, N.R., Swanson, S.M. and Kinghorn, A.D. (2009) Cytotoxic xanthone constituents of the stem bark of *Garcinia mangostana* (Mangosteen). *Journal of Natural Products* 72, 2028–2031.

Han, Q., Yang, L., Liu, Y., Qiao, C., Song, J., Xu, L., Yang, D. and Xu, H. (2006) Gambogic acid and epigambogic acid, C-2 epimers with novel anticancer effects from *Garcinia hanburyi. Planta Medica* 72, 281–284.

Han, Q.B., Tian, H.L., Yang, N.Y., Qiao, C.F., Song, J.Z., Chang, D.C., Luo, K.Q. and Xu, H.X. (2008a) Polyprenylated xanthones from *Garcinia lancilimba* showing apoptotic effects against HeLa-C3 cells. *Chemistry and Biodiversity* 5, 2710–2717.

Han, Q.B., Yang, N.Y., Tian, H., Qiao, C.F., Song, J.Z., Chang, D.C., Chen, S.L., Luo, K.Q. and Xu, H.X. (2008b) Xanthones with growth inhibition against HeLa cells from *Garcinia xipshuanbannaensis. Phytochemistry* 69, 2187–2192.

Harbone, J.B., Baxter, H. and Moss, G.P. (eds) (1999*) Phytochemical Dictionary: A Handbook of Bioactive Compounds from Plants*, 2nd edn, Taylor and Francis, London, UK, p. 587.

Harrison, L.J., Leong, L.S., Sia, G.L., Sim, K.Y. and Tan, H.T.W. (1993) Xanthones from *Garcinia forbesii. Phytochemistry* 33, 727–728.

Hartati, S., Soemiati, A., Wang, H.B., Kardono, L.B., Hanafi, M., Kosela, S. and Qin, G.W. (2008) A novel polyisoprenyl bezophenone derivative from *Garcinia eugeniafolia. Journal of Asian Natural Product Research* 10, 509–513.

Hasani-Ranjbar, S., Nayebi, N., Larijani, B. and Abdollahi, M. (2009) A systematic review of the efficacy and safety of herbal medicines used in the treatment of obesity. *World Journal of Gastroenterology* 15, 3073–3085.

Hay, A.E., Aumond, M.C., Mallet, S., Dumontet, V., Litaudon, M., Rondeau, D. and Richomme, P. (2004a) Antioxidant xanthones from *Garcinia vieillardii. Journal of Natural Products* 67, 707–709.

Hay, A.E., Helesbeux, J.J., Duval, O., Labaied, M., Grellier, P. and Richomme, P. (2004b) Antimalarial xanthones from *Calophyllum caledonicum* and *Garcinia vieillardii. Life Sciences* 75, 3077–3085.

Hay, A.E., Merza, J., Landreau, A., Litaudon, M., Pagniez, F., Le Pope, P. and Richomme, P. (2008) Antileishmanial polyphenols from *Garcinia vieillardii. Fitoterapia* 79, 42–46.

Hayamizu, K., Hirakawa, H., Oikawa, D., Nakanishi, T., Takagi, T., Tachibana, T. and Furuse, M. (2003) Effect of *Garcinia cambogia* extract on serum leptin and insulin in mice. *Fetoterapia* 74, 267–273.

Heymsfield, S.B., Allison, D.B., Vasselli, J.R., Pietrobelli, A., Greenfield, D. and Nunez, C. (1998) *Garcinia cambogia* (hydroxycitric acid) as a potential antiobesity agent: A randomized controlled trial. *Journal of the American Medical Association* 280, 1596–1600.

Ho, C.K., Huang, Y.L. and Chen, C.C. (2002) Garunone E, a xanthone derivative, has potent cytotoxic effect against hepatocellular carcinoma cell lines. *Planta Medica* 68, 975–979.

Hsu, J., Skover, G. and Goldman, M.P. (2007) Evaluating the efficacy in improving facial photo-damage with a mixture of topical antioxidants. *Journal of Drugs in Dermatology* 6, 1141–1148.

Huang, S.X., Feng, C., Zhou, Y., Xu, G., Han, Q.B., Qiao, C.F., Chang, D.C., Luo, K.Q. and Xu, H.X. (2009) Bioassay-guided isolation of xanthones and polycyclic prenylated acylphloroglucinols from *Garcinia oblongifolia. Journal of Natural Products* 72, 130–135.

Hui, M.X. (2005) Chemical Constituents and Biological Activities of *Gracinia cuneifolia, Mesua beccariana* and *Mesua ferrea*, Master of Science Thesis, University of Putra, Malaysia.

Hussain, R.A., Owegby, A.G., Parimoo, P. and Waterman, P.G. (1982) Kolanone, a novel polyisoprenylated benzophenone with antimicrobial properties from the fruit of *Garcinia kola. Planta Medica* 44, 78–81.

Iinuma, M., Tosa, H., Ito, T., Tanaka, T. and Riswan, S. (1996a) Three new benzophenone-xanthone dimers from the root of *Garcinia dulcis. Chemical and Pharmaceutical Bulletin* 44, 1744–1747.

Iinuma, M., Tosa, H., Tanaka, T. and Riswan, S. (1996b) Three new xanthones from the bark of *Garcinia dioica. Chemical and Pharmaceutical Bulletin* 44, 232–234.

Iinuma, M., Tosa, H., Tanaka, T., Shimano, R., Asai, F. and Yonemori, S. (1994) Two xanthones from root bark of *Garcinia subelliptica. Phytochemistry* 35, 1355–1360.
Iinuma, M., Tosa, H., Tanaka, T., Asai, F., Kobayashi, Y., Shimano, R. and Miyauchi, K. (1996c) Antibacterial activity of xanthones from Guttiferaeous plants against methicillin-resistant *Staphylococcus aureus. Journal of Pharmacy and Pharmacology* 48, 861–865.
Iinuma, M., Tosa, H., Tanaka, T., Kanamaru, S., Asai, F., Kobayashi, Y., Miyauchi, K. and Shimano, R. (1996d) Antibacterial activity of some *Garcinia* benzophenone derivatives against methicillin-resistant *Staphylococcus aureus. Biological and Pharmaceutical Bulletin* 19, 311–314.
Ilyas, M., Kamil, M., Parveen, M. and Khan, M.S. (1994) Isoflavones from *Garcinia nervosa. Phytochemistry* 36, 807–809.
Ito, C., Itoigawa, M., Mishina, Y., Tomiyasu, H., Litaudon, M., Coson, J.-P., Mukainaka, T., Tokuda, H., Nishino, H. and Furukawa, H. (2001) Cancer chemopreventive agents. New depsidones from *Garcinia* plants. *Journal of Natural Products* 64, 147–150.
Ito, C., Itoigawa, M., Miyamoto, Y., Onoda, S., Rao, K.S., Mukinaka, T., Tokuda, H., Nishino, H. and Furukawa, H. (2003a) Polyprenylated benzophenones from *Garcinia assigu* and their potential cancer chemopreventive activities. *Journal of Natural Products* 66, 206–209.
Ito, C., Itoigawa, M., Takakura, T., Ruangrungsi, N., Enjo, F., Tokuda, H., Nishino, H. and Fukuwa, H. (2003b) Chemical constituents of *Garcinia fusca*: Structure elucidation of eight new xanthones and their cancer chemopreventive activity. *Journal of Natural Products* 66, 200–205.
Iwu, M.M. (1985) Antihepatotoxic constituents of *Garcinia kola* seeds. *Experientia* 41, 699–700.
Iwu, M.M. (1986) Biflavanones of *Garcinia:* Pharmacological and biological activities. *Progress in Clinical and Biological Research* 213, 485–488.
Iwu, M.M. (1993) *CRC Handbook of African Medicinal Plants*, CRC Press, Boca Raton, Florida, USA.
Iwu, M.M., Igboko, O.A., Onwuchekwa, U.A. and Okunji, C.O. (1987) Evaluation of the antihepatotoxic activity of the biflavonoids of *Garcinia kola* seeds. *Journal of Ethnopharmacology* 21, 127–138.
Iwu, M.M., Igboko, O.A., Okunji, C.O. and Tempesta, M.S. (1990) Antidiabetic and aldose reductase activities of biflavones of *Garcinia kola. Journal of Pharmacy and Pharmacology* 42, 290–292.
Jabit, M.L., Khalid, R., Abas, F., Shaari, K., Hui, L.S., Stanslas, J. and Lajis, N.H. (2007) Cytotoxic xanthones of *Garcinia penangiana* Pierre. *Zeitschrift fur Naturforschung C* 62, 786–792.
Jabit, M.L., Wahyuni, F.S., Khalid, R., Israf, D.A., Shaari, K., Lajis, N.H. and Stanslas, J. (2009) Cytotoxic and nitric oxide inhibitory activities of methanol extracts of *Garcinia* species. *Pharmaceutical Biology* 47, 1019–1026.
Jantan, I. (2004) Medicinal plant research in Malaysia: scientific interests and advances. *Journal Sains Kesihatan Malaysia* 2, 27–46.
Jayanthi, V., Anand, L., Ashok, L. and Srinivasan, V. (2005) Dietary factors in pathogenesis of gallstone disease in Southern India – a hospital-based case-control study. *Indian Journal of Gastroenterolgy* 24, 97–99.
Jayaprakasha, G.K. and Sakariah, K.K. (2002) Determination of organic acids in leaves and fruit rinds of *Garcinia indica* (Desr.) by LC. *Journal of Pharmaceutical and Biomedicinal Analysis* 28, 379–384.
Jayaprakasha, G.K., Jena, B.S. and Sakariah, K.K. (2003) Improved liquid chromatographic method for determination of organic acids in leaves, pulp, fruits, and rinds of *Garcinia pedunculata. Journal of the Association of Official Agricultural Chemists* 86, 1063–1068.
Jena, B.S., Jayaprakasha, G.K. and Sakariah, K.K. (2002a) Organic acids from leaves, fruits and rinds of *Garcinia cowa. Journal of Agricultural and Food Chemistry* 50, 3431–3434.
Jena, B.S., Jayaprakasha, G.K., Singh, R.P. and Sakariah, K.K. (2002b) Chemistry and biochemistry of (-)-hydroxycitric acid from *Garcinia. Journal of Agricultural and Food Chemistry* 50, 10–22.
Joseph, G.S., Jayaprakasha, G.K., Selvi, A.T., Jena, B.S. and Sakariah, K.K. (2005) Antiaflatoxigenic and antioxidant activities of *Garcinia* extracts. *International Journal of Food Microbiology* 25, 153–160.
Jung, H.A., Su, B.N., Keller, W.J., Mehta, R.G. and Kinghorn, A.D. (2006) Antioxidant xanthones from the pericarp of *Garcinia mangostana* (Mangosteen). *Journal of Agricultural and Food Chemistry* 54, 2077–2082.
Kaikabo, A.A., Samuel, B.B. and Eloff, J.N. (2009) Isolation and activity of two antibacterial biflavonoids from leaf extracts of *Garcinia livingstonei* (Clusiaceae). *Natural Product Communications* 4, 1363–1366.
Kamat, N.M. (2006) *Non-traditional products from Kokum: Inland and global opportunities. Proceedings of the 2005 Second National Seminar on Kokum (Garcinia indica Choisy)*, Western Ghats Kokum Foundation and University of Goa, Taleigao, Goa, India.

Kamdemwaffo, A.F., Mulholland, D., Wansi, J.D., Mbaze, L.M., Powo, R., Mpondo, T.N., Fomum, Z.T., Konig, W. and Nkengfack, A.E. (2006) Afzeliixanthones A and B, two new prenylated xanthones from *Garcinia afzelii* ENGL. (Guttiferae). *Chemical and Pharmaceutical Bulletin* 54, 448–451.
Kapadia, G.J., Oguntimein, B. and Shukla, Y.N. (1994) High-speed counter-current chromatographic separation of biflavonoids from *Garcinia kola* seeds. *Journal of Chromatography A* 673, 142–146.
Khan, M.H. and Yadava, P.S. (2010) Herbal remedies of asthma in Thoubal District of Manipur in North East India. *Indian Journal of Natural Products and Resources* 1, 80–84.
Kijjoa, A., Gonzalez, M.J., Pinto, M.M., Nascimento, M.S.J., Campos, N., Mondranondra, I.O., Silva, A.M.S., Eaton, G. and Herz, W. (2008) Cytotoxicity of prenylated xanthones and other constituents from the wood of *Garcinia merguensis*. *Planta Medica* 74, 864–866.
Komguem, J., Meli, A.L., Manfouo, R.N., Lontsi, D., Ngounou, F.N., Kuete, V., Kamdem, H.W., Tone, P., Ngadjui, B.T., Sondengam, B.L. and Connolly, J.D. (2005) Xanthones from *Garcinia smeathmanni* (Oliver) and their antimicrobial activity. *Phytochemistry* 66, 1713–1717.
Kondo, M., Zhang, L., Ji, H., Kou, Y. and Ou, B. (2009) Bioavailability and antioxidant effects of a xanthone-rich Mangosteen (*Garcinia mangostana*) product in humans. *Journal of Agricultural and Food Chemistry* 57, 8788–8792.
Konoshima, M. and Ikeshiro, Y. (1970) Fukugiside, the first biflavonoid glycoside from *Garcinia spicata* Hook.f. *Tetrahedron Letters* 20, 1717–1720.
Koshy, A.S. and Vijayalakshmi, N.R. (2001) Impact of certain flavonoids on lipid profiles – potential action of *Garcinia cambogia* flavonoids. *Phytotherapy Research* 15, 395–400.
Kumar, S., Sharma, S. and Chattopadhyay, S.K. (2009) High-performance liquid chromatography and LC-ESI-MS method for identification and quatification of two isomeric polyisoprenylated benzophenones isoanthochymol and camboginol in different extracts of *Garcinia* species. *Biomedical Chromatography* 23, 888–897.
Lannang, A.M., Komguem, J., Ngninzeko, F.N., Tangmouo, J.G., Ajaz, A., Choudhary, M.I., Ranjit, R., Devkota, K.P. and Sondengam, B.L. (2005) Bangangxanthone A and B, two xanthones from the stem bark of *Garcinia polyantha* Oliv. *Phytochemistry* 66, 2351–2355.
Lannang, A.M., Louh, G.N., Lontsi, D., Sprecht, S., Sarite, S.R., Florke, U., Hassain, H., Hoerauf, A. and Krohn, K. (2008) Antimalarial compounds from the root bark of *Garcinia polyantha* Oliv. *Journal of Antibiotics* 61, 518–523.
Lenz, T.L. and Hamilton, W.R. (2003) Supplemental products used for weight loss. *Journal of the American Pharmaceutical Association* 44, 59–67.
Lewis, Y.S. and Neelakantan, S. (1965) (-)-Hydroxycitric acid – the principal acid in the fruits of *Garcinia cambogia* desr. *Phytochemistry* 4, 619–625.
Li, S.L., Song, J.Z., Han, Q.B., Qiao, C.F. and Xu, H.X. (2008) Improved high-performance liquid chromatographic method for simultaneous determination of 12 cytotoxic caged xanthones in gamboges, a potential anticancer resin from *Garcinia hanburyi*. *Biomedical Chromatography* 22, 637–644.
Liao, C.H., Sang, S., Hp, C.T. and Lin, J.K. (2005) Garcinol modulates tyrosine phosphorylation of FAK and subsequently induces apoptosis through down-regulation of Src, ERK, and Akt survival signaling in human colon cancer cells. *Journal of Cell Biology* 96, 155–169.
Likhitwitayawuid, K., Chanmahasathien, W., Ruangrungsi, N. and Krungkrai, J. (1998a) Xanthones with antimalarial activity from *Garcinia dulcis*. *Planta Medica* 64, 281–282.
Likhitwitayawuid, K., Phadungcharoen, T. and Krungkrai, J. (1998b) Antimalarial xanthones from *Garcinia cowa*. *Planta Medica* 64, 70–72.
Lin, C-N., Kiang, C-W., Lu, C-M., Wu, R-R. and Lee, K-K. (1996) Garcinielliptin oxide, a triterpenoid with a novel skeleton, isolated from *Garcinia subelliptica*. *Chemical Communications* 1996, 1315–1316.
Lin, Y.M., Anderson, H., Flavin, M.T., Pai, Y.H.S., Mata-Greenwood, E., Pengsuparp, T., Pezzuto, J.M., Schinazi, R.F., Hughes, S.H. and Chen, F.C. (1997) *In vitro* anti-HIV activity of biflavonoids isolated from *Rhus succedanea* and *Garcinia multiflora*. *Journal of Natural Products* 60, 884–888.
Lin, Y.M., Flavin, M.T., Schure, R., Chen, F.C., Sidwell, R., Barnard, D.L., Huffman, J.H. and Kern, E.R. (1999) Antiviral activities of biflavonoids. *Planta Medica* 65, 120–125.
Louh, G.N., Lannang, A.M., Mbazoa, C.D., Tamgmouo, J.G., Komguen, J., Castilho, P., Ngninzeko, F.N., Qamar, N., Lontsi, D., Choudhary, M.I. and Sondengam, B.L. (2008) Polyanxanthone A, B and C, three xanthones from the wood trunk of *Garcinia polyantha* Oliv. *Phytochemistry* 69, 1013–1017.
Mackeen, M.M., Ali, A.M., Lajis, N.H., Kawazu, K., Kikuzaki, H. and Nakatani, N. (2002) Antifungal garcinia acid esters from the fruits of *Garcinia atroviridis*. *Zeitschrift fur Naturforschung* C 57, 291–295.

MacLeod, A.J. and Pieris, N.M. (1982) Volatile flavor components of *Garcinia mangostana. Phytochemistry* 21, 117–119.

Magadula, J.J., Kapingu, M.C., Bezabih, M. and Abegaz, B.M. (2008) Polyisoprenylated benzophenones from *Garcinia semseii* (Clusiaceae). *Phytochemistry Letters* 1, 215–218.

Mahabusarakam, W., Chairerk, P. and Taylor, W.C. (2005) Xanthones from *Garcinia cowa* Roxb. latex. *Phytochemistry* 66, 1148–1153.

Marcason, W. (2006) What are the facts and myths about mangosteen? *Journal of the American Dietetic Association* 106, 986.

Martins, F.T., Doriguetto, A.C., de Souza, T.C., de Souza, K.R., Dos Santos, M.H., Moreira, M.E. and Barbosa, L.C. (2008) Composition, and anti-inflammatory and antioxidant activities of the volatile oil from the fruit peel of *Garcinia brasiliensis. Chemical and Biodiversity* 5, 252–258.

Matsumoto, K., Akao, Y., Hong, Y., Ohguchi, K., Ito, T., Tanaka, T., Kobayashi, E., Iinuma, M. and Nozawa, Y. (2004) Preferential target is mitochondria in alpha-magostin-induced apoptosis in human leukemia HL60 cells. *Bioorganic and Medicinal Chemistry* 12, 5799–5806.

Mbafor, J.T., Fomum, Z.T., Promsattha, R., Sanson, D.R. and Tempensta, M.S. (1989) Isolation and characterization of taxifolin 6-C-glucoside from *Garcinia epunctata. Journal of Natural Products* 52, 417–419.

Mbwambo, Z.H., Kapingu, M.C., Moshi, M.J., Machumi, F., Apers, S., Cos, P., Ferreira, D., Marais, J.P.J., Vanden Berghe, D., Maes, L., Vlietinck, A. and Pieters, L. (2006) Antiparasitic activity of some xanthones and biflavonoids from the root bark of *Garcinia livingstonei. Journal of Natural Products* 69, 369–372.

Merza, J., Mallet, S., Dumontet, V., Seraphin, D. and Richomme, P. (2006) New cytotoxic guttiferone analogues from *Garcinia virgata* from New Caledonia. *Planta Medica* 72, 87–89.

Minami, H., Kinoshita, M., Fukuyama, Y., Kodama, M., Yoshizawa, T., Sugiura, M., Nakagawa, K. and Tago, H. (1994) Antioxidant xanthones from *Garcinia subelliptica. Phytochemistry* 36, 501–506.

Minami, H., Takahashi, E., Fukuyama, Y., Kodama, M., Yoshizawa, T. and Nakagawa, K. (1995) Novel xanthones with superoxide scavenging activity from *Garcinia subelliptica. Chemical and Pharmaceutical Bulletin* 43, 347–349.

Nakatani, K., Atsumi, M., Arakawa, T., Oosawa, K., Shimura, S., Nakahata, N. and Ohizumi, Y. (2002) Inhibitions of histamine release and prostaglandin E2 synthesis by mangosteen, a Thai medicinal plant. *Biological and Pharmaceutical Bulletin* 25, 1137–1141.

Naldoni, F.J., Caludino, A.L., Cruz, J.W., Jr., Chavasco, J.K., Faria e Silva, P.M., Veloso, M.P. and Dos Santos, M.H. (2009) Antimicrobial activity of benzophenones and extracts from the fruits of *Garcinia brasiliensis. Journal of Medicinal Food* 12, 403–407.

Nanditha, B. and Prabhasankar, P. (2009) Antioxidants in bakery products: A review. *Critical Reviews in Food Science and Nutrition* 49, 1–27.

Ndukwe, K.C., Okeke, I.N., Lamikanra, A., Adsina, S.K. and Aboderin, O. (2005) Antibacterial activity of aqueous extracts of selected chewing sticks. *Journal of Contemporary Dental Practice* 15, 86–94.

Negi, P.S., Jayaprakasha, G.K. and Jena, B.S. (2008) Antibacterial activity of extracts from the fruit rinds of *Garcinia cowa* and *Garcinia pedunculata* against food borne pathogens and spoilage bacteria. *Food Science and Technology* 41, 1857–1861.

Neves, J.S., Coelho, L.P., Cordeiro, R.S., Veloso, M.P., Rodrigues e Silva, P.M., dos Santos, M.H. and Martins, M.A. (2007) Antianaphylactic properties of 7-epiclusianone, a tetraprenylated benzophenone isolated from *Garcinia brasiliensis. Planta Medica* 73, 644–649.

Ngoupayo, J., Tabopda, T.K., Ali, M.S. and Tsamo, E. (2008) Alpha-glucosidase inhibitors from *Garcinia brevipedicellata* (Clusiaceae). *Chemical and Pharmaceutical Bulletin* 56, 1466–1469.

Ngoupayo, J., Tabopda, T.K. and Ali, M.S. (2009) Antimicrobial and immunomodulatory properties of prenylated xanthones from twigs of G*arcinia staudtii. Bioorganic and Medicinal Chemistry* 17, 5688–5695.

Nguyen, L.H.D. and Harrison, L.J. (2000) Xanthones and triterpenoids from the bark of *Garcinia vilersiana. Phytochemistry* 53, 111–114.

Nguyen, L.H.D., Vo, H.T., Pham, H.D., Connolly, J.D. and Harrison, L.J. (2003) Xanthones from the bark of *Garcinia merguensis. Phytochemistry* 63, 467–470.

Nguyen, L.H.D., Venkataraman, G., Sim, K.Y. and Harrison, L.J. (2005) Xanthones and benzophenones from *Garcinia griffithii* and *Garcinia mangostana. Phytochemistry* 66, 1718–1723.

Nilar, H.L.J. (2002) Xanthones from the heartwood of *Garcinia mangostana. Phytochemistry* 60, 541–548.

Niwa, M., Terashima, K., Ito, J. and Aqil, M. (1994) Two novel arylbenzofurans, garcifuran-A and garcifuran-B from *Garcinia kola. Heterocycles* 38, 1071–1076.
Obolskiy, D., Pischel, I., Siriwatanametanon, N. and Hienrich, M. (2009) *Garcinia mangostana* L: A phytochemical and pharmacological review. *Phytochemistry Research* 23, 1047–1065.
Okoko, T. (2009) *In vitro* antioxidant and free radical scavenging activities of *Garcinia kola* seeds. *Food and Chemical Toxicology* 47, 2620–2623.
Okunji, C.O. and Iwu, M.M. (1991) Molluscidal activity of *Garcinia kola* biflavones. *Fitoterapia* LXII, 74–76.
Okunji, C., Komarnytsky, S., Fear, G., Poulev, A., Ribnicky, D.M., Awachie, P.I., Ito, Y. and Raskin, I. (2007) Preparative isolation and identification of tyrosinase inhibitors from the seeds of *Garcinia kola* by high-speed counter-current chromatography. *Journal of Chromatography A* 1151, 45–50.
Okunji, C.O., Ware, T.A., Hicks, R.P., Iwu, M.M. and Skanchy, D.J. (2002) Capillary electrophoresis determination of biflavonones from *Garcinia kola* in three traditional African medicinal formulations. *Planta Medica* 68, 440–444.
Orie, N.N. and Ekon, E.U. (1993) The bronchodilator effect of *Garcinia kola. East African Medical Journal* 70, 143–145.
Padhye, S., Ahmad, A., Oswal, N. and Sarkar, F.H. (2009) Emerging role of garcinol, the antioxidant chalcone from *Garcinia indica* Choisy and its synthetic analogs. *Journal of Hematology and Oncology* 2, 38–51.
Panthong, A., Norkaew, P., Kajanapothi, D., Taesotikul, T., Anantachoke, N. and Reutrakul, V. (2007) Anti-inflammatory, analgesic and antipyretic activities of the extract of gamboges from *Garcinia hanburyi* Hook f. *Journal of Ethnopharmacology* 111, 335–340.
Parveen, M., Ilyas, M., Mushfiq, M., Busudan, O.A. and Muhaisen, H.M. (2004) A new biflavonoid from leaves of *Garcinia nervosa. Natural Product Research* 18, 269–275.
Pattalung, P., Thongtheeraparp, W., Wiriyachitra, P. and Taylor, W.C. (1994) Xanthones of *Garcinia cowa. Planta Medica* 60, 365–368.
Pedraza-Chaverri, J., Cardenas-Rodriguez, N., Orzco-Ibarra, M. and Perez-Rojas, J.M. (2008) Medicinal properties of mangosteen (*Garcinia mangostana*). *Food and Chemical Toxicology* 46, 3227–3239.
Pereira, I.O., Marques, M.J., Pavan, A.L., Codonho, B.S., Barbieri, C.L., Beijo, L.A., Doriguetto, A.C., D'Martin, E.C. and dos Santos, M.H. (2009) Leishmanicidal activity of benzophenones and extracts from *Garcinia brasiliensis* Mart. fruits. *Phytomedicine* 17, 339–345.
Permanaa, D., Lajis, N.H., Mackeen, M.M., Ali, A.M., Aimi, N., Kitajima, M. and Takayama, H. (2001) Isolation and bioactivities of constituents of the roots of *Garcinia atroviridis. Journal of Natural Products* 64, 976–979.
Permanaa, D., Abas, F., Maulidiani, Shaari, K., Stanslas, J., Ali, A.M. and Lajis, N.H. (2005) Atrovirisidone B, a new prenylated depsidone with cytotoxic property from the roots of *Garcinia atroviridis. Zeitschrift fur Naturforschung C* 60, 523–526.
Phongpaichit, S., Rungjindamai, N., Rukachaisirikul, V. and Skayaroj, J. (2006) Antimicrobial activity in cultures of endophytic fungi isolated from *Garcinia* species. *Immunology and Medical Microbiology* 48, 367–372.
Pittler, M.H. and Ernst, E. (2004) Dietary supplements for body-weight reduction: A systematic review. *American Journal of Clinical Nutrition* 79, 529–536.
Preus, H.G., Rao, C.V., Garis, R., Bramble, J.D., Ohia, S.E., Bagchi, M. and Bagchi, D. (2004) An overview of the safety and efficacy of a novel, natural (-)-hydroxycitric acid (HCA-SX) for weight management. *Journal of Medicine* 35, 33–48.
Rao, A.V.R., Venkataswamy, G. and Pendse, D. (1980) Camboginol and cambogin. *Tetrahedron Letters* 21, 1975–1978.
Rassameemasmaung, S., Sirikulsathean, A., Amornchat, C., Hirunrat, K., Rojanapanthu, P. and Gritsanapan, W. (2007) Effects of herbal mouthwash containing the pericarp extract of *Garcinia mangostana* L. on halitosis, plaque and papillary bleeding index. *Journal of International Academy of Periodontology* 9, 19–25.
Rassameemasmaung, S., Sirikulsathean, A., Amornchat, C., Maungmingsook, P., Rojanapanthu, P. and Gritsanapan, W. (2008) Topical application of *Garcinia mangostana* L. pericarp gel as an adjunct to periodontal treatment. *Complementary Therapies in Medicine* 16, 262–267.
Roux, D., Hadi, H.A., Thoret, S., Guenard, D., Thoison, O., Pais, M. and Sevenet, T. (2000) Structure-activity relationship of polyisoprenyl benzophenones from *Garcinia pyrifera* on the tubulin/microtubule system. *Journal of Natural Products* 63, 1070–1076.
Rukachaisirikul, V., Kamkaew, M., Sukavisit, D., Phongpaichit, S., Sawangchote, P. and Taylor, W.C.

(2003a) Antibacterial xanthones from the leaves of *Garcinia nigrolineata. Journal of Natural Products* 66, 1531–1535.

Rukachaisirikul, V., Pailee, P., Hiranrat, A., Tuchinda, P., Kasisit, J., Taylor, W.C. and Reutrakul, V. (2003b) Anti-HIV-1 protostane triterpenes and digeranylbenzophenone from trunk bark of *Garcinia speciosa. Planta Medica* 69, 1141–1146.

Rukachaisirikul, V., Phainuphong, P., Sukpondma, Y., Koysomboon, S., Sawangchote, P. and Taylor, W.C. (2003c) Caged-triprenylated and –tetraprenylated xanthones from the latex of *Garcinia scortechinii. Journal of Natural Products* 66, 933–938.

Rukachaisirikul, V., Ritthiwigrom, T., Pinsa, A., Sawangchote, P. and Taylor, W.C. (2003d) Xanthones from the stem bark of *Garcinia nigrolineata. Phytochemistry* 64, 1149–1156.

Rukachaisirikul, V., Naklue, W., Sukpondma, Y. and Phongpaichit, S. (2005a) An antibacterial biphenyl derivative from *Garcinia bancana* MIQ. *Chemical and Pharmaceutical Bulletin* 53, 342–343.

Rukachaisirikul, V., Phainuphong, P., P., Sukpondma, Y., Phongpaichit, S. and Taylor, W.C. (2005b) Antibacterial caged-tetraprenylated xanthones from the stem bark of *Garcinia scortechinii. Planta Medica* 71, 165–170.

Rukachaisirikul, V., Saelim, S., Karnsomchoke, P. and Phongpaichit, S. (2005c) Friedolanostanes and lanostanes from the leaves of *Garcinia hombroniana. Journal of Natural Products* 68, 1222–1225.

Rukachaisirikul, V., Tadpetch, K., Wattanaphanit, A., Saengsanae, N. and Phongpaichit, S. (2005d) Benzopyran, biphenyl, and tetraoxygenated xanthone derivatives from the twigs of *Garcinia nigrolineata. Journal of Natural Products* 68, 1218–1221.

Rukachaisirikul, V., Naklue, W., Phongpaichit, S., Towantana, N.H. and Maneenoon, K. (2006) Phloroglucinols, depsidones and xanthones from the twigs of *Garcinia parvifolia. Tetrahedron* 62, 6578–6585.

Rukachaisirikul, V., Trisuwan, K., Sukpondma, Y. and Phongpaichit, S. (2008) A new benzoquinone derivative from the leaves of *Garcinia parvifolia. Archives of Pharmacal Research* 31, 17–20.

Saito, M., Ueno, M., Kubo, K., Nagata, J. and Takeuchi, M. (2005) High dose of *Garcinia cambogia* is effective in suppressing fat accumulation in developing male Zucker rats, but highly toxic to the testis. *Food and Chemical Toxicology* 43, 411–419.

Sang, S., Liao, C-H., Pan, M-H., Rosen, R.T., Lin-Shiau, S-Y., Lin, J-K. and Ho, C-T. (2002) Chemical studies on antioxidant mechanism of garcinol: Analysis of radical reaction products of garcinol with peroxy radicals and their antitumor activities. *Tetrahedron* 58, 10095–10102.

Sani, B.P. and Rao, P.L. (1966) Antibiotic principles of *Garcinia morella.* VII. Antiprotozoal activity of morellin, neomorellin and other insoluble neutral phenols of the seed coat of *Garcinia morella. Indian Journal of Experimental Biology* 4, 27–28.

Santhanam, K. and Rao, P.L. (1969) Antibiotic principles of *Garcinia morella.* XIII. Antimicrobial activity and toxicity of alpha 1 and γ-guttiferins and their derivatives. *Indian Journal of Experimental Biology* 7, 34–36.

Schmid, W. (1855) Uber das mangostin. *Justus Liebigs Annalen der Chemie* 93, 83–88.

Selvi, A.T., Joseph, G.S. and Jayaprakasha, G.K. (2003) Inhibition of growth and aflatoxin production in *Aspergillus flavus* by *Garcinia indica* extract and its antioxidant activity. *Food Microbiolgy* 20, 455–460.

Seth, M.K. (2004) Trees and their economic importance. *Botanical Review* 69, 321–376.

Shadid, K.A., Shaari, K., Abas, F., Israf, D.A., Hamzah, A.S., Syakroni, N., Saha, K. and Lajis, N.H. (2007) Cytotoxic caged-polyprenylated xanthonoids and a xanthone from *Garcinia cantleyana. Phytochemistry* 68, 2537–2544.

Shankaranarayanan, D., Gopalakrishnan, C. and Kameswaran, L. (1979) Pharmacological profile of mangostin and its derivatives. *Archives of International Pharmacodynamics and Therapeutics* 239, 257–269.

Siva, R. (2007) Status of natural dyes and dye-yielding plants in India. *Current Science* 92, 916–925.

Sordat-Diserens, I., Marston, A., Hamburger, M., Hostettmann, K. and Rogers, C. (2004) Novel prenylated xanthones from *Garcinia gerrardii* Harvey. *Helvetica Chimica Acta* 72, 1001–1007.

Sordat-Diserens, I., Rogers, C., Sordat, B. and Hostettman, K. (1992) Prenylated xanthones from *Garcinia livingstonei. Phytochemistry* 31, 313–316.

Stevens, T., Qadri, A. and Zein, N.N. (2005) Two patients with acute liver injury associated with use of the herbal weight-loss supplement hydroxycut. *Annals of Internal Medicine* 42, 477–478.

Stohs, S.J., Preuss, H.G., Ohia, S.E., Kaats, G.R., Keen, C.L., Williams, L.D. and Burdock, G.A. (2009) No evidence of demonstrating hepatotoxicity of hydroxycitric acid. *World Journal of Gastroenterology* 15, 4087–4089.

Suffredini, I.B., Paciencia, M.L.B., Nepomuceno, D.C., Younes, R.N. and Varella, A.D. (2006) Antibacterial and cytotoxic activity of Brazilian plant extracts – Clusiaceae. *Memoriasdo Instituto Oswaldo Cruz, Rio de Janiero* 101, 287–290.

Sukpondma, Y., Rukachaisirikul, V. and Phongpaichit, S. (2005) Xanthone and sesquiterpene derivatives from the fruits of *Garcinia scrotechinii. Journal of Natural Products* 68, 1010–1017.

Suksamram, S., Suwannapoch. N., Phakhodee, W., Thanuhiranlert, J., Ratnanukul, P., Chimnoi, N. and Suksamram, A. (2003) Antimycobacterial activity of prenylated xanthones from the fruits of *Garcinia mangostana. Chemical and Pharmaceutical Bulletin* 51, 857–859.

Suksamram, S., Komutiban, O., Ratananukul, P., Chimnoi, N., Lartpornmatulee, N. and Suksamram, A. (2006) Cytotoxic prenylated xanthones from the young fruit of *Garcinia mangostana. Chemical and Pharmaceutical Bulletin* 54, 301–305.

Sundaram, B.M., Gopalakrishnan, C., Subramanian, S., Shankaranarayanan, D. and Kameswaran, L. (1983) Antimicrobial activities of *Garcinia mangostana. Planta Medica* 48, 59–60.

Taiwo, O., Xu, H.X. and Lee, S.F. (1999) Antibacterial activities of extracts from Nigerian chewing sticks. *Phytotherapy Research* 13, 675–679.

Tao, S-J., Guan, S-H., Wang, W., Lu, Z-Q., Chen, G-T., Sha, N., Yue, Q-X., Liu, X. and Guo, D-A. (2009) Cytotoxic polyprenylated xanthones from the resin of *Garcinia hanburyi. Journal of Natural Products* 72, 117–124.

Thadhani, V.M., Jansz, E.R. and Peiris, H. (2002) Effect of exogenous histidine and *Garcinia cambogia* on histamine formation in skipjack (*Katsowonus pelamis*) homogenates. *International Journal of Food Sciences and Nutrition* 53, 29–34.

Thoison, O., Fehy, J., Dumontet, V., Chiaroni, A., Riche, C., Tri, M.V. and Sevenet, T. (2000) Cytotoxic prenylxanthones from *Garcinia bracteata. Journal of Natural Products* 63, 441–446.

Tona, L., Ngimbi, N.P., Tsakala, M., Mesia, K., Cimanga, K., Apers, S., De Bruyne, T., Pieters, L., Totte, J. and Vlietinck, A.J. (1999) Antimalarial activity of 20 crude extracts from nine African medicinal plants used in Kinshasa, Congo. *Journal of Ethnopharmacology* 68, 193–203.

Toromanyan, E., Aslanyan, G., Amroyan, E., Gabrielyan, E. and Panossian, A. (2007) Efficacy of Slim 339 in reducing body weight of overweight and obese human subjects. *Phytotherapeutical Research* 21, 1177–1181.

Upadhyaya, A., Nayak, P.S., Agarwal, V.K. and Tiwari, A.B. (2009) *Healing Herbs of India,* Satish Serial Publishing House, New Delhi, India.

Vieira, L.M., Kijjoa, A., Wilairat, R., Nascimento, M.S.J., Gales, L., Damas, A.M., Silva, A.M.S., Mondranondra, I.O. and Herz, W. (2004) Bioactive friedolanostanes and 11(10 $\rightarrow$ 8)-abeolanostanes from the bark of *Garcinia speciosa. Journal of Natural Products* 67, 2043–2047.

Waterman, P.G. and Crichton, E.G. (1980) Xanthones, benzophenones and triterpenes from the stem bark of *Garcinia ovalifolia. Planta Medica* 40, 351–355.

Wang, L.L., Li, Z.L., Hua, H.M. and Liu, X.Q. (2008a) Chemical constituents from barks of *Garcinia tetralanta. China Journal of Chinese Materia Medica* 33, 2350–2352.

Wang, L.L., Li, Z.L., Song, D.D., Sun, L., Pei, Y.H., Jing, Y.K. and Hua, H.M. (2008b) Two novel triterpenoids with antiproliferative and apoptotic activities in human leukemia cells isolated from the resin of *Garcinia hanburyi. Planta Medica* 74, 1735–1740.

Weng, J.R., Tsao, L.T., Wang, J.P., Wu, R.R. and Lin, C.N. (2004) Anti-inflammatory phloroglucinols and terpenoids from *Garcinia subelliptica. Journal of Natural Products* 67, 1796–1799.

Wiart, C. (2009) *Medicinal Plants from the East.* Nottingham University Press, Nottingham, UK.

Williams, P., Ongsakul, M., Proudfoot, J., Croft, K. and Beilin, L. (1995) Mangostin inhibits the oxidative modification of human low density lipoprotein. *Free Radical Research* 23, 175–184.

Williams, R.B., Hoch, J., Glass, T.E., Evans, R., Miller, J.S., Wisse, J.H. and Kinghorn, D.G. (2003) A novel cytotoxic guttiferone analogue from *Garcinia macrophylla* from the Suriname rainforest. *Planta Medica* 69, 846–866. Erratum (2005) *Planta Medica* 71, 354.

Wolfrom, M.L., Dickey, E.E., McWain, P., Thompson, A., Looker, J.H., Windrath, O.M. and Komitsky, Jr., F. (1964) Osage orange pigments XIII. Isolation of three new pigments from the bark. *Journal of Organic Chemistry* 29, 689–691.

Wong, L.P. and Klemmer, P.J. (2008) Severe lactic acidosis associated with juice of the mangosteen fruit *Garcinia mangostana. Journal of the National Kidney Foundation* 51, 829–833.

Wu, C.C., Weng, J.R., Won, S.J. and Lin, C.N. (2005) Constituents of the pericarp of *Garcinia subelliptica. Journal of Natural Products* 68, 1125–1127.

Xu, G., Feng, C., Zhou, Y., Han, Q.B., Qiao, C.F., Huang, S.X., Chang, D.C., Zhao, Q.S., Luo, K.Q. and Xu, H.X. (2008) Bioassay and ultra-performance liquid chromatography/mass spectrometry guided isolation of apoptosis-inducing benzophenones and xanthone from the pericarp of *Garcinia yunnanensis* Hu. *Journal of Agricultural and Food Chemistry* 56, 11144–11150.

Xu, Y.J., Lai, Y.H., Imiyabir, Z. and Goh, S.H. (2001) Xanthones from *Garcinia parvifolia. Journal of Natural Products* 64, 1191–1195.

Yamada, T., Hida, H. and Yamada, Y. (2007) Chemistry, physiological properties and microbial production of hydroxycitric acid. *Applied Microbiology and Biotechnology* 75, 977–982.

Yamaguchi, F., Saito, M., Ishikawa, H., Kataoka, S. and Ariga, T. (1999) Healthy foods and cosmetics. US Patent No. 5972357.

Yamaguchi, F., Ariga, T., Yoshimura, Y. and Nakazawa, H. (2000) Antioxidative and antiglycination activity of garcinol from *Garcinia indica* fruit rind. *Journal of Agricultural and Food Chemistry* 48, 180–185.

Yang, N.Y., Han, Q.B., Cao, X.W., Qiao, C.F., Song, J.Z., Chen, S.L., Yang, D.J., Yiu, H. and Xu, H.X. (2007) Two new xanthones isolated from the stem bark of *Garcinia lancilimba. Chemical and Pharmaceutical Bulletin* 55, 950–952.

Ye, D. and Kong, L. (1996) Selection of technology for processing steamed *Garcinia hunburyi* with high pressure. *China Journal of Chinese Materia Medica* 21, 472–473.

Zhong, F., Chen, Y., Wang, P., Feng, H. and Yang, G. (2009) Xanthones from the bark of *Garcinia xanthochymus* and their DPPH radical-scavenging activity. *Chinese Journal of Chemistry* 27, 74–80.

Zhou, Y., Han, Q.B., Song, J.Z., Qiao, C.F. and Xu, H.X. (2008) Characterization of polyprenylated xanthones in *Garcinia xipshuanbannaensis* using liquid chromatography coupled with electrospray ionization quadrupole time-of-flight tandem mass spectrometry. *Journal of Chromatography A* 1206, 131–139.

21 Predictive Modelling of Antimicrobial Effects of Natural Aromatic Compounds in Model and Food Systems

Nicoletta Belletti,* Sylvain Sado Kamdem, Rosalba Lanciotti and Fausto Gardini

Introduction

Essential oils (EOs) or the so-called volatile or ethereal oils (Guenther, 1948) are aromatic oily liquids obtained from plant organs, i.e. flower, bud, seed, leaf, twig, bark, herb, wood, fruit and root. The term 'essential oil' has been derived from the words *Quinta essentia* with medical use attributed to Paracelsus. According to Bakkali *et al.* (2008) they are volatile, natural, complex compounds characterized by a strong odour and are formed by plant as secondary metabolites. They are usually obtained by stream or hydro-distillation (nowadays performed at high or low pressure), first developed in the Middle Ages by Arabs. Liquid carbon dioxide or microwaves can also be used to extract EOs.

Known for their antiseptic properties, i.e. bactericidal, viricidal and fungicidal, and for their medicinal properties and their fragrance, they are used in the flavouring of foods and as antimicrobial, analgesic, sedative, anti-inflammatory, spasmolytic and traditional anaesthetic remedies (Bakkali *et al.*, 2008). However, although empirically recognized for centuries, the antimicrobial properties of EOs only came to scientific attention recently (Appendini and Hotchkiss, 2002; Burt, 2004; Serrano *et al.*, 2008). In fact, besides their use as food flavouring agents, EOs are now, together with their pure components, gaining increasing interest for their antimicrobial and antioxidant properties. Furthermore, most EOs are generally recognized as safe (GRAS) substances, which means that they could be widely accepted by consumers as natural alternatives (Cowan, 1999; Newberne *et al.*, 2000). The use of natural antimicrobials as substitutes for those from chemical synthesis has encouraged the search for new sources including EOs (Ruberto and Baratta, 2000; Capecka, *et al.*, 2005; Sacchetti *et al.*, 2005).

Spices (woody-stemmed plants) which have strong antimicrobial activity include allspice, cinnamon, clove, mustard and vanilla. Among herbs (green-stemmed plants), the higher antimicrobial activities were found in basil, oregano, rosemary, sage and thyme. Generally, the phenolic components of the EOs exhibit the antimicrobial activity (Deans *et al.*, 1995; Kim *et al.*, 1995a,b). Exceptions include: mustard where allyl groups and related isothiocyanates are responsible for antimicrobial action; and allicin in garlic and onion. These are both non-phenolic, aliphatic compounds. Phenolic compounds in olive oil (oleuropein) and tea-tree oil (terpenes), which are not classified as either spices or herbs, also show antimicrobial activity

* Corresponding author.

(Davidson and Naidu, 2000). Spices and herbs which show limited antimicrobial activity include: anise, bay (laurel), black pepper, cardamom, cayenne (red pepper), celery seed, chilli powder, coriander, cumin, curry powder, dill, fenugreek, ginger, juniper oil, mace, marjoram, mint, nutmeg, orris root, paprika, sesame, spearmint, tarragon and white pepper (Davidson and Naidu, 2000).

The literature concerning the antimicrobial activity of EOs against many microorganisms including several pathogens has notably increased in the past decade (Burt *et al.*, 2004; Bakkali *et al.* 2008; Tiwari *et al.*, 2009; Tajkarimi *et al.*, 2010). Recently, their potential role in food preservation has also been exploited (Vázquez *et al.*, 2001; Lanciotti *et al.*, 2004; Tajkarimi *et al.*, 2010).

The chemical profiles of the EO products differ not only in the number of molecules but also in the stereochemical types of molecules extracted, according to the type of extraction which in its turn is chosen according to the purpose of the use. The extraction product can vary in quality, quantity and in composition according to climate, soil composition, plant organ, age and vegetative cycle stage (Angioni *et al.*, 2006). Most of the commercialized EOs are chemotyped by gas chromatography and mass spectrometry analyses.

The composition, structure as well as functional groups, of the EOs have an important role in determining their antimicrobial activity. Usually compounds with phenolic groups are most effective (Deans *et al.*, 1995; Dorman and Deans, 2000; Holley and Patel, 2005). The oils of clove, oregano, rosemary, thyme, sage and vanilla have been often found to be most effective against microorganisms. They are generally more inhibitory against Gram-positive than against Gram-negative bacteria (Zaika, 1988; Mangena and Muyima, 1999; Marino *et al.*, 2001). While this is true for many EOs, there are some which are effective against both groups (oregano, clove, cinnamon and citral) (Kim *et al.*, 1995a; Sivropoulou *et al.*, 1996; Skandamis *et al.*, 2002). There are also some non-phenolic constituents of oils which are more effective, such as allyl isothiocyanate (Ward *et al.*, 1998) or quite effective against Gram-negative bacteria (garlic oil; Yin and Cheng, 2003). In addition, allyl isothiocyanate is effective against many fungi (Nielsen and Rios, 2000).

Plant EOs are usually mixtures of several components which are often synthesized by plants with protective purposes. For combating infectious or parasitic agents, plants synthesize secondary metabolites which may be present constitutively (Cowan, 1999; Rauha *et al.*, 2000) or generated from inactive precursors in response to stress (Sofos *et al.*, 1998). Preformed substances (constitutively inhibiting metabolites) in plant tissue include phenolic compounds, flavonols, flavonoids, glycosides, alkaloids, and even polyacetylenes. Stress-generated metabolites are stored as inactive precursors which are activated by hydrolases or oxidases, usually in the plant tissue. Examples are onion sulfoxides and mustard glucosinolates. In onion and garlic, the precursor alliin is converted by alliinase to yield allicin (which is antimicrobial) plus pyruvate and ammonia. In mustard and horseradish, precursor glucosinolates are converted by the enzyme myrosinase to yield a variety of isothiocyanates (as well as thiocyanate, nitriles and glucose) including the allyl form which is strongly antimicrobial (Delaquis and Mazza, 1995).The oils with high levels of eugenol (allspice, clove bud and leaf, bay, and cinnamon leaf), cinnamamic aldehyde (cinnamon bark and cassia oil), thymol and carvacrol (oregano and thyme) and citral (citrus fruits and lemongrass) are usually strong antimicrobials (Lis-Balchin *et al.*, 1998; Davidson and Naidu, 2000). The activity of sage and rosemary is due to borneol and other phenolics in the terpenic fraction. The volatile terpenes carvacrol, p-cymene and thymol are responsible for the antimicrobial activity of oregano, thyme and savory. In sage EO the antimicrobial activity is due to the presence of the terpene thejone while in rosemary it depends on a group of terpenes (borneol, camphore, 1,8 cineole, alpha-pinene, camphone, verbenonone and bornyl acetate) (Davidson and Naidu, 2000).

Some of the compounds involved in the plant defence mechanisms are produced throughout the lipoxygenase pathway that catalyses the oxygenation of unsaturated fatty acids, forming fatty acid hydroperoxides. In particular, aldehydes and the related alcohols are produced by the action of hydroperoxide lyases, isomerases and dehydrogenases. Many of the natural aromas of fruits and vegetables responsible for their 'green notes', such as hexanal, hexanol, 2-(*E*)-hexenal and 3-(Z)-hexenol, are six carbon compounds formed through this pathway. In addition, they are important constituents of the aroma of tomatoes, tea, strawberry, olive oil, grape, apples and pear (Gardini, *et al*, 2002). Some of these molecules are endowed with a strong antimicrobial activity both in model (Caccioni *et al.*, 1997; Gardini *et al.*, 2002) and real systems (Lanciotti *et al.*, 1999).

The pronounced antimicrobial activity of citrus EOs has been mainly explained with the presence of C10 and C15 terpenes (mono- and sesquiterpenes) having aromatic rings and a phenolic hydroxylic group able to form hydrogen bonds with active sites of target enzymes (Farag *et al.*, 1989; Juven *et al.*, 1994; Daferera *et al.*, 2000; Dorman and Deans, 2000). Nevertheless, other active terpenes, as well as alcohols, aldehydes and esters, can contribute to the overall antimicrobial effects of the EOs (Sikkema *et al.*, 1995). The relevant antimicrobial activity of citron (*Citrus medica*) and lemon myrtle (*Backhousia citriodora*) EOs was ascribed to their high citral content (Hayes and Markovic, 2002; Belletti *et al.*, 2004). Citral (3,7-dimethyl-2-7-octadienal), an acyclic α,β-unsaturated monoterpene aldehyde, is an isoprenoid compound with two isomers, geranial and neral, naturally occurring in citrus EOs and characterized by a broad-spectrum antimicrobial activity (Hayes and Markovic, 2002; Wuryatmo *et al.*, 2003).

However, the complexity and variability of EOs make it difficult, in general, to correlate their action to a specific component, as well as the possibility of antagonistic and synergistic effects. For this reason, Caccioni *et al.* (1998) proposed a holistic approach to explain the antimicrobial capabilities of citrus EOs, whose performances could be the result of a certain quantitative balance of various components.

While their antimicrobial activity *in vitro* is well demonstrated, the use of EOs to inhibit microbial growth in foods is less consolidated. In fact, the use of EOs to control microbial growth in food has been proposed for beverages (Fitzgerald *et al.*, 2004; Belletti *et al.*, 2007), meats (Singh *et al.*, 2003; Rižnar *et al.*, 2006; Chouliara *et al.*, 2007; Solomakos *et al.*, 2008), bakery products (Guynot *et al.*, 2003), fresh fruits and vegetables (Roller and Seedhar, 2002; Singh *et al.*, 2002; Belletti *et al.*, 2008), and salads (Skandamis and Nychas 2000). However, few are applied at the industrial level. This is mainly due to: the sensorial impact of the EOs that requires compatibility with food; the variability of their composition, which can be reflected in their antimicrobial potential; the need for a deeper knowledge of the mechanisms of action of their bioactive constituents; and their interactions with the food component (Gutierrez *et al.*, 2009) that do not guarantee a constant antimicrobial activity (Burt, 2004; Lanciotti *et al.*, 2004).

Predictive Microbiology

Safety and quality characteristics of food can be biased by the presence and the possible proliferation of pathogenic and/or spoilage microorganisms during the life cycle of the product (i.e. from the raw ingredients until the moment of consumption). The prediction of the microbial evolution in foods can be achieved through mathematical modelling, in the field of predictive microbiology (Van Impe *et al.*, 2005).

A general definition of the term predictive microbiology (PM) has been provided by McMeekin *et al.* (2002), who stated that predictive microbiology can be defined as the quantitative microbial ecology of foods. The fundamental concept starts with the assumption that both microbial growth and inactivation are reproducible responses. The responses of microorganisms depend on relatively few parameters (temperature, pH, water availability, and presence of salt or other food preservatives). In this view, a detailed comprehension of microbial responses to environmental and technological conditions can lead to an objective evaluation of the effect of processing, distribution and storage conditions on the dynamics of microbial growth. On the basis of this objective evaluation, it is possible to predict the consequences of microbial activities on the shelf life and/or safety of foods. In other words, the modelling process can be seen as a distillation of the existing microbiological knowledge into reliable mathematical models (Ross, 1999).

Maybe the first example of a predictive model can be found in the thermal processing of foods where heat-process conditions are set in order to destroy 10^{12} spores of *Clostridium botulinum* type A. This predictive model (based on a first-order death kinetic of microorganisms) was developed by Esty and Meyer (1922). Nowadays this approach is still widely accepted and applied by the canning industry in spite of the fact that this approach is only a 'simplification of reality'. In the past couple of decades it has been widely documented that thermal inactivation kinetics can be complicated by the presence of 'shoulders' and 'tails' (Peleg and Cole, 1998).

During the 1980s, some important food-poisoning outbreaks affected food production. Those outbreaks were related to the presence in food of both 'traditional' pathogens of foods (e.g. *Salmonella* in eggs) as well as emerging pathogens (e.g. *Listeria monocytogenes* able to grow in refrigerated foods) (Membré and Lambert, 2008). The increasing trend of food-poisoning outbreaks has augmented the interest for predictive microbiology that finally has been recognized as a scientific discipline. As reported previously, the typical application of PM is the assessment of bacterial growth, inactivation or survival as a function of intrinsic parameters (e.g. pH, water activity, acids, salt and preservatives), extrinsic parameters (chilling, modified atmosphere) or processing parameters (heat treatment, pressurization and irradiation) or their combinations.

The need to get an objective evaluation of the effect of the modulation of those parameters on the microbial growth has become more and more important. Recently, with the need for quantitative risk assessment applications, the demand to predict probabilities of survival and growth at low cell numbers also began to increase. The understanding of the effect of the combination of various factors on microbial behaviour is essential for the optimization of both process conditions and food formulation. In fact, the PM approach can help not only in the objective evaluation of the effect of a single parameter but can explain the effect of the combination of different parameters. Several predictive models for microbial inactivation have been developed with the specific objective to become a tool for food/process optimization. The use of predictive models as tools at an industrial level can be achieved only after a correct validation with independent data within the domain of the tested parameters. Where protocols based on predictions are selected to minimize processing and the safety factor is reduced, more rigorous selection and validation of models are required (McMeekin *et al.*, 2002). The validation procedures followed to evaluate/validate kinetic models can be performed by comparison of observed and predicted values. The comparison can be mathematically described by performance criteria such as the bias and accuracy factors defined by Ross (1996).

The need to predict microbial behaviour has become pressing as the preservation techniques have become milder or shifted to more 'natural' compounds. The past decade has been characterized by an increasing consumer demand for more natural-tasting food products, which means additive-free, fresher and minimally treated protocols but with a maintained microbiological safety (Gould, 1996). This trend has forced the search for new preservative techniques or for the combination of preservatives in the philosophy of the 'Hurdle Technology' concept in food safety proposed by Leistner (1995). The use of natural antimicrobial compounds and EOs, as preservation compounds, is thus of utmost interest in the food industry.

An efficient exploitation of such compounds at the industrial level is still under study. To speed up this process, as well as to help in the scaling up of the usage of EOs in food, the predictive modelling approach can be seen as a useful tool. However, at present, few predictive models of EO-induced microbial inactivation can be found. Similarly, few works aiming to study the inhibitory effect of EOs or their constituents, in combination with other barrier factors such as temperature and pH, on microorganisms can be found (Skandamis *et al.*, 2002; Belletti *et al.*, 2007; Char *et al.*, 2009).

Thus, although the philosophy of the 'hurdle concept' is widely accepted, its optimization requires the understanding of microbial physiology, which can also be achieved by the accumulation of quantitative information (e.g. predictive models) on microbial behaviour in foods (Koutsoumanis *et al.*, 2004). In addition, the practical exploitation of new antimicrobials, alone or in combination with different hurdles, is subordinate to answering the following questions:

- Product formulation: the microbial profile of the raw materials, pH and water activity, the presence of natural and other hurdles: are these factors capable of preventing bacterial growth, at least within shelf life?
- Process steps: how does each step influence the contamination, survival, inhibition or growth of microorganisms?
- Distribution and consumption conditions: how do they modify the quality and the safety of the products?
- How is it possible to control the process and the contamination level?

Numerous investigations have confirmed the antimicrobial action of EOs in model and real foods throughout a modelling approach (Lanciotti *et al.*, 2001; Skandamis *et al.*, 2002; Char *et al.*, 2009). The comprehension of the mechanisms of action of EOs can be better achieved in model systems, where it is quite easy to control and modulate single or multiple factors.

As already observed, many *in vitro* studies report a high efficacy of EOs against food-borne pathogens and spoilage bacteria (Smith-Palmer *et al.*, 1998; Dorman and Deans, 2000; Hammer *et al.*, 2002). Most of the models regarding the application of aroma compounds have been developed in liquid laboratory media. However, many authors report discrepancies between a model prediction and observed growth in real foods. In fact, it is generally accepted that a higher concentration of EO is needed to reach the same effect in food as *in vitro* (Burt, 2004). It is likely that models obtained *in vitro* often do not take into account significant factors for microbial growth such as food structure and composition (i.e. fat content), microbial interaction (pathogens and lactic acid bacteria) and physiological state of the microbial cells (Robins and Wilson, 1994; Gram and Melchiorsen, 1996; Pin, *et al.*, 1999; Gutierrez *et al.* 2009). Thus, the application of predictive models developed in laboratory media to foods requires extensive 'product validation' studies in order to account for the impact of the above factors on their performance (McMeekin *et al.*, 1993).

Furthermore, another aspect for the optimized application of EOs in foods is the impact on sensory acceptability. In many published research works the concentrations to be used to obtain a useful antimicrobial activity can result in unacceptable or inappropriate flavours and odours if applied to a food product. Therefore, research in this area should be focused on optimizing EO combinations and applications to obtain effective antimicrobial activity without adversely affecting the sensorial quality of foods.

The Modelling Process

A modelling approach is usually performed through the collection of growth or inactivation data for a defined target microorganism. The modulation of the controlling factors is usually defined by an experimental design that traces the domain of validity of the model. In 1993, Whiting and Buchanan proposed the following predictive models classification scheme, which applies both to models that describes microbial growth or inactivation.

Primary models are the mathematical description of the microbial evolution (growth, survival or inactivation) as a function of time. Microbial evolution can be measured directly as colony-forming units or biomass, and indirectly as absorbance measurements, impedance, metabolite accumulation or substrate–product depletion (see, e.g. Whiting, 1995). The Gompertz function and its modifications or the Baranyi model are examples of primary models for describing microbial growth. First-order kinetic models, the Baranyi model and the Weibull equation can be used to describe microbial inactivation.

Secondary models are often based on data acquired through primary modelling or deriving from measures carried out in definite conditions (e.g. toxin production) which are modelled as a function of environmental parameters such as temperature, pH, water activity, and so on, frequently under the conditions defined by suitable experimental designs.

Tertiary models are software applications comprising primary and secondary models based on laboratory or literature data (see some reference software in Membré and Lambert, 2008) and can represent very valuable resources. Those applications are created to be easy to use and allow experts, and should also allow non-experts, to obtain valuable predictions on microbial behaviour. Additionally, some versions intended to estimate the 'remaining shelf life' of a food product can include the possibility to introduce a temperature history.

However, between secondary modelling and tertiary modelling a model validation is essential. A model can be viewed as a simplification of reality and this validation step can prove that this simplification is not adversely affecting predictions, or, in other words, that the simplification can satisfactorily describe the reality.

Kinetic modelling

The past decade has been characterized by a considerable effort to generate predictive microbiology databases for food-borne pathogens or spoilage microorganisms. The majority of the works has focused on developing kinetic models for the prediction of the effect of environmental factors (temperature, pH, water activity or salt concentration, preservatives, etc.) on microbial dynamic parameters (maximum specific growth rate, duration of the lag phase) (McMeekin *et al.*, 1997; 2000, 2002).

As defined by Ratkowsky and Ross (1995), kinetic models allow the user to calculate the shelf life of foods or to predict the timespan in which significant microbial growth (and toxin or undesirable substance accumulation) might occur. Some of the developed kinetic models for spoilage and pathogenic organisms have been incorporated into publicly available applications software, described before as tertiary models.

Several studies have been focused on modelling the influence of the environmental conditions on the specific growth rate of the microorganisms. The modelling is usually carried out in two steps. First, growth curves are obtained at different levels of the studied environmental factors (temperature, pH, water activity, preservatives, etc.) and fitted with a primary model to estimate the bacterial growth rate. Then, the growth rates are fitted with secondary models as a function of the environmental conditions.

Predictions from the kinetic models should not be made outside the interpolation region. Baranyi *et al.* (1996) demonstrated that extrapolation can lead to serious problems. Within the use of EOs and the quantification of their antimicrobial activity it is important to report that the inactivation or inhibition pattern of microorganisms exposed to lethal or inhibitory agents or environments can be described by a dose–response curve. In those models the microbial response is plotted as a function of the concentration (or intensity) of the preservation factor (Peleg *et al.*, 1997). The dose–response curve has a sigmoid shape, in which a region of unnoticeable inhibition (or mortality) is followed by a linear region of exponential decay (Peleg, 1996).

Probability modelling

In 'probability modelling', the interest of the modeller is directed to deciding whether a microorganism might or might not grow or produce toxins (Ratkowsky and Ross, 1995; Ratkowsky, 2002). In 1995, Ratkowsky and Ross proposed the application of a modified logistic regression for modelling the edge, defined as boundary interface, between growth and no growth. They modified a kinetic model in order to predict the probability of growth using linear logistic regression with the cardinal parameters treated as fixed values.

This approach allowed the convergence between predictive microbiology and hurdle technology, integrating probabilistic and kinetic aspects of predictive microbiology. In fact, data for no growth are incorporated into the model as well as data for growth, including the variability in the response. This approach is of particular importance when low contamination levels are involved, in the case, for example, of pathogen contamination. In addition, the knowledge of the combination (or the combinations) of factors (intrinsic and extrinsic) within the no growth region of a microorganism can help manufacturers in guaranteeing the safety and stability of their products. Furthermore, growth/no growth interface models may be important for deciding food safety regulations (Schaffner and Labuza, 1997).

Recently, modelling the behaviour of microorganisms at the growth/no growth interface has been acknowledged as an important element of 'modern' predictive microbiology (McMeekin *et al.*, 1997, 2000, 2002).

Applications of Primary Models in Antimicrobial Assessment of Essential Oil Effects

Primary models

Primary models in biology studies are equations obtained from experimental data that describe an output of choice as a dependence of time and/or a set of inputs. These models, which are generally empirical and deterministic, always give descriptions that are reliable only within the range of the analysed data. In fact, these models only describe data under experimental conditions in the form of a convenient mathematical relationship (Gibson *et al.*, 1987). Primary models used in studies on the application of EOs as antimicrobial can be grouped in growth and survival models.

Growth models

The general trend in the wide literature regarding the antimicrobial use of EOs is their application in order to obtain a bactericidal effect and, in the worst case, a bacteriostatic action. These are the main aims behind the assessment of the minimum bactericidal concentration (MBC) and minimum inhibitory concentration (MIC). While the MBC in theory assures the absence of living or cultivable cells after the application of the EO, the MIC gives information on the EO quantity that can prevent the growth (intended as an increase in cell density) in a defined period of time. It is this last aspect of the EOs' antimicrobial effects that gives scope to models describing the growth in their presence. In fact, during the assessment of the EOs MICs, especially when optical density (OD) devices are used, the time for the cell load variation (generally a difference in OD > 0.1 is used) that defines the end of the bacteriostatic effect masks two different possible events: (i) the EO have effectively had a bacteriostatic effect until the time of OD variation detected, and so no cell was inactivated, but only delayed in the growth; (ii) the EO had a partial bactericidal effect, which means that part of the population was inactivated while the other part could grow even if with a lower rate than in the absence of EO. This second event, generally not taken into consideration, is on the contrary very important in the MIC definition. This is the reason why the assessment of the influence of EOs on the growth parameters (lag and exponential phases) becomes important. The fact that the less the initial cell load is, the longer is the lag phase and, hence, the detection time has already been stressed in the literature (Pin and Baranyi, 2006) and it is reliable that the same influence exists even in the presence of EOs.

Unfortunately, there are not many papers that deal with the assessment of growth dynamics in the presence of EOs. However, the comparison between measures of MIC and MBC is always difficult because the operative condition are not standardized (different from antibiotic tests) and variation in the medium, the initial inoculum, the incubation condition and the way in which EO or their components are added can dramatically change the final result.

In this context, an interesting tool was proposed by Lambert and Pearson (2000) who developed a model, inspired by the Gompertz equation, to evaluate the dose–response effect in the presence of inhibitors, including EO components. According to this model, *fa* is calculated, which is the fractional area, i.e. the ratio of inhibited growth to uninhibited growth measured with an indirect method (for instance, optical density, impedance, etc.):

$$fa = \exp\left[-\left(\frac{x}{P_1}\right)^{P_2}\right]$$

where *fa* (ranging from 0 to 1) is the fractional area, x is the inhibitor concentration, P_1 is the concentration at maximum slope (of a log x versus *fa* plot) and P_2 is a slope parameter.

The application of this approach is useful to understand the interactions among different molecules that can affect the antimicrobial effect achieved, which can be antagonistic, additive or synergistic. Koblinsky *et al.* (2007), using a factorial experimental design, demonstrated that the effects of many antimicrobials (among which were several EO constituents) have to be considered additive rather than synergistic. The possible misinterpretation of the results depends on the definition of synergy (the combination of two components is more effective than each compound alone), which is based on the hypothesis that the inhibitory effect of each compound varies linearly with concentration when it acts alone. This is not the case for EO constituents, which generally show a threshold level under which scarce or no antimicrobial activity is present and beyond which the inhibition strongly increases.

In general there are few papers based on the modellization of the growth curve of microorganisms in the presence of EO and their constituents. Obviously they are mainly focused on pathogenic and spoilage organisms. However, Hayouni *et al.* (2008) observed the lack of information about the effects of EOs on useful bacteria. In this perspective, they tested different concentrations of *Melaleuca armillaris* EO on six lactic acid bacteria species often used as starter cultures (*Lactobacillus pentosus, L. rhamnosus, L. plantarum, L. casei, L. delbrueckii* subsp. *bulgaricus* and *L. sakei*). The OD values recorded were plotted over time and fitted with the Baranyi equation (Baranyi and Roberts, 1994). The results obtained indicated that the detection times ('lag time') were longer as the EOs' concentration increased. For the highest concentrations, they observed for some strains a bacteriostatic effect exceeding the 72 h of the experiment. More interestingly, they assessed the physiological state of the lactic acid bacteria, through flow cytometry, to detect a stress response induced by the presence of EOs. According to their results, the population could be divided into viable, stressed (injured) and dead cells. The ratio within the three groups changed during the exposure to EOs with the reduction of the first group and the increase of the last two. The increased percentage of stressed cells and dead cells explained the longer detection time observed.

Primary models such as the Gompertz equation modified by Zwietering *et al.* (1990) were used to evaluate the effects of hexanal, used as a component of a modified atmosphere packaging. In particular, the shelf life and the evolution of naturally occurring microbial populations in fresh apple slices during storage at 4 and 15°C was modelled (Lanciotti *et al.*, 1999). The models obtained allowed the calculation of the shelf life extension in relation to hexanal concentration and storage temperatures (4 and 15°C). In fact, the presence of hexanal in the storage atmosphere at 4°C totally inhibited mesophilic bacteria and considerably prolonged the lag phase of psychrotrophic bacteria. Also, at 15°C, hexanal strongly inhibited moulds, yeasts, and mesophilic and psychrotrophic bacteria. Moreover, hexanal led to a yeast selection favouring species having a reduced spoilage potential due to their prevalent respiratory activity. When added to a modified atmosphere (70% N_2 and 30% CO_2), this molecule was also very effective in preventing browning reactions for at least 16 days at 15°C.

Lanciotti *et al.* (2003) studied the effects of different concentrations of hexanal, (*E*)-2-hexenal, hexyl acetate, and their mixtures on the fate of pathogenic species such as *Escherichia coli, Salmonella enteritidis* and *L. monocytogenes* inoculated in model systems and fresh-sliced apples. In order to evaluate the effects of the chosen molecules on microbial growth kinetics and product shelf life, the data collected during the incubation of the model systems or during the storage of the products, were modelled according to the Gompertz equation modified by Zwietering *et al.* (1990). The result obtained in this work pointed out the potential use of compounds such as hexanal, (*E*)-2-hexenal, and hexyl acetate for both the extension of shelf life and the improvement of hygienic safety of 'minimally processed foods'. In fact, hexanal, (*E*)-2-hexenal and hexyl acetate had a significant inhibitory effect against pathogen microorganisms frequently isolated from raw materials (*E. coli, S. enteritidis* and *L. monocytogenes*) when inoculated both in a model system and food. In this latter condition, these compounds, at the levels used (150, 150 and 20 ppm for hexanal, hexyl acetate and (*E*)-2-hexenal, respectively), displayed a bactericide effect on *L. monocytogenes* and they exhibited significant extensions of lag phase of *E. coli* and *S. enteritidis* inoculated at levels of 10^4–10^5 CFU/g.

Corbo *et al.* (2000) developed a central composite design (CCD) to assess the individual and interactive effects of hexanal and (*E*)-2-hexenal and storage temperatures on (i) the growth of the naturally occurring microflora; (ii) the evolution over time of an inoculated spoilage yeast (*Pichia subpelliculosa*); and (iii) the enzymatic browning reaction in minimally processed apples. The cell loads of *P. subpelliculosa* in the different runs of CCD recorded over storage were modelled by using the Gompertz equation as modified by Zwietering *et al.* (1990). The lag phase lengths were analysed in relation to the independent variables of the CCD to obtain a polynomial equation describing the main, interactive, and quadratic effects of hexanal and *E*-2-hexenal concentrations and temperature levels on the dependent variables considered (i.e. lag phase extension). The models obtained allowed the calculation of the effect of inclusion of different hexanal and (*E*)-2-hexenal concentrations on the extension of shelf life, also when *P. subpelliculosa* was inoculated at levels of 10^3 cfu/g and abusive storage temperatures were used.

Another example of a primary model used to fit data obtained from the assessment of EO activity on microorganisms is reported by Gardini *et al.* (2009). These authors assessed the composition of four EOs extracted by hydrodistillation from plants of common use in Cameroon (*Curcuma longa, Xylopia aethiopica, Zanthoxylum leprieurii* L. and *Zanthoxylum xanthoxyloides*) and their effect on the growth dynamics of *L. monocytogenes, S. enteritidis* and *Staphylococcus aureus*. This was performed through OD values recorded using a MicroFoss-32. Because the EOs had a very low bactericidal activity, they were tested for their bacteriostatic activity also as function of the initial cell inoculum. The detection times, obtained both by the instrument automatic calculation as well as by fitting the data to the Baranyi and Roberts (1994) model, increased with the EO concentration increase and the inoculum cell load decrease.

A similar paper (Belletti *et al.*, 2004) described the effects of nine citrus EOs used for soft-drink production on the growth of a *Saccharomyces cerevisiae* strain isolated from spoiled orangeade. The growth was indirectly assessed in model systems by measuring the carbon dioxide produced by the yeast metabolism (Gardini *et al.*, 1997), which was modelled over time using the Gompertz equation as modified by Zwietering *et al.* (1990). The differences in the parameters of the equations obtained (in particular growth rate and lag phase) indicated that some of these EOs were characterized by significant antimicrobial effects. In particular, the citron (*Citrus medica*) EO showed the most pronounced antimicrobial activity which was related to its high content of citral. Citral is a mixture of two isomers (neral and geranial) whose antimicrobial properties have been studied by several authors (Caccioni and Deans 1993; Wuryatmo *et al.*, 2003; Rivera-Carriles *et al.*, 2005; Bellettti *et al.*, 2007, 2008). The same work also showed differences in the antimicrobial effects of the EO when added to the head space of the model system or solubilized directly in the medium.

Periago *et al.* (2006) used the Baranyi model to investigate the effects on *Bacillus megaterium* spore germination and growth inhibition caused by the presence of EO components (thymol and carvacrol) in combination with a heat treatment (90°C for 25 min). They demonstrated that 0.6 mM thymol increased the lag phase duration almost three times, but it did not change the growth rate, while carvacrol at 0.6 mM reduced the growth rate but did not increase the lag phase duration significantly, resulting in a quite similar growth curve. The combination of thymol and carvacrol at the same concentrations caused a further increase of the lag phase and a significant decrease of the growth rate. When carvacrol and/or thymol were combined with the thermal treatment previously mentioned (able to kill 90% of the population), the growth of the survivors was inhibited for at least 7 days.

Survival models

In several points of view, primary models describing the survival of microorganisms in the presence of EOs are more interesting for food-safety purposes than the growth models. They give information on the EO concentration alone or in combination with other antimicrobial compounds and/or physical stresses that are needed to reduce the microbial population completely or to a certain extent. These models are usually applied to pathogens but also to spoilage bacteria.

An example of a survival model applied to EO bactericidal activity was presented by Delgado *et al.* (2004). These authors studied the bactericidal action of thymol and cymene on two *Bacillus cereus* strains whose vegetative cells were grown up to the exponential phase at pH 7. The survival curves did not follow a first-order kinetic and were fitted with the Weibull probability distribution model (Peleg and Cole, 1998). This approach requires the estimation of a shape parameter, related to the inactivation behaviour described by the curve (concave, convex or linear) and a scale parameter, related to the death kinetic of the first log unit of cells. The survival kinetics were strain dependent and were studied at different concentrations of the compounds (0.2–1.0 mM for thymol and 0.2–2.0 mM for cymene) considered alone or in combination. The good fitting of this model allowed the calculation of the time necessary to achieve a 6 log cfu/ml reduction and, hence, to compare the efficacy of the compounds to other deactivation techniques. In addition to the enhanced antimicrobial effect at increasing concentration of the two terpenes added alone, the authors also stressed a synergistic effect of the combination of the tested natural antimicrobials on *B. cereus* viability.

Similarly, Periago *et al.* (2004) used the Weibull model to study the antibacterial action of carvacrol and cymene on the cells of two *L. monocytogenes* strains collected during lag and exponential phase. The effects were evaluated in a culture medium (brain heart infusion) and in carrot juice. The model accurately predicted the inactivation kinetic achieved. The authors underlined a remarkable synergistic effect obtained in the presence of a low concentration (0.75 mM) of each component, which was promising for an eventual practical exploitation in minimally processed foods.

Aragao *et al.* (2007) elaborated the data

presented by Skandamis and Nychas (2000) to compare the Weibull and the Log-normal models for fitting the survivor data of *E. coli* inoculated in Greek salads in presence of different concentrations of oregano EO. Both the models were able to give a good fit of the experimental data, without significant differences. These authors also investigated the models when plotted in their probability density function (PDF) form. This approach allowed comparing the effect of oregano EO concentration on the survival of *E. coli* using the mathematic characteristics of this distribution. In fact, increased lethality determined by increasing EO concentration resulted in shorter mean and mode, smaller standard deviation and increased overall symmetry (skewness) of the cell resistance distribution.

This latter approach was also used by Char *et al.* (2009) who combined vanillin (4-hydroxy-3-methylbenzaldheyde) with a thermal treatment in order to reduce the severity of the thermal treatment. A milder treatment can have a reduced impact on the organoleptic and nutritional quality of the final product. The added value of their work relied on the fact that they performed their experiment in real food (fruit juice). In this perspective, they assessed the response of *Listeria innocua* to combined treatments involving moderate temperatures (57–61°C) and addition of different levels of vanillin (0–1100 ppm) to find the most effective inactivation treatment in orange juice. For the modelling of these data, the authors tested two different equations: a modified Gompertz equation and the Weibull equation in the form described previously. The modified Gompertz equation (Linton *et al.*, 1995) used here needed the estimation of three parameters, such as *A*, *B* and *C*, which represent the different regions in the survival curve: *A* is the initial shoulder [min]; *B* is the maximum death rate [min^{-1}]; and *C* is the overall change in the number of survivors. They also used the Weibull distribution based on the assumption that in the population there is a spectrum of resistances to the treatments. This means that each organism is destroyed sooner or later than others; the shapes of the survival curves are reflections of treatment resistance distributions having different mode, variance and skewness values, according to the PDF distribution. As a result, the survival curve is the cumulative form of a temporal distribution of lethal events where each individual organism is inactivated at a specific time, thus generating a spectrum of heat resistances (Peleg and Cole, 1998). With this distribution approach, the authors were able to demonstrate that increasing the vanillin concentration in orange juice as well as the heating temperature produced a strong reduction in frequency spread (lower variance), mean and mode. This fact indicated that cells became more sensitive to heat with increasing vanillin concentrations.

Also Knight and McKellar (2007) studied the effect of EOs (cinnamon, clove and lemon oils) or their constituents, such as (*R*)-(–)-carvone, (*S*)-(–)-perillaldehyde, citral, geraniol, methyljasmonate and *p*-anisaldehyde, against *E. coli* O157:H7. The study was carried out in acidic (pH 4.5) and neutral (pH 7.2) conditions. The data were elaborated with the traditional first-order kinetic model and the resulting D-values were compared. Cinnamon and clove EOs (tested at concentration ranging between 0.01 and 0.1%) strongly inhibited *E. coli* O157:H7 independently of the pH, while (*R*)-(–)-carvone and (*S*)-(–)-perillaldehyde were moderately active at both pH values. By contrast, citral and geraniol were active, under the adopted conditions, only in the acidic environment. Synergistic effects were recorded between the EOs and the lower pH. Moreover, cinnamon and clove oils (0.01%) were further tested in apple cider inoculated with *E. coli* O157:H7 in combination with mild heat treatments. The concomitant presence of the two EOs resulted in lower D-values than those for single oils, suggesting a synergistic effect and the potential efficacy of a mild heat treatment for apple cider.

Interesting synergies against *L. monocytogenes* were also found among S-carvone (5 mM) and a mild thermal treatment (45°C for 30 min) (Karatzas *et al.*, 2000). Significant diminutions of survivors were observed by combining the two treatments, which alone had no effects on *L. monocytogenes* viability. Similar results were obtained using carvacrol (1.7 mM), cinnamaldehyde (2.5 mM), thymol (1.5 mM) and decanal (2 mM).The results induced the authors to hypothesize an important role of such a strategy in the stabilization of minimally processed foods, in the framework of the hurdle technology philosophy. With the same aim, Sado *et al.* (2009) applied the Weibull equation to model the survivor data of a strain of *L. monocytogenes* subjected to thermal treatments ranging from 53 to 68°C in the presence of sublethal concentrations of carvacrol, (*E*)-2-hexenal and citral. Using the models obtained for each treatment, the time required to kill 5 log units of *L. monocytogenes* were calculated. The presence of the aroma compounds was able to reduce to about one third this time in comparison with the treatment carried out in the absence of these molecules at temperatures from 53 and 60°C. However, also at higher temperature, although to a lower extent, the thermal treatment efficacy was significantly enhanced by the three molecules.

Applications of Secondary Models in Essential Oil Studies

As already stated, secondary modelling relates the primary model parameters or a result of microbial activity (i.e. toxin accumulation) to environmental or intrinsic variables such as temperature or pH. In this perspective, secondary models have been widely employed to describe the changes of microbial growth parameters as a function of traditional antimicrobial compound concentration and other intrinsic and extrinsic parameters (Rosso *et al.*, 1995; Presser *et al.*, 1997, 1998; Le Marc *et al.*, 2002; Coroller *et al.*, 2005). However, fewer models for the prediction of growth and survival parameters in the presence of EOs are available. Among these, surface response models for the growth/no growth approach prevail. Bacteriostatic/bactericidal probabilistic models are not clearly classified as primary or secondary models. Anyway, if the probability of growth/no growth is expressed as a function of several intrinsic and extrinsic variables, they can be seen as a prediction of a parameter of cell density variation and, in this case, we can speak of a secondary model.

In many cases, secondary models are obtained by fitting data collected according an experimental plan. For example, Gardini *et al.* (1997) considered hexanal as a model molecule to study the relationship between the antifungal activity of a volatile compound and its vapour pressure. Hexanal is an aliphatic aldehyde produced by plants through the lipoxygenase pathway usually as a response to mechanical stresses and is responsible for the 'green notes' of fruits and vegetables. The antimicrobial activity of this molecule was tested *in vitro* against *Aspergillus niger.* A Central Composite Design (3 variables, 5 levels) was developed to evaluate the effect of water activity (between 0.998 and 0.964), temperature (between 15 and 25°C) and hexanal concentration (between 0 and 3.256 μmol) on the inhibition of the growth of *A. niger.* Also the changes of the hexanal vapour pressure during incubation in model systems were evaluated. The results obtained in each run of the experimental design were fitted with a polynomial quadratic equation. The model obtained indicated that the indices taken into consideration evidenced that increasing hexanal vapour pressure determined higher antimicrobial effects. Consequently, the temperature increase enhanced the hexanal antifungal activity, due to its effect on vapour pressure. This behaviour allowed the authors to state that, under the adopted conditions, the hexanal can easily solubilize in the fungal cell membrane when it is present in its hydrophobic (i.e. vapour) state. Similar results were obtained for (*E*)-2-hexenal, another aldehyde produced by the lipoxygenase pathway, tested against *A. niger* (Gardini *et al.*, 2001). The same dependent variables were taken into consideration with the same approach. In general, the use of this unsaturated aldehyde allowed noticeable antimicrobial activity at lower concentrations compared with hexanal. Moreover, important increases in the antimicrobial activity were observed at the lowest water activity tested. This behaviour can be caused by the 'salting out' effect which enhanced the vapour pressure of (*E*)-2-hexenal.

One of the first applications of secondary modelling to EO effect on the microbial stability of foods can be found in Skandamis and Nychas (2000), who performed an experiment where the survival over time of *E. coli* O157:H7 in different combinations of temperature, pH and *Origanum vulgare* EO concentration was assessed in Greek aubergine salad. The various survival data were firstly fitted by the Baranyi and Roberts equation (as reported in the primary modelling section) to fit the survivor data in relation to the condition adopted. Then, the shoulder (or survival) period (lag in the death kinetic), as well as the death rate, were estimated. In the first part of the work a three-way analysis of variance experiment was carried out by considering four storage temperatures (ranging from 0 to 15°C), three pH levels (from 4.0 to 5.0) and four oregano EO concentrations (from 0.0, to 2.1% expressed as vol/wt) were studied. This approach demonstrated that all the variables considered significantly affected the *E. coli* O157:H7 survival. The results (lag phase and death rate) of the primary modelling or their log transformation were fitted with a polynomial quadratic equation which allowed obtaining surface response models. Besides the traditional diagnostics of this procedure (R^2 and standard error of fit), the authors also checked the bias and the accuracy factor of the models. The values of all the dignostics indicated, in addition to the goodness of fit, that the models produced 'fail-safe' predictions. Finally, the predictions were validated with new data obtained by a new batch of aubergine salad. The proposed models were critically discussed by the authors who found that they were more accurate for the prediction of the early stage of inactivation where the cell load was still high and several data points were recorded. The fact that the experiment was performed in a real system also increased the prediction difficulty. The authors also admitted that the application of such a model would have to take into account that the EO composition is rarely the same and that most of the food parameters are not constant during storage. These authors concluded that the use of oregano EO can be considered an important further hurdle for the safety of this food product. They found an increase of EO antimicrobial activity that was associated to the higher hydrophobicity of the EO constituents at the lower pH. It is, in fact, known that the increase of hydrophobicity causes an increase in the antimicrobial effect because it is easier for the solubilization of such molecules in the cytoplasmic membranes of microbial cells.

The same approach has been already used by Koutsoumanis *et al.* (1999) to evaluate the survival of *S. enteritidis* in home-made taramasalata, a traditional Greek appetizer, added with different concentrations of oregano EO (from 0 to 2%), stored at different temperatures (5–20°C) and at different pH (from 4.3 to 5.3 adjusted with lemon juice). The primary modelling with the Baranyi model was followed by the use of a polynomial equation to model the parameters. Among the variables considered, the oregano EO concentration was the most important in speeding up the *S. enteritidis* death rate. In this work the authors stressed the importance of having models established on real foods. These models can take into account factors such as food structure and composition and the interaction between the components of the microbiota, which cannot be considered in laboratory media.

Another approach based on an adaptation of the Fermi function was used by López-Malo *et al.* (2002) to predict the effects of selected concentrations of antimicrobials from natural (vanillin, thymol, eugenol, carvacrol or citral) or synthetic (potassium sorbate or sodium benzoate) origin on *A. flavus* lag time. The strain was inoculated in laboratory media formulated at water activity (a_w) 0.99 and pH 4.5 or 3.5. The lag time was defined as the time to detect a colony with a diameter > 0.5 mm.

The Fermi equation used in this work was modified as follows:

$$S(P) = \left[\frac{1}{1+\exp(\frac{P-P_c}{k})} \right]$$

where $S(P)$ is the mould lag time in the control samples (without antimicrobial) divided by the lag time in the growth medium formulated with a selected antimicrobial concentration, P the antimicrobial concentration used (ppm), P_c a critical level of P where $S(P)$ is 0.5, and k is a constant (ppm) indicating the steepness of the dose–response curve around P_c. Since about 90% of the inhibition occurs within $P_c \pm 3k$ (Peleg, 1996), a large value of k means a wide span, while a small value a very steep decline. The authors observed that the generated equations obtained by non-linear regression, predicted *A. flavus* lag time with an error lower than 10% for the tested antimicrobials and that these errors increased as the standard errors on the estimates for P_c and k increased due to error propagation.

In addition to the previous secondary models presented, another approach used in the assessment of EO effect on microorganisms in association with intrinsic and extrinsic parameters is the growth/no growth probabilistic approach. This approach uses large experimental plans obtained mostly with the central composite design or the Box–Behnken design is aimed at obtaining response surface models applicable in the range of the parameters values tested. The data obtained in such experiments are transformed in a binary result that describes the probability that an event (growth/no growth, toxin produced/no toxin produced) occurs. This approach makes the exploitation of the boundary possible. The logit model is particularly appropriate for this elaboration. For each observation of the experimental design the values 1 or 0 are assigned depending on the occurrence or not of the event (i.e. growth or no growth). Several observations for each condition tested make the model more reliable.

The logit model derives from the logistic equation in its form asymptotic to 0 and 1. Under these premises, it can describe the probability (P) that the event occurs in relation to the independent variable x:

$$P = \frac{e^{f(x)}}{1+e^{f(x)}}$$

This equation, which has a sigmoidal shape, can be easily linearized as follows:

$$Logit(P) = \ln\left(\frac{P}{1-P}\right) = f(x)$$

where Logit (P) is the natural logarithm of the ratio between the probability that the event occurs (P) and that it does not occur ($1–P$).

This approach can be easily extended to describe the effect of several variables on P, as demonstrated by Belletti *et al.* (2007) who studied the combined effects of a mild heat treatment (55°C) and the presence of three aroma compounds (citron EO, citral and (*E*)-2-hexenal) on the spoilage of non-carbonated beverages inoculated with different amounts of a *S. cerevisiae* strain. In this work, the logistic regression proved to be an important tool to study alternative hurdle strategies for the stabilization of non-carbonated beverages. The experimental design was set up using ten repetitions for each run (experimental conditions), and a total of 20 runs were considered. The results, expressed as growth/no growth (reflecting the ability of the yeast to spoil or not the beverage), were elaborated using a logistic regression in order to assess the probability of beverage spoilage as a function of thermal treatment length, concentration of flavouring agents, and initial yeast inoculum. The logit models obtained for the three substances were extremely precise. The thermal treatment alone, even if prolonged for 20 min, was not able to prevent yeast growth. However, the presence of increasing concentrations of aroma compounds improved the stability of the products. The inhibiting effect of the compounds was enhanced by a prolonged thermal treatment. In fact, it influenced the vapour pressure of the molecules, which can easily interact with microbial membranes when they are in gaseous form. (*E*)-2-hexenal showed a threshold level, related to initial inoculum and thermal treatment length, over which yeast growth was rapidly inhibited, with the concomitant cell death. Concentrations over 100 ppm of citral and thermal treatment longer than 16 min allowed a 90% probability of stability for bottles inoculated with 10^5 CFU/bottle of *S. cerevisiae*. Citron EO (which contained high amount of citral) gave the most interesting responses: beverages with 500 ppm of EO needed only 3 min of treatment to prevent yeast growth.

Similarly, Belletti *et al.* (2010) evaluated the antimicrobial activity of three terpenes (citral, linalool and β-pinene), in combination with a mild heat treatment (55°C, 15 min). The study was carried out on an orange-based soft drink inoculated using a strain of *S. cerevisiae*. While the temperature here was fixed with respect to the previous work on beverage, the aroma compounds were employed in order to assess their combined effect. The experimental design used was a CCD with 3 variables (citral, linalool and β-pinene concentration) at 5 levels reinforced by the addition of 5 combinations, located in the boundaries of growth/no-growth of yeast. A model comprising only significant individual parameters ($p \leq 0.05$) and describing the relationships between terpene concentrations and the probability of having stable (unspoiled) beverages was obtained. Interesting interactive effects among the terpenes were observed. For example, when citral and β-pinene were combined, the citral concentration required to achieve a 50% probability of having stable bottles (P=0.5) dropped from 100.9 μl/l alone to 49.3 μl/l in the presence of 20 μl/l of β-pinene. The mixture of citral and linalool was less effective and the same probability was obtained combining 60 μl/l of linalool with 35.1 μl/l of citral. The addition of 20 μl/l of linalool and β-pinene reinforced citral bioactivity and the concentration of citral needed to reach P = 0.5 fell from 100.9 μl/l in the presence of citral alone to 42.0 μl/l. The authors concluded that the antimicrobial potential of the three terpenes alone can be strengthened by combining appropriate concentrations of each of them. Their study confirmed also the potentiating effect of a mild temperature treatment on the antimicrobial efficacy of the molecules. Neither the thermal treatment alone nor the presence of the terpenes at their maximum concentrations (without thermal treatment) was able to guarantee the microbial stability of the beverages.

Conclusions and Future Perspectives

The use of EOs and aroma compounds as antimicrobials has been known since ancient times. Nowadays the need to find natural alternatives to replace chemical preservatives has renewed interest in the research for the exploitation of the use of EOs or their components as antimicrobials in foods.

The main limitations for their use in food processing have been summarized by Lanciotti *et al.* (2004), who stated that possible ways to reduce the organoleptic impact include:

1. Minimizing perception of the presence of spices/herbs and EOs in food by optimizing food formulation.
2. The application of combined methods. This could be done with the use of EOs (or their most bioactive compound) in combination with other preservatives (such as pH, heat treatment, high pressure treatment, etc.) within the philosophy of the 'hurdle technology'.
3. Enhancing a calibrated vapour pressure capacity in order to increase interaction between EO and the bacterial cell membrane.

Furthermore, as reported by Koutsoumanis *et al.* (1999), the evaluation of new preservatives such as natural antimicrobials in food, the evaluation of food structure composition and the study of the interaction between natural microflora and food-borne disease agents could be made much more precise by the application of predictive models.

Predictive modelling represents a useful tool in order to predict the effect of the combination of different hurdles. The variability of extracted EOs combined with the food components variability can explain the scarcity and the difficulty of having reliable models for EOs activities in food. Even though there are still only a few reliable models referring to the use of essential oils, the variability of these oils will always limit a general application of any model. A deterministic approach oriented to single components seems to be the road for more reliable models and their application in industries. Neural network modelling could also, in future, be an approach to help obtain reliable models. The existence of good predictive models could be the key for the future standardized application of EOs in foods.

References

Angioni, A., Barra, A., Coroneo, V., Dessi, S. and Cabras, P. (2006) Chemical composition, seasonal variability, and antifungal activity of *Lavandula stoechas* L. ssp. *stoechas* essential oils from stem/leaves and flowers. *Journal of Agriculture and Food Chemistry* 54, 4364–4370.

Appendini, P. and Hotchkiss, J.H. (2002) Review of antimicrobial food packaging. *Innovative Food Science & Emerging Technologies* 3, 113–126.

Aragao, G.M.F., Corradini, M.G., Normand, M.D. and Peleg, M. (2007) Evaluation of the Weibull and log normal distribution functions as survival models of *Escherichia coli* under isothermal and non isothermal conditions. *International Journal of Food Microbiology* 119, 243–257.

Bakkali, F., Averbeck, S., Averbeck, D. and Idaomar, M. (2008) Biological effects of essential oils — a review. *Food and Chemical Toxicology* 46, 446–475.

Baranyi, J. and Roberts, T.A. (1994) A dynamic approach to predicting bacterial growth in food. *International Journal of Food Microbiology* 23, 277–294.

Baranyi, J., Ross, T., McMeekin, T.A. and Roberts, T.A. (1996) Effects of parameterisation on the performance of empirical models used in predictive microbiology. *Food Microbiology* 13, 83–91.

Belletti, N., Ndagijimana, M., Sisto, C., Guerzoni, M.E., Lanciotti, R., and Gardini, F. (2004) Evaluation of the antimicrobial activity of citrus essence on *Saccharomyces cerevisiae. Journal of Agricultural and Food Chemistry* 52, 6932–6938.

Belletti, N., Sado Kamdem, S., Patrignani, F., Lanciotti, R., Covelli, A. and Gardini, F., (2007) Antimicrobial activity of aroma compounds against *Saccharomyces cerevisiae* and improvement of microbiological stability of soft drinks as assessed by logistic regression. *Applied and Environmental Microbiology* 73, 5580–5586.

Belletti, N., Lanciotti, R., Patrignani, F. and Gardini, F. (2008) Antimicrobial efficacy of citron essential oil on spoilage and pathogenic microorganisms in fruit-based salads. *Journal of Food Science* 73, 331–338.

Belletti, N., Sado Kamdem, S., Tabanelli, G., Lanciotti, R. and Gardini, F. (2010) Modeling of combined effects of citral, linalool and β-pinene used against *Saccharomyces cerevisiae* in citrus-based beverages subjected to a mild heat treatment. *International Journal of Food Microbiology* 136, 283–289.

Burt, S. (2004) Essential oils: their antibacterial properties and potential applications in foods, a review. *International Journal of Food Microbiology* 94, 223–253.

Caccioni, D.R.L. and Deans, S.G. (1993) Action of citrus fruits essential oils on germination of *Penicillium digitatum* and *Penicillium italicum. International Symposium on Industrial Crops and Products*, Pisa, Italy, 22–24 November.

Caccioni, D.R.L., Gardini, F., Lanciotti, R., and Guerzoni, M.E. (1997) Antifungal activity of natural volatile compounds in relation to their vapour pressure. *Science des Aliments* 17, 21–34.

Caccioni, D.R.L., Guizzardi, M., Biondi, D.M., Renda, A. and Ruberto, G. (1998) Relationships between volatile components of citrus fruit essential oils and antimicrobial action on *Penicillium digitatum* and *Penicillium italicum. International Journal of Food Microbiology* 43, 73–79.

Capecka, E., Mareczek, A. and Leja, M. (2005) Antioxidant activity of fresh and dry herbs of some Lamiaceae species. *Food Chemistry* 93, 223–226.

Char, C., Guerrero, S. and Alzamora, S.M. (2009) Survival of *Listeria innocua* in thermally processed orange juice as affected by vanillin addition. *Food Control* 20, 67–74.

Chouliara,E., Karatapanis, A., Savvaidis, I.N. and Kontominas, M.G. (2007) Combined effect of oregano essential oil and modified atmosphere packaging on shelf-life extension of fresh chicken breast meat, stored at 4°C. *Food Microbiology* 24, 607–617.

Corbo, M.R., Lanciotti, R., Gardini, F., Sinigaglia, M. and Guerzoni, M.E. (2000) Effects of hexanal, (*E*)-2-hexenal, and storage temperature on shelf life of fresh sliced apples. *Journal of Agriculture and Food Chemistry* 48, 2401–2408.

Coroller, L., Guerrot, V., Huchet, V., Le Marc, Y., Mafart, P., Sohier, D. and Thuault, D. (2005) Modelling the influence of single acid and mixture on bacterial growth. *International Journal of Food Microbiology* 100, 167–178.

Cowan, M.M. (1999) Plant products as antimicrobial agents. *Clinical Microbiology Reviews* 12, 564–582.

Daferera, D.J., Ziogas, B.N., and Polissiou, M.G. (2000) GC-MS analysis of essential oils from some Greek aromatic plants and their fungitoxicity on *Penicillium digitatum*. *Journal of Agriculture and Food Chemistry* 48, 2576–2581.

Davidson, P.M. and Naidu, A.S. (2000) Phytophenols. In Naidu, A.S. (ed.), *Natural Food Antimicrobial Systems*. CRC Press, Boca Raton, Florida, pp. 265–295.

Deans, S.G., Noble, R.C., Hiltunen, R., Wuryani, W. and Pénzes L.G. (1995) Antimicrobial and antioxidant properties of *Syzygium aromaticum* (L.) Merr. & Perry: Impact upon bacteria, fungi and fatty acid levels in ageing mice. *Flavour and Fragrance Journal* 10, 323–328.

Delaquis, P.J. and Mazza, G. (1995) Antimicrobial properties of isothiocyanate in food preservation. *Food Technology* 49, 73–84.

Delgado, B., Fernández, P.S., Palop, A. and Periamo, P.M. (2004) Effect of thymol and cymene on *Bacillus cereus* vegetative cells evaluated through the use of frequency distributions. *Food Microbiology* 21, 327–334.

Dorman, H.J.D. and Deans, S.G. (2000) Antimicrobial agents from plants: antibacterial activity of plant volatile oils. *Journal of Applied Microbiology* 88, 308–316.

Esty, J.R. and Meyer, K.F. (1922) The heat resistance of the spores of *B. botulinus* and allied anaerobes. XI. *Journal of Infectious Diseases* 31, 656–663.

Farag, R.S., Daw, Z.Y., Hewedy, F.M. and El-Baroti, G.S.A. (1989) Antimicrobial activity of some Egyptian spice essential oils. *Journal of Food Protection* 52, 665–667.

Fitzgerald, D.J., Stratford, M., Gasson, M.J., Ueckert, J., Bos, A. and Arbad, A. (2004) Mode of antimicrobial action of vanillin against *Escherichia coli*, *Lactobacillus plantarum* and *Listeria innocua*. *Journal of Applied Microbiology* 97, 104–113.

Gardini, F., Belletti, N., Ndagijimana, M., Guerzoni, M.E., Tchoumbougnang, F., Amvam Zollo, P.H., Micci, C., Lanciotti, R. and Sado Kamdem, S.L. (2009) Composition of four essential oils obtained from plants from Cameroon, and their bactericidal and bacteriostatic activity against *Listeria monocytogenes*, *Salmonella enteritidis* and *Staphylococcus aureus*. *African Journal of Microbiology Research* 3, 264–271.

Gardini, F., Lanciotti, R., Sinigaglia, M. and Guerzoni, M.E. (1997) A headspace gas chromatographic approach for the monitoring of the microbial cell activity and the cell viability evaluation. *Journal of Microbiological Methods* 29, 103–114.

Gardini, F., Lanciotti, R. and Guerzoni, M.E. (2001) Effect of (E)-2-hexenal on the growth of Aspergillus flavus in relation to its concentration, temperature and water activity. *Letters in Applied Microbiology* 33, 50–55.

Gardini, F., Lanciotti, R., Belletti, N. and Guerzoni, M.E. (2002) Use of natural aroma compounds to control microbial growth in foods. In: Mohan, R. (ed.) *Research Advances in Food Science*. Global Research Network, pp. 63–78.

Gibson, A.M., Bratchell, N. and Roberts, T.A. (1987) The effect of sodium chloride and temperature on the rate and extent of growth of *Clostridium botulinum* type A in pasteurised pork slurry. *Journal of Applied Bacteriology* 62, 479–490.

Gould, G.W. (1996) Industry perspectives on the use of natural antimicrobials and inhibitors for food applications. *Journal of Food Protection* Supplement 82–86.

Gram, L. and Melchiorsen, J. (1996) Interaction between fish spoilage bacteria *Pseudomonas* sp. and *S. putrefaciens* in fish extracts and on fish tissue. *Journal of Applied Bacteriology* 80, 589–595.

Guenther, E. (1948) *The Essential Oils.* Vol. 1 D. van Nostrand Co., Inc. Toronto, New York, London, p. 87.

Gutierrez, J., Barry-Ryan, C. and Bourke, P. (2009) Antimicrobial activity of plant essential oils using food model media: Efficacy, synergistic potential and interactions with food components. *Food Microbiology* 26, 142–150.

Guynot, M.E., Ramos, A.J., Setó, L., Purroy, P., Sanchis, V. and Marín, S. (2003) Antifungal activity of volatile compounds generated by essential oils against fungi commonly causing deterioration of bakery products. *Journal of Applied Microbiology* 94, 893–899.

Hammer, K.A., Carson, C.F. and Riley, T.V. (2002) In vitro activity of *Melaleuca alternifolia* (tea tree) oil against dermatophytes and other filamentous fungi. *Journal of Antimicrobial Chemotherapy* 50, 195–199.

Hayes, A.J. and Markovic, B. (2002) Toxicity of Australian essential oil *Backhousia citriodora* (lemon myrtle). Part 1. Antimicrobial activity and *in vitro* cytotoxicity. *Food Chemistry and Toxicology* 40, 535–543.

Hayouni, E.A., Bouix, M., Abedrabba, M., Leveau, J-Y. and Hamdi, M. (2008) Mechanism of action of *Melaleuca armillaris* (Sol. Ex Gaertu) Sm. essential oil on six LAB strains as assessed by multiparametric flow cytometry and automated microtiter-based assay. *Food Chemistry* 111, 707–718.

Holley, R.A. and Patel, D. (2005) Improvement in shelf-life and safety of perishable foods by plant essential oils and smoke antimicrobials. *Food Microbiology* 22, 273–292.

Juven, B.J., Kanner, J., Sched, F. and Weisslowicz, H. (1994) Factors that interact with the antibacterial of thyme essential oil and its active constituents. *Journal of Applied Bacteriology* 76, 626–631.

Karatzas, A.K., Bennik, M.H.J., Smid, E.J. and Kets, E.P.W. (2000) Combined action of S-carvone and mild heat treatment on *Listeria monocytogenes* Scott A. *Journal of Applied Microbiology* 89, 296–301.

Kim, J., Marshall, M.R. and Wei, C. (1995a) Antibacterial activity of some essential oil components against five foodborne pathogens. *Journal of Agricultural and Food Chemistry* 43, 2839–2845.

Kim, J.M., Marshall, M.R., Cornell, J.A., Preston, J.F., and Wei, C.I. (1995b) Antibacterial activity of carvacrol, citral, and geraniol against *Salmonella typhimurium* in culture medium and on fish cubes. *Journal of Food Science* 60, 1364–1374.

Knight, K.P. and McKellar, R.C. (2007) Influence of cinnamon and clove essential oils on the dand z-values of *Escherichia coli* O157:H7 in apple cider. *Journal of Food Protection* 70, 2089–2094.

Kobilinsky A., Nazer, A.I., Dubois-Brissonnet, F. (2007) Modeling the inhibition of *Salmonella typhimurium* growth by combination of food antimicrobials. *International Journal of Food Microbiology,* 115, 95–109.

Koutsoumanis, K., Lambropoulou, K. and Nychas, G-J.E. (1999) A predictive model for the non-thermal inactivation of *Salmonella enteritidis* in a food model system supplemented with a natural antimicrobial. *International Journal of Food Microbiology* 49, 63–74.

Koutsoumanis, K., Kendall, P.A. and Sofos, J.N. (2004) Modeling the boundaries of growth of *Salmonella typhimurium* in broth as a function of temperature, water activity, and pH. *Journal of Food Protection* 67, 53–59.

Lambert, R.J.W. and Pearson, J. (2000) Susceptibility testing: accurate and reproducible minimum inhibitory concentration (MIC) and non-inhibitory concentration (NIC) values. *Journal of Applied Microbiology* 88, 784–790.

Lanciotti, R., Corbo, M.R., Gardini, F., Sinigaglia, M. and Guerzoni, M.E. (1999) Effect of hexanal on the shelf-life of fresh apple slices. *Journal of Agriculture and Food Chemistry* 47, 4769–4776.

Lanciotti, R., Sinigaglia, M., Gardini, F., Vannini, L. and Guerzoni, M.E. (2001) Growth/no growth interfaces of *Bacillus cereus*, *Staphylococcus aureus* and *Salmonella enteritidis* in model systems based on water activity, pH, temperature and ethanol concentration. *Food Microbiology* 18, 659–668.

Lanciotti, R., Belletti, N., Patrignani, F., Gianotti, A., Gardini, F. and Guerzoni, M.E. (2003) Application of hexanal, (E)-2-hexenal, and hexyl acetate to improve the safety of fresh-sliced apples. *Journal of Agriculture and Food Chemistry* 51, 2958–2963.

Lanciotti, R., Gianotti, A., Patrignani, F., Belletti, N., Guerzoni, M.E. and Gardini, F. (2004) Use of natural aroma compounds to improve shelflife and safety of minimally processed fruits. *Trends in Food Science and Technology* 15, 201–208.

Leistner, L. (1995) Principles and applications of hurdle technology. In: Gould, G.W. (ed.) *New Methods of Food Preservation.* Blackie Academic & Professional, New York, pp. 1–21.

Le Marc, Y., Huchet, V., Bourgeois, C., Guyonnet, J., Mafart, P. and Thuault, D. (2002) Modelling the growth kinetics of *Listeria* as a function of temperature, pH and organic acid concentration. *International Journal of Food Microbiology* 73, 219–237.

Linton, R.H., Carter, W.H., Pierson, M.D. and Hackney, C.R. (1995) Use of a modified Gompertz equation to model nonlinear survival curves for *Listeria monocytogenes* Scott A. *Journal of Food Protection* 58, 946–954.

Lis-Balchin, M., Deans, S.G. and Eaglesham, E. (1998) Relationship between bioactivity and chemical composition of commercial essential oils. *Flavour and Fragrance Journal* 13, 98–104.

López-Malo, A., Alzamora, S.M. and Palou, E. (2002) *Aspergillus flavus* dose response curves to selected natural and synthetic antimicrobials. *International Journal of Food Microbiology* 73, 213–218.

Mangena, T. and Muyima, N.Y.O. (1999) Comparative evaluation of the antimicrobial activities of essential oils of *Artemisia afra, Pteronia incana* and *Rosmarinus officinalis* on selected bacteria and yeast strains. *Letters in Applied Microbiology* 28, 291–296.

Marino, M., Bersani, C. and Comi, G. (2001) Impedance measurements to study the antimicrobial activity of essential oils from *Lamiaceae* and *Compositae. International Journal of Food Microbiology* 67, 187–195.

McMeekin, T.A., Olley, J., Ross, T. and Ratkowsky. D.A. (1993) *Predictive Microbiology: Theory and Application.* Research Studies Press Ltd, Taunton, UK.

McMeekin, T.A., Brown, J.L., Krist, K., Miles, D., Nuemeyer, K., Nichols, D.S., Olley, J., Presser, K., Ratkowsky, D.A., Ross, T., Salter, M., and Soontranon, S. (1997) Quantitative microbiology: a basis for food safety. *Emerging Infectious Diseases* 3, 541– 549.

McMeekin, T.A., Presser, K., Ratkowsky, D.A,. Ross, T., Salter, M. and Tienungoon, S. (2000) Quantifying the hurdle concept by modelling the bacterial growth/no growth interface. *International Journal of Food Microbiology* 55, 93–98.

McMeekin, T.A., Olley, J., Ratkowsky, D.A. and Ross, T. (2002) Predictive microbiology: towards the interface and beyond. *International Journal of Food Microbiology* 73, 395–407.

Membré, J.M. and Lambert, R.J.W (2008) Application of predictive modelling techniques in industry: from food design up to risk assessment. *International Journal of Food Microbiology* 128, 10–15.

Newberne, P., Smith, R.L., Doull, J., Feron, V.J., Goodman, J.I., Murno, I.C., Portoghese, P.S., Waddel, W.J., Wagner, B.M., Weil, C.S., Adams, T.B. and Hallagan, J.B. (2000) GRAS flavouring substances. *Food Technology* 54, 66–83.

Nielsen, P.V. and Rios, R. (2000) Inhibition of fungal growth on bread by volatile components from spices and herbs, and their possible application in active packaging, with special emphasis on mustard essential oil. *International Journal of Food Microbiology* 60, 219–229.

Peleg, M. (1996) Evaluation of the Fermi equation as a model of dose–response curves. *Applied Microbiology and Biotechnology* 46, 303–306.

Peleg, M. and Cole, M.B. (1998) Reinterpretation of microbial survival curves. *Critical Reviews in Food Science* 38, 353–380.

Peleg, M., Normand, M.D. and Damrau, E. (1997) Mathematical interpretation of dose–response curves. *Bulletin of Mathematical Biology* 59, 747–761.

Periago P.M., Delgado, B., Fernández, P.S. and Palop, A. (2004) Use of carvacrol and cymene to control growth and viability of *Listeria monocytogenes* cells and predictions of survivors using frequency distribution functions. *Journal of Food Protection* 67, 1408–1416.

Periago P.M., Conesa, R., Delgado, B., Fernández, P.S. and Palop, A. (2006) *Bacillus megaterium* spore germination and growth inhibition by a treatment combining heat with natural antimicrobials. *Food Technology and Biotechnology* 44, 17–23.

Pin, C. and Baranyi, J. (2006) Kinetics of single cells: observation and modelling of a stochastic process. *Applied and Environmental Microbiology* 72, 2163–2169.

Pin, C., Sutherland, J.P. and Baranyi, J. (1999) Validating predictive models of food spoilage organisms. *Journal of Applied Microbiology* 87, 491–499.

Presser, K., Ratkowsky, D. and Ross, T. (1997) Modelling the growth rate of *Escherichia coli* as a function of pH and lactic acid concentration. *Applied Environmental Microbiology* 63, 2355–2360.

Presser, K.A., Ross, T. and Ratkowsky, D.A. (1998) Modelling the growth limits (growth/no growth interface) of *Escherichia coli* as a function of temperature, pH, lactic acid concentration, and water activity. *Applied Environmental Microbiology* 64, 1773–1779.

Ratkowsky, D.A. (2002) Some examples of, and some problems with, the use of nonlinear logostic regression in predictive food microbiology. *International Journal of Food Microbiology* 73, 119–125.

Ratkowsky, D.A. and Ross. T. (1995) Modelling the bacterial growth/no growth interface. *Letters in Applied Microbiology* 20, 29–33.

Rauha, J.P., Remes, S., Heinonen, M., Hopia, A., Kahkonen, M., Kujala, T., *et al.* (2000) Antimicrobial effects of Finnish plant extracts containing flavonoids and other phenolic compounds. *International Journal of Food Microbiology* 56, 3–12.

Rivera-Carriles, K., Argaiz, A., Palou, E. and López-Malo, A. (2005) Synergistic inhibitory effect of citral with selected phenolics against *Zygosaccharomyces bailii. Journal of Food Protection* 68, 602–606.

Rižnar, K., Čelan, S., Knez, Ž., Škerget, M., Bauman, D. and Glaser, R. (2006) Antioxidant and antimicrobial activity of rosemary extract in chicken frankfurters. *Journal of Food Science* 71, 425–429.

Robins, M.M., and Wilson.P.D.G. (1994) Food structure and microbial growth. *Trends in Food Science and Technology* 5, 289–293.

Roller, S. and Seedhar, P. (2002) Carvacrol and cinnamic acid inhibit microbial growth in fresh-cut melon and kiwifruit at 4°C and 8°C. *Letters in Applied Microbiology* 35, 390–394.

Ross, T. (1999) *Predictive Microbiology for the Meat Industry.* Meat and Livestock Australia, North Sydney, 196 pp.

Ross, T. (1996) Indices for performance evaluation of predictive models in food microbiology. *Journal of Applied Microbiology* 81, 501–508.

Rosso, L., Lobry, J.R., Bajard, S. and Flandrois, J.P. (1995) Convenient model to describe the combined effects of temperature and pH on microbial growth. *Applied and Environmental Microbiology*, 610–616.

Ruberto, G. and Baratta, M.T. (2000) Antioxidant activity of selected essential oil components in two lipid model systems. *Food Chemistry* 69, 167–174.

Sacchetti, G., Maietti, S., Muzzoli, M., Scaglianti, M., Manfredini, S., Radice, M. and Bruni, R. (2005) Comparative evaluation of 11 essential oils of different origin as functional antioxidants, antiradicals and antimicrobials in food. *Food Chemistry* 91, 621–632.

Sado Kamdem, S.L., Belletti, N., Lanciotti, R. and Gardini, F. (2009) Effects of citral, carvacrol and (E)-2-hexenal on growth inactivation of *Listeria monocytogenes* during heat treatment. 6th International Conference on Predictive Modeling in Foods (ICPMF) in: Proceedings, Extended abstracts, pp. 306–310, Washington (USA) 8–12 September 2009.

Schaffner, D.W. and Labuza, T.P. (1997) Predictive microbiology: where are we and where are we going? *Food Technology* 51, 95– 99.

Serrano, M., Martínez-Romero, D., Guilléb, F., Valverde, J.M., Zapata, P.J., Castello, S. and Valero, D. (2008) The addition of essential oils to MAP as a tool to maintain the overall quality of fruits. *Trends in Food Science and Technologies* 19, 464–471.

Sikkema, J., de Bont, J.A.M. and Poolman, B. (1995) Mechanisms of membrane toxicity of hydrocarbons. *Microbiology Reviews* 59, 201–222.

Singh, A., Singh, R.K., Bhunia, A.K. and Singh, N. (2003) Efficacy of plant essential oils as antimicrobial agents against *Listeria monocytogenes* in hotdogs. *LWT Food Science and Technology* 36, 787–794.

Singh, G., Kapoor, I.P., Pandey, S.K., Singh, U.K. and Singh, R.K. (2002) Studies on essential oils: part 10; antibacterial activity of volatile oils of some spices. *Phytotherapy Research* 16, 680–682.

Sivropoulou, A., Papanikolaou, E., Nikolaou, C. and Kokkini, S. (1996) Antimicrobial and cytotoxic activities of *Origanum* essential oils. *Journal of Agricultural and Food Chemistry* 44, 1202–1205.

Skandamis, P. and Nychas, G-J. E. (2000) Development and evaluation of a model predicting the survival of *Escherichia coli* O157:H7 NCTC 12900 in homemade eggplant salad at various temperatures, pHs, and oregano essential oil concentrations. *Applied and Environmental Microbiology* 66, 1646–1653.

Skandamis, P., Davies, K.W., McClure, P.J., Koutsoumanis, K. and Tassou, C. (2002) Avitalistic approach for non-thermal inactivation of pathogens in traditional Greek salads. *FoodMicrobiology* 19, 405–421.

Smith-Palmer, A., Stewart, J. and Fyfe, L. (1998) Antimicrobial properties of plant essential oils and essences against five important food-borne pathogens. *Letters in Applied Microbiology*, 26, 118–122.

Sofos, J.N., Beuchat, L.R., Davidson, P.M. and Johnson, E.A. (1998) Naturally occurring antimicrobials in food. *Regulatory Toxicology and Pharmacology* 28, 71–72.

Solomakos, N., Govaris, A., Koidis, P. and Botsoglou, N. (2008) The antimicrobial effect of thyme essential oil, nisin, and their combination against *Listeria monocytogenes* in minced beef during refrigerated storage. *Food Microbiology* 25, 120–127.

Tajkarimi, M.M., Ibrahim, S.A. and Cliver, D.O. (2010) Antimicrobial herb and spice compounds in food. *Food Control* 21, 1199–1218.

Tiwari, B.K., Valdramidis, V.P., Donnell C.P.O., Muthukumarappan, K., Bourke, P. and Cullen, P.J. (2009) Application of natural antimicrobials for food preservation. *Journal of Agriculture and Food Chemistry* 57, 5987–6000.

Van Impe, J.F., Poschet, F., Geeraerd, A.H. and Vereecken, K.M. (2005) Towards a novel class of predictive microbial growth models. *International Journal of Food Microbiology* 100, 97–105.

Vázquez, B.I., Fente, C., Franco, C.M., Vázquez, M.J. and Cepeda, A. (2001) Inhibitory effects of eugenol and thymol on *Penicillium citrinum* strains in culture media and cheese. *International Journal of Food Microbiology* 67, 157–163.

Ward, S.M., Delaquis, P.J., Holley, R.A. and Mazza, G. (1998) Inhibition of spoilage and pathogenic bacteria on agar and pre-cooked roast beef by volatile horseradish distillates. *Food Research International* 31, 19–26.

Whiting, R.C. (1995) Microbial modeling in foods. *Critical Reviews in Food Science* 35, 467–494.

Whiting, R.C. and Buchanan, R.L. (1993) A classification of models in predictive microbiology—a reply to K.R. Davey. *Food Microbiology* 10, 175–177.

Wuryatmo, E., Klieber, A. and Scott, E.S. (2003) Inhibition of citrus postharvest pathogens by vapor of citral and related compounds in culture. *Journal of Agricultural and Food Chemistry* 51, 2637–2640.

Yin, M. and Cheng, W. (2003) Antioxidant and antimicrobial effects of four garlic-derived organosulfur compounds in ground beef. *Meat Science* 63, 23–28.

Zaika, L.L. (1988) Spices and herbs: their antimicrobial activity and its determination. *Journal of Food Safety* 9, 97–118.

Zwietering, M.H., Jongenberger, J., Rombouts, F.M. and van't Riet, K. (1990) Modeling of the bacterial growth curve. *Applied and Environmental Microbiology* 56, 1785–1881.

22 Database Mining for Bacteriocin Discovery

Riadh Hammami,* Abdelmajid Zouhir, Christophe Le Lay, Jeannette Ben Hamida and Ismail Fliss*

Introduction

Bacteriocins are a very diverse group of antimicrobial peptides produced by a wide range of bacteria and known for their inhibitory activity against various human and animal pathogens. Three main features distinguish the majority of bacteriocins from conventional antibiotics: bacteriocins are ribosomally synthesized, are only toxic to bacteria and have a relatively narrow killing spectrum (Riley and Wertz, 2002). These characteristics have raised considerable interest for bacteriocin application in food preservation. A survey with the key 'bacteriocin' revealed more than 6000 papers recorded in PubMed (December 2009). This increasing interest for bacteriocins is well illustrated by Fig. 22.1, which shows the important growth of papers dealing with them in PubMed.

Bacteriocins were first identified almost 100 years ago. These toxins have been found among most families of bacteria and many actinomycetes and described as universally produced, including by some members of the Archaea (Riley and Wertz, 2002; Shand and Leyva, 2008). Klaenhammer estimates that 99% of all bacteria probably produce at least one bacteriocin and the only reason we have not isolated more is that few researchers are looking for them (Klaenhammer, 1988). Bacteriocins make up a highly diverse family of proteins in terms of size, microbial target, mode of action and release and mechanism of immunity and can be divided into two broad groups: those produced by Gram-negative bacteria and those produced by Gram-positive bacteria (Gordon *et al.*, 2007; Heng *et al.*, 2007). Bacteriocins of Gram-positive bacteria are more abundant and more diverse than those found in Gram-negative bacteria (Hammami *et al.*, 2010). Two main features distinguish Gram-positive from Gram-negative bacteriocins. First, bacteriocin production is not necessarily the lethal event it is for Gram-negative bacteria (Riley and Wertz, 2002). In addition, the Gram-positive bacteria have evolved bacteriocin-specific regulation, whereas bacteriocins of Gram-negative bacteria rely solely on host regulatory networks (Riley and Wertz, 2002).

The bacteriocin action starts with entry into the target cell by recognizing specific cell-surface receptors. Then, microbial cell killing occurs through various mechanisms: formation of ion-permeable channels in the cytoplasmic membrane, non-specific degradation of cellular DNA, inhibition of protein synthesis through the specific cleavage of 16s rRNA, or by cell lysis resulting from

* Corresponding authors.

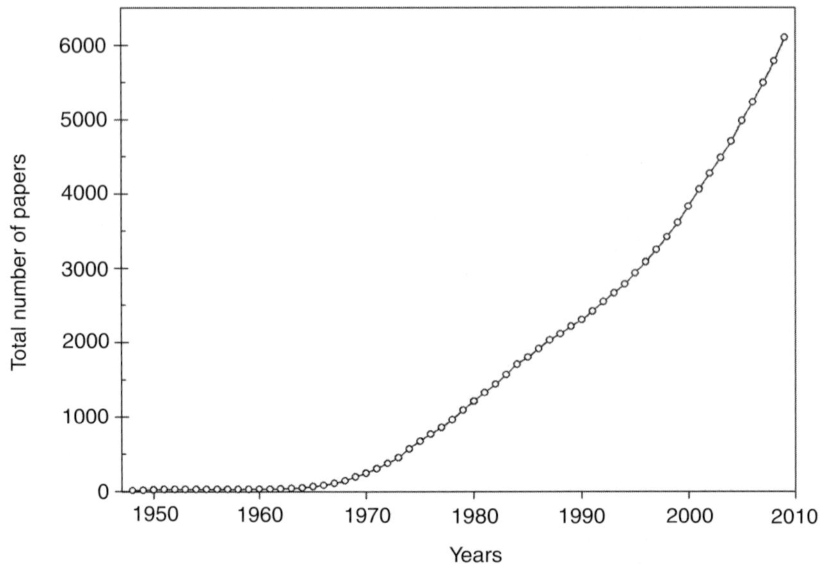

Fig. 22.1. Evolution of the total number of papers dealing with bacteriocins in PubMed.

inhibition of peptidoglycan synthesis (Vriezen *et al.*, 2009). According to their biochemical and genetic properties, bacteriocins are categorized into three different classes (for a review see Cintas *et al.*, 2001). Class I bacteriocins are the lantibiotics, which are small (18–39 residues), post-translationally modified peptides that contain unusual amino acids such as lanthionine or b-methyllanthionine (Willey and van der Donk, 2007). Class II includes unmodified bacteriocins which are subdivided into three subclasses, namely, class IIa (pediocin-like bacteriocins), class IIb (two-peptide bacteriocins) and class IIc (other [i.e. non-pediocin-like], one-peptide bacteriocins) (Drider *et al.*, 2006). Class III bacteriocins are large (>30 kDa) and heat-labile proteins. The Archaea produce their own distinct family of bacteriocin-like antimicrobials, known as archaeocins (Shand and Leyva, 2007). A new structure-based classification of bacteriocins has been recently proposed by our team (Zouhir *et al.*, 2010). This resourceful and consistent classification approach allows the classification of more than 70% of Gram-positive bacteriocins, including those remained unclassified until recently.

BACTIBASE database

Why?

Since their discovery by Gartia in 1925, nearly 300 bacteriocins have been identified and some of them have been used successfully for inhibiting both animal and human pathogens (Snelling, 2005; Kirkup, 2006). A few of these have been well characterized and information such as amino acid sequence and spectra of antimicrobial activity are now available. However, for many other bacteriocins, this type of information is still missing or is scattered in the scientific literature and therefore unavailable to potential users. This situation could be improved by a central resource such as a database in which data could be collected, analysed and used to generate new and useful information about bacteriocins. Thus, we developed BACTIBASE, a database dedicated to bacteriocins produced by both Gram-positive and Gram-negative bacteria (Hammami *et al.*, 2007) (http://bactibase.pfba-lab-tun.org). The microbial, physico-chemical and structural proprieties provided in this database would allow better and more comprehensive structural and functional analysis of this special group of antimicrobial peptides.

The core content and web interface

Bacteriocin sequences were collected from the UniProt database (The UniProt Consortium, 2009) and from the scientific literature using PubMed. Since not all known bacteriocin sequences were present in the ExPASy (http://www.expasy.org/srs/) SRS server or NCBI server (http://www.ncbi.nlm.nih.gov/entrez/), a literature search was used to complete the BACTIBASE sequence database. Sequences were retrieved in SciDBMaker (Hammami *et al.*, 2008) and manually curated. The peptides collected in this version of BACTIBASE are mainly from natural sources. Precursor sequences were removed to keep only mature peptide sequences. A physicochemical dataset was designed using SciDBMaker including empirical formula, mass, length, isoelectric point, net charge, the number of basic, acidic, hydrophobic and polar residues, amino acids content, absent and most prevalent amino acids, hydropathy index, binding potential index, instability index, aliphatic index, half-life in mammalian cells, yeast and *Escherichia coli*, extinction coefficient, absorbance at 280 nm, secondary (α-helix or β-strand) and tertiary structure (when available), physical method used for structural determination (e.g. NMR spectroscopy or X-ray diffraction) and critical residues for activity, whenever information was available (Hammami *et al.*, 2010). The resulting tables were then exported from SciDBMaker to the MySQL server (http://www.mysql.com). Since information about target organisms (spectrum of activity) are dispersed in the scientific literature, microbiological information was collected from the literature by PubMed search. Analysis of data about producer and target organisms would certainly be useful in applications domains such as food preservation. The current release of the BACTIBASE database (version 2, July 2009) contains 177 bacteriocin sequences, of which 156 are the products of Gram-positive organisms and 18 of Gram-negative organisms. Additionally, the database includes 'BACTIBASE references', a cross-linked sub-database which lists all published scientific articles consulted on the subject of each bacteriocin (Hammami *et al.*, 2010).

BACTIBASE runs on a LAMP platform (Linux + Apache + MySQL + PHP). The public interface of BACTIBASE has recently been reengineered to improve usability and performance. The website uses JavaScript to ensure interactivity. Figure 22.2 shows a screen capture of the web interface with all menu navigation elements such as statistics, advanced search, browse and tools. The information page for bacteriocin is organized into five tabs which contain all the information for a bacteriocin including its description, taxonomy, spectrum of activity (when available), cross-link to other databases (such as UniProt, EMBL, InterPro, etc.), references, structural data, physicochemical data and comments (Fig. 22.3). The latter tab may serve as a forum (via notes) allowing registered users to exchange information or ask questions regarding each bacteriocin present in the database.

In addition to the usual data-browsing interface, BACTIBASE offers a powerful, easy-to-use, query interface for identifying bacteriocins of interest. The provided search interface is intuitive for novice users. The advanced search page presents an interactive and dynamic user interface allowing the search of all possible criteria such as bacteriocin name, producer organism, target organisms, sequence motif, reference and physicochemical parameters (Fig. 22.4). The following example describes how to find bacteriocins in the BACTIBASE database with particular characteristics. An example: How to find a list of bacteriocins produced by *Lactobacillus* and active against *Listeria* or *Salmonella*? First, go to page search (http://bactibase.pfba-lab-tun.org/search.php?view=GeneralView). Then, type '*Lactobacillus*' in the producer organism input box. As MySQL server allows different type of comparison, you should choose the correct comparison operator (in our case 'start with' or 'contains'). Next, go to target organisms

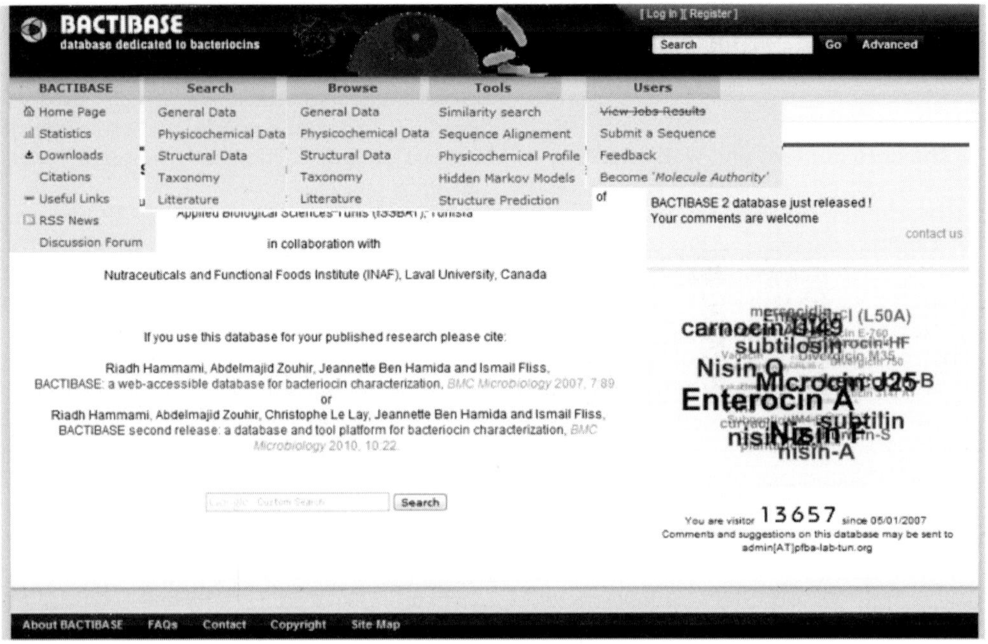

Fig. 22.2. The web interface of the BACTIBASE database. A screen capture of the web interface with all menu navigation elements such as statistics, advanced search, browse and tools.

region, and type '*Listeria*' and '*Salmonella*' separately in the first and second input boxes. Now, set the comparison operator to 'contains' for the previous keywords and finish by clicking on the 'Search' button. The current example should result 18 entries found in the BACTIBASE database. Clicking on 'MENU:Browse->physicochemical data' will show physicochemical data about these 18 bacteriocin sequences. Only, two entries (with these parameters) have a resolved tri-dimensional structure as shown in 'MENU:Browse->structural data'. The data in all resulting tables can be filtered according to your needs by clicking on column headers, which appear at the top of the table's columns. Beside this, data can be easily saved as a file in various formats (XML, FASTA and XLS) by clicking the corresponding icons in the upper right corner of the page, allowing further analysis of the data result.

The BACTIBASE website also provides a section to explore bacteriocin taxonomy (http:// bactibase.pfba-lab-tun.org/taxonomy.php). The user interface shows an integrated phylogenetic tree view which was designed to facilitate data retrieval via bacterial species name (Hammami *et al.*, 2010). The tree is displayed on the left and the corresponding bacteriocins are listed in tabulated form on the right. All bacteriocins associated with the selected genus are summarized in the table and a report can be generated in PDF format for further analysis. Clicking on the provided link displays the detailed entry for each bacteriocin. Alternatively, the information contained within the BACTIBASE database is exposed as XML (extensible markup language), ensuring thereby an automated database interaction through web-based application programming interfaces (API). When certain URLs are visited, an XML file with the requested data is returned, following the representational state transfer (REST) interface for data exchange. For example, calling the URL: http://bactibase.pfba-lab-tun.org/bacteriocinslist.php?RecPerPage=ALL&x_Strain=lactobacillus&export=xml serves an XML file with

Fig. 22.3. The detailed information page for bacteriocins: the page is organized into five tabs which contain all the information for a bacteriocin including its description, taxonomy, spectrum of activity (when available), cross-link to other databases (such as UniProt, EMBL, InterPro, etc.), references, structural data, physicochemical data and comments.

detailed information of all bacteriocins produced by the genus *Lactobacillus*.

Integrated tools for bacteriocin characterization

BACTIBASE offers several tools to assist biologist users: (i) different homology search engines, such as BLAST (Altschul *et al.*, 1997) and FASTA (Pearson and Lipman, 1988); (ii) multiple sequence alignments, including CLUSTALW (Larkin *et al.*, 2007), MUSCLE (Edgar, 2004) and T-COFFEE (Notredame *et al.*, 2000); (iii) visualization of sequence alignment (JalView: Waterhouse *et al.*, 2009) and structure (Jmol: http://jmol.sourceforge.net/); (iv) Hidden Markov Models (HMMER; Durbin *et al.*, 1998); (v) physicochemical profile; and (vi) taxonomy Browser. Besides this, a pipeline has been developed for automatic homology modelling from user bacteriocin sequence. Thus, the program MODELLER (Sali and Blundell, 1993) has

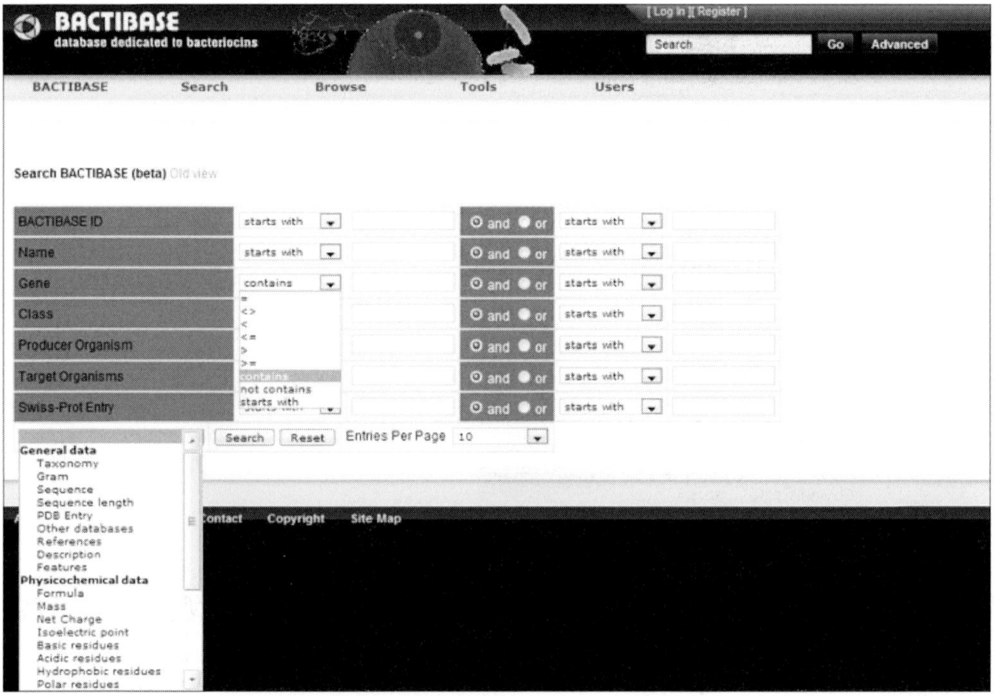

Fig. 22.4. An advanced search page presenting an interactive and dynamic user interface allowing a search by all possible criteria, such as name, producer organism, target organisms, sequence motif, reference and physicochemical parameters.

been incorporated into the platform, which implements comparative protein structure modelling by satisfaction of spatial restraint. This feature should be very useful for the *in silico* design of novel bacteriocins. The ability to develop novel bacteriocin-based drugs that target prokaryotic as well as eukaryotic cells may open new possibilities for the design of improved antibiotics possessing refined characteristics (Hammami *et al.*, 2010). Figure 22.5 shows the tools page (http://bactibase.pfba-lab-tun.org/toolsui.php); the page is organized into five tabs which contain various tools for sequence analysis including similarity search, multiple alignment, physicochemical profile, Hidden Markov Models and molecular modelling. The user navigates between tools by first selecting the tab with the corresponding tool category and then clicking the particular tool.

Microbiological and physicochemical statistics on bacteriocins found in BACTIBASE

The BACTIBASE database lists 177 entries including 45 lanthionine-containing bacteriocins (Class I), 47 non-lanthionine-containing bacteriocins (Class II) and other classified/unclassified entries. Due to their 'generally recognised as safe' (GRAS) status, lactic acid bacteria (LAB) have attracted great interest as food preservatives and several works have focused on the isolation and development of new strains of bacteriocin-producing bacteria. As consequence, the LAB (order *Lactobacillales*) makes up the predominant group of producers as found in BACTIBASE database, with 113 bacteriocins. Figure 22.6 illustrates the distribution of bacteriocins among the producer genera in the BACTIBASE database. The majority of

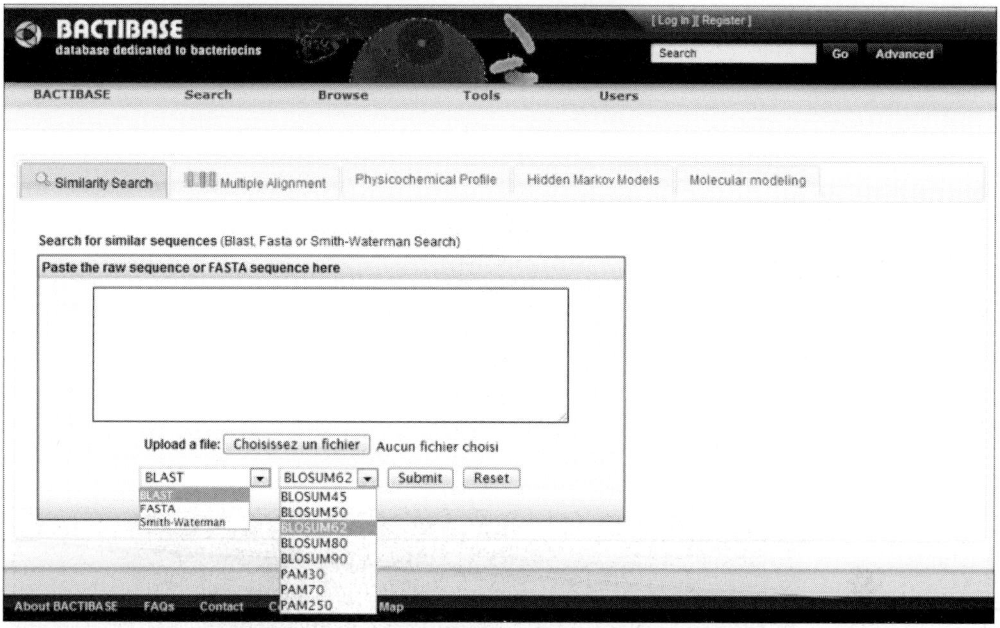

Fig. 22.5. A screen capture of the tools interface. The page is organized into five tabs which contain various tools for sequence analysis including similarity search, multiple alignment, physicochemical profile, Hidden Markov Models and molecular modelling.

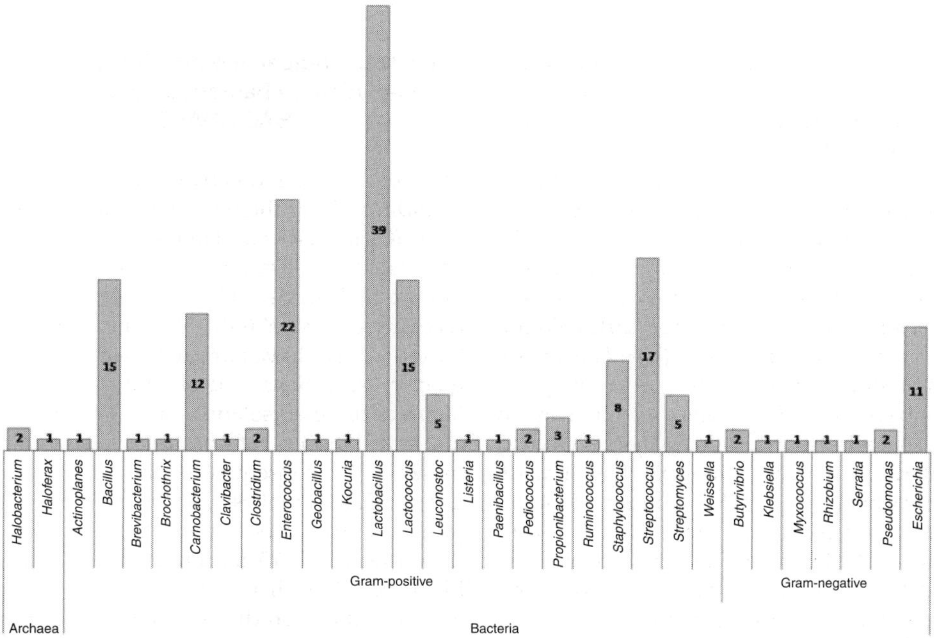

Fig. 22.6. The distribution of bacteriocins among the producer genera in the BACTIBASE database.

bacteriocins found in our database are produced by *Bacillus* (15), *Carnobacterium* (12), *Enterococcus* (22), *Lactobacillus* (39), *Lactococcus* (15), *Streptococcus* (17) and *Escherichia* (11). Except for the *Pediococcus* genus with the Pediocin PA-1 bacteriocin, the Lactobacillaceae family is exclusively composed of the *Lactobacillus* genus. This genus produces the greatest number of bacteriocins, some of which are classified in the literature as lantibiotics, class IIa, class IIb and the rest are unclassified (Zouhir *et al.*, unpublished data). In general, class I bacteriocins have a large inhibitory spectrum, such as nisin and thermophilin 13, which inhibit various species from the genera *Enterococcus, Lactobacillus, Lactococcus, Leuconostoc, Pediococcus, Streptococcus, L. monocytogenes, Staphylococcus, Bacillus* and *Clostridium*. Comparatively, most number of class IIa bacteriocins have a limited spectrum of inhibition including *Enterococcus, Lactobacillus, Pediococcus* and *Listeria*.

A statistical analysis was made that correlated various physicochemical parameters of bacteriocin sequences. As can be seen in Table 22.1, the length of sequence significantly influences the amino acid content and the protein-binding potential index (Boman Index; Boman, 2003). Also, the Boman Index negatively correlates with amino acid content (basic, acidic, hydrophobic and polar residues, etc.), as demonstrated by Pearson's correlation coefficient (r). Conversely, calculated Pearson coefficients revealed a positive correlation between sequence length and number of hydrophobic, polar, basic and acidic residues, indicating that their content is fairly constant (Fig. 22.7).

Conclusion

Due to their GRAS status, LAB have attracted great interest as food preservatives and several works have focused on the isolation and development of new strains of bacteriocin-producing bacteria. Bacteriocins are a heterogeneous group of antibacterial toxins that vary in spectrum of activity, genetic origin, biochemical properties and mechanism of action. Over the past few decades, numerous bacteriocins from both Gram-positive and Gram-negative bacteria have been isolated and their biochemical and genetic characteristics have been characterized. BACTIBASE is an integrated open-access database designed for the characterization of bacteriocins. The purpose of the database is to serve the research community by organizing information relevant to all types of bacteriocins from all groups of bacteria. The BACTIBASE database brings together physicochemical, structural, taxonomic, spectrum of activity and literature data for bacteriocins produced by both Gram-positive and Gram-negative bacteria. The provided features should make BACTIBASE a useful tool in food preservation or food-safety applications and could have implications for the development of new drugs for medical use.

Table 22.1. Pearson correlation between various physicochemical parameters of bacteriocins sequences.

	Net charge	Acidic	Length	Glycine	Hydrophobic	Polar	C. Extinct	B. Index
pI	0.651	–0.081	–0.002	0.064	–0.003	–0.005	0.101	–0.016
Basic		0.963	0.972	0.780	0.962	0.925	0.765	–0.982
Acidic			0.968	0.758	0.961	0.913	0.759	–0.979
Length				0.862	0.991	0.974	0.810	–0.960
Glycin					0.843	0.915	0.784	–0.746
Hydrophobic						0.946	0.786	–0.939
Polar							0.835	–0.920
C. Extinct								–0.759

C. Extinct, coefficient of extinction; B. Index, Boman Index.

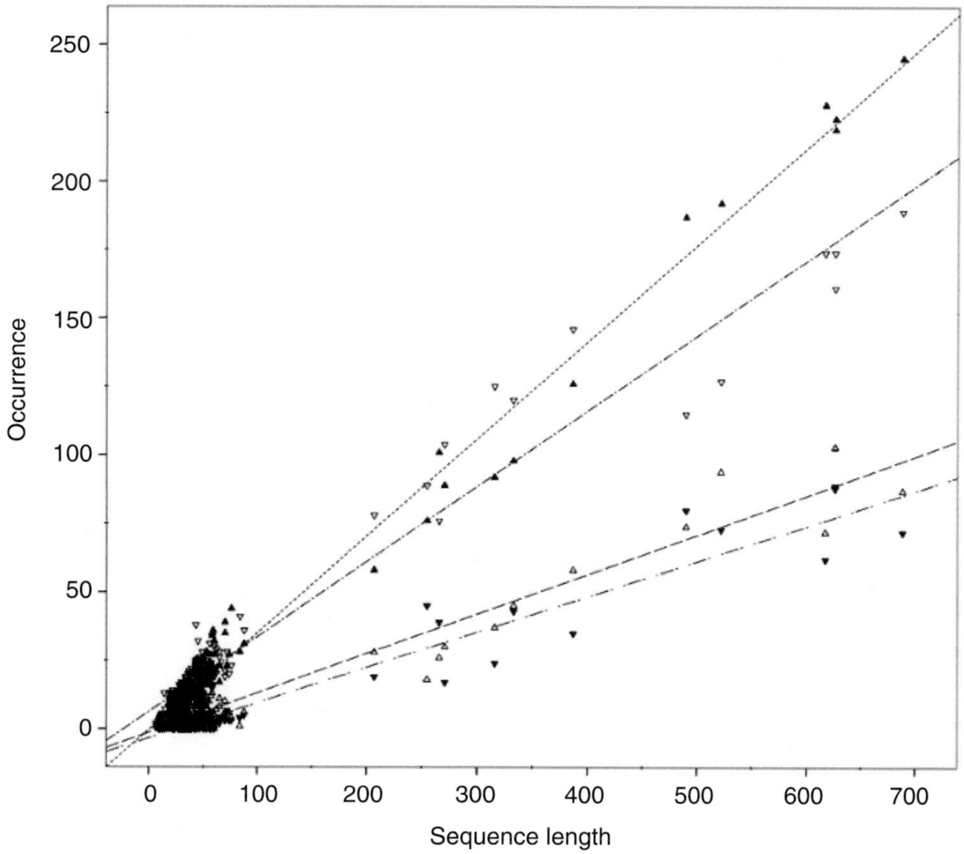

Fig. 22.7. Correlation between sequence length (x axis), hydrophobic (▲), polar (▽), basic (△) and acidic (▼) residues.

References

Altschul, S., Madden, T., Schaffer, A., Zhang, J., Zhang, Z., Miller, W. and Lipman, D. (1997) Gapped BLAST and PSI-BLAST: a new generation of protein database search programs. *Nucleic Acids Research* 25, 3389–3402.

Boman, H.G. (2003) Antibacterial peptides: basic facts and emerging concepts. *Journal of Internal Medicine* 254, 197–215.

Cintas, L.M., Casaus, M.P., Herranz, C., Nes, I.F. and Hernandez, P.E. (2001) Review: Bacteriocins of lactic acid bacteria. *Food Science and Technology International* 7, 281–305.

Drider, D., Fimland, G., Hechard, Y., McMullen, L. and Prevost, H. (2006) The continuing story of class IIa bacteriocins. *Microbiology and Molecular Biology Reviews* 70, 564–582.

Durbin, R., Eddy, S., Krogh, A. and Mitchison, G. (1998) *Biological Sequence Analysis: Probabilistic Models of Proteins and Nucleic Acids.* Cambridge University Press, Cambridge.

Edgar, R.C. (2004) MUSCLE: multiple sequence alignment with high accuracy and high throughput. *Nucleic Acids Research* 32, 1792–1797.

Gartia, A. (1925) Sur un remarquable exemple d'antagonisme entre deux souches de colibacille. *Comptes Rendus des Séances de la Société de Biologie et de ses Filiales,* 93, 1040–1041.

Gordon, D.M., Oliver, E. and Littlefield-Wyer, J. (2007) The diversity of bacteriocins in Gram-negative bacteria. In: Riley, M.A. and Chavan, M. (eds) *Bacteriocins: Ecology and Evolution.* Springer, Berlin, pp. 5–18.

Hammami, R., Zouhir, A., Ben Hamida, J. and Fliss, I. (2007) BACTIBASE: a new web-accessible database for bacteriocin characterization. *BMC Microbiology* 7, 89.

Hammami, R., Zouhir, A., Naghmouchi, K., Ben Hamida, J. and Fliss, I. (2008) SciDBMaker: new software for computer-aided design of specialized biological databases. *BMC Bioinformatics* 9, 121.

Hammami, R., Zouhir, A., Le Lay, C., Ben Hamida, J. and Fliss, I. (2010) BACTIBASE second release: a database and tool platform for bacteriocin characterization. *BMC Microbiology* 10, 22.

Heng, N.C.K., Wescombe, P.A., Burton, J.P., Jack, R.W. and Tagg, J.R. (2007) The diversity of bacteriocins in Gram-positive bacteria. In Riley, M.A. and Chavan, M. (eds) *Bacteriocins: Ecology and Evolution.* Springer, Berlin, pp. 45–92.

Kirkup, B. (2006) Bacteriocins as oral and gastrointestinal antibiotics: theoretical considerations, applied research, and practical applications. *Current Medical Chemistry* 13, 3335–3350.

Klaenhammer, T.R. (1988) Bacteriocins of lactic acid bacteria. *Biochimie* 70, 337–349.

Larkin, M.A., Blackshields, G., Brown, N.P., Chenna, R., McGettigan, P.A., McWilliam, H., Valentin, F., Wallace, I.M., Wilm, A., Lopez, R., Thompson, J.D., Gibson, T.J. and Higgins, D.G. (2007) Clustal W and Clustal X version 2.0. *Bioinformatics,* 23, 2947–2948.

Notredame, C., Higgins, D.G. and Heringa, J. (2000) T-coffee: a novel method for fast and accurate multiple sequence alignment. *Journal of Molecular Biology,* 302, 205–217.

Pearson, W.R. and Lipman, D. J. (1988) Improved tools for biological sequence comparison. *Proceedings of the National Academy of Sciences of the United States of America,* 85, 2444–2448.

Riley, M.A. and Wertz, J.E. (2002) Bacteriocins: Evolution, ecology and application. *Annual Review of Microbiology* 56, 117–137.

Sali, A. and Blundell, T.L. (1993) Comparative protein modelling by satisfaction of spatial restraints. *Journal of Molecular Biology* 234, 779–815.

Shand, R. and Leyva, K. (2007) Peptide and protein antibiotics from the domain Archaea: Halocins and Sulfolobicins. *In: Bacteriocins: Ecology and Evolution*, Springer, Berlin, pp. 93–109.

Shand, R.F. and Leyva, K.J. (2008) Archaeal antimicrobials: an undiscovered country. In: Norfolk, B.P. (ed.), *Archaea: New Models for Prokaryotic Biology.* Caister Academic, pp. 233–242.

Snelling, A. (2005) Effects of probiotics on the gastrointestinal tract. *Current Opinion in Infectious Diseases* 18, 420–426.

The UniProt Consortium (2009) The Universal Protein Resource (UniProt) 2009. *Nucleic Acids Research* 37, D169–174.

Vriezen, J.A.C., Valliere, M. and Riley, M.A. (2009) The evolution of reduced microbial killing. *Genome Biology and Evolution* 2009, 400–408.

Waterhouse, A.M., Procter, J.B., Martin, D.M.A., Clamp, M. and Barton, G.J. (2009) Jalview Version 2–a multiple sequence alignment editor and analysis workbench. *Bioinformatics* 25, 1189–1191.

Willey, J.M. and van der Donk, W.A. (2007) Lantibiotics: peptides of diverse structure and function. *Annual Review of Microbiology* 61, 477–501.

Zouhir, A., Hammami, R., Fliss, I. and Hamida, J.B. (2010) A new structure-based classification of gram-positive bacteriocins. *Protein Journal* 29, 432–439.

Index

Page numbers in **bold** type refer to figures and tables.